Lecture Notes in Mathematics

AF867373

Volume 2390

Editors-in-Chief

Jean-Michel Morel, City University of Hong Kong, Kowloon Tong, China

Bernard Teissier, IMJ-PRG, Paris, France

Series Editors

Karin Baur, University of Leeds, Leeds, UK

Michel Brion, UGA, Grenoble, France

Annalisa Buffa, Institute of Mathematics, École Polytechnique Fédérale de Lausanne, Lausanne, Switzerland

Adrian Constantin, Department of Mathematics, University of Vienna, Wien, Austria

Rupert Frank, LMU, Munich, Germany

Annette Huber, Albert Ludwig University, Freiburg, Germany

Davar Khoshnevisan, The University of Utah, Salt Lake City, UT, USA

Ioannis Kontoyiannis, University of Cambridge, Cambridge, UK

Angela Kunoth, University of Cologne, Cologne, Germany

Ariane Mézard, IMJ-PRG, Paris, France

Mark Podolskij, University of Luxembourg, Esch-sur-Alzette, Luxembourg

Mark Policott, Mathematics Institute, University of Warwick, Coventry, UK

László Székelyhidi, MPI for Mathematics in the Sciences, Leipzig, Germany

Gabriele Vezzosi, UniFI, Florence, Italy

Anna Wienhard, MPI for Mathematics in the Sciences, Leipzig, Germany

This series reports on new developments in all areas of mathematics and their applications - quickly, informally and at a high level. Mathematical texts analysing new developments in modelling and numerical simulation are welcome. The type of material considered for publication includes:

1. Research monographs
2. Lectures on a new field or presentations of a new angle in a classical field
3. Summer schools and intensive courses on topics of current research.

Texts which are out of print but still in demand may also be considered if they fall within these categories. The timeliness of a manuscript is sometimes more important than its form, which may be preliminary or tentative. Please visit the LNM Editorial Policy (https://drive.google.com/file/d/1MOg4TbwOSokRnFJ3ZR3ciEeKs9hOnNX_/view?usp=sharing)

Titles from this series are indexed by Scopus, Web of Science, Mathematical Reviews, and zbMATH.

Nicholas D. Alikakos • Cédric Villani
Editors

Analysis Techniques for Mathematical Physics and Geometry – and Music

Festum Pi, Chania, 2024

Springer

FESTUM π
αιγαιοπέλαγος

Editors
Nicholas D. Alikakos
Department of Mathematics
University of Athens (EKPA)
Athens, Greece

Cédric Villani
Institut Camille Jordan
Université Claude Bernard Lyon 1
Lyon, France

ISSN 0075-8434 ISSN 1617-9692 (electronic)
Lecture Notes in Mathematics
ISBN 978-3-032-11395-5 ISBN 978-3-032-11396-2 (eBook)
https://doi.org/10.1007/978-3-032-11396-2

© The Editor(s) (if applicable) and The Author(s), under exclusive license to Springer Nature Switzerland AG 2026

This work is subject to copyright. All rights are solely and exclusively licensed by the Publisher, whether the whole or part of the material is concerned, specifically the rights of translation, reprinting, reuse of illustrations, recitation, broadcasting, reproduction on microfilms or in any other physical way, and transmission or information storage and retrieval, electronic adaptation, computer software, or by similar or dissimilar methodology now known or hereafter developed.
The use of general descriptive names, registered names, trademarks, service marks, etc. in this publication does not imply, even in the absence of a specific statement, that such names are exempt from the relevant protective laws and regulations and therefore free for general use.
The publisher, the authors and the editors are safe to assume that the advice and information in this book are believed to be true and accurate at the date of publication. Neither the publisher nor the authors or the editors give a warranty, expressed or implied, with respect to the material contained herein or for any errors or omissions that may have been made. The publisher remains neutral with regard to jurisdictional claims in published maps and institutional affiliations.

This Springer imprint is published by the registered company Springer Nature Switzerland AG
The registered company address is: Gewerbestrasse 11, 6330 Cham, Switzerland

If disposing of this product, please recycle the paper.

Foreword

Festum Pi started with an encounter and a dream, when Cédric Villani and I decided to create a festival that would express our mixed French-Greek culture, reviving the marriage of mathematics and music in their common cradle of ancient Greece. Mathematics and music do share a lot in their fabric, their insistence on variations and combinations, but also the common culture which flows through these arts. After all, children learning to play a musical instrument learn about accuracy, order, comparison, symmetry, fractions, resonance, years before they learn these concepts in mathematics course. To serve our purpose, the event had to be not just a revival of the past or the exploration of the rich interface of mathematics and music: It had to attract mathematical researchers, exchanging and debating about most recent advances, and contemporary musicians, musicologists, and composers as well, exploring and sharing their passion; the interaction had to be not just an intellectual one, but also one pertaining to the senses of culture through the creation of a music festival and a festival orchestra, Orchestre. In 2024 it was the third edition already, and the project was coming to maturity. Over two weeks in Chania, the dozens of participants, from Greece and France and elsewhere, could exchange the rhythm of ideas and melodies. The melodies of music, declined in concerts and masterclasses. The melodies of mathematics, through summer school and research conference and work sessions. The melodies of sharing ideas and emotions, through interdisciplinary explorations and public concerts. The help from the Crete Region, especially President Arnaoutakis and Vice-President Kalogeris, and from the MAICH Institute of Agronomics and its director George Baourakis have been invaluable and that conference was also the occasion of an extraordinary international collaboration. The particularly high level of the mathematics for that third session, devoted to analysis applied to mathematical physics and geometry, made it clear that we had to turn the courses into a lecture notes volume. The speakers made great efforts so that their lectures would remain as valuable pieces for various audiences, and Nikos Alikakos and Cédric Villani worked hard on the editing, with the fantastic help of Habiba Kalantarova from Springer. It is our ambition that the next edition will explore the shores of artificial intelligence and machine learning, in their interface both with mathematical analysis and algorithmic sciences, and with music. Both these interfaces have been the subject of intense activity lately, and will lead us to a richer dialogue between tradition and research. For the musical part I'm aiming to incorporate the richness of Cretan traditional music to projects commissioned to contemporary classical composers, we are currently working together with Karol Beffa in view of an orchestral work for strings and Cretan lyra. It is also our dream that a permanent activity of research through the creation of a Pythagorean Research Center at the interface of those beautiful arts will be established in Crete, as a truly interdisciplinary European project. But for now, here is already a gem of art born in the festival, in the form of this rich set of carefully polished original lecture notes.

Festum Pi founder and artistic director
Paris and Chania
April 2025

Dionysios Dervis-Bournias

Introduction

The mathematics summer school and conference Festum π (MAΘHMATA) was held in Chania, in July 2024. This was an opportunity to present several up-to-date advances in various areas of mathematical analysis, encompassing both mathematical physics and differential geometry. The present volume collects some of these contributions in the form of lecture notes, intended both for researchers and for students.

The first part of this volume is devoted to modeling and derivation; this covers in particular limits from microscopic physics to statistical and fluid models.

Nathalie Ayi (Sorbonne Université, Paris) opens this volume with the general issue of large-number limits for non-exchangeable particle systems, a limit regime in which statistics is present, but individuality is not completely lost. At stake is the possibility to model situations in which global patterns arise and evolve rapidly in large assemblies, as in the flocking phenomenon, even in the absence of a centralized decision center; from the mathematical point of view, the challenge is to go beyond the time-honored but limited mean-field limit in which all agents are indistinguishable. Since the early models of Cucker and Smale two decades ago, there has been a lot of activity in this theme. Ayi surveys three main approaches that have been applied to this problem: adaptive dynamical networks, which capture evolving interaction structures; random graphs, in which interactions between individuals are partly determined by laws of chance; and hypergraphs, which capture higher-order interactions between individuals. This area combines elements of analysis, probability, and graph theory with observations in physical or biological crowds.

In the second contribution, one of us (Nicholas Alikakos, University of Athens) lectures on phase transitions based on the Allen-Cahn equation, a specific phase field model, that now extends to a system since the focus is on three or more phases. The vectorial case relates to minimal partitions, with the new phenomenon being the appearance of junctions—sets of higher codimension where three or more phases meet. The lectures address the triple junction, the simplest possible junction. A key point in the considerations is that no a priori symmetry hypotheses are imposed on the solution. Finally, an analog of the De Giorgi conjecture on the reduction of variables is stated.

The second part of the volume is devoted to a review of tools for certain classes of equations, with specific structures.

This starts with the lecture by Clément Mouhot (University of Cambridge), teaming up with Giovanni Brigati (Institute of Science and Technology Austria) for the written version of the course, on the famous De Giorgi–Nash–Moser iteration schemes for regularity of elliptic and parabolic equations. This emblematic subject has gone through many variations and has proven its versatility. More than 60 years after its birth, it is still being improved and adapted to various new settings, for instance, in the kinetic theory of gases and plasmas. In their contribution Brigati and Mouhot blend historical remarks with a systematic review and present the main techniques there.

Next, Filippo Santambrogio (Université Claude Bernard Lyon 1) surveys another important set of structural tools, namely the gradient flow structure for optimal transport theory, and the associated estimates. The combination of gradient flow structure and optimal transport (Monge–Kantorovich theory, Wasserstein distances) has become hugely popular after their introduction by Jordan, Kinderlehrer, and Otto at the end of the 1990s. A range of linear and nonlinear diffusion equations are covered by that approach. Santambrogio reviews the main results in that field which is nearly 30 years old now, addressing regularity, stability, long-time behavior, and numerical analysis.

Finally, Yvain Bruned (Université de Lorraine), Frédéric Rousset (Université Paris-Saclay), and Katharina Schratz (Sorbonne Université) together consider an original and very promising topic, the study of singularities for numerical simulations of equations exhibiting blow-up phenomena. Common wisdom has it that resonances are associated to blow-up and severely disrupt the numerical scheme for any partial differential equation; but the authors show that in many cases one can build numerical schemes on the underlying structure of resonances of the solution at the level of a Duhamel formula, rather than just discretizing the equation. To control higher-order oscillations, a decorated tree formalism is used, in the spirit of Hairer's work on regularity structures. Promising new schemes for low-regularity dispersive equations are proposed with that strategy.

The third part of this volume, more focused, is devoted to recent research on the analytic properties of certain models of interest, namely the Ricci flow in geometric analysis and the Boltzmann equation in statistical physics. In both cases, this includes a detailed introduction of the model, suitable for non-experts, and then surveys the fine analysis of a certain feature.

First, Panagiotis Gianniotis (University of Athens) introduces the Ricci flow of Hamilton and exposes Perelman's bounded diameter conjecture (bounding the diameter of the horns which may appear in the singularities of the 3d Ricci flow), then he explains the concept of splitting and almost splitting maps, in the particular case of Type I singularities. The results presented here have been part of the fascinating set of questions and problems inaugurated by Perelman's famous work on the Poincaré conjecture and Thurston's geometrization program. The notes by Gianniotis also provide a self-contained introduction to that great theme of geometric analysis.

Finally, one of us (Cédric Villani, Université Claude Bernard Lyon I) introduces the theory of the Boltzmann equation and covers recent advances showing that Boltzmann collisions tend to decrease the Fisher information, a fact which has been crucial in establishing regularity estimates for very singular potentials in the spatially homogeneous regime. The recent results obtained along those lines by Nestor Guillen, Luis Silvestre, Cyril Imbert, and the author have solved longstanding open problems in kinetic theory and are leading the way to the completion of the program of well-posedness for the spatially homogeneous Boltzmann equation, initiated by Carleman in the 1930s. By establishing the monotonicity of Fisher information for all Boltzmann interactions, they also provide a far-reaching extension of the pioneering work by McKean in the 1960s, and challenge statistical physics with an interpretation of that striking property. Villani's long review is also an opportunity to recall the use of Fisher information in various problems related to statistics and diffusion processes.

A glimpse of the interdisciplinary spirit of the Festum π Conference is conveyed by the inclusion of two lectures bridging music analysis and mathematics, making up the fourth part of this volume.

First, Franck Jedrzejewski (Universite Paris Saclay) presents the concept of microtone, in other words minute variations of frequencies, which made their way into music at the beginning of the twentieth century in Europe and America, and surveys their use across various musical genres to the present day.

Next, Claude Ledoux (Conservatoire National Supérieur de Musique de Paris) takes the example of Messiaen's music and his depiction of bird songs, to reflect on the parallels and synergies between mathematics and music.

Both contributions illustrate how musical analysis and mathematical analysis, though distant, remain connected by subtle threads; at times through numbers, techniques, and variations of rhythm and frequencies; at times through their inner beauty and spirit, mixing rules, variations, and imagination.

We cannot, in this introduction, do justice to or fully capture the remarkable atmosphere of intellectual exchange and curiosity—engaging both students and professors—that made Festum π 2024 a gem of a mathematical event.

Athens, Greece
Lyon, France
March 14, 2025

Nicholas D. Alikakos
Cédric Villani

Contents

Part I
Modelling and Derivation

Graph and Mean-Field Limits for Interacting Particle Systems on Weighted Deterministic and Random Graphs

Nathalie Ayi

1 Introduction

In recent years, models of collective dynamics have garnered considerable attention. From studying flocking and swarming behaviors in animal groups [1, 8, 34] to analyzing how opinions spread within populations linked by social networks [25], mathematical tools from analysis, probability and graph theory have played a significant role in advancing our understanding of these phenomena.

A general mathematical model is to consider a population of N agents (also referred to as particles) that interact through binary interactions via the following system:

$$\frac{d}{dt}x_i^N(t) = \frac{1}{N}\sum_{j=1}^{N} a_{ij}\phi(x_i^N(t), x_j^N(t)),\ t \geq 0,\ i = 1, \ldots, N, \tag{1}$$

where $x_i^N \in \mathbb{R}^d$, $d \geq 1$, is the state variable, representing for instance the opinion or the position of agent i and ϕ is the interaction function. The constant $a_{ij} \in \mathbb{R}$, called the interaction coefficient, captures the idea that not all our interactions have the same impact on us, depending on the individual we are interacting with. A typical example is the Hegselmann-Krause dynamics [25] where $a_{ij} = a(|x_i^N - x_j^N|)$ with $a : \mathbb{R}^+ \to \mathbb{R}^+$ being called the influence function. This type of model conveys the idea that individuals with opinions significantly different from ours are usually

N. Ayi (✉)
Sorbonne Université, Université Paris Cité, CNRS, Inria, Laboratoire Jacques-Louis Lions (LJLL), Institut Universitaire de France, Paris, France
e-mail: nathalie.ayi@sorbonne-universite.fr

© The Author(s), under exclusive license to Springer Nature Switzerland AG 2026
N. D. Alikakos, C. Villani (eds.), *Analysis Techniques for Mathematical Physics and Geometry – and Music*, Lecture Notes in Mathematics 2390,
https://doi.org/10.1007/978-3-032-11396-2_1

unable to influence our perspective. In this case, the function a can be typically compactly supported.

When studying these models, a wide range of relevant directions can be explored. One key question concerns their long-time behavior. A distinctive feature of these models is their ability to naturally give rise to global patterns despite the absence of centralized intelligence. For instance, Hegselmann-Krause-type models inherently lead to phenomena such as consensus, where the entire population eventually shares the same opinion, or clustering, where people settle into a finite number of distinct opinions. In second-order models, like Cucker-Smale-type for instance, comparable phenomena, such as flocking and swarming behaviors in animal groups, can also be observed [18, 24].

Another significant class of questions, and the central focus of these lecture notes, is the study of the large-population limit. More specifically, we aim to understand how the particle system can be described as N, the number of agents, goes to infinity. A classic and historical approach is the mean-field limit. Initially introduced in the context of gas dynamics (see [15] for instance), this approach involves a shift in perspective. Instead of tracking the trajectories of each individual particle, the focus changes to understanding how opinions are distributed within the population. Thus, the quantity of interest becomes the measure $\mu_t(x)$ which represents the probability of finding an agent with opinion x at time t. For instance, it has been rigorously established that, for the Hegselmann-Krause dynamics, when passing to the mean-field limit, we recover the following non-local transport equation of Vlasov-type [19]:

$$\partial_t \mu_t + \nabla \cdot (V[\mu_t]\mu_t) = 0, \qquad V[\mu_t](x) = \int_{\mathbb{R}^d} a(\|x-y\|)(y-x)d\mu_t(y).$$

However, this classic approach has a major limitation. Indeed, it can only be applied to systems exhibiting the property of indistinguishability. Essentially, this means that swapping the labels of particles does not affect the dynamics. Therefore, it means that the identity of individual particles is not relevant for those models. Consequently, this imposes a significant restriction on the ones that can be considered. In the context of opinion dynamics, it becomes evident that disregarding the identity of individuals and the specific nature of the relationships they have with those they interact with leads to overlook an essential facet.

The use of graph theory, which accounts for these aspects, has become a prominent approach and has known significant progress in recent years. Indeed, natural objects from it appears as it can be seen with the following example, which takes place in the setting of Kuramoto-type models. In [55], they study a family of interacting oscillators via the following nonlocally coupled system

$$\frac{d}{dt}x_i^N = \frac{1}{N}\sum_{j=i-\ell}^{i+\ell} \sin(x_j^N - x_i^N), \qquad i = 1, \ldots, N, \tag{2}$$

$$\begin{pmatrix} 0 & 1 & 1 & 0 & 0 & 0 & 0 & 0 & 1 & 1 \\ 1 & 0 & 1 & 1 & 0 & 0 & 0 & 0 & 0 & 1 \\ 1 & 1 & 0 & 1 & 1 & 0 & 0 & 0 & 0 & 0 \\ 0 & 1 & 1 & 0 & 1 & 1 & 0 & 0 & 0 & 0 \\ 0 & 0 & 1 & 1 & 0 & 1 & 1 & 0 & 0 & 0 \\ 0 & 0 & 0 & 1 & 1 & 0 & 1 & 1 & 0 & 0 \\ 0 & 0 & 0 & 0 & 1 & 1 & 0 & 1 & 1 & 0 \\ 0 & 0 & 0 & 0 & 0 & 1 & 1 & 0 & 1 & 1 \\ 1 & 0 & 0 & 0 & 0 & 0 & 1 & 1 & 0 & 1 \\ 1 & 1 & 0 & 0 & 0 & 0 & 0 & 1 & 1 & 0 \end{pmatrix}$$

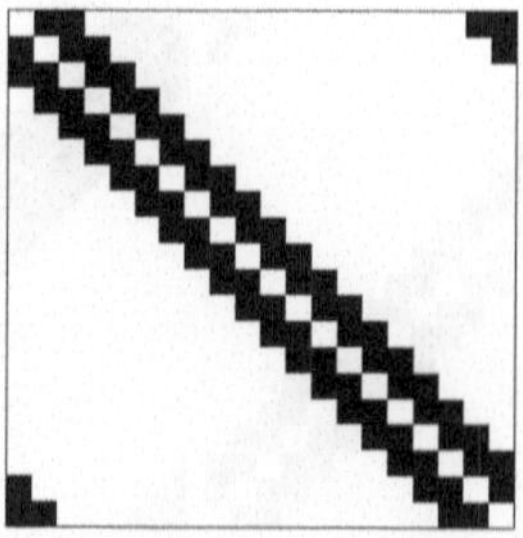

Fig. 1 Adjacency matrix of the graph G_N (left) and pixel representation of it built using the function w^{G_N} (right)

with $\ell = \lfloor rN \rfloor, r \in [0, 1]$ and where x_i represents the phase of oscillator i, the oscillators being located on a ring and indexed by integers from $\mathbb{Z}/N\mathbb{Z}$. The key point lies in the fact that the above model can be seen as a system of differential equations on a graph

$$G_N =< V(G_N), E(G_N) >$$

with the vertex set $V(G_N) = \{1, \ldots, N\}$ and the set of edges

$$E(G_N) = \{(i, j) \in \{1, 2, \ldots, N\}^2 | \, 0 < dist(i, j) \leq \ell\}$$

where $dist(i, j) = \min\{|i - j|, N - |i - j|\}$. In full generality, a graph can be characterized by its adjacency matrix, i.e. a square $N \times N$ matrix $A = (a_{ij})$ such that its element a_{ij} is 1 when there is an edge from vertex j to i, and 0 when there is no edge. A pixel version of this matrix (see Fig. 1) can be built via the following map, let $w^{G_N} : [0, 1]^2 \to \{0, 1\}$ be defined as

$$w^{G_N}(\xi, \zeta) = \begin{cases} 1 & \text{if } (i, j) \in E(G_N) \text{ and } (\xi, \zeta) \in \left[\frac{i-1}{N}, \frac{i}{N}\right) \times \left[\frac{j-1}{N}, \frac{j}{N}\right) \\ 0 & \text{else.} \end{cases}$$

The advantage of this method is that it allows for the visualization of an object that emerges as the natural limit associated with a convergent sequence of graphs. In Fig. 2, when representing the graph associated with Eq. (2), we can visually observe the convergence of the associated function w^{G_N} towards the function $w(\xi, \zeta) = \chi_{[0,r]}(|\xi - \zeta|)$. In fact, it can be rigorously established using concepts from graph theory, where the limiting object, the function w, is referred to as a *graphon*, short for *graph function*.

The crucial object graphon appears when considering limits of sequences of convergent *dense* graphs $(G_N)_{N \geq 1}$, i.e. sequences in which $|E(G_N)| \approx C|V(G_N)|^2$ for some constant $C \in (0, 1)$. Let us quickly mention here the existence of alternatives to graphons such as L^p-graphons for which $|V(G_N)| \ll |E(G_N)| \lesssim$

Fig. 2 Pixel representation of the graph G_N built using the function w^{G_N} with $N = 20$ (left) and $N = 100$ (center). Plot of the graphon $w(\xi, \zeta) = \chi_{[0,r]}(|\xi - \zeta|)$, limit of the sequence of graphs (right)

$|V(G_N)|^2$ [14] and graphons for which $|V(G_N)| \lesssim |E(G_N)| \lesssim |V(G_N)|^2$ [7]. We refer an interested reader to the associated references. If we go back to the graphon setting, the convergence of the graph sequence $\{G_N\}_{N\geq 0}$ is defined in terms of the homomorphism densities (see [35]), i.e. the sequence is called convergent if

$$t(F, G_N) = \frac{\hom(F, G_N)}{|V(G_N)|^{|V(F)|}}$$

is convergent for every simple graph F, where $\hom(F, G_N)$ is the number of homomorphisms (i.e. adjacency preserving maps $V(F) \to V(G_N)$) and where we call simple a graph without loops and multiple edges. The appearance of the graphon as a limit object comes from the following result: for every convergent sequence of simple graphs, there is w a graphon, i.e. a measurable symmetric function from $[0, 1]^2$ to $[0, 1]$ such that

$$t(F, G_N) \to t(F, w) := \int_{I^{|V(F)|}} \prod_{(i,j)\in E(F)} w(\xi_i, \xi_j) d\xi_1 \dots d\xi_N$$

for every simple graph F. A key feature, which will be of great use in the following, is that we can endow the space of graphons $\mathcal{G}_W := \{w \in L^\infty_+([0, 1]^2) : \|w\|_{L^\infty} \leq W, \text{ and } w \text{ is symmetric}\}$ with a distance. For any two graphons $w, \bar{w} \in \mathcal{G}_W$, we define the labeled cut distance

$$d_\square(w, \bar{w}) := \sup_{S,T\subset[0,1]} \left| \iint_{S\times T} (w(\xi, \zeta) - \bar{w}(\xi, \zeta))\, d\xi\, d\zeta \right|,$$

as well as the (unlabeled) cut distance $\delta_\square(w, \bar{w}) = \inf_\Phi d_\square(w, \bar{w}^\Phi)$, where Φ ranges over all bijective measure-preserving maps $\Phi : [0, 1] \longrightarrow [0, 1]$, and $\bar{w}^\Phi \in \mathcal{G}_W$ represents the rearranged graphon $\bar{w}^\Phi(\xi, \zeta) = \bar{w}(\Phi(\xi), \Phi(\zeta))$. The distance can be used to study the convergence of sequence of graphs, the key point being

$\delta_\square(w_N, w) \to 0$ if, and only if, $t(F, G_N) \to t(F, w)$. Besides, $(\mathcal{G}_W, \delta_\square)$ has the property of compactness. We refer an interested reader to [35] for a more complete view on the graph theory.

Finally, it is worth noticing that the labeled cut-distance is strictly weaker than the L^1-norm, i.e.

$$d_\square(w, \bar{w}) \leq \|w - \bar{w}\|_{L^1},$$

which explains why many of the proofs in the rest of the notes will be done using the L^1-convergence of graphs. However, the space of graphons associated with the L^1-norm does not inherently exhibit the compactness property that arises with the cut-distance. As a result, this approach requires the weights to be well-prepared, as we will see. Moreover, let us note that, since the L^1-norm is strictly stronger than the cut-distance, there exist sequences of graphs that converge with respect to the cut-distance but not with respect to the L^1-norm. Therefore, working with the cut-distance naturally allows for a broader range of cases to be covered.

Thus, graphons naturally emerge as limiting objects associated with sequences of (dense) graphs G_N as N goes to infinity. Since our primary goal is to study the large-population limit, it seems clear that these quantities will appear in the limit equations.

The notes are organized as follows: in Sect. 2, we introduce the two different large-population limits that will be studied, the graph limit and the mean-field limit. We provide a rigorous proof of the first limit in the case of dense graphs. Regarding the mean-field limit, after giving the proof in the indistinguishable case, we explain the key idea on the non-exchangeable case before concluding this section by clarifying the links between these two approaches. The remainder of the notes is devoted to studying these two limits in contexts that are variations of the classic framework. In Sect. 3, we consider adaptive dynamical networks while in Sect. 4, we discuss random graphs. Finally, in Sect. 5, we focus on incorporating higher-order interactions through hypergraphs.

2 Large-population Limits

In this section, we investigate the large-population limits of interacting particle systems on graphs. The particle systems under consideration can involve several thousand individuals (see [8] for instance), and in the case of neurons, populations can reach up to 10^{11} particles [26]. From a computational standpoint, simulating such systems becomes quickly too expensive for very large numbers of particles N. To overcome this challenge, two approaches are commonly employed, the graph limit (or continuum limit) where the system of ODEs is replaced by a integro-partial differential equation, and the mean-field limit which focuses on the evolution of the population density, instead of the individual trajectories.

2.1 Graph Limits (or Continuum Limits)

Let us denote $I := [0, 1]$, we define the subintervals $I_i^N := [\frac{i-1}{N}, \frac{i}{N})$ for all $i \in \{1, \cdots, N\}$. Let w be a bounded function from I^2 to $\mathbb{R}$. We introduce a new type of graph that has not been discussed yet. A *weighted graph* is a graph for which, additionally to the information that an edge is present or not, a weight, proportional to the intensity of the connection between the nodes it links, is associated. This is actually an essential feature to fully capture the complexity of many real networks (see [2, 9, 10, 32, 44] for instance).

Let us then define a sequence of weighted graphs $G_N =< \{1, \ldots, N\}, \{1, \ldots, N\}^2, \bar{w}^N >$ with:

$$\bar{w}_{ij}^N = N^2 \iint_{I_i^N \times I_j^N} w(\xi, \zeta) d\xi \, d\zeta.$$

where $\bar{w}_{ij}^N$ is the weight associated to the edge (i, j). We take an interest in the following systems of ODEs

$$\frac{d}{dt} x_i^N = \frac{1}{N} \sum_{j=1}^{N} \bar{w}_{ij}^N \phi(x_j^N - x_i^N), \quad x_i^N \in \mathbb{R}^d, \quad i \in \{1, \cdots, N\}. \tag{3}$$

We start by noticing that it can be seen as a system of ODEs on the graph G_N where the weights are built by averaging the function w on subintervals. Such a sequence of graphs happens to be convergent and w is then the limit graphon. Indeed, using the Lebesgue differentiation theorem and the dominated convergence theorem, we can prove that the function

$$w_N(\xi, \zeta) = \begin{cases} \bar{w}_{ij}^N & \text{if } (i, j) \in E(G_N) \text{ and } (\xi, \zeta) \in \left[\frac{i-1}{N}, \frac{i}{N}\right) \times \left[\frac{j-1}{N}, \frac{j}{N}\right) \\ 0 & \text{else.} \end{cases}$$

converges to w in the L^1-norm which implies the convergence in cut-distance and then the convergence in terms of homomorphism densities. The graph limit, also called continuum limit, consists in deriving rigorously the following equation

$$\partial_t x(t, \xi) = \int_I w(\xi, \zeta) \phi(x(t, \zeta) - x(t, \xi)) d\zeta. \tag{4}$$

We start by stating the well-posedness of the equation (see [38]).

Theorem 1 *Suppose that ϕ is Lipschitz-continuous, $w \in L^\infty(I^2)$ and the initial data satisfy $x(0, .) = u$ with $u \in L^\infty(I)$. Then, there exists a unique $x \in C^1(\mathbb{R}; L^\infty(I))$ solution to Eq. (4).*

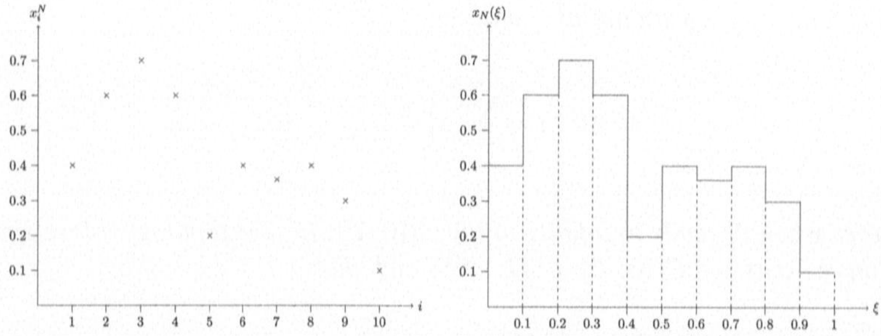

Fig. 3 Plot of the opinions x_i^N of the different agents (left) and associated piecewise-constant function x_N (right)

The proof, quite classic, relies on the use of the Banach fixed point theorem (see [38] for more details). Similar arguments allow to obtain the existence of a solution to the particles system.

Theorem 2 *Suppose that ϕ is Lipschitz-continuous, $w \in L^\infty(I^2)$ and let the initial data $(x_1^{0,N}, \dots, x_N^{0,N})$ be bounded in $(\mathbb{R}^d)^N$. Then, there exists a unique $(x_1^N, \dots, x_N^N) \in C^1(\mathbb{R}; (\mathbb{R}^d)^N)$ solution to Eq. (3).*

At this stage, let us remember that we wish to compare a solution to the particle system to one of the graph limit Eq. (4). However, they do not live in the same functional spaces. Thus, we need to transform vectors into piecewise-constant functions (see Fig. 3). In order to do so, we define the function $x_N : \mathbb{R} \times I \to \mathbb{R}$ as

$$x_N(t, \xi) := \sum_{i=1}^{N} x_i^N(t) \mathbf{1}_{I_i^N}(\xi).$$

Let $T > 0$, we introduce the functional space $C([0, T]; L^2(I))$ to which f belongs if the associated norm satisfies

$$\sup_{t \in [0,T]} \|f\|_{L^2(I)} < +\infty.$$

Then we can state the main result of this section established in [38].

Theorem 3 (Medvedev [38]) *Suppose that the function ϕ is Lipschitz-continuous and w belongs to $L^\infty(I^2)$. Let the initial data of the graph limit equation x_0 belong to the space $L^\infty(I; \mathbb{R}^d)$ and the initial data of the particle system $(x_1^{0,N}, \dots, x_N^{0,N})$ satisfy*

$$(x_1^{0,N}, \dots, x_N^{0,N}) = \left(N \int_{\frac{i-1}{N}}^{\frac{i}{N}} x_0(s) ds \right)_{i \in \{1, \dots, N\}} \in (\mathbb{R}^d)^N.$$

Then the function x_N with initial conditions

$$x_N(0,\xi) = \sum_{i=1}^{N} x_i^{0,N} \mathbf{1}_{[\frac{i-1}{N},\frac{i}{N})}(\xi)$$

converges when N tends to infinity in the $C([0,T];L^2(I))$ topology. More specifically, there exists $x \in C([0,T];L^2(I,\mathbb{R}^d))$ such that

$$\|x - x_N\|_{C([0,T];L^2(I,\mathbb{R}^d))} \xrightarrow[N\to+\infty]{} 0. \tag{5}$$

Furthermore, the limit function x is solution to the graph limit Eq. (4) *supplemented by the initial conditions $x(0,\cdot) = x_0$.*

Sketch of the Proof The proof relies on the observation that $(x_1^N,\ldots,x_N^N)$ is solution of Eq. (1) if, and only if, x_N is solution of the following equation

$$\partial_t x_N(t,\xi) = \int_I w_N(\xi,\zeta)\,\phi(x_N(t,\zeta) - x_N(t,\xi))d\zeta \tag{6}$$

where we recall

$$w_N(\xi,\zeta) = \sum_{i=1}^{N}\sum_{j=1}^{N} \bar{w}_{ij}^N \mathbf{1}_{I_i^N}(\xi)\mathbf{1}_{I_j^N}(\zeta).$$

Indeed, for $\xi \in I_i^n$, the above equation becomes

$$\begin{aligned}
\frac{d}{dt}x_i^N(t) &= \sum_{j=1}^{N} \int_{I_j^N} w_N(\xi,\zeta)\phi(x_N(t,\zeta) - x_N(t,\xi))d\zeta \\
&= \sum_{j=1}^{N} \bar{w}_{ij}^N \phi(x_j^N(t) - x_i^N(t)) \int_{I_j^N} d\zeta \\
&= \frac{1}{N}\sum_{j=1}^{N} \bar{w}_{ij}^N \phi(x_j^N(t) - x_i^N(t)).
\end{aligned}$$

Let us define $\alpha_N := x_N - x$. Then, we have

$$\begin{aligned}
\partial_t \alpha_N(t,\xi) &= \int_I w_N(\xi,\zeta)\,\phi(x_N(t,\zeta) - x_N(t,\xi))d\zeta \\
&\quad - \int_I w(\xi,\zeta)\,\phi(x(t,\zeta) - x(t,\xi))d\zeta \\
&= \int_I w_N(\xi,\zeta)\,[\phi(x_N(t,\zeta) - x_N(t,\xi)) - \phi(x(t,\zeta) - x(t,\xi))]\,d\zeta \\
&\quad + \int_I [w_N(\xi,\zeta) - w(\xi,\zeta)]\,\phi(x(t,\zeta) - x(t,\xi))d\zeta.
\end{aligned}$$

Then, we multiply the above equation by $\alpha_N(t,\xi)$ and we integrate over I. Using the Lipschitz-continuity of ϕ, we get

$$\begin{aligned}
&\frac{1}{2}\frac{d}{dt}\|\alpha_N(t)\|^2_{L^2(I)} \\
&\leq \int_{I^2} |w_N(\xi,\zeta)|\,|\alpha_N(t,\xi)|L_\phi|\alpha_N(t,\zeta)-\alpha_N(t,\xi)|d\xi d\zeta \\
&\quad + \left|\int_{I^2} \alpha_N(t,\xi)\,[w_N(\xi,\zeta)-w(\xi,\zeta)]\,\phi(x(t,\zeta)-x(t,\xi))d\zeta\right|.
\end{aligned}$$

Since $w \in L^\infty(I^2)$, there exists $\bar{M} > 0$ such that for all $(\xi,\zeta) \in I^2$,

$$|w(\xi,\zeta)| \leq \bar{M}.$$

Thus, using the triangular and the Cauchy-Schwarz inequalities, we deduce that

$$\begin{aligned}
&\frac{1}{2}\frac{d}{dt}\|\alpha_N(t)\|^2_{L^2(I)} \\
&\leq 2\bar{M}L_\phi\|\alpha_N(t)\|^2_{L^2(I)} \\
&\quad + \left|\int_{I^2} \alpha_N(t,\xi)\,[w_N(\xi,\zeta)-w(\xi,\zeta)]\,\phi(x(t,\zeta)-x(t,\xi))d\zeta\right|.
\end{aligned}$$

Let us now deal with the second term of the above right-hand side. We know that

$$\sup_{t\in[0,T]}\ \operatorname*{ess\,sup}_{(\xi,\zeta)\in I^2} |\phi(x(t,\zeta)-x(t,\xi))| \leq 2L_\phi\|x\|_{C([0,T]);L^\infty(I))} =: \bar{C}$$

Thus we deduce, using the Cauchy-Schwarz inequality, that

$$\begin{aligned}
&\left|\int_{I^2} \alpha_N(t,\xi)\,[w_N(\xi,\zeta)-w(\xi,\zeta)]\,\phi(x(t,\zeta)-x(t,\xi))d\zeta\right| \\
&\leq \bar{C}\|w_N - w\|_{L^2(I)}\|\alpha_N\|_{L^2(I)}.
\end{aligned}$$

Thus, finally, for all $t \in [0,T]$,

$$\frac{1}{2}\frac{d}{dt}\|\alpha_N(t)\|^2_{L^2(I)} \leq 2\bar{M}L_\phi\|\alpha_N\|^2_{L^2(I)} + \bar{C}\|w_N - w\|_{L^2(I)}\|\alpha_N\|_{L^2(I)}.$$

We divide by $\|\alpha_N(t)\|_{L^2(I)}$ and we get

$$\frac{d}{dt}\|\alpha_N(t)\|_{L^2(I)} \leq 2\bar{M}L_\phi\|\alpha_N\|_{L^2(I)} + \bar{C}\|w_N - w\|_{L^2(I)}.$$

Then, applying Gronwall's lemma, we obtain that

$$\|\alpha_N(t)\|_{L^2(I)} \leq \left(\|\alpha_N(0)\|_{L^2(I)} + \bar{C}\|w_N - w\|_{L^2(I)}t\right) e^{2\bar{M}L_\phi t}$$

from which we deduce that

$$\sup_{t\in[0,T]} \|\alpha_N(t)\|_{L^2(I)} \leq \left(\|\alpha_N(0)\|_{L^2(I)} + \bar{C}\|w_N - w\|_{L^2(I)}T\right) e^{2\bar{M}L_\phi T}$$

Finally, we use the Lebesgue differentiation theorem and the dominated convergence theorem to prove the convergence of the above right-hand side to 0 which concludes the proof. □

Remark 1 As mentioned in the introduction, this framework, leveraging graph theory in the context of graphons, is applicable only to dense graphs. However, alternative theories have been developed to address this limitation, such as the L^p-graphon framework [14], thereby allowing for the treatment of sparse graphs. Building on this approach, derivation of continuum limits have also been established, extending the previous result to a wider class of graphs (see [27, 40]).

Remark 2 It should be noted that, in this subsection, when dealing with the graph limit for deterministic graphs, we only have worked with deterministic initial data. In the next sections, where we will consider the mean-field limit, or the graph limit but this time for random graphs, we will allow the initial data to be random as well.

2.2 *Mean-Field Limits*

We now turn our attention to mean-field limits. As mentioned previously, they focus on the evolution of the population density, instead of the individual trajectories. We begin by recalling the theory in the case of indistinguishable particles.

2.2.1 Indistinguishable Case

A particle system is said to be indistinguishable if any permutation function σ : $\{1, \ldots, N\} \to \{1, \ldots, N\}$ of the sets of particle labels does not affect the dynamics. Let us consider the following particle system

$$\frac{d}{dt}x_i^N(t) = \frac{1}{N}\sum_{j=1}^{N} \phi(x_j^N(t) - x_i^N(t)), \quad i = 1, \ldots, N, \tag{7}$$

which enters this framework. We start by introducing the empirical measure, a crucial quantity, that we define as

$$\mu_t^N(x) := \frac{1}{N}\sum_{i=1}^{N} \delta_{x_i^N(t)}(x) \tag{8}$$

Let us consider a test function $f \in C_c^\infty(\mathbb{R}^d)$, then we can do the following computation

$$\begin{aligned}
&\frac{d}{dt}\int f(x)d\mu_t^N(x) \\
&= \frac{d}{dt}\left[\frac{1}{N}\sum_{i=1}^{N} f(x_i^N)\right] = \frac{1}{N}\sum_{i=1}^{N}\frac{d}{dt}x_i^N \cdot \nabla f(x_i^N) \\
&= \frac{1}{N^2}\sum_{i=1}^{N}\sum_{j=1}^{N}\phi(x_j^N(t) - x_i^N(t)) \cdot \nabla f(x_i^N) \\
&= \int_{\mathbb{R}^d}\int_{\mathbb{R}^d}\phi(y-x)\cdot\nabla f(x)d\mu_t^N(x)d\mu_t^N(y).
\end{aligned}$$

Therefore, we deduce that μ_t^N is a solution to the Vlasov-type equation

$$\partial_t \mu_t(x) + \nabla \cdot \left(\left(\int_{\mathbb{R}^d}\phi(y-x)d\mu_t(y)\right)\mu_t(x)\right) = 0. \tag{9}$$

This observation is actually the starting point of the rigorous derivation of the mean-field limit which relies on stability estimates obtained for the Vlasov-type equation. Let us start by saying a word on the well-posedness of such an equation. We endow the space $\mathcal{P}_1(\mathbb{R}^d)$, i.e. the space of probability measure μ satisfying

$$\int_{\mathbb{R}^d}|x|d\mu(x) < +\infty,$$

with the Monge-Kantorovich distance, which is defined as follows, for $\mu, \nu \in \mathcal{P}_1(\mathbb{R}^d)$,

$$W_1(\mu,\nu) := \sup_{[\varphi]_{\mathrm{Lip}}\le 1}\int_{\mathbb{R}^d}\varphi(x)d(\mu-\nu)(x) \tag{10}$$

where $[\varphi]_{\mathrm{Lip}}$ stands for the Lipschitz constant of φ.

Theorem 4 (Well-posedness of the Vlasov Equation) *Assume that the function ϕ is Lipschitz-continuous. Then, for every $\mu_0 \in \mathcal{P}_1(\mathbb{R}^d)$, there exists a unique $\mu \in C(\mathbb{R}_+, \mathcal{P}_1(\mathbb{R}^d))$ solution to Eq.* (9).

Let us say a word about the proof. We first define the push-forward of a measure. Let $T : \mathbb{R}^d \to \mathbb{R}^d$ be a map, for every Borel set A, for $\nu = T\#\mu$, it means

$$\nu(A) = \mu(T^{-1}(A)).$$

We introduce the characteristics associated with the Vlasov equation

$$\frac{d}{dt}X[\mu](t,x) = F[\mu_t](X[\mu](t,x)) \tag{11}$$

where the force is defined as

$$F[\mu_t](x) := \int_{\mathbb{R}^d} \phi(y-x)d\mu_t(y).$$

The existence of a solution to the characteristics equation is a direct consequence of the Cauchy-Lipschitz theorem. Let us denote

$$T_t[\mu](x) := X[\mu](t,x)$$

for all $t \geq 0$, $x \in \mathbb{R}^d$. Then, the proof of Theorem 4 actually relies on the fact that the solution to the Vlasov equation (9) is given by

$$\mu_t = T_t[\mu]\#\mu_0.$$

Indeed, we can prove that there exists a solution to this fixed point equation thanks to the fact that $\mathcal{F}[\mu] := T_t[\mu]\#\mu_0$, is actually a contraction mapping.

Theorem 5 (Stability Estimates for the Vlasov Equation) *Assume that the function ϕ is Lipschitz-continuous, with its Lipschitz constant denoted L_ϕ. For any initial data $\mu_0, \bar{\mu}_0 \in \mathcal{P}_1(\mathbb{R}^d)$, we have, for $t \geq 0$,*

$$W_1(\mu_t, \bar{\mu}_t) \leq e^{2L_\phi t} W_1(\mu_0, \bar{\mu}_0). \tag{12}$$

The proof relies on the two following lemmas.

Lemma 1 *For every family of probability measures $(\mu_t)_{t\geq 0}$, the map $x \in \mathbb{R}^d \mapsto T_t[\mu](x)$ is Lipschtiz-continuous and the Lipschitz constant satisfies*

$$[T_t[\mu]]_{Lip} \leq e^{L_\phi t}. \tag{13}$$

Proof We notice that

$$\begin{aligned}
&|T_t[\mu](x) - T_t[\mu](y)| \\
&\quad \leq \left| x + \int_0^t F[\mu_s]\left(X[\mu](s,x)\right) ds - y - \int_0^t F[\mu_s]\left(X[\mu](s,y)\right) ds \right| \\
&\quad \leq |x-y| + \int_0^t L_\phi \left|X[\mu](s,x) - X[\mu](s,y)\right| ds \\
&\quad = |x-y| + L_\phi \int_0^t |T_s[\mu](x) - T_s[\mu](y)|\, ds.
\end{aligned}$$

Then, by Gronwall's lemma, we obtain

$$|T_t[\mu](x) - T_t[\mu](y)| \le |x - y| e^{L_\phi t}$$

which concludes the proof. □

Lemma 2 *The solution to the characteristics Eq.* (11) *satisfies, for every* $t \geq 0$, $x \in \mathbb{R}^d$,

$$|X[\mu](t,x) - X[\bar{\mu}](t,x))| \le L_\phi e^{L_\phi t} \int_0^t W_1(\mu_s, \bar{\mu}_s) ds. \tag{14}$$

Proof Using the triangular inequality and the Lipschitz character of ϕ, we get

$$\begin{aligned}
&|X[\mu](t,x) - X[\bar{\mu}](t,x))| \\
&\quad = \left| \int_0^t \int_{\mathbb{R}^d} \phi(y, X[\mu](s,x)) d\mu_s(y) ds - \int_0^t \int_{\mathbb{R}^d} \phi(y, X[\mu](s,x)) d\bar{\mu}_s(y) ds \right| \\
&\quad + \left| \int_0^t \int_{\mathbb{R}^d} \phi(y, X[\mu](s,x)) d\bar{\mu}_s(y) ds - \int_0^t \int_{\mathbb{R}^d} \phi(y, X[\bar{\mu}](s,x)) d\bar{\mu}_s(y) ds \right| \\
&\quad \le L_\phi \int_0^t W_1(\mu_s, \bar{\mu}_s) ds + L_\phi \int_0^t |X[\mu](s,x) - X[\bar{\mu}](s,x)| ds.
\end{aligned}$$

Using Gronwall's lemma, we conclude the proof. □

Proof of Theorem 5 Applying the triangular inequality, we obtain

$$\begin{aligned}
W_1(\mu_t, \bar{\mu}_t) &= W_1(T_t[\mu]\#\mu_0, T_t[\bar{\mu}]\#\bar{\mu}_0) \\
&\le W_1(T_t[\mu]\#\mu_0, T_t[\mu]\#\bar{\mu}_0) + W_1(T_t[\mu]\#\bar{\mu}_0, T_t[\bar{\mu}]\#\bar{\mu}_0).
\end{aligned}$$

Let us deal with the first term. We notice that

$$\begin{aligned}
&W_1(T_t[\mu]\#\mu_0, T_t[\mu]\#\bar{\mu}_0) \\
&\quad = \sup_{[\varphi]_{\mathrm{Lip}} \le 1} \int_{\mathbb{R}^d} \varphi \, d\left(T_t[\mu]\#\mu_0 - T_t[\mu]\#\bar{\mu}_0\right) \\
&\quad = \sup_{[\varphi]_{\mathrm{Lip}} \le 1} \int_{\mathbb{R}^d} \varphi \circ T_t[\mu] \, d\left(\mu_0 - \bar{\mu}_0\right).
\end{aligned}$$

Using Lemma 1, we deduce that

$$[\varphi \circ T_t[\mu]]_{\mathrm{Lip}} \le [\varphi]_{\mathrm{Lip}} [T_t[\mu]]_{\mathrm{Lip}} \le e^{L_\phi t} [\varphi]_{\mathrm{Lip}}$$

and then, we obtain

$$\begin{aligned}
&W_1(T_t[\mu]\#\mu_0, T_t[\mu]\#\bar{\mu}_0) \\
&= e^{L_\phi t} \sup_{[\varphi]_{\mathrm{Lip}} \le 1} \int_{\mathbb{R}^d} \frac{\varphi \circ T_t[\mu]}{e^{L_\phi t}} \, d\left(\mu_0 - \bar{\mu}_0\right) \\
&\le e^{L_\phi t} W_1(\mu_0, \bar{\mu}_0).
\end{aligned}$$

Let us deal with the second term. We have

$$\begin{aligned} W_1(T_t[\mu]\#\bar{\mu}_0, T_t[\bar{\mu}]\#\bar{\mu}_0) &= \sup_{[\varphi]_{\mathrm{Lip}}\leq 1} \int_{\mathbb{R}^d} (\varphi\circ T_t[\mu] - \varphi\circ T_t[\bar{\mu}])\, d\bar{\mu}_0 \\ &\leq \int_{\mathbb{R}^d} |X[\mu](t,x) - X[\bar{\mu}](t,x))|\, d\bar{\mu}_0(x) \\ &\leq L_\phi e^{L_\phi t} \int_0^t W_1(\mu_s, \bar{\mu}_s) ds \end{aligned}$$

where the last inequality is obtained using Lemma 2. Finally, we have

$$W_1(\mu_t, \bar{\mu}_t) \leq e^{L_\phi t} W_1(\mu_0, \bar{\mu}_0) + L_\phi e^{L_\phi t} \int_0^t W_1(\mu_s, \bar{\mu}_s) ds$$

which concludes the proof applying again Gronwall's lemma. □

Corollary 1 (Mean-field Limit) *Let μ_t^N be the empirical measure defined in* (8) *where $(x_1^N, \dots, x_N^N)$ is a solution to* (7) *with initial data $(x_1^{0,N}, \dots, x_N^{0,N}) \in (\mathbb{R}^d)^N$. Then, if $W_1(\mu_0^N, \mu_0) \to 0$ as N goes to ∞ for some $\mu_0 \in \mathcal{P}_1(\mathbb{R}^d)$, it holds $W_1(\mu_t^N, \mu_t) \to 0$ as N goes to ∞ for all $t \geq 0$ where μ_t is the solution to the Vlasov equation* (9) *with initial data μ_0.*

An alternative proof can be done by introducing an intermediate particle system [51]. A sketch of the proof, following this path, in the context of non-exchangeability, will be given in Sect. 5.

2.2.2 Non-exchangeable Case

The approach outlined in the previous subsection was initially introduced in the context of gas dynamics (see, for instance, [15]). However, its primary limitation lies in its reliance on the indistinguishability of the particles in the system. Consequently, it is unsurprising that this framework fails to yield an equation such as (9) in the non-exchangeable setting. Indeed, the key quantity in (9), $\mu_t(x)$, represents the probability of finding an agent with opinion x at time t. In the non-exchangeable case, however, the identities of the particles play a crucial role. This suggests that the framework lacks an essential parameter to capture this individuality. And indeed, when introducing the parameter $\xi \in [0, 1]$ appearing in the previous sections, the new quantity of interest becomes $\mu_t^\xi(x)$, the probability of finding an agent with identity ξ and opinion x at time t. Then, in that new setting, a non-exchangeable mean-field limit can be obtained (see [16, 26, 28–30] for instance) and leads to the Vlasov-type equation

$$\partial_t \mu_t^\xi(x) + \nabla_x \cdot \left(\left(\int_I \int_{\mathbb{R}^d} w(\xi, \zeta) \phi(y - x) \mu_t^\zeta(dy) d\zeta \right) \mu_t^\xi(x) \right) = 0. \qquad (15)$$

We notice, as expected, the presence of the graphon w in the equation and we refer the reader to Sect. 5 for a sketch of the proof of such a derivation starting from a particle system in the more complex case of hypergraphs.

2.3 Link Between The Approaches

Both of the limits discussed above are large-population limits. A relevant point is to understand how they are connected to one another. Indeed, it is possible to derive a solution to the Vlasov-type equation from a solution of the graph limit equation, and in some cases, to do the reverse (see Fig. 4).

Indeed, as explained in [5], we can reach the Vlasov-type equation (15) from the graph limit Eq. (4). Actually, taking advantage of the graph limit, this provides an alternative proof for the non-exchangeable mean-field limit. Let $x(t, \xi)$ denote the solution to the graph limit Eq. (4). Let $\overline{\mu}_t$ denote a "continuous" empirical measure defined by

$$\overline{\mu}_t^{\xi}(x) = \int_I \delta_{x(t,\zeta)}(x)\delta_{\zeta}(\xi)d\zeta.$$

For all test functions $f \in C_c^{\infty}(I \times \mathbb{R}^d)$, we have

$$\begin{aligned}
\frac{d}{dt}\int_{I\times\mathbb{R}^d} f(\xi, x)d\overline{\mu}_t^{\xi}(x)d\xi &= \frac{d}{dt}\int_I f(\xi, x(t, \xi))d\xi \\
&= \int_I \nabla_x f(\xi, x(t, \xi)) \cdot \left(\int_I w(\xi, \zeta)\phi(x(t, \xi), x(t, \zeta))d\zeta\right) d\xi \\
&= \int_{I\times\mathbb{R}^d} \nabla_x f(\xi, x) \cdot \left(\int_{I\times\mathbb{R}^d} w(\xi, \zeta)\phi(x, y)d\overline{\mu}_t^{\zeta}(y)d\zeta\right) d\overline{\mu}_t^{\xi}(x)d\xi,
\end{aligned}$$

which means that $\overline{\mu}_t^{\xi}(x)$ is indeed a solution to Eq. (15).

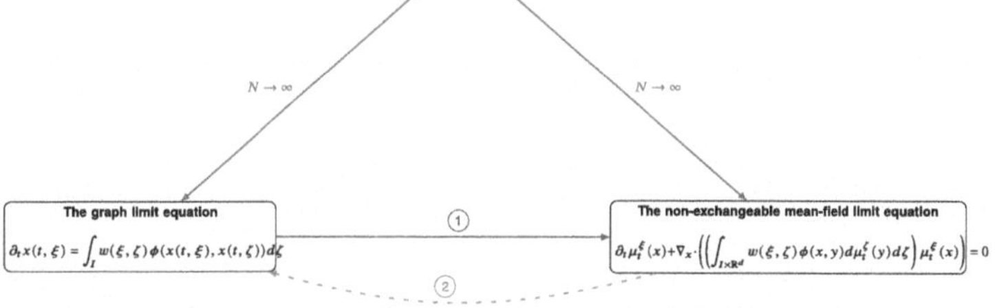

Fig. 4 Links between the different equations. The *red arrows* corresponds to large-population limits, respectively graph limit and non-exchangeable mean-field limit. The non-exchangeable mean-field limit can be reached through the graph limit while the inverse is true for specific interaction functions

Let us now address the passage from the non-exchangeable mean-field limit to the graph limit in the case of dimension $d = 1$. We denote

$$\bar{x}(t,\xi) := \int_{\mathbb{R}} x \, d\mu_t^\xi(x)$$

the first moment of the measure $\mu_t^\xi(x)$ solution of Eq. (15). Then, by computations, we get

$$\begin{aligned}
\partial_t \bar{x}(t,\xi) =& \partial_t \int_{\mathbb{R}} x \, d\mu_t^\xi(x) \\
=& \int_{\mathbb{R}} \partial_x(x) \left(\int_{I\times\mathbb{R}} w(\xi,\zeta)\phi(x,y) d\mu_t^\zeta(y) d\zeta \right) d\mu_t^\xi(x) \\
=& \int_{\mathbb{R}} \left(\int_{I\times\mathbb{R}} w(\xi,\zeta)\phi(x,y) d\mu_t^\zeta(y) d\zeta \right) d\mu_t^\xi(x).
\end{aligned}$$

At this stage, we notice that, in all generality, we cannot close the equation. Let us introduce an hypothesis on the interaction function which will allow it.

Hypothesis 1 *We suppose that*

$$\phi(x,y) = (\lambda_1 x + \lambda_2 y),$$

with $(\lambda_1, \lambda_2) \in \mathbb{R}^2$.

Remark 3 Although highly restrictive, this condition is not without interest as many models used in the literature satisfy it, such as the Hegselmann-Krause model, for instance for which the interaction function corresponds to $(y - x)$.

Therefore, under this assumption, we have

$$\begin{aligned}
\partial_t \bar{x}(t,\xi) &= \int_{\mathbb{R}} \left(\int_{I\times\mathbb{R}} w(\xi,\zeta)(\lambda_1 x + \lambda_2 y) d\mu_t^\zeta(y) d\zeta \right) d\mu_t^\xi(x) \\
&= \int_I w(\xi,\zeta) \left(\lambda_1 \int_{\mathbb{R}} x d\mu_t^\xi(x) + \lambda_2 \int_{\mathbb{R}} y d\mu_t^\zeta(y) \right) d\zeta \\
&= \int_I w(\xi,\zeta) \left(\lambda_1 \bar{x}(t,\xi) + \lambda_2 \bar{x}(t,\zeta) \right) d\zeta \\
&= \int_I w(\xi,\zeta)\phi(\bar{x}(t,\xi), \bar{x}(t,\zeta)) d\zeta.
\end{aligned}$$

which concludes obtaining the graph limit equation. Let us emphasize here that it is not clear that a closed equation can be obtained through this path in the general (nonlinear) case (for further comments, see [45]).

Remark 4 Note however that, in the particular case where the initial datum of the Vlasov equation is taken as a Dirac measure, i.e. $\mu_0^\xi = \delta_{x_0(\xi)}$ with $x_0 \in L^\infty(I; \mathbb{R}^d)$, then we recover the graph limit equation. Indeed, the unique solution μ to the Vlasov equation must have the form $\mu_t^\xi = \delta_{x(t,\xi)}$ where x is the unique solution to the continuum limit equation.

3 Adaptive Dynamical Networks (ADN)

Adaptive dynamical networks (ADN) are particularly relevant for modeling real-life interactions. For instance, in the context of opinion dynamics, relationships influence our opinions, but our opinions, in turn, also reshape these relationships. This mutual feedback induces changes in the network's structure over time, making the network's connectivity dynamic and state-dependent. This phenomenon is not limited to social interactions but extends to a variety of fields, including neural and neuronal networks [46, 47], machine learning applications [33, 48], power grid models [43], and more (see [12] for a review on the matter). Let us give a formal definition.

Definition 1 A network is said to be adaptive if the evolution of an edge (i, j) explicitly depends on the states of nodes i and j.

As mentioned in [12], there are two main classes for adaptive dynamical networks:

- *event-based adaptation*, where the network structure evolves at discrete points in time, with the triggers for these changes being dependent on the state of the system itself,
- *continuous adaptation*, which can be presented in its general form as

$$\begin{cases} \dfrac{d}{dt} x_i^N(t) = \displaystyle\sum_{j=1}^N w_{ij}^N(t)\phi\left(x_i^N(t), x_j^N(t), t\right) & \text{for all } i \in \{1, \cdots, N\} \\ \dfrac{d}{dt} w_{ij}^N(t) = h_{ij}(w^N(t), x^N(t), t) \end{cases} \tag{16}$$

  where $w^N := (w_{ij}^N)_{i,j\in\{1,\cdots,N\}}$ with w_{ij}^N represents the time-evolving weight of the connection from node j to i.

We will focus on the latter.

3.1 The Graph Limit for a Particular ADN

Let us go back a moment to the indistinguishable case

$$\frac{d}{dt}x_i^N(t) = \frac{1}{N}\sum_{j=1}^{N}\phi(x_j^N(t) - x_i^N(t)), \quad x_i^N(t) \in \mathbb{R}^d, \quad i \in \{1, \cdots, N\}.$$

and focus on an extended model where, in addition to holding an opinion, each agent is assigned an influence weight that evolves over time in a coupled way with the opinions. This weight represents a social hierarchy within the group, where agents with higher weights exert greater influence on the group's behavior. Therefore, the weight can be interpreted as a measure of charisma or popularity. This can be formulated as a system of 2N ODEs, as follows:

$$\begin{cases} \frac{d}{dt}x_i^N(t) = \frac{1}{N}\sum_{j=1}^{N} m_j^N(t)\phi(x_j^N(t) - x_i^N(t)) \\ \frac{d}{dt}m_i^N(t) = \psi_i(x^N(t), m^N(t)) \end{cases} \tag{17}$$

with $x^N = (x_1^N, \ldots, x_N^N)$, $m^N = (m_1^N, \ldots, m_N^N)$ and where $x_i^N \in \mathbb{R}^d$ is the opinion of agent i, $m_i^N \in \mathbb{R}^+$ the agent's weight. We initialize the system with the total weight

$$\sum_{i=1}^{N} m_i^N(0) = N$$

and we suppose that ψ_i, which dictates the weight dynamics, satisfy $\sum_i \psi_i \equiv 0$. It implies that the weight actually shifts between the agents. This model can actually be viewed as a system of ODEs defined on a specific adaptive dynamical network such that all edges originating from the same node share the same weight (see Fig. 5). This model has been initially introduced in [37] with specific choices for the weight dynamics.

Example 1 In [37], they study a model based on the idea that influence grows by persuading already influential individuals. Let us denote $e_{jk}^N := m_k^N a(|x_j^N -$

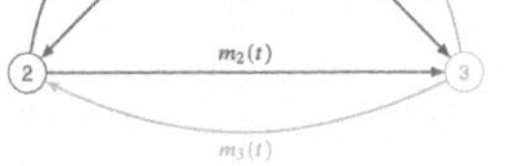

Fig. 5 Representation of the graph associated with the weight-varying microscopic dynamics (17). We notice that this graph is non-symmetric. Moreover, it fits the framework of adaptive dynamical networks but it is a very specific one. Indeed, all edges originating from the same node share the same weight

$x_k^N|)|x_j^N - x_k^N|$ the influence of k over j, $I_j^N := \underset{k\in\{1,\dots,N\}}{\operatorname{argmax}}\ e_{jk}^N$ the set of agents that influence j the most and $|I_j^N|_{m^N} := \sum_{k\in I_j^N} m_k^N$ the weight of the agents that influence j the most, this model can be expressed as follows:

$$\begin{cases} \dfrac{d}{dt}x_i^N(t) = \dfrac{1}{N}\displaystyle\sum_{j=1}^{N} m_j^N(t)a(|x_j^N - x_i^N|)(x_j^N - x_i^N) \\ \dfrac{d}{dt}m_i^N(t) = m_i^N\left(\displaystyle\sum_{j|i\in I_j} \dfrac{m_j^N}{|I_j|_m} - 1\right). \end{cases} \tag{18}$$

We observe that, in the case where there exists j such that $I_j^N = \{i\}$ with $m_j^N > m_i^N$, i.e if we can find an agent j, for which i is the most influential agent and such that its weight is greater than the one of i, then the weight of i indeed increases.

However, this previous example does not fit all the following regularity assumptions, that will be needed to rigorously obtain the graph limit result established in [3].

Hypothesis 2 *The interaction function ϕ satisfies $\phi \in \text{Lip}(\mathbb{R}^d;\mathbb{R})$, with $\|\phi\|_{\text{Lip}} = L_\phi$, and $\psi(0) = 0$.*

Hypothesis 3 *The function $\psi : I \times L^\infty(I;\mathbb{R}^d) \times L^\infty(I;\mathbb{R})$ is assumed to satisfy the following Lipschitz properties: there exists $L_\psi > 0$ such that for all $(x_1, x_2, m_1, m_2) \in L^2(I)^4$,*

$$\begin{cases} \|\psi(\cdot, x_1, m_1) - \psi(\cdot, x_2, m_1)\|_{L^2(I)} \le L_\psi \|x_1 - x_2\|_{L^2(I)} \\ \|\psi(\cdot, x_1, m_1) - \psi(\cdot, x_1, m_2)\|_{L^2(I)} \le L_\psi \|m_1 - m_2\|_{L^2(I)}. \end{cases} \tag{19}$$

Assume also that there exists $C_\psi > 0$ such that for all $(x, m) \in L^\infty(I, \mathbb{R}^d \times \mathbb{R})$, for all $\xi \in I$,

$$|\psi(\xi, x, m)| \le C_\psi(1 + \|m\|_{L^\infty(I)}). \tag{20}$$

Remark 5 We notice that m^N plays the same role in the Eq. (17) as the weights $\bar{w}^N$, built with the graphon w, did in Eq. (3) in Sect. 2.1. Thus, the sublinear growth bound (20) is actually there to ensure that the weights do not blow up in finite time, placing us in a context similar to the previous one which required the graphon w to be bounded.

Example 2 In [3], we consider a situation in which the agents are divided into K groups $I_k := [\frac{k-1}{K}, \frac{k}{K})$, with $k \in \{1, \cdots, K\}$. Each group is composed of leaders I_k^L and followers I_k^F as follows: for $r \in (0, 1)$, $I_k^L = [\frac{k-1}{K}, \frac{k-1}{K} + \frac{r}{K})$ and $I_k^F = [\frac{k-1}{K} + \frac{r}{K}, \frac{k}{K})$. The weights dynamics is built such as, within each group, the weight of each leader increases proportionally to itself and to the total weight of all the followers of the group. Conversely, the weight of each follower decreases proportionally to itself and to the total weight of all the leaders. Provided that $n := N/K \in \mathbb{N}$ and that $rn \in \mathbb{N}$, the corresponding microscopic dynamics can be written as:

$$\forall k \in \{1, \cdots, K\}, \quad \frac{d}{dt} m^N_{(k-1)n+i}$$
$$= \begin{cases} \frac{\beta}{N} m^N_{(k-1)n+i} \sum_{j=rn+1}^{n} m^N_{(k-1)n+j} & \text{if } i \in \{1, \cdots, rn\}, \\ -\frac{\beta}{N} m^N_{(k-1)n+i} \sum_{j=1}^{rn} m^N_{(k-1)n+j} & \text{if } i \in \{rn+1, \cdots, n\}. \end{cases} \tag{21}$$

Contrary to the first example, the associated ψ function fits the above regularity assumptions.

Then we can state the main result of this section, i.e. the convergence to the graph limit equation

$$\begin{cases} \partial_t x(t,\xi) = \int_I m(t,\zeta)\phi(x(t,\zeta) - x(t,\xi))d\zeta \\ \partial_t m(t,\xi) = \psi(\xi, x(t,\cdot), m(t,\cdot)), \end{cases} \tag{22}$$

for a weight dynamics defined as

$$\forall i \in \{1, \cdots, N\}, \qquad \psi_i(x^N(t), m^N(t)) = N \int_{\frac{i-1}{N}}^{\frac{i}{N}} \psi(\xi, x_N(t,\xi), m_N(t,\xi))d\xi, \tag{23}$$

with

$$x_N(t,\xi) := \sum_{i=1}^{N} x_i^N(t) \mathbf{1}_{I_i^N}(\xi), \quad m_N(t,\xi) := \sum_{i=1}^{N} m_i^N(t) \mathbf{1}_{I_i^N}(\xi)$$

Theorem 6 (Ayi and Pouradier Duteil [3]) *Suppose that the function ϕ satisfies Hypothesis 2 and the function ψ Hypothesis 3. Let (x_0, m_0) belongs to the space $L^\infty(I; \mathbb{R}^d) \times L^\infty(I; \mathbb{R})$ with*

$$\int_I m_0(\xi)d\xi = 1.$$

We consider the functions $x_N : I \times \mathbb{R} \to \mathbb{R}^d$ and $m_N : I \times \mathbb{R} \to \mathbb{R}$ defined as above and such that (x^N, m^N) is a solution to the particle system (17) *with initial data satisfying*

$$
\begin{aligned}
(x_1^{0,N}, \dots, x_N^{0,N}) &= \left(N \int_{\frac{i-1}{N}}^{\frac{i}{N}} x_0(s) ds \right)_{i \in \{1,\dots,N\}} &\in (\mathbb{R}^d)^N, \\
(m_1^{0,N}, \dots, m_N^{0,N}) &= \left(N \int_{\frac{i-1}{N}}^{\frac{i}{N}} m_0(s) ds \right)_{i \in \{1,\dots,N\}} &\in (\mathbb{R})^N.
\end{aligned} \tag{24}
$$

Then, we have

$$
\|x - x_N\|_{C([0,T];L^2(I,\mathbb{R}^d))} \xrightarrow[N \to +\infty]{} 0 \quad \textit{and} \quad \|m - m_N\|_{C([0,T];L^2(I,\mathbb{R}))} \xrightarrow[N \to +\infty]{} 0 \tag{25}
$$

where (x, m) is solution to the integro-differential system (22) *supplemented by the initial conditions $x(0, \cdot) = x_0$ and $m(0, \cdot) = m_0$.*

Sketch of the Proof As previously, we can prove that (x_N, m_N) satisfy the following system of integro-differential equations:

$$
\begin{cases}
\partial_t x_N(t, \xi) = \displaystyle\int_I m_N(t, \zeta) \phi(x_N(t, \zeta) - x_N(t, \xi))) \, d\zeta \\
\partial_t m_N(t, \xi) = N \displaystyle\int_{\frac{1}{N}\lfloor \xi) N \rfloor}^{\frac{1}{N}(\lfloor \xi) N \rfloor + 1)} \psi(\zeta, x_N(t, \cdot), m_N(t, \cdot)) \, d\zeta.
\end{cases} \tag{26}
$$

We denote $\alpha_N := x_N - x$ and $\beta_N := m_N - m$. Exploiting the sublinear growth assumption on ψ in Eq. (20), we can actually prove that there exists $\bar{M} > 0$ such that, for all $i \in \{1, \dots, N\}$, for all $t \in [0, T]$, we have $|m_i^N(t)| \le \bar{M}$. Then, doing similar computations as the one performed in Sect. 2.1, we deduce that

$$
\frac{1}{2} \frac{d}{dt} \|\alpha_N(t)\|_{L^2(I)}^2 \le 2\bar{M} L \|\alpha_N(t)\|_{L^2(I)}^2 + L_\phi \, \|\alpha_N(t)\|_{L^2(I)} \, \|\beta_N(t)\|_{L^2(I)}. \tag{27}
$$

Let us deal with the new term β_N. We rewrite the equation it satisfies as

$$
\begin{aligned}
\frac{\partial}{\partial t} \beta_N(t, \xi) =& N \int_{\frac{1}{N}\lfloor \xi N \rfloor}^{\frac{1}{N}(\lfloor \xi N \rfloor + 1)} \psi(\zeta, x_N(t), m_N(t)) \, d\zeta - \psi(\xi, x(t), m(t)) \\
=& N \int_{\frac{1}{N}\lfloor \xi N \rfloor}^{\frac{1}{N}(\lfloor \xi N \rfloor + 1)} [\psi(\zeta, x_N(t), m_N(t)) - \psi(\zeta, x(t), m(t))] \, d\zeta \\
&+ N \int_{\frac{1}{N}\lfloor \xi N \rfloor}^{\frac{1}{N}(\lfloor \xi N \rfloor + 1)} \psi(\zeta, x(t), m(t) \, d\zeta - \psi(\xi, x(t), m(t)).
\end{aligned}
$$

We multiply by $\beta_N(t, \xi)$ and integrate over I to get, using Cauchy-Schwarz inequality,

$$\frac{1}{2}\int_I \frac{\partial}{\partial t}\beta_N(t,\xi)^2\,d\xi = \int_I N\int_{\frac{1}{N}\lfloor\xi)N\rfloor}^{\frac{1}{N}(\lfloor\xi N\rfloor+1)} [\psi(\zeta, x_N(t), m_N(t))$$
$$-\psi(\zeta, x(t), m(t))]\,d\zeta\ \beta_N(t,\xi)\,d\xi$$
$$+\int_I\left[N\int_{\frac{1}{N}\lfloor\xi N\rfloor}^{\frac{1}{N}(\lfloor\xi N\rfloor+1)} \psi(\zeta, x(t), m(t))\,d\zeta\right.$$
$$-\psi(\xi, x(t), m(t))]\,\beta_N(t,\xi)\,d\xi$$
$$\leq \|h_N(t)\|_{L^2(I)}\|\beta_N(t)\|_{L^2(I)} + \|g_N(t)\|_{L^2(I)}\|\beta_N(t)\|_{L^2(I)}, \tag{28}$$

where we denoted by h_N and g_N the functions

$$h_N : (t,\xi) \mapsto N\int_{\frac{1}{N}\lfloor\xi N\rfloor}^{\frac{1}{N}(\lfloor\xi N\rfloor+1)} [\psi(\zeta, x_N(t), m_N(t)) - \psi(\zeta, x(t), m(t))]\,d\zeta$$

and

$$g_N : (t,\xi) \mapsto N\int_{\frac{1}{N}\lfloor\xi N\rfloor}^{\frac{1}{N}(\lfloor\xi N\rfloor+1)} \psi(\zeta, x(t), m(t))\,d\zeta - \psi(\xi, x(t), m(t)).$$

We can bound the first function by applying Cauchy-Schwarz inequality and obtain

$$\|h_N(t)\|^2_{L^2(I)} \leq \|\psi(\cdot, x_N(t), m_N(t)) - \psi(\cdot, x(t), m(t))\|^2_{L^2(I)},$$

which, using Eq. (19), leads to

$$\|h_N(t)\|_{L^2(I)} \leq \|\psi(\cdot, x_N(t), m_N(t)) - \psi(\cdot, x(t), m_N(t))\|_{L^2(I)}$$
$$+ \|\psi(\cdot, x(t), m_N(t)) - \psi(\cdot, x(t), m(t))\|_{L^2(I)}$$
$$\leq L_\psi(\|x_N(t) - x(t)\|_{L^2(I)} + \|m_N(t) - m(t)\|_{L^2(I)})$$
$$= L_\psi(\|\alpha_N(t)\|_{L^2(I)} + \|\beta_N(t)\|_{L^2(I)}).$$

Regarding the second term of (28), according to Lebesgue's differentiation theorem, for almost every $\xi \in I$, we have

$$\lim_{N\to+\infty} N\int_{\frac{1}{N}\lfloor\xi N\rfloor}^{\frac{1}{N}(\lfloor\xi N\rfloor+1)} \psi(\zeta, x(t), m(t))\,d\zeta = \psi(\xi, x(t), m(t)).$$

which implies

$$\lim_{N\to+\infty} \|g_N(t)\|_{L^2(I)} = 0. \tag{29}$$

We deduce from (28), summing up the contributions of h_N and g_N, that

$$\frac{1}{2}\frac{d}{dt}\|\beta_N(t)\|^2_{L^2(I)} \leq L_\psi \left(\|\alpha_N(t)\|_{L^2(I)} + \|\beta_N(t)\|_{L^2(I)}\right)\|\beta_N(t)\|_{L^2(I)} + \|g_N(t)\|_{L^2(I)}\|\beta_N(t)\|_{L^2(I)}. \tag{30}$$

Applying Gronwall's lemma, we finally get

$$\sup_{t\in[0,T]} \left(\|\alpha_N(t)\|_{L^2(I)} + \|\beta_N(t)\|_{L^2(I)}\right) \leq \left(\|\alpha_N(0)\|_{L^2(I)} + \|\beta_N(0)\|_{L^2(I)} + \int_0^T \|g_N(\tau)\|_{L^2(I)}d\tau\right)e^{KT},$$

and given the choice of initial data, the fact that for all t, $\lim_{N\to+\infty}\|g_N(t)\|_{L^2(I)} = 0$ and using the dominated convergence theorem for the last term, this concludes the proof. □

We finish this subsection by mentioning here other results in the literature in the context of adaptive dynamical networks. In [23], the authors obtain the continuum limit for the following model

$$\begin{cases} \dfrac{d}{dt}x_i^N(t) = \dfrac{1}{N}\displaystyle\sum_{j=1}^{N} w_{ij}^N(t)\phi(x_j^N(t) - x_i^N(t)) \\ \dfrac{d}{dt}w_{ij}^N(t) = -\varepsilon w_{ij}^N + \varepsilon H(x_i^N(t), x_j^N(t)) \end{cases} \tag{31}$$

in the more general context of graphops, a graphop being a bounded self-adjoint and positivity-preserving operator $A : L^\infty(I) \to L^1(I)$ which can be viewed as a generalized concept for the adjacency matrix. Graphops can also be defined using a family of finite measures $(\eta^\xi)_{\xi\in I}$ called fiber measures thanks to the Riesz representation theorem. In the case where η^ξ has a density, i.e.

$$d\eta^\xi(\zeta) = w(\xi,\zeta)d\zeta,$$

then w plays the role of the usual graphon.

In [52], the author obtains a similar result as Theorem 6, under similar regularity assumptions, but without the restriction to have all the edges originating from the same node having the same weight. Finally, if we get back to this restricted case, there exists a result in the presence of a singularity in the weight dynamics. This one can present a jump discontinuity at the origin (see [11] for more details).

3.2 The Mean-Field Limit in the Indistinguishable Case

We introduce a new empirical measure defined by

$$\tilde{\mu}_t^N(x) := \frac{1}{N}\sum_{i=1}^N m_i^N(t)\delta_{x_i^N(t)}(x). \tag{32}$$

We notice that $\tilde{\mu}_t^N$ is invariant by relabeling of the indices, as the classic empirical measure μ_t^N defined in (8), but is also invariant by grouping of the agents, i.e. for every $(x^N, m^N) \in (\mathbb{R}^d)^N \times \mathbb{R}^N$, for every $J \subset \{1, \cdots, N\}$, such that $x_i^N = x_J$ for all $i \in J$, we have

$$\frac{1}{N}\sum_{i=1}^N m_i^N \delta_{x_i^N}(x) = \frac{1}{N}\left[\left(\sum_{i\in J} m_i^N\right)\delta_{x_J}(x) + \sum_{i\in\{1,\cdots,N\}\setminus J} m_i^N \delta_{x_i^N}(x)\right].$$

Therefore, we state the adapted notion of indistinguishability developed in [3].

Definition 2 (Ayi and Pouradier Duteil [3]) We say that the particle system (17) preserves indistinguishability if for all $J \subset \{1, \cdots, N\}$, for all initial conditions $(x^{0,N}, m^{0,N}) \in \mathbb{R}^{dN} \times \mathbb{R}^N$ and $(y^{0,N}, p^{0,N}) \in \mathbb{R}^{dN} \times \mathbb{R}^N$ satisfying

$$\begin{cases} x_i^{0,N} = y_i^{0,N} = x_j^{0,N} = y_j^{0,N} & \text{for all } (i,j) \in J^2 \\ x_i^{0,N} = y_i^{0,N} & \text{for all } i \in \{1, \cdots, N\} \\ m_i^{0,N} = p_i^{0,N} & \text{for all } i \in J^c \\ \sum_{i\in J} m_i^{0,N} = \sum_{i\in J} p_i^{0,N}, \end{cases}$$

the solutions $t \mapsto (x^N(t), m^N(t))$ and $t \mapsto (y^N(t), p^N(t))$ to (17) with respective initial conditions $(x^{0,N}, m^{0,N})$ and $(y^{0,N}, p^{0,N})$ satisfy for all $t \geq 0$,

$$\begin{cases} x_i^N(t) = y_i^N(t) = x_j^N(t) = y_j^N(t) & \text{for all } (i,j) \in J^2 \\ x_i^N(t) = y_i^N(t) & \text{for all } i \in \{1, \cdots, N\} \\ m_i^N(t) = p_i^N(t) & \text{for all } i \in J^c \\ \sum_{i\in J} m_i^N(t) = \sum_{i\in J} p_i^N(t). \end{cases}$$

Naturally, not every weight dynamics preserves this property. A general class of functions for which it stays true can be written as

$$\begin{aligned} \psi_i(m^N(t), x^N(t)) = m_i^N(t)\frac{1}{N^k}\sum_{j_1=1}^N \cdots \sum_{j_k=1}^N m_{j_1}^N(t)\cdots m_{j_k}^N(t) \\ \times S(x_i^N(t), x_{j_1}^N(t), \cdots x_{j_k}^N(t)), \end{aligned} \tag{33}$$

see [3] for a rigorous proof. For conciseness, let us restrict to the case $k = 1$. The graph limit can also be performed for this special class of weights and leads to the convergence of the particle systems to the following integro-differential system

$$\begin{cases} \partial_t x(\xi, t) = \int_I m(\zeta, t)\phi(x(\xi, t) - x(\zeta, t))d\zeta \\ \partial_t m(\xi, t) = m(\xi) \int_I m(\zeta) S(x(\xi), x(\zeta))\, d\zeta. \end{cases} \tag{34}$$

Let us now take an interest into the classic mean-field limit. Similarly as in Sect. 2.2.1, let f be a test function in $C_c^\infty(\mathbb{R}^d)$, then we have

$$\begin{aligned} &\frac{d}{dt}\int f(x)d\tilde{\mu}_t^N(x) \\ &= \frac{d}{dt}\left[\frac{1}{N}\sum_{i=1}^N m_i^N f(x_i^N)\right] = \frac{1}{N}\sum_{i=1}^N \frac{d}{dt}m_i^N f(x_i^N) + \frac{1}{N}\sum_{i=1}^N m_i^N \frac{d}{dt}x_i^N \cdot \nabla f(x_i^N) \\ &= \frac{1}{N^2}\sum_{i=1}^N\sum_{j=1}^N m_i^N m_j^N(t) S(x_i^N(t), x_j^N(t)) f(x_i^N) \\ &\quad + \frac{1}{N^2}\sum_{i=1}^N\sum_{j=1}^N m_i^N m_j^N(t)\phi(x_j^N(t) - x_i^N(t)) \cdot \nabla f(x_i^N) \\ &= \int\int S(x, y)d\tilde{\mu}_t^N(y) f(x)d\tilde{\mu}_t^N(x) + \int\int \phi(y - x)\cdot\nabla f(x)d\tilde{\mu}_t^N(x)d\tilde{\mu}_t^N(y). \end{aligned}$$

Therefore, we deduce that $\tilde{\mu}_t^N$ is a solution to the equation

$$\partial_t \mu_t(x) + \nabla\cdot(V[\mu_t](x)\mu_t(x)) = h[\mu_t](x) \tag{35}$$

with $h[\mu](x) = \int S(x, y)d\mu_t(y)\mu(x)$ and $V[\mu](x) = \int \phi(y - x)d\mu(y)$, a Vlasov-type equation where we notice that the adaptive characteristic of the network appears in the source term. Let $(x, m) \in C([0, T]; L^2(I; \mathbb{R}^d)) \times C([0, T]; L^2(I; \mathbb{R}))$ be the solution to the graph limit Eq. (34) and define the "continuous" empirical measure $\tilde{\mu} \in \mathcal{P}(\mathbb{R}^d)$ as

$$\tilde{\mu}_t(x) := \int_I m(\xi, t)\delta(x - x(\xi, t))d\xi,$$

Then, by similar computations, we get that $\tilde{\mu} \in \mathcal{P}(\mathbb{R}^d)$ satisfies Eq. (35). Then, we can prove the convergence to the Vlasov-type equation by taking advantage of the rigorously established graph limit.

Theorem 7 (Ayi and Pouradier Duteil [3]) *Let (x, m) be the solutions to the graph limit system* (34) *with initial conditions given by $x(0, \cdot) = x_0 \in L^\infty(I; \mathbb{R}^d)$ and $m(0, \cdot) = m_0 \in L^\infty(I; \mathbb{R}^d)$. Let $\tilde{\mu}_t^N$ be the new empirical measure defined* *in* (32) *associated with the solution of the particle system* (17) *with initial conditions given by* (24) *and special weight dynamics*

$$\psi_i(x^N(t), m^N(t)) = \frac{1}{N}\sum_{j=1}^{N} m_i^N(t) m_j^N(t) S(x_i^N(t), x_j^N(t)).$$

Then, for all test function $\varphi \in C_c^\infty(\mathbb{R}^d)$, *and all* $t \in [0, T]$, *it holds*

$$\lim_{N\to\infty} \int_{\mathbb{R}^d} \varphi(x)(d\tilde{\mu}_t^N(x) - d\tilde{\mu}_t(x)) = 0$$

and $\tilde{\mu}_t$ *is solution to the Vlasov-type equation with source term* (35).

The key idea of the proof is to notice that, for all test function $\varphi \in C_c^\infty(\mathbb{R}^d)$,

$$\int_{\mathbb{R}^d} \varphi(x) d\tilde{\mu}_t^N(x) = \int_{\mathbb{R}^d} \varphi(x) d\hat{\mu}_t^N(x),$$

with $\hat{\mu}_t^N \in \mathcal{P}(\mathbb{R}^d)$ being defined as

$$\hat{\mu}_t^N(x) := \int_I m_N(t,\xi)\delta(x - x_N(t,\xi))d\xi.$$

The convergence result follows then directly from the regularity assumptions and from the graph limit

$$\|x - x_N\|_{C([0,T];L^2(I,\mathbb{R}^d))} \xrightarrow[N\to+\infty]{} 0 \quad \text{and} \quad \|m - m_N\|_{C([0,T];L^2(I,\mathbb{R}))} \xrightarrow[N\to+\infty]{} 0.$$

4 Random Graphs

The aim of this section is simply to introduce the reader to the concept of random graphs. Accordingly, we will focus on highlighting their relevance and illustrating them with classic examples and more recent ones. We will conclude by stating what graph limits become in this context.

As explained in [20], random graphs hold significant interest as they provide realistic models for generating graphs, enabling us to capture, understand, and analyze patterns observed in real-world networks. By mimicking the structural characteristics of these graphs, random graph models help evaluate how well they replicate key features, offering insights into the underlying mechanisms of complex systems. Another advantage is their use in extrapolation. Social networks, for instance, are continually growing, and random graph models allow us to

Fig. 6 Pixel representation of realizations of Erdös-Rényi graphs for $p = 0.5$ from left to right for $N = 20$, $N = 100$ and $N = 1000$. Plot of the constant graphon $w(\xi, \zeta) = p$ (right)

scale an existing graph to a larger size while preserving its essential properties. A final point is the utility of randomization for privacy preservation. In cases where anonymization of datasets is insufficient to ensure privacy, randomizing existing graphs becomes a valuable tool for safeguarding sensitive information while retaining the dataset's structural essence for analysis (see [20] for more details).

Definition 3 We call random a graph which is generated by a random process.

- Erdös-Rényi graph [21]

Let $p \in (0, 1)$, we consider a sequence of random graphs

$$G(N, p) = < V(G(N, p)), E(G(N, p)) >$$

with the vertex set $V(G(N, p)) = \{1, \dots, N\}$. Regarding the set of edges, an edge is inserted between a pair of distinct nodes with probability p, i.e. for every $i, j \in \{1, \dots, N\}, i \neq j$,

$$\mathbb{P}\{(i, j) \in E(G(N, p))\} = p.$$

The sequence $G(N, p)$ converges to the constant graphon p as it can be visually observed in Fig. 6.

- Small world graph [54]

We start from the ℓ-nearest graph $G_N = < V(G_N), E(G_N) >$ mentioned in the introduction where $V(G_N) = \{1, \dots, N\}$ and the set of edges is

$$E(G_N) = \{(i, j) \in \{1, 2, \dots, N\}^2 \,|\, 0 < dist(i, j) \leq \ell\}$$

Then, we rewire each edge at random with probability p. This construction can be seen as an interpolation graph between a deterministic graph, for $p = 0$, and a completely random one, for $p = 1$.

- W-Random graph [39]

Let us denote $\overline{\xi} = (\xi_1, \xi_2, \xi_3, \dots)$ and $\overline{\xi}^N = (\xi_1, \xi_2, \dots, \xi_N)$ where for $i \in \mathbb{N}$, ξ_i are independent identical distributed random variables, following a uniform law on $[0, 1]$.

Definition 4 (Medvedev [38]) A W-random graph on N nodes generated by the random sequence $\overline{\xi}$, denoted $G_N = \mathbb{G}(\overline{\xi}_N, W)$ is such that the edges of G_N are selected at random and

$$\mathbb{P}\{(i, j) \in E(G_N)\} = W(\xi_i, \xi_j) \text{ for each } (i, j) \in \{1, \dots, N\}^2 \text{ for } i \neq j.$$

The decision whether to include a pair $(i, j) \in \{1, \dots, N\}^2$ is made independently as for the decisions of other pairs.

Let us now return to the study of particle systems, this time set on a W-random graph, i.e. we consider the following particle system

$$\frac{d}{dt} x_i^N(t) = \frac{1}{N} \sum_{j=1}^{N} \sigma_{ij} \phi(x_j^N(t) - x_i^N(t)) \tag{36}$$

where the law of σ_{ij} knowing $\bar{\xi}$ is nothing else than a Bernoulli of parameter $W(\xi_i, \xi_j)$, i.e. $\mathcal{L}(\sigma_{ij}|\overline{\xi}) = \mathcal{B}(W(\xi_i, \xi_j))$. Then, in [39, 41], the author established the convergence to the following integro-differential equation

$$\partial_t x(t, \xi) = \int_I W(\xi, \zeta) \phi(x(t, \zeta) - x(t, \xi)) d\zeta. \tag{37}$$

Theorem 8 (Medvedev [40]) *Suppose $W : I^2 \to I$, a symmetric measurable function, ϕ a Lipschitz continuous function on $\mathbb{R}$ and g a function belonging to $L^\infty(I) \cap W^{1,2}(0, 1)$. Assume also that $W(\cdot, \zeta) \in W^{1,2}(0, 1)$ for almost every $\zeta \in (0, 1)$.*

Then, for any $T > 0$, the solutions of (36) and (37) satisfy the following relation

$$\lim_{N \to +\infty} N^{\frac{1-\alpha}{4}} \sup_{t \in [0,T]} \|x^N(t) - \mathbf{P}_{\overline{\xi}^N} x(t, \xi)\|_{2,N} = 0 \textit{ almost surely,}$$

where $0 < \alpha < 1$ arbitrary and with $\mathbf{P}_{\overline{\xi}^N} x(t, \xi) = (x(t, \xi_1), x(t, \xi_2), \dots, x(t, \xi_N))$ and

$$(x, y)_N := \frac{1}{N} \sum_{i=1}^{N} x_i y_i, \textit{ and the corresponding norm } \|x\|_{2,N} := \sqrt{(x, x)_N}.$$

Remark 6 We observe that, although the initial particle system incorporates randomness, the limiting equation is entirely deterministic. In this sense, the result can be interpreted as a law of large numbers.

- q-weighted random graphs [4]

As previously, we denote $\overline{\xi} = (\xi_1, \xi_2, \xi_3, \dots)$ and $\overline{\xi}^N = (\xi_1, \xi_2, \dots, \xi_N)$ where for $i \in \mathbb{N}$, ξ_i are i.i.d. random variables with uniform law on $[0, 1]$.

Definition 5 (Ayi and Pouradier Duteil [4]) A q-weighted random graph on N nodes generated by the random sequence $\overline{\xi}$, denoted G_N, is such that the weight of an edge of G_N is randomly attributed. More precisely, the law for the weight of the edge (i, j) is $q(\xi_i, \xi_j, .)$ where

$$q : I \times I \to \mathcal{P}(\mathbb{R}_+)$$

$$(\xi, \zeta) \mapsto q(\xi, \zeta; .).$$

The decision of the attribution of the weight of a pair $(i, j) \in \{1, \dots, N\}^2$ is made independently from the decision for other pairs.

Remark 7 The advantage of this framework is that it provides a unified way to express existing models by simply specifying the law q. For instance, for W-random graphs, we just write

$$q(\xi, \zeta; \cdot) = (1 - W(\xi, \zeta))\delta_0 + W(\xi, \zeta)\delta_1, \qquad \text{for all } \xi, \zeta \in \mathbb{R}.$$

Another example is the Garlaschelli model [22]. It is a weighted random graph model in which the probability of drawing an edge of discrete weight $w \in \mathbb{N}$ between vertices i and j is given by

$$\mathbb{P}(\sigma_{ij} = w) = q_{ij}(w) = p^w(1 - p).$$

We can write it as

$$q(\xi, \zeta; \cdot) = (1 - p)\sum_{i=0}^{+\infty} p^i \delta_i, \qquad \text{for all } \xi, \zeta \in \mathbb{R}.$$

The main result that we want to state in this section is the convergence of the particle system (36) on a q-weighted random graph, i.e. with $\mathcal{L}(\sigma_{ij}|\overline{\xi}) = q(\xi_i, \xi_j; \cdot)$ towards the continuum limit

$$\partial_t x(t, \xi) = \int_I \left(\int_{\mathbb{R}_+} w q(\xi, \zeta; dw) \right) \phi(x(t, \zeta) - x(t, \xi)) d\zeta \tag{38}$$

In order to establish it, some assumptions and bounds are needed.

Hypothesis 4 *Let* $\phi \in L^\infty(\mathbb{R})$ *be bounded and Lipschitz continuous, with* $\|\phi\|_{\text{Lip}} := L$ *and* $\|\phi\|_{L^\infty(\mathbb{R})} := K$.

Hypothesis 5 *There exists* $M > 0$ *such that for all* $(\xi, \zeta) \in I^2$*, for all* $k \in \{1, \cdots, 4\}$,

$$\left(\int_{\mathbb{R}_+} w^k q(\xi, \zeta; dw) \right)^{1/k} \leq M,$$

i.e. the first four moments of the probability measure $q(\xi, \zeta; \cdot)$ *are bounded uniformly in* ξ *and* ζ.

We can now conclude this section by properly state the convergence result.

Theorem 9 (Ayi and Pouradier Duteil [4]) *Let* ϕ *satisfy Hypothesis 1, let* $g \in L^\infty(I)$ *and let* q *be a weighted random graph law satisfying Hypothesis 2. Then, as* N *goes to infinity, solution* x^N *to the discrete system* (36) *with* $\mathcal{L}(\sigma_{ij}|\overline{\xi}) = q(\xi_i, \xi_j; \cdot)$ *and initial data* $x_i^N(0) = g(\xi_i)$ *for* $i \in \{1, \ldots, N\}$ *converges to the solution* x *of the continuous model* (38) *with initial data* $x(0, \xi) = g(\xi)$ *for* $\xi \in I$. *More precisely,*

$$\mathbb{P}\left[\sup_{t \in [0,T]} \|x^N(t) - \mathbf{P}_{\overline{\xi}^N} x(\cdot, t)\|_{2,N} \geq \frac{C_1(T)}{\sqrt{N}} \right] \leq \frac{\tilde{C}_1}{N}$$

where the constants $C_1(T)$ *and* $\tilde{C}_1$ *are respectively defined by* $C_1(T) := \sqrt{T}\sqrt{1 + M^2K^2} e^{(\frac{1}{2}+4ML)T}$ *and* $\tilde{C}_1 := 3M^4K^4 + 6$.

Remark 8 The first-order moment $\int_{\mathbb{R}_+} wq(\xi, \zeta; dw)$ plays the role of the graphon. Therefore, in the case of a Bernoulli distribution with parameter $W(\xi, \zeta)$, i.e. for W-random graphs, Theorem 8 is indeed recovered.

Remark 9 We do not wish to get into the details of the proof but we can mention here what we consider like some of the keys. The proof mainly relies on exploiting the independence of the involved random variables and a clever application of the Bienaymé-Chebyshev inequality. We refer the reader to the paper [4] for seeing the full picture and discovering where those keys come into play.

5 Hypergraphs

Many existing models focus on binary interactions, yet real-life dynamics often involve group interactions that go beyond pairs of individuals, such as physical meetings or virtual group chats. In these scenarios, opinions can be shaped collectively, and decisions are often the result of joint influences. To address this, we adopt the natural framework of hypergraphs. Unlike traditional graphs that connect pairs of nodes, hypergraphs generalize this concept by allowing connections among multiple nodes. In recent years, the study of opinion dynamics on hypergraphs [13, 17, 36, 42, 49, 53, 56–58] has gained increasing attention, highlighting the growing recognition of hypergraphs as a valuable tool for capturing the complexities and subtleties of group interactions with greater accuracy and realism.

5.1 Hypergraph Theory

We start by recalling the definition of an hypergraph (see Fig. 7 for an example).

Definition 6 (Hypergraph) A weighted hypergraph (or simply hypergraph) consists in a triple $H = (V, E, W)$, where V is a finite set, $E \subset 2^V$ is a finite collection of subsets of V, and $W : 2^V \longrightarrow \mathbb{R}_+$ satisfies $W(e) > 0$ if, and only if, $e \in E$.

Elements in $V = \{1, \dots, N\}$ are called *nodes (or vertices)*, elements in E are called *hyperedges*, and W is called the *weight function*.

For each $\ell \in \{1, \dots, N-1\}$, we define the $(\ell+1)$*-order adjacency tensor*

$$(w^{\ell,N}_{ij_1\cdots j_\ell})_{1\le i,j_1,\dots,j_\ell\le N} : w^{\ell,N}_{ij_1\cdots j_\ell} := W(i, j_1, \dots, j_\ell), \quad i, j_1, \cdots, j_\ell \in \{1, \dots, N\}.$$

We introduce the space of *hypergraphons of unbounded rank* [6]. Given $M > 0$, it is defined as

$$\mathcal{H}_M := \left\{ w = (w_\ell)_{\ell\in\mathbb{N}} : \begin{array}{l} w_\ell \in L^\infty_+([0,1]^{\ell+1}),\ \|w_\ell\|_{L^\infty} \le M, \\ \text{and } w_\ell \text{ is symmetric for all } \ell \in \mathbb{N} \end{array} \right\}.$$

This space becomes a metric space when equipped with the cut-distance, defined as follows: for any $w, \bar{w} \in \mathcal{H}_M$, for all $\ell \in \mathbb{N}$, the ℓt-th order labeled cut distance is

$$d_{\square,\ell}(w_\ell, \bar{w}_\ell) := \sup_{S, S_1, \dots, S_\ell \subset [0,1]} \left| \int_{S\times S_1\times\cdots\times S_\ell} (w_\ell - \bar{w}_\ell)\, d\xi\, d\xi_1 \dots d\xi_\ell \right|,$$

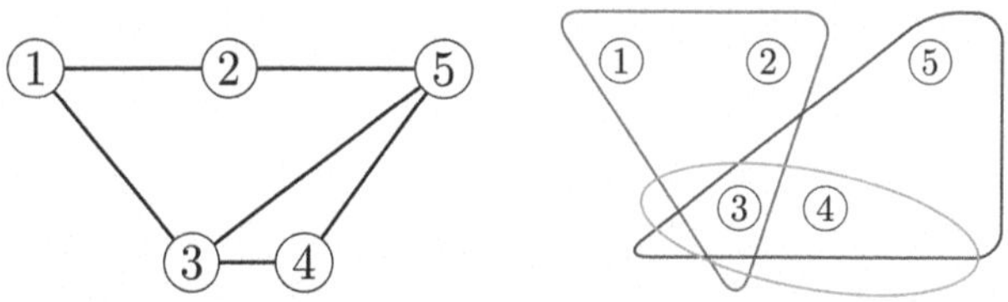

Fig. 7 Examples of graph (left) and hypergraph (right). The hyperedges of this hypergraph are {1,2,3}, {3,4,5} and {3,4}

and for any strictly positive summable sequence $(\alpha_\ell)_{\ell\in\mathbb{N}}$, the labeled cut distance is

$$d_\square(w,\bar{w};(\alpha_\ell)_{\ell\in\mathbb{N}}) := \sum_{\ell=1}^{\infty} \alpha_\ell\, d_{\square,\ell}(w_\ell,\bar{w}_\ell),$$

and the unlabeled cut distance is $\delta_\square(w,\bar{w};(\alpha_\ell)_{\ell\in\mathbb{N}}) = \inf_\Phi d_\square(w,\bar{w}^\Phi;(\alpha_\ell)_{\ell\in\mathbb{N}})$ where $\Phi : [0,1] \longrightarrow [0,1]$ is a bijective, measure-preserving map and

$$\bar{w}_\ell^\Phi(\xi,\xi_1,\ldots,\xi_\ell) = \bar{w}_\ell(\Phi(\xi),\Phi(\xi_1),\ldots,\Phi(\xi_\ell)).$$

Similarly as in the binary case, for any sequence of hypergraphs $(H_N)_{N\in\mathbb{N}}$, we can build a piecewise-constant function which is associated. For all $\ell \in \mathbb{N}$, for all $(\xi,\xi_1,\cdots,\xi_\ell) \in [0,1]^{\ell+1}$, we define

$$w_\ell^{H_N}(\xi,\xi_1,\cdots,\xi_\ell) := \sum_{i,j_1,\cdots j_\ell=1}^{N} w_{ij_1\cdots j_\ell}^{\ell,N} N^\ell \mathbb{1}_{I_i^N\times I_{j_1}^N\times\cdots I_{j_\ell}^N}(\xi,\xi_1,\cdots,\xi_\ell), \tag{39}$$

if $\ell \le N-1$ and

$$w_\ell^{H_N}(\xi,\xi_1,\cdots,\xi_\ell) := 0$$

if $\ell \ge N$.

Definition 7 (Convergence of a Sequence of Hypergraphs) The sequence of hypergraphs $(H_N)_{N\in\mathbb{N}}$ is said to converge to a UR-hypergraphon w when

$$\lim_{N\to\infty} \delta_\square(w,w^{H_N};(\alpha_\ell)_{\ell\in\mathbb{N}}) = 0$$

for some positive and summable sequence $(\alpha_\ell)_{\ell\in\mathbb{N}}$.

Example 3 Consider the sequence of hypergraphs $(H_N)_{N\in\mathbb{N}}$ for homogeneous groups defined for some radius $\theta \in (0,1)$ by the hyperedge weights $(w_{ij_1\cdots j_\ell}^{\ell,N})_{i,j_1,\ldots,j_\ell\in[\![1,N]\!]}$ given by

$$w_{ij_1\cdots j_\ell}^{\ell,N} = \begin{cases} \dfrac{1}{N^\ell}, & \text{if} \quad \max_{k_1,k_2\in\{i,j_1,\cdots,j_\ell\}} |k_1-k_2| \le \theta N, \\ 0, & \text{otherwise,} \end{cases}$$

for each $N \in \mathbb{N}$ and $\ell \in \{1,\ldots,N-1\}$. Then, $(H_N)_{N\in\mathbb{N}}$ converges in the labeled cut-distance $d_\square$ (and hence also in the unlabeled cut-distance $\delta_\square$) to the UR-hypergraph $w = (w_\ell)_{\ell\in\mathbb{N}}$ given by

$$w_\ell(\xi_0, \ldots, \xi_\ell) = \begin{cases} 1, & \text{if} \quad \max_{k_1,k_2\in\{1,\ldots,\ell\}} |\xi_{k_1} - \xi_{k_2}| \leq \theta, \\ 0, & \text{otherwise,} \end{cases}$$

for all $\ell \in \mathbb{N}$.

A crucial property of this metric space is its compactness.

Proposition 1 (Compactness of UR-hypergraphons) *The unlabeled cut distance $\delta_\square$ induces topologically equivalent pseudometrics on the space of UR-hypergraphons $\mathcal{H}_M$ which do not depend on the summable sequence $(\alpha_\ell)_{\ell\in\mathbb{N}}$. When induced on the quotient $\mathcal{H}_M/\sim$ that identifies UR-hypergraphons identical modulo rearrangement, the space becomes a compact metric space.*

Thus, we deduce from the above property that any sequence of hypergraphs such that there exists $M > 0$ for which we have the bound

$$\sup_{N\in\mathbb{N}} \max_{1\leq\ell\leq N-1} \max_{1\leq i,j_1,\cdots,j_\ell\leq N} N^\ell w_{ij_1\cdots j_\ell}^{\ell,N} \leq M, \tag{40}$$

converges toward some limiting UR-hypergraphon up to a subsequence. Besides, similarly as in the binary case, UR-hypergraphons can be suitably approximated in the cut distance by sequences of finite hypergraphs through L^1 and pointwise approximations.

Proposition 2 *Let $w = (w_\ell)_{\ell\in\mathbb{N}} \in \mathcal{H}_M$ be a UR-hypergraphon and $(H_N)_{N\in\mathbb{N}}$ be the sequence of hypergraphs whose adjacency tensors $(w^{\ell,N})_{\ell\in\mathbb{N}}$ are defined by L^1-approximations of the UR-hypergraphon w, i.e. for all $(i, j_1 \cdots, j_\ell) \in [\![1, N]\!]^{\ell+1}$,*

$$w_{ij_1\cdots j_\ell}^{\ell,N} = \begin{cases} N \displaystyle\int_{I_i^N\times I_{j_1}^N\times\cdots\times I_{j_\ell}^N} w_\ell(\xi, \xi_1, \cdots, \xi_\ell)\, d\xi\, d\xi_1 \cdots d\xi_\ell, , & \textit{if } \ell \in [\![1, N-1]\!], \\ 0, & \textit{otherwise.} \end{cases}$$

Then, the sequence of hypergraphs $(H_N)_{N\in\mathbb{N}}$ converges as N tends to infinity to the UR-hypergraphon w in the labeled cut-distance $d_\square$ (and hence also in the unlabeled cut-distance $\delta_\square$).

Proposition 3 *Let $w = (w_\ell)_{\ell\in\mathbb{N}} \in \mathcal{H}_M$ be a UR-hypergraphon, and suppose that w_ℓ is continuous for all $\ell \in \mathbb{N}$.*
Let $(\bar{H}_N)_{N\in\mathbb{N}}$ be the sequence of hypergraphs whose adjacency tensors $(\bar{w}^{\ell,N})_{\ell\in\mathbb{N}}$ are defined by evaluating the UR-hypergraphon w on the grid points $(\bar{\xi}_i^N)_{i\in[\![1,N]\!]}^{\ell+1}$ where $\bar{\xi}_i^N \in \left[\frac{i-1}{N}, \frac{i}{N}\right)$ for all $i \in [\![1, N]\!]$, i.e.

$$\bar{w}_{ij_1\cdots j_\ell}^{\ell,N} = \begin{cases} \dfrac{1}{N^\ell} w_\ell(\bar{\xi}_i^N, \bar{\xi}_{j_1}^N, \cdots, \bar{\xi}_{j_\ell}^N), & \textit{if } \ell \in [\![1, N-1]\!], \\ 0, & \textit{otherwise.} \end{cases}$$

Then, the sequence of hypergraphs $(\bar{H}_N)_{N\in\mathbb{N}}$ converges as N tends to infinity to the UR-hypergraphon w in the labeled cut-distance $d_\square$ (and hence also in the unlabeled cut-distance $\delta_\square$).

5.2 Mean-Field Limit in the Presence of Hypergraphs

We are interested in the following multi-agent system with high-order interactions:

$$\begin{cases} \dfrac{dX_i^N(t)}{dt} = \displaystyle\sum_{\ell=1}^{N-1} \sum_{j_1,\dots,j_\ell=1}^{N} w_{ij_1\cdots j_\ell}^{\ell,N} K_\ell(X_i^N(t), X_{j_1}^N(t), \dots, X_{j_\ell}^N(t)), \\ X_i^N(0) = X_{i,0}^N, \qquad i \in \{1, \cdots, N\}. \end{cases} \tag{41}$$

We can find several examples of higher-order multi-agent dynamics in the literature.

Example 4 In [50], they study an extension of the Kuramoto-Saraguchi model on hypergraphs,

$$\begin{aligned} \frac{d}{dt}x_i^N &= \sum_{j_1=1}^{N} w_{ij_1}^{N,1} \sin(x_{j_1}^N - x_i^N) + \sum_{j_1=1}^{N}\sum_{j_2=1}^{N} w_{ij_1j_2}^{N,2} \sin(2x_{j_1}^N - x_{j_2}^N - x_i^N) \\ &+ \sum_{j_1=1}^{N}\sum_{j_2=1}^{N}\sum_{j_3=1}^{N} w_{ij_1j_2j_3}^{N,3} \sin(x_{j_1}^N + x_{j_2}^N - x_{j_3}^N - x_i^N). \end{aligned}$$

Example 5 In [42], they propose to study a model for higher-order opinion dynamics on a uniform hypergraph of rank 2,

$$\frac{d}{dt}x_i = \sum_{j_1=1}^{N}\sum_{j_2=1}^{N} w_{ij_1j_2}^{N,2} e^{\lambda|x_{j_1}-x_{j_2}|}\left(\frac{x_{j_1}+x_{j_2}}{2} - x_i\right).$$

The main result of this section is the mean-field limit to the Vlasov-type equation

$$\begin{cases} \partial_t \mu_t^\xi + \mathrm{div}_x(F_w[\mu_t](\cdot,\xi)\,\mu_t^\xi) = 0, \quad t \geq 0,\ x \in \mathbb{R}^d,\ \xi \in [0,1], \\ \mu_{t=0}^\xi = \mu_0^\xi. \end{cases} \tag{42}$$

where

$$F_w[\mu_t](x,\xi) := \sum_{\ell=1}^{\infty} \int_{[0,1]^\ell} w_\ell(\xi, \xi_1, \dots, \xi_\ell) \times \left(\int_{\mathbb{R}^{d\ell}} K_\ell(x, x_1, \dots, x_\ell)\, d\mu_t^{\xi_1}(x_1) \cdots d\mu_t^{\xi_\ell}(x_\ell) \right) d\xi_1, \dots d\xi_\ell, \tag{43}$$

and $w = (w_\ell)_{\ell \in \mathbb{N}}$ is a UR-hypergraphon established in [6].

Theorem 10 (Ayi et al. [6]) *Assume that the kernels K_ℓ satisfy some regularity assumptions and the coupling weights the scaling (40). Suppose additionally that both of them satisfy some symmetries. For any $(X_{1,0}^N, \dots, X_{N,0}^N)$ with i.i.d. $X_{i,0}^N$ (but N dependent law) such that there exists $p \in [1, 2]$ for which $X_{i,0}^N$ satisfies*

$$\sup_{N \in \mathbb{N}} \max_{1 \le i \le N} \mathbb{E}|X_{i,0}^N|^p < \infty,$$

consider the unique solutions $(X_1^N, \dots, X_N^N)$ to the microscopic dynamics. Then, there is a subsequence $N_k \to \infty$ such that the mean-field limit of the multi-agent system is characterized in a suitable sense by a solution to the Vlasov-type equation for some $(\mu_t^\xi)_{\xi \in [0,1]} \subset \mathcal{P}(\mathbb{R}^d)$ and some $w = (w_\ell)_{\ell \in \mathbb{N}}$ such that $\sup_{\ell \in \mathbb{N}} \|w_\ell\|_{L^\infty} \le M$.

We refer the reader to [6] for more precise statement on the regularity and symmetry assumptions.

Remark 10 Regarding the regularity assumptions on the kernels, we actually not only need them to be fulfilled, i.e. in our case, for the kernels to be bounded Lipschitz and to belong to the Sobolev space $H^{\frac{d(\ell+1)}{2}+\varepsilon}$ for some $\varepsilon > 0$, but we need a stronger control on the associated norms as ℓ goes to ∞. For instance, we will need to have

$$\sum_{\ell=1}^{\infty} \frac{4^\ell \pi^{\frac{d\ell}{4}}}{\sqrt{\Gamma(\frac{d\ell}{2})}} \|K_\ell\|_{H^{\frac{d(\ell+1)}{2}+\varepsilon}} < \infty, \tag{44}$$

where Γ represents the Gamma function $\Gamma(a) := \int_0^\infty s^{a-1} e^{-s}\, ds$ for $a > 0$.

Remark 11 We also highlight the existence of a mean-field limit result in the context of directed hypergraph measures [31], a generalization of the notion of digraph measures introduced in [30]. It is worth noting that this concept differs from the hypergraph limits commonly defined in the (hyper)graph theory literature. Specifically, it is not naturally linked to classic notions such as homomorphism density or cut-distance. However, unlike our framework, it allows for handling graphs that are sparse or have intermediate densities.

5.3 Functional Setting

Solutions to the Vlasov equation (42)–(43) will be parametrized families $(\mu^\xi)_{\xi\in[0,1]} \subset \mathcal{P}(\mathbb{R}^d)$ of probability densities. Thus, we introduce the following proper definition.

Definition 8 (Borel Family of Probability Measures) Consider any $\nu \in \mathcal{P}([0,1])$, and let $(\mu^\xi)_{\xi\in[0,1]} \subset \mathcal{P}(\mathbb{R}^d)$ be a parametrized family of probability measures defined for ν-a.e. $\xi \in [0,1]$. We say that $(\mu^\xi)_{\xi\in[0,1]}$ is a Borel family if the map $\xi \in [0,1] \longmapsto \mu^\xi(B)$ is Borel-measurable for every Borel set $B \subset \mathbb{R}^d$.

It is known that Borel families of probability measures can be alternatively regarded, hence the following reformulation.

Definition 9 (Fibered Probability Measures) Consider any $\nu \in \mathcal{P}([0,1])$. We define the space of fibered probability measures by

$$\mathcal{P}_\nu(\mathbb{R}^d \times [0,1]) := \{\mu \in \mathcal{P}(\mathbb{R}^d \times [0,1]) : \pi_{\xi\#}\mu = \nu\},$$

where $\pi_\xi(x,\xi) = \xi$ is the projection on the second component, and therefore $\pi_{\xi\#}\mu$ stands for the marginal of μ in the second component.

Indeed, the link between the two above definitions follows from the following classic result.

Theorem 11 (Disintegration Theorem) *Consider any $\nu \in \mathcal{P}([0,1])$ and $\mu \in \mathcal{P}_\nu(\mathbb{R}^d \times [0,1])$, then there exists a ν-a.e. uniquely defined Borel family $(\mu^\xi)_{\xi\in[0,1]} \subset \mathcal{P}(\mathbb{R}^d)$ so that*

$$\iint_{\mathbb{R}^d\times[0,1]} \varphi(x,\xi)\,d\mu(x,\xi) = \int_0^1 \left(\int_{\mathbb{R}^d} \varphi(x,\xi)\,d\mu^\xi(x)\right) d\nu(\xi),$$

for every bounded Borel-measurable map $\varphi : \mathbb{R}^d \times [0,1] \longrightarrow \mathbb{R}$. Conversely, given any Borel family $(\mu^\xi)_{\xi\in[0,1]}$, then we can associate a unique fibered probability measure $\mu \in \mathcal{P}_\nu(\mathbb{R}^d \times [0,1])$ so that the above formula holds true, and for simplicity, we shall write $\mu(x,\xi) = \mu^\xi(x) \otimes \nu(\xi)$.

Remark 12 We will not state it here but, since we work with a full curve $t \in [0,T] \mapsto \mu_t \in \mathcal{P}_\nu(\mathbb{R}^d \times [0,1])$ and not an isolated measure $\mu \in \mathcal{P}_\nu(\mathbb{R}^d \times [0,1])$, fortunately, a time-dependent disintegration can also be obtained (see [6] for more details).

We endow the space of fibered probability measures with the following metric.

Definition 10 Consider any $\nu \in \mathcal{P}([0,1])$ and any $p \in [1,\infty]$, we define

$$\mathcal{P}_{p,\nu}(\mathbb{R}^d \times [0,1]) := \left\{\mu \in \mathcal{P}_\nu(\mathbb{R}^d \times [0,1]) : \int_0^1 d_{\mathrm{BL}}^p(\mu^\xi, \delta_0)\,d\nu(\xi) < \infty\right\},$$

and

$$d_{p,\nu}(\mu_1, \mu_2) := \left(\int_0^1 d_{\mathrm{BL}}^p(\mu_1^\xi, \mu_2^\xi)\, d\nu(\xi) \right)^{1/p}, \quad \mu_1, \mu_2 \in \mathcal{P}_{p,\nu}(\mathbb{R}^d \times [0,1]).$$

Thus, we can reformulate (42) as follows:

$$\begin{cases} \partial_t \mu_t + \mathrm{div}_x(F_w[\mu_t]\,\mu_t) = 0, & t \geq 0,\ x \in \mathbb{R}^d,\ \xi \in [0,1], \\ \mu_{t=0} = \mu_0, \end{cases} \tag{45}$$

with $\mu_t \in \mathcal{P}_\nu(\mathbb{R}^d \times [0,1])$.

5.4 *Strategy of The Proof*

We go back to the proof of the mean-field limit. The key idea is to introduce the following auxiliary multi-agent system:

$$\begin{cases} \dfrac{d\bar{X}_i^N}{dt} = \displaystyle\sum_{\ell=1}^{N-1} \sum_{j_1,\ldots,j_\ell=1}^{N} w_{ij_1\cdots j_\ell}^{\ell,N}\, \mathbb{E}_i^N\, K_\ell(\bar{X}_i^N, \bar{X}_{j_1}^N, \ldots, \bar{X}_{j_\ell}^N), \\ \bar{X}_i^N(0) = X_{i,0}^N. \end{cases} \tag{46}$$

where $\mathbb{E}_i^N = \mathbb{E}[\,\cdot\,|\bar{\mathcal{F}}_i^N]$ denotes the expectation conditioned to the natural filtration $\bar{\mathcal{F}}_i^N$ of $\bar{X}_i^N$, i.e. the family of time-dependent sub-σ-algebras defined by

$$\bar{\mathcal{F}}_i^N(t) = \sigma(\{\bar{X}_i^N(s) : 0 \leq s \leq t\}). \tag{47}$$

We can prove that this system actually approximates the original multi-agent system (41) as N goes to ∞ with the following result.

Lemma 3 (Error Estimate) *Assume that the kernels K_ℓ satisfy some regularity assumptions and the coupling weights the scaling* (40). *Suppose additionally that both of them satisfy some symmetries. For any $(X_{1,0}^N, \ldots, X_{N,0}^N)$ with independent $X_{i,0}^N$ such that $\mathbb{E}|X_{i,0}^N|^p < \infty$ for some $p \in [1,2]$, consider the unique solutions $(X_1^N, \ldots, X_N^N)$ to* (41) *and $(\bar{X}_1^N, \ldots, \bar{X}_N^N)$ to* (46). *Then,*

$$\left(\frac{1}{N} \sum_{i=1}^{N} \mathbb{E}|X_i^N(t) - \bar{X}_i^N(t)|^p \right)^{1/p} \leq A_1^N e^{A_2^N t}, \tag{48}$$

for all $t \geq 0$, where the constants A_1^N, A_2^N only depend on the regularity, bounds and scaling assumptions.

We denote the laws associated to the intermediate particle system

$$\bar{\lambda}_t^{N,i} := \mathrm{Law}(\bar{X}_i^N(t)), \quad t \geq 0,\ 1 \leq i \leq N.$$

Proposition 4 *Then, $(\bar{\lambda}^{N,i})_{1\leq i\leq N}$ is a solution in distributional sense to the following coupled PDE system*

$$\begin{cases} \partial_t \bar{\lambda}_t^{N,i} + \mathrm{div}_x(F_i^N[\bar{\lambda}_t^{N,1}, \cdots, \bar{\lambda}_t^{N,N}]\bar{\lambda}_t^{N,i}) = 0, & t \geq 0,\ x \in \mathbb{R}^d,\ 1 \leq i \leq N, \\ \bar{\lambda}_0^{N,i} = \mathrm{Law}(X_{i,0}^N), \end{cases}$$

where

$$F_i^N[\bar{\lambda}_t^{N,1}, \cdots, \bar{\lambda}_t^{N,N}](x)$$
$$= \sum_{\ell=1}^{N-1} \sum_{j_1,\cdots,j_\ell=1}^{N} w_{ij_1\cdots j_\ell}^{\ell,N} \int_{\mathbb{R}^{d\ell}} K_\ell(x, x_1, \ldots, x_\ell)\, d\bar{\lambda}_t^{N,1}(x_1)\cdots d\bar{\lambda}_t^{N,N}(x_N).$$

We can finally introduce a graphon reformulation to make the connection with the Vlasov-type equation. For every $N \in \mathbb{N}$, and $t \in \mathbb{R}_+, \xi \in [0, 1]$ we define

$$\bar{\mu}_t^{N,\xi} := \sum_{i=1}^{N} \mathbb{1}_{I_i^N}(\xi)\, \bar{\lambda}_t^{N,i}.$$

Lemma 4 *Under the previous assumptions, consider the unique solution $(\bar{X}_1^N, \ldots, \bar{X}_N^N)$ to the intermediate particle system, their associated laws $(\bar{\lambda}^{N,i})_{1\leq i\leq N}$ and the graphon reformulation $(\bar{\mu}^N, w^N)$ with $w^N := w^{H_N}$ defined in (39). Then, $\bar{\mu}^N$ is a distributional solution to the Vlasov equations (45)–(43) with hypergraphon $w^N = (w_\ell^N)_{\ell\in\mathbb{N}}$ and initial datum $\bar{\mu}_{t=0}^{N,\xi} = \sum_{i=1}^N \mathbb{1}_{I_i^N}(\xi)\,\mathrm{Law}(X_{i,0})$.*

We state the final main tool that we will be using to do a sketch of the proof, the stability estimate for the Vlasov equation.

Theorem 12 (Stability Estimate for the Vlasov Equation) *For any initial data $\mu_0, \bar{\mu}_0 \in \mathcal{P}_{p,\nu}(\mathbb{R}^d \times [0, 1])$ with $p \in [1, \infty)$, let $\mu, \bar{\mu} \in C(\mathbb{R}_+, \mathcal{P}_{p,\nu}(\mathbb{R}^d \times [0, 1]))$ be the unique global-in-time distributional solutions issued at μ_0 with given w (respectively, $\bar{\mu}_0$ and $\bar{w}$). Then, we have*

$$d_{p,\nu}(\mu_t, \bar{\mu}_t) \leq e^{(C_p+L_F)t}\left(d_{p,\nu}(\mu_0, \bar{\mu}_0) + \frac{D_\infty^{1/q}}{L_F} d_\square(w, \bar{w}; (4^\ell \|\hat{K}_\ell\|_{L^1})_{\ell\in\mathbb{N}})^{1/p}\right),$$

for every $t \geq 0$, where $\frac{1}{p} + \frac{1}{q} = 1$, C_p ,L_F, D_∞ are bounded by constants which only depend on the regularity, bounds and scaling assumptions and $(4^\ell \|\hat{K}_\ell\|)_{\ell\in\mathbb{N}}$ is summable for the same reasons.

We conclude this section by giving a sketch of the proof of the main result, the mean-field limit to the Vlasov equation. We define

$$\mu_t^{N,\xi} := \sum_{i=1}^{N} \mathbb{1}_{I_i^N}(\xi)\delta_{X_i^N(t)}.$$

Sketch of the Proof of Theorem 10 Thanks to the bounds on the weights and the initial data, there exists $w = (w_\ell)_{\ell\in\mathbb{N}} \in \mathcal{H}_M$, $\mu_0 \in \mathcal{P}_{p,\nu}(\mathbb{R}^d \times [0,1])$ and $\Phi_k : [0,1] \longrightarrow [0,1]$ measure-preserving maps such that

$$d_\square(w^{N_k,\Phi_k}, w) \to 0, \tag{49}$$

and

$$d_{p,\nu}(\bar{\mu}_0^{N_k,\Phi_k}, \mu_0) \to 0. \tag{50}$$

for a suitable subsequence N_k (see [6] for more details). Let us define $\mu \in C(\mathbb{R}_+, \mathcal{P}_{p,\nu}(\mathbb{R}^d \times [0,1]))$ to be the distributional solution to (45)–(43) with given w and initial datum μ_0.

- **Step 1**: Comparing μ^N and $\bar{\mu}^N$.
  Let ξ be in $[0,1]$, we fix $i \in [\![1, N]\!]$ such that $\xi \in I_i^N$. This implies that $\mu_t^{N,\xi} = \delta_{X_i^N(t)}$ and $\bar{\mu}_t^{N,\xi} = \bar{\lambda}_t^{N,i}$. Thus, we deduce that

$$\begin{aligned}
d_{\mathrm{BL}}(\mathbb{E}\mu_t^{N,\xi}, \bar{\mu}_t^{N,\xi}) &= d_{\mathrm{BL}}(\mathbb{E}\delta_{X_i^N(t)}, \bar{\lambda}_t^{N,i}) \\
&= \sup_{\|\phi\|_{\mathrm{BL}}\leq 1} \int_{\mathbb{R}^d} \phi(x)\, d(\mathbb{E}\delta_{X_i^N(t)}(x) - \bar{\lambda}_t^{N,i}(x)) \\
&= \sup_{\|\phi\|_{\mathrm{BL}}\leq 1} \int_{\mathbb{R}^d} \phi(x)\, d(\mathbb{E}\delta_{X_i^N(t)}(x) - \mathbb{E}\delta_{\bar{X}_i^N(t)}(x)) \\
&= \sup_{\|\phi\|_{\mathrm{BL}}\leq 1} \mathbb{E}\left[\phi(X_i^N(t)) - \phi(\bar{X}_i^N(t))\right] \\
&\leq \mathbb{E}\left|X_i^N(t) - \bar{X}_i^N(t)\right|,
\end{aligned}$$

  for all $1 \leq i \leq N$. We deduce that

$$d_{p,\nu}(\mathbb{E}\mu_t^N, \bar{\mu}_t^N) \le \left(\frac{1}{N} \sum_{i=1}^N \mathbb{E}|X_i^N(t) - \bar{X}_i^N(t)|^p \right)^{1/p}.$$

- **Step 2**: Comparing $\bar{\mu}^{N_k,\Phi_k}$ and μ.
By Lemma 4, we know that $\bar{\mu}^{N_k}$ is a distributional solution to (45)–(43) given w^{N_k}. Moreover, since Φ_k are measure-preserving maps, then $\bar{\mu}^{N_k,\Phi_k}$ is also a distributional solution to (45)–(43) given w^{N_k,Φ_k}. Therefore, we can use the stability estimate in Theorem 12 to compare it to the distributional solution μ introduced above. More precisely, we have

$$\begin{aligned} \sup_{t\in[0,T]} d_{p,\nu}(\bar{\mu}_t^{N_k,\Phi_k}, \mu_t) &\le e^{(C_p+L_F)T} \Big(d_{p,\nu}(\bar{\mu}_0^{N_k,\Phi_k}, \mu_0) \\ &+ \frac{D_\infty^{1/q}}{L_F} d_\square(w^{N_k,\Phi_k}, w; (2^\ell \|\hat{K}_\ell\|_{L^1})_{\ell\in\mathbb{N}})^{1/p} \Big), \end{aligned} \tag{51}$$

for all $T \in \mathbb{R}_+$.
- **Step 3**: Comparing μ^{N_k,Φ_k} and μ.
To conclude, we use the triangle inequality and take expectations and supremum over $t \in [0, T]$ to get

$$\begin{aligned} &\sup_{t\in[0,T]} d_{p,\nu}(\mathbb{E}\mu^{N_k,\Phi_k}, \mu_t) \\ &\le \sup_{t\in[0,T]} d_{p,\nu}(\mathbb{E}\mu_t^{N_k,\Phi_k}, \bar{\mu}_t^{N_k,\Phi_k}) + \sup_{t\in[0,T]} d_{p,\nu}(\bar{\mu}^{N_k,\Phi_k}, \mu_t) \\ &= \sup_{t\in[0,T]} d_{p,\nu}(\mathbb{E}\mu_t^{N_k}, \bar{\mu}_t^{N_k}) + \sup_{t\in[0,T]} d_{p,\nu}(\bar{\mu}^{N_k,\Phi_k}, \mu_t) \\ &\le \sup_{t\in[0,T]} \left(\frac{1}{N} \sum_{i=1}^N \mathbb{E}|X_i^N(t) - \bar{X}_i^N(t)|^p \right)^{1/p} + e^{(C_p+L_F)T} \Big(d_{p,\nu}(\bar{\mu}_0^{N_k,\Phi_k}, \mu_0) \\ &+ \frac{D_\infty^{1/q}}{L_F} d_\square(w^{N_k,\Phi_k}, w; (2^\ell \|\hat{K}_\ell\|_{L^1})_{\ell\in\mathbb{N}})^{1/p} \Big), \end{aligned}$$

where we also used that L^p norms are invariant under measure-preserving rearrangements and **Step 1** and **Step 2**. We conclude by using the convergence of initial data in (50) and the convergence of UR-hypergraphons in (49) and the control of the error of the approximation by the intermediate particle system obtained in Lemma 3. □

Acknowledgments NA has received funding from the Emergence grant of Sorbonne University.

References

1. I. Aoki, A simulation study on the schooling mechanism in fish. Nippon Suisan Gakkaishi **48**(8), 1081–1088 (1982)
2. S. Aref, D. Friggens, S. Hendy, Analysing scientific collaborations of new zealand institutions using scopus bibliometric data, in *Proceedings of the Australasian Computer Science Week Multiconference, ACSW '18, New York, NY, USA, 2018* (Association for Computing Machinery, 2018)
3. N. Ayi, N. Pouradier Duteil, Mean-field and graph limits for collective dynamics models with time-varying weights. J. Differ. Equ. **299**, 65–110 (2021)
4. N. Ayi, N. Pouradier Duteil, Graph limit for interacting particle systems on weighted random graphs. Math. Models Methods Appl. Sci. **36**, 1129–1174 (2026)
5. N. Ayi, N. Pouradier Duteil, Large-population limits of non-exchangeable particle systems. *Active Particles*, vol. 4 (Springer, Berlin, 2024)
6. N. Ayi, N.P. Duteil, D. Poyato, Mean-field limit of non-exchangeable multi-agent systems over hypergraphs with unbounded rank. arXiv:2406.04691
7. G. Backhausz, B. Szegedy, Action convergence of operators and graphs. Can. J. Math. **74**(1), 72–121 (2022)
8. M. Ballerini, N. Cabibbo, R. Candelier, A. Cavagna, E. Cisbani, I. Giardina, V. Lecomte, A. Orlandi, G. Parisi, A. Procaccini, et al., Interaction ruling animal collective behavior depends on topological rather than metric distance: Evidence from a field study. Proc. Natl. Acad. Sci. **105**(4), 1232–1237 (2008)
9. M. Bellingeri, D. Bevacqua, F. Sartori, M. Turchetto, F. Scotognella, R. Alfieri, N.K.K. Nguyen, T.T. Le, Q. Nguyen, D. Cassi, Considering weights in real social networks: a review. Front. Phys. **11** (2023)
10. A. Barrat, M. Barthélemy, R. Pastor-Satorras, A. Vespignani, The architecture of complex weighted networks. Proc. Natl. Acad. Sci. **101**(11), 3747–3752 (2004)
11. I. Ben-Porat, J.A. Carrillo, P.-E. Jabin, The graph limit for a pairwise competition model. J Differ. Equ. **413**, 329–69 (2024)
12. R. Berner, T. Gross, C. Kuehn, J. Kurths, S. Yanchuk, Adaptive dynamical networks. Phys. Rep. **1031**, 1–59 (2023)
13. S. Boccaletti, P. De Lellis, C. del Genio, K. Alfaro-Bittner, R. Criado, S. Jalan, M. Romance, The structure and dynamics of networks with higher order interactions. Phys. Rep. **1018**, 1–64 (2023)
14. C. Borgs, J. Chayes, H. Cohn, Y. Zhao, An l^p theory of sparse graph convergence i: Limits, sparse random graph models, and power law distributions. Trans. Am. Math. Soc. **372**, 01 (2014)
15. W. Braun, K. Hepp, The Vlasov dynamics and its fluctuations in the $1/N$ limit of interacting classical particles. Commun. Math. Phys. **56**(2), 101–113 (1977)
16. H. Chiba, G.S. Medvedev, The mean field analysis of the Kuramoto model on graphs i. the mean field equation and transition point formulas. Discrete Contin. Dyn. Syst. **39**(1), 131–155 (2019)
17. E. Cruciani, E.L. Giacomelli, J. Lee, On the convergence of nonlinear averaging dynamics with three-body interactions on hypergraphs. SIAM J. Appl. Dyn. Syst. **23**(3), 2364–2406 (2021)
18. F. Cucker, S. Smale, Emergent behavior in flocks. IEEE Trans. Automatic Control **52**, 852–862 (2007)
19. R.L. Dobrushin, Vlasov equations. Funct. Anal. Appl. **13**(2), 115–123 (1979)
20. M. Drobyshevskiy, D. Turdakov, Random graph modeling: A survey of the concepts. ACM Comput. Surv. **52**(6) (2019)
21. P. Erdös, A. Rényi, On random graphs. i. Publ. Math. **6**(3–4), 290–297 (1959)
22. D. Garlaschelli, The weighted random graph model. New J. Phys. **11**(7), 073005 (2009)
23. M.A. Gkogkas, C. Kuehn, C. Xu, Continuum limits for adaptive network dynamics. Commun. Math. Sci. **21**, 83–106 (2023)

24. S.Y. Ha, J.-G. Liu, A simple proof of the cucker-smale flocking dynamics and mean-field limit. Commun. Math. Sci. **7**, 297–325 (2009)
25. R. Hegselmann, U. Krause, Opinion Dynamics and Bounded Confidence Models, Analysis and Simulation. J. Artif. Soc. Soc. Simul. **5** (2002)
26. P.-E. Jabin, D. Poyato, J. Soler, Mean-field limit of non-exchangeable systems. Comm. Pure Appl. Math. **78** 651–741 (2025). https://doi.org/10.1002/cpa.22235
27. D. Kaliuzhnyi-Verbovetskyi, G. Medvedev, The semilinear heat equation on sparse random graphs. SIAM J. Math. Anal. **49** (2016)
28. D. Kaliuzhnyi-Verbovetskyi, G.S. Medvedev, The mean field equation for the Kuramoto model on graph sequences with non-lipschitz limit. SIAM J. Math. Anal. **50**(3), 2441–2465 (2018)
29. C. Kuehn, Network dynamics on graphops. New J. Phys. **22**(5), 053030 (2020)
30. C. Kuehn, C. Xu, Vlasov equations on digraph measures. J. Differ. Equ. **339**, 261–349 (2022)
31. C. Kuehn, C. Xu, Vlasov equations on directed hypergraph measures. Partial Differ. Equ. Appl. **6**, 9 (2025). https://doi.org/10.1007/s42985-025-00313-6
32. J. Kunegis, Konect: the koblenz network collection, in *Proceedings of the 22nd International Conference on World Wide Web, WWW '13 Companion, page 1343–1350, New York, NY, USA, 2013* (Association for Computing Machinery, 2013)
33. Y. LeCun, Y. Bengio, G. Hinton, Deep learning. Nature **521**, 436–44 (2015)
34. U. Lopez, J. Gautrais, I.D. Couzin, G. Theraulaz, From behavioural analyses to models of collective motion in fish schools. Interface Focus **2**(6), 693–707 (2012)
35. L. Lovâsz, *Large Networks and Graph Limits*. Colloquium Publications, vol. 60 (American Mathematical Society, Providence, 2012)
36. M. Lucas, G. Cencetti, F. Battiston, Multiorder Laplacian for synchronization in higher-order networks. Phys. Rev. Res. **2**, 033410 (2020)
37. S. McQuade, B. Piccoli, N. Pouradier Duteil, Social dynamics models with time-varying influence. Math. Models Methods Appl. Sci. **29**(4), 681–716 (2019)
38. G.S. Medvedev, The nonlinear heat equation on dense graphs and graph limits. SIAM J. Math. Anal. **46**, 2743–2766 (2014)
39. G.S. Medvedev, The nonlinear heat equation on W-random graphs. Arch. Ration. Mech. Anal. **212**, 781–803 (2014). https://doi.org/10.1007/s00205-013-0706-9
40. G. Medvedev, The continuum limit of the kuramoto model on sparse random graphs. Commun. Math. Sci. **17**, 883–898 (2019)
41. G.S. Medvedev, Correction to: The nonlinear heat equation on w-random graphs. Arch. Ration. Mech. Anal. **231**(2), 1305–1308 (2019)
42. L. Neuhäuser, R. Lambiotte, M.T. Schaub, Consensus dynamics and opinion formation on hypergraphs, in *Higher-Order Systems*. Understanding Complex Systems, ed. by F. Battiston, G. Petri (Springer, Cham, 2022), pp. 347–376
43. T. Nishikawa, A.E. Motter, Comparative analysis of existing models for power-grid synchronization. New J. Phys. **17**(1), 015012 (2015)
44. J.-P. Onnela, J. Saramäki, J. Hyvönen, G. Szabó, D. Lazer, K. Kaski, J. Kertész, A.-L. Barabási, Structure and tie strengths in mobile communication networks. Proc. Natl. Acad. Sci. **104**(18), 7332–7336 (2007)
45. T. Paul, E. Trélat, From microscopic to macroscopic scale equations: mean field, hydrodynamic and graph limits. arXiv:2209.08832
46. O. Popovych, M. Xenakis, P. Tass, The spacing principle for unlearning abnormal neuronal synchrony. PLoS ONE **10** (2015)
47. V. Rohr, R. Berner, E. Lameu, O. Popovych, S. Yanchuk, Frequency cluster formation and slow oscillations in neural populations with plasticity. PLOS ONE **14**, e0225094 (2019)
48. D.E. Rumelhart, R.J. Williams, G.E. Hinton, Learning representations by back-propagating errors. Nature **323**, 533–536 (1986)
49. H. Schawe, L. Hernández, Higher order interactions destroy phase transitions in deffuant opinion dynamics model. Commun. Phys. **5**(1), 32 (2022)
50. P.S. Skardal, A. Arenas, Higher order interactions in complex networks of phase oscillators promote abrupt synchronization switching. Commun. Phys. **3**(1), 218 (2020)

51. A.S. Sznitman, in *Topics in propagation of chaos*, ed. by P.L. Hennequin. Ecole d'Eté de Probabilités de Saint-Flour XIX — 1989. Lecture Notes in Mathematics, vol 1464 (Springer, Berlin, Heidelberg, 1991). https://doi.org/10.1007/BFb0085169
52. S. Throm, Continuum limit for interacting systems on adaptive networks. Eur. J. Appl. Math. 1–15 (2024). Published online. https://doi.org/10.1017/S0956792524000354
53. X. Wang, C. Xu, Z. Zheng, Phase transition and scaling in kuramoto model with high-order coupling. Nonlinear Dyn. **103**, 2721–2732 (2021)
54. D.J. Watts, S.H. Strogatz, Collective dynamics of small-world networks. Nature **393**(6684), 440–442 (1998)
55. D.A. Wiley, S.H. Strogatz, M. Girvan, The size of the sync basin. Chaos: Interdisciplinary J. Nonlinear Sci. **16**(1) (2006) 015103.
56. C. Xu, X. Wang, P.S. Skardal, Bifurcation analysis and structural stability of simplicial oscillator populations. Phys. Rev. Res. **2**, 023281 (2020)
57. C. Xu, P.S. Skardal, Spectrum of extensive multiclusters in the kuramoto model with higher-order interactions. Phys. Rev. Res. **3**, 013013 (2021)
58. Y. Zhang, M. Lucas, F. Battiston, Higher-order interactions shape collective dynamics differently in hypergraphs and simplicial complexes. Nat. Commun. **14**, 1605 (2023)

Four Lectures on The Mathematical Theory of (Multi) Phase Transitions

Nicholas D. Alikakos

Festum π Conference in Chania, 2024

1 Introduction

1. We begin with the physical context. The basic quantity is the *Free Energy*

$$J_\varepsilon(u, \Omega) = \int_\Omega \left(\frac{\varepsilon^2}{2} |\nabla u|^2 + W(u) \right) dx \, ,$$

$$\nabla u = \left(\frac{\partial u_i}{\partial x_j} \right), \; i = 1, \ldots, m, \; j = 1, \ldots, n \tag{1.1}$$

where $\Omega \subset \mathbb{R}^n$, an open bounded, smooth set in $\mathbb{R}^n$, $u : \Omega \to \mathbb{R}^m$, a Sobolev map. The potential W is a nonnegative C^2 function $W : \mathbb{R}^m \to \mathbb{R}^+$, with a discrete set of nondegenerate zeroes $\{W = 0\} = \{a_1, \ldots, a_N\}$, $\partial^2 W(a_i)$ positive definite. We refer to the $a_i's$ as the *Phases*, $u = u(x)$ is the *Order Parameter*, that is a macroscopic indicator of the microscopic structure of the matter contained in Ω. For instance in the case of a ferromagnetic material u can be identified with the local density of magnetization, while in other situations u may stand for the concentration. $\varepsilon > 0$ is a small parameter. In the physical context considered, ε gives the order of magnitude of the surface tension with respect to the bulk energy measured by $W(u)$. Although a 2nd order effect energetically, the surface tension is the primary factor in determining the morphology of the interface. The gradient tern in the free energy penalizes the appearance of *Diffuse Interfaces* that is regions in space where u changes abruptly from values close to a phase to values close to another phase. The origin

N. D. Alikakos (✉)
University of Athens, EKPA, Athens, Greece
e-mail: nalikako@math.uoa.gr

© The Author(s), under exclusive license to Springer Nature Switzerland AG 2026
N. D. Alikakos, C. Villani (eds.), *Analysis Techniques for Mathematical Physics and Geometry – and Music*, Lecture Notes in Mathematics 2390,
https://doi.org/10.1007/978-3-032-11396-2_2

of the physical theory goes back to Van der Waals in the 1870's. It was refined and further developed by the material scientist Cahn and his collaborators in the 1970's [19].
Independently and at about the same time De Giorgi [23, 24] introduced (1.1) the scalar bistable case ($m = 1$, $N = 2$) as a means of approximating the constrained perimeter functional (Volume constraint) with a quadratic functional. The relationship of the associated gradient flows to motion by mean curvature was discovered by the physicists [2, 39] and established rigorously for the scalar bistable case by mathematicians in varying degree of generality in [14, 21, 25, 29, 38], while the vector multiphase case ($m \geq 2$, $N \geq 3$) has been treated very recently in [31]. Motion by mean curvature was discovered earlier by the material scientist Mullins [43] in the 1950's.

2. *The Gradient Flows*

We consider

$$u_t = -grad_{L^2(\Omega)} J_\varepsilon(u) := J_\varepsilon' u) \tag{1.2}$$

Note that

$$J_\varepsilon'(u)v = \frac{d}{dh}|_{h=0} J_\varepsilon(u+hv) = \int_\Omega (\varepsilon \nabla u \nabla v + W_u(u)v)\,dx$$

$$= \int_\Omega \Big(-\varepsilon^2 \Delta u + W_u(u)\Big)\, v dx = \Big(-\varepsilon^2 \Delta u + W_u(u), v\Big)_{L^2(\Omega)},$$

provided $\frac{\partial u}{\partial \nu}|_{\partial\Omega} = 0.$
Thus

$$\frac{d}{dt} J_\varepsilon(u(\cdot,t)) = \big(J_\varepsilon'(u), u_t\big)_{L^2(\Omega)} = -||u_t||^2_{L^2(\Omega)} \leq 0. \tag{1.3}$$

Notice that (1.2) via the Riesz representation in $L^2(\Omega)$ is equivalent to the *Allen-Cahn system*

$$u_t = \varepsilon^2 \Delta u - W_u(u) \quad , \quad W_u(u) = (\nabla_u W(u))^T, \tag{1.4}$$

with homogeneous Neumann boundary conditions. Equation (1.3) states that the free energy is decreasing in time away from steady states, with the fastest possible rate in $L^2(\Omega)$. If instead of $L^2(\Omega)$ we consider $H^{-1}(\Omega)$, then we obtain the *Cahn-Hilliard system*

$$\begin{cases} u_t = -\Delta(\varepsilon^2 \Delta u - W_u(u)) \\ \frac{\partial u}{\partial \nu}|_{\partial\Omega} = \frac{\partial}{\partial \nu}(\varepsilon^2 \Delta u - W_u(u))|_{\partial\Omega} = 0. \end{cases} \tag{1.5}$$

Fig. 1 Double-well potential

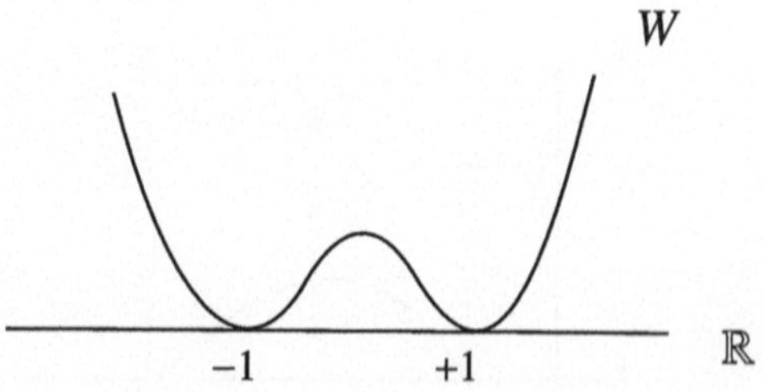

Equation (1.5) is the simplest *local* equation that preserves the mass $\int_\Omega u(x,t)dx$, and reduces the free energy, and it is physically more relevant than (1.4). In these lectures we are interested in solutions that describe the coexistence of different phases, that are time independent, and for these there are no essential differences between (1.4) and (1.5).

3. *The Scalar Case* ($m = 1$, $N = 2$)

(A) We begin with the simplest equation. We take $m = 1$, $n = 1$ and $N = 2$, two phases. For definiteness we take

$$W(u) = \frac{1}{4}(u^2 - 1)^2 \quad , \ \{W = 0\} = \{\pm 1\}. \text{ (Fig. 1)}$$

$$\begin{cases} u_t = \varepsilon^2 u_{xx} - W'(u) \ , \ x \in (a,b) \ , \ t > 0 \\ u_x = 0 \ , \quad x = a, b \\ u(x,0) = u_0(x) \end{cases} \tag{1.6}$$

At the beginning, the evolution is governed mainly by the ODE (x is a parameter)

$$u_t = -W'(u) \tag{1.7}$$

which *separates* the solution, and makes it *layered*, that is a function with values close to ± 1, together with sharp transitions at the interfaces ξ_1, ξ_2, ξ_3 (graph below) (Fig. 2). This happens at the time T, when $u_{xx} \sim \frac{1}{\varepsilon^2}$, i.e. the diffusion becomes significant and balances the reaction. At this time the layers start moving, and near the interfaces $\xi_1(t)$, $\xi_2(t)$, $\xi_3(t)$, u satisfies (approximately)

$$\varepsilon^2 u_{xx} - W'(u) = 0. \tag{1.8}$$

Rescaling x via $\eta = \frac{x - \xi_i}{\varepsilon}$ we obtain for $U(\eta) = u(\varepsilon\eta)$

$$\begin{cases} \ddot{U} - W'(U) = 0 \quad , \quad U : \mathbb{R} \to \mathbb{R} \\ U(\pm\infty) = \pm 1 \end{cases} \tag{1.9}$$

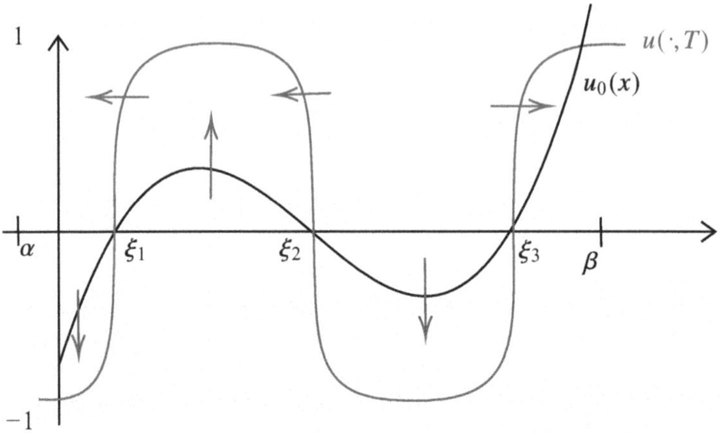

Fig. 2 We indicate the evolution of an initial condition u_0 with three zeroes ξ_1 , ξ_2, ξ_3 in (a, b), for small $\varepsilon > 0$ ([20, 33])

Equation (1.9) has a *unique* solution modulo translations. It can be characterized variationally via

$$\sigma = \min_V \left\{ \int_{\mathbb{R}} \left(\frac{1}{2} |\dot{V}|^2 + W(V) \right) d\eta \ , \quad V(\pm\infty) = \pm 1 \right\} \tag{1.10}$$

where σ is the *Surface Tension Coefficient*, that measures the interface energy. This is the simplest example of the kind of solution that is the main focus of these lectures. It describes the profile of the solution u in the transition zone. We call it the *Heteroclinic Connection*. Multiplying the Eq. (1.9) by $\dot{U}$, integrating and utilizing the b.c.'s at infinity we obtain

$$\frac{1}{2} |\dot{U}|^2 = W(U) \ , \quad U(\pm\infty) = \pm 1, \tag{1.11}$$

which can be integrated easily to give the explicit

$$U(\eta) = tanh \ \eta \tag{1.12}$$

(B) We continue with the higher dimensional $n \geq 2$, still scalar case. Rescaling time in (1.4) we obtain

$$\begin{cases} u_t = \varepsilon \Delta u - \frac{1}{\varepsilon} W'(u) \ , \quad x \in \Omega \ , \ t > 0, \\ \frac{\partial u}{\partial \nu} |_{\partial \Omega} = 0 \end{cases} \tag{1.13}$$

$\Omega \subset \mathbb{R}^n$, open, bounded, with smooth boundary.

We consider now the following formal computation which is based on the natural assumption that the transition near the interface should be 1−dim, i.e.

$$u(x,t) \cong U(\frac{d(x,\Gamma(t))}{\varepsilon}) \quad \text{(approximately, } d \text{ is the distance function)) ,} \tag{1.14}$$

where now $\Gamma(t)$ is the evolving interface (hypersurface in $\mathbb{R}^n$, playing the role of $\xi(t)$).
We calculate

$$u_t = \dot{U}(\frac{d(x,\Gamma(t)}{\varepsilon})\frac{1}{\varepsilon}d_t(x,\Gamma(t))$$
$$u_{,i} = \dot{U}(\frac{d(x,\Gamma(t)}{\varepsilon})\frac{1}{\varepsilon}d_{,i}$$
$$u_{,ii} = \ddot{U}(\frac{d(x,\Gamma(t)}{\varepsilon})\frac{1}{\varepsilon}d_{,i}d_{,i} + \frac{1}{\varepsilon^2}\dot{U}(\frac{d(x,\Gamma(t)}{\varepsilon})d_{,ii}$$

Substituting in (1.13), utilizing that $|\nabla d| = 1$ we obtain

$$\frac{1}{\varepsilon}\dot{U}d_t = \varepsilon[\ddot{U}\frac{1}{\varepsilon^2} + \frac{1}{\varepsilon}\dot{U}\Delta d] - \frac{1}{\varepsilon}W'(U)$$
$$= \frac{1}{\varepsilon}\dot{U}\Delta d \quad \text{(by (1.9))}$$

Hence

$$d_t(x,\Gamma(t)) = \Delta d(x,\Gamma(t)) \tag{1.15}$$

Evaluating now (1.15) for $x \in \Gamma(t)$ we obtain (see [37] for information on the distance function)

$$V = -H \tag{1.16}$$

where V is the normal velocity of the evolving hypersurface $\Gamma(t)$ at x, and H is the mean curvature of $\Gamma(t)$ at x. Thus for this class of layered solutions of (1.13) we see the relationship to the geometric evolution law. The interfaces in the context of (1.13) are not sharp, but are instead transition zones of width roughly $O(\varepsilon)$, where u changes abruptly between phases. The heteroclinic U describes the profile in the diffuse interface.

4. *The Vector case* ($m \geq 2$, $N \geq 3$)
So far the physical context we have considered is binary systems, for example atoms of two different species A , B, with the phases corresponding to two stable ways that the atoms can combine. $u(x)$ could stand for the ratio of the percentages of A over B, at each point x in space. In the case however of three species A , B , C , $m = 2$ since we need a 2−vector order parameter for

Fig. 3 triple-well potential

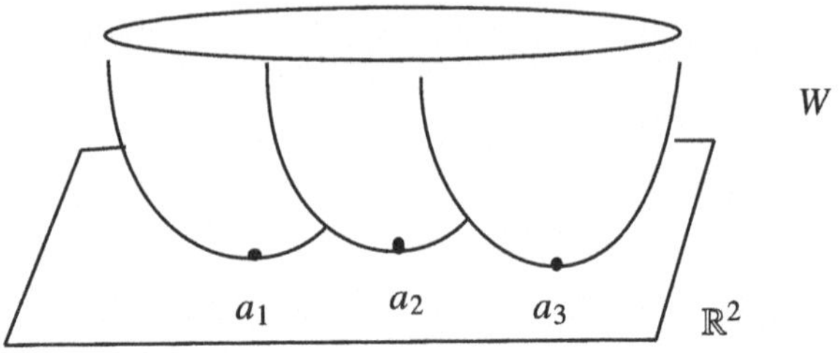

describing the three binary systems AB , BC , CA. So here W is a triple-well potential $W : \mathbb{R}^2 \to \mathbb{R}^+$
$\{W = 0\} = \{a_1, a_2, a_3\}$ (Fig. 3). In this new set-up we have binary transitions where two phases meet, and ternary transitions where all three phases meet.
In the vector case (1.9) is replaced by the vector equation

$$\begin{cases} \ddot{U}_{ij} - W_u(U_{ij}) = 0 \ , \quad U_{ij} : \mathbb{R} \to \mathbb{R}^2 \\ U_{ij}(\pm\infty) = a_\pm \in \{a_1, a_2, a_3\} \ , \ \ a_+ = a_i \ , \ a_- = a_j \end{cases} \tag{1.17}$$

The solutions of (1.17) we are interested in are obtained via minimization

$$\sigma_{ij} := \min \left\{ \int_{\mathbb{R}} \left(\frac{1}{2} |\dot{U}_{ij}|^2 + W(U_{ij}) \right) d\eta \ , \quad \lim_{\eta \to \pm\infty} U(\eta) = a_\pm \right\} \tag{1.18}$$

and unlike (1.9), they are not generally unique ([9]).
The heteroclinic connection U can be viewed as a curve that connects two of the minima of W in the most efficient way. We notice that the analog of (1.11) holds also for (1.17). Thus

$$\int_{\mathbb{R}} \left(\frac{1}{2} |\dot{U}|^2 + W(U) \right) d\eta = \sqrt{2} \int_{\mathbb{R}} |\dot{U}| \sqrt{W(U)} d\eta \tag{1.19}$$

and so by (1.18) U can be viewed as a geodesic connecting a_+ to a_- with $\mathbb{R}^2$ equipped with the metric $g_{kl}(u) = \sqrt{W(u)} du_k du_l$, $k, l = 1, 2$.
On the other hand at a ternary point the transition is two-dimensional and is captured by the PDE

$$\Delta u - W_u(u) = 0 \quad , \quad u : \mathbb{R}^2 \to \mathbb{R}^2 \tag{1.20}$$

with $u(x) \to a_i$ as $|x| \to \infty$, $x \in D_i$ (see Fig. 4b).
We call this solution the *Triple Junction Solution*.
The corresponding dynamic problem, the analog of (1.13), is

$$u_t = \varepsilon \Delta u - \frac{1}{\varepsilon} W_u(u) \ , \ u : \Omega \to \mathbb{R}^2 \tag{1.21}$$

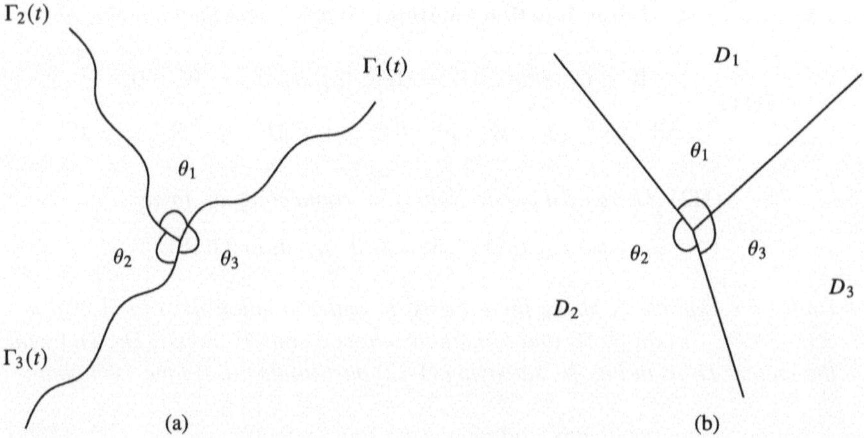

Fig. 4 In (**a**) we show three curves evolving by curvature, $V_i - K_i$, intersecting at the junction ([15]). In (**b**) we show the blow-down of the triple junction solution of (1.20). It is the unique (up to translation and rotations) singular minimal cone, partitioning the plane into three sectors D_1, D_2, D_3. It coincides with the blow-up at the junction in (**a**)

with Neumann boundary conditions, which has associated to it a sharp interface problem, the analog of (1.16), which is motion by curvature. However now we have a new feature. A point, the *Junction*, where three curves meet at angles determined by Young's law

$$\frac{sin\alpha_1}{\sigma_{23}} = \frac{sin\alpha_2}{\sigma_{13}} = \frac{sin\alpha_3}{\sigma_{12}} \quad , \quad \sigma_{ij} < \sigma_{ik} + \sigma_{kj} \quad , \; i \neq j \neq k \; , \; i, j, k \in \{1, 2, 3\} \tag{1.22}$$

We seek the solution of (1.20) as a *De Giorgi Minimizer* of the free energy functional,

$$J(u, V) = \int_V \left(\frac{1}{2} |\nabla u|^2 + W(u) \right) dx \quad , \quad V \text{ open, bounded in } \mathbb{R}^2 \tag{1.23}$$

that is a $u(x)$ in $W^{1,2}_{loc}$, $u : \mathbb{R}^2 \to \mathbb{R}^2$, such that

$$J(u, V) \leq J(u + v, V) \quad , \; \forall \text{ bounded, open set } V \subset \mathbb{R}^2 \; , \; \forall v \in C^1_0(V). \tag{1.24}$$

The reason behind this definition is that for $n > 1$ the condition

$$\int_{\mathbb{R}^n} (\frac{1}{2} |\nabla u|^2 + W(u)) dx < \infty \; , \; u \in W^{1,2}_{loc}(\mathbb{R}^n; \mathbb{R}^m) \cap L^{\infty}_{loc}(\mathbb{R}^n; \mathbb{R}^m),$$

for any $W \geq 0$, $\{W = 0\} \neq \emptyset$, implies that $u \equiv$constant ([3]).
We have the following

Theorem 1 (The Triple Junction Solution) *Suppose that the following hold:*

$$\textbf{(H1)} \quad \begin{array}{l} W \in C^2(\mathbb{R}^2)\ ,\ \{W=0\}=\{a_1,a_2,a_3\}\ ,\ W \geq 0 \\ \partial^2 W(a_i) \geq cI\ ,\ \ W_u \cdot u > 0 \ \ if\ |u|>M\ ,\ \ u \in \mathbb{R}^2\ ,\ c>0 \end{array}$$

$$\textbf{(H2)}\ \ \textit{There exist connections}\ \ U_{ij}\ \ \textit{connecting}\ \ a_i\ \ \textit{to}\ \ a_j\ ,$$
$$a_i \neq a_j\ \ i,j \in \{1,2,3\}\ ,\ \textit{with}\ \ \sigma_{ij}\ \ \textit{as in}\ \ (1.18).$$

Under the hypothesis above there exists an entire solution $u(x)$ to (1.20)*, u : $\mathbb{R}^2 \to \mathbb{R}^2$, $|u(x)| \leq M$, and a singular minimal cone $\mathcal{P} = \{D_1, D_2, D_3\}$ with the sectors D_i as in Fig. 4b, satisfying* (1.22) *partitioning the plane, such that*

$$\lim_{r \to +\infty} ||u(r\cdot) - u_0(\cdot)||_{L^1_{loc}(\mathbb{R}^2)} = 0\ , \tag{1.25}$$

where

$$u_0(x) = \sum_{i=1}^{3} a_i \chi_{D_i}(x)\ \ ,\ \ u(r\cdot)\ \ \textit{stands for}\ \ u(rx).$$

The theorem above was established, almost concurrently with entirely different proofs, in [8] and [47], for some sequence $\{r_k\} \to +\infty$, leaving open the possibility that the cone could depend on the sequence. It was subsequently refined to the statement above in [35].
The theorem has a long history. The first existence proof was given in 1996 [16] in the equivariant class of maps with the six symmetries and minimality of the solution in this symmetry class. Consequently the possibility of the solution being unstable under general perturbations remained open. Later in 2022 [32] an existence proof was given in the class of rotations of the equivariant triangle, thus eliminating the reflections. However, despite being less restrictive this result faces the same stability criticisms.

5. *The De Giorgi insight*
So far we have emphasized the role of the triple junction solution in the mathematical physics of phase transitions. However there is also an interesting relationship to geometric analysis. In 1978 [24] De Giorgi stated a profound conjecture about bounded solutions of the scalar, bistable Allen-Cahn equation,

$$\Delta u - W'(u) = 0\ ,\ \ u : \mathbb{R}^n \to \mathbb{R}\ ,\ \ W(u) = \frac{1}{4}(u^2-1)^2. \tag{1.26}$$

Conjecture Let $u \in C^2(\mathbb{R}^n)$ be a solution to (1.26) such that

1. $|u| < 1$
2. $\dfrac{\partial u}{\partial x_n} > 0$ for $x \in \mathbb{R}^n$.

Is it true that all the level sets of u are hyperplanes, at least for $n \leq 8$, or equivalently $u(x) = U(x \cdot \nu)$ for some $\nu \in \mathbb{R}^n$, where U is the heteroclinic connection in (1.9)?

The analogy with the Bernstein problem for minimal graphs is behind this conjecture. In low dimensions the only minimal graphs are hyperplanes. This suggests the restriction $n \leq 8$. As far as we know the conjecture in its full generality has been settled only for $n = 2$ in [36] and $n = 3$ in [10]. However under the additional hypothesis

$$\lim_{x_n \to \pm\infty} u(x', x_n) = \pm 1 \ , \ x = (x', x_n) \in \mathbb{R}^n \tag{1.27}$$

it has been established in [45] for $4 \leq n \leq 8$. Finally in [26] a counterexample was given for $n > 8$. Previous related work, partial results, etc. were obtained in [13, 18, 41, 43] among others. We refer to the expository papers [30] and [46] for a detailed account.

In [1] it was established that the second hypothesis $\frac{\partial u}{\partial x_n} > 0$ in the conjecture together with (1.27) imply that the solution is a De Giorgi minimizer as in (1.24).

In light of De Giorgi's insight, it is natural to expect that in the vector case for minimizers, there should be an analogous result of variable reduction, where the heteroclinic representing (the transition across) the plane, is replaced by the triple junction solution representing the singular minimal cone.

The following is a result in this spirit but under symmetry hypothesis.

Theorem 2 ([7]) *Let u be a C^2 solution of*

$$\Delta u - W_u(u) = 0 \ , \ u : \mathbb{R}^3 \to \mathbb{R}^3 \tag{1.28}$$

Assume

- *W is a triple-well potential under the symmetry group of the equilateral triangle acting on $\mathbb{R}^3$, leaving the x_3–axis fixed,*
- *u equivariant and minimizing in the corresponding equivariant class,*
- *the heteroclinic connections are unique, and hyperbolic (zero is not in the spectrum of the leariged operator in the symmetry class)*
- *Assume that the triple junction solution $u_{tr} : \mathbb{R}^2 \to \mathbb{R}^3$ is unique and hyperbolic in its equivariant class.*

Then

$$u(x_1, x_2, x_3) = u_{tr}(x_1, x_2) \tag{1.29}$$

We expect that the following general result

$$u(x_1, x_2, x_3) = u_{tr}(\pi x) \ , \pi \text{ an orthogonal projection to an unknown 2-plane in } \mathbb{R}^3, \tag{1.30}$$

should be true under uniqueness and appropriate hypotheses on the zero eigenvalue of the linearized operator about the heteroclinic connections and the triple junction, *but without any symmetry assumptions* on the De Giorgi minimizer u, the triple junction and the connections.

The rest of these lecture notes is structured as follows:

In Sect. 2 we give an existence proof for heteroclinic connections (1.17). In Sect. 3 we state and prove a variational maximum principle which is a tool for clearing out the diffuse interface.

In Sect. 4 we state a density estimate [5] that extends to the vector case the result of [17] that converts integral estimates to pointwise estimates and is useful for localizing the diffuse interface.

Finally in Sect. 5 we give the major step in the proof of Theorem 1, together with a brief outline of the rest of the proof based on the tools developed in Sects. 2, 3 and 4.

For the scalar Allen-Cahn equation we refer the reader to the lectures [22], while for a detailed account for the vector equation we refer to the book [7].

2 One Dimensional Solutions: Heteroclinic Connections

We consider the existence of solutions for

$$u'' - W_u(u) = 0 \quad , \quad u(\pm\infty) = a^{\pm} \quad , \; a^+ \neq a^- \tag{2.1}$$

$a^{\pm} \in A = \{W = 0\}$, $|A| \geq 2$.

$$J_{(s_1,s_2)}(u) := \int_{s_1}^{s_2} \left(\frac{1}{2}|u'|^2 + W(u) \right) dx \quad , \quad J(u) = J_{(-\infty,\infty)}(u) \tag{2.2}$$

Hypotheses

(h1) $W \in C^2$, $c_2|\xi|^2 \geq \xi^T \partial^2 W_u(a)\xi \geq c_1|\xi|^2$, $a \in A$

(h2)

(a) $\liminf\limits_{|u|\to+\infty} W(u) > 0$

or the *weaker*

(b) $\sqrt{W(u)} \geq \gamma(|u|)$, $\gamma : (0, +\infty) \to \mathbb{R}$,

$$\int_0^\infty \gamma(r)dr = +\infty$$

or the *stronger*

(c) $W_u(u) \cdot u > 0$ if $|u| > M$.

Theorem 3 (Existence) *Under* **(h1)**, **(h2)**(b), *given* $a^- \in A$, *there exist* $a^+ \in A \setminus \{a^-\}$ *and* u *a classical solution to* (2.1) *satisfying*

$$\frac{1}{2}|u'|^2 - W(u) = 0 \qquad (equipartition) \tag{2.3}$$

which minimizes J *on*

$$\mathscr{A} = \left\{ u \in W^{1,2}_{loc}(\mathbb{R};\mathbb{R}^m) \mid \lim_{x\to-\infty} u(x) = a^- \;,\; \lim_{x\to+\infty} u(x) \in A \setminus \{a^-\} \right\} \tag{2.4}$$

Note

(1) $W(u(x)) > 0 \;,\; x \in \mathbb{R}$, by uniqueness for (2.3).

(2) If $|A|$ is even the theorem provides at least $\dfrac{|A|}{2}$ connections, while if $|A|$ is odd at least $\dfrac{|A|+1}{2}$ connections.

Example 6 $m = 1 \;,\; W(u) = \frac{1}{2}(u^2-1)^2$, bistable, then

$$u(x) = tanh\, x \;,\; u^{+} = \pm 1 \quad , \text{ unique modulo translations}$$

Remark 1 For $m = 2$ and potentials of the form $W(u_1, u_2) = |f(z)|^2 \;,\; z = u_1 + iu_2$, for holomorphic $f's$ (or even meromorphic) e.g. $f(z) = \prod_{i=1}^{N}(z - a_i)$, one can get $u(x)$ in closed form and construct a variety of explicit examples exhibiting unexpected behavior. For this class of potentials given $a^+ \neq a^-$ the solution to (2.1) is unique. On the other hand there are examples of meromorphic $f's$ which exhibit non uniqueness. We refer the reader to [9] for all these phenomena.

Remark 2 ([7]) *Sufficient condition for existence of connection* $a_i \to a_j \;,\; \{W = 0\} = \{a_1, \ldots, a_N\}$

Let

$$\sigma_{ij} = \inf_{\mathscr{A}_{ij}} J(u) \;,$$

$$\mathscr{A}_{ij} := \left\{ u \in W^{1,2}_{loc}(\mathbb{R};\mathbb{R}^m) \mid \lim_{x\to-\infty} u(x) = a_i \;,\; \lim_{x\to+\infty} u(x) = a_j \right\} \tag{2.5}$$

$i, j = 1, \ldots, N$. Then the condition

$$\sigma_{ij} < \sigma_{ih} + \sigma_{hj} \;\; \forall a_h \in A \setminus \{a_i, a_j\} \quad \text{(Triangle Inequality)} \tag{2.6}$$

is sufficient for the existence of an orbit connecting a_i to a_j.

(For $W(z) = |(z - z_1)(z - z_2)(z - z_3)|^2$ it is also necessary.)

Remark 3 There are various proofs of variants of Theorem 3 by several authors ([4, 12, 42, 44, 48–50]). The proof we present here is from [34].

Proof of Theorem 3

1. *Removing noncompactness due to translations*
   Choose $r_0 > 0$ small such that

$$\min_{a\in A\ ,\ |u-a|<r_0} W(u) = W_0. \tag{2.7}$$

   Note $\forall\ u(\cdot) \in \mathscr{A}\ ,\ \exists\ x_0$ such that $W(u(x_0)) = W_0$. Consider then the translates of $u \in \mathscr{A}$ with $W(u(0)) = W_0$. We restrict ourselves to this modified $\mathscr{A}$, which we still denote by $\mathscr{A}$.
2. $\mathscr{A} \neq \emptyset$ Indeed given a^- consider the element $\tilde{a} \neq a^-$ in A closest to a^-.
   Let

$$\tilde{u}(x) = (1 - (x + x_0))a^- + (x + x_0)\tilde{a}$$

   Choose $x_0 \in (0, 1)$ s.t. $W(\tilde{u}(0)) = W_0$.
   Let

$$J(\tilde{u}) = \sigma \tag{2.8}$$

3. L^∞ *bound*
   $\exists\ M > 0$ depending on γ in **(h2)**(b) such that for $u \in \mathscr{A}$ with

$$J(u) \leq \sigma \Rightarrow ||u||_{L^\infty(\mathbb{R};\mathbb{R}^m)} \leq M. \tag{2.9}$$

   Indeed let $u(\overline{x}) = M$, some $\overline{x}$

$$\sigma \geq J_{(-\infty,\overline{x})}(u) \geq \int_{-\infty}^{\overline{x}} \sqrt{2W(u(x))}|u'(x)|dx \geq \sqrt{2}\int_{|a^-|}^{M} \gamma(r)dr$$

   and we conclude with **(h2)**(b).
4. *Compactness-Direct Method*
   Let $\{u_j\} \subset \mathscr{A}$ be a minimizing sequence,

$$J(u_j) \to \inf_{\mathscr{A}} J(u) =: \sigma_0 \leq \sigma$$

   By (2.9), and the Cauchy-Schwarz inequality

$$|u_j(x_1) - u_j(x_2)| \leq |\int_{x_1}^{x_2} |u_j'(\xi)|d\xi| \leq \sqrt{2}\sigma|x_1 - x_2|^{1/2} \tag{2.10}$$

By the Ascoli-Arzela theorem and a diagonal argument there exists subsequence $\{u_j\}$,

$$u_j \to u^* \quad , \ \text{uniformly on compacts of } \mathbb{R}. \tag{2.11}$$

Note that $\{u_j\}$ is bounded in $W^{1,2}_{loc}(\mathbb{R};\mathbb{R}^m)$: Given $l_1 < l_2$

$$\frac{1}{2}\int_{l_1}^{l_2} |u_j'|^2 dx \leq J(u_j) \leq \sigma \ , \quad \int_{l_1}^{l_2} |u_j|^2 dx < C \ , \ \text{by} \quad (2.9).$$

Hence

$$u_j \overset{w}{\rightharpoonup} u^* \ \text{ in } \ W^{1,2}_{loc}(\mathbb{R};\mathbb{R}^m) \tag{2.12}$$

Therefore, by weak lower semicontinuity

$$\frac{1}{2}\int_{l_1}^{l_2} |u_x^*|^2 dx \leq \liminf_{j\to+\infty} \frac{1}{2}\int_{l_1}^{l_2} |u_j'|^2 dx \leq \liminf_{j\to+\infty} \frac{1}{2}\int_{\mathbb{R}} |u_j'|^2 dx \tag{2.13}$$

Also by Fatou,

$$\lim_{j\to+\infty} \int_{\mathbb{R}} W(u^*) \leq \liminf_{j\to+\infty} \int_{\mathbb{R}} W(u_j) dx \tag{2.14}$$

Thus

$$J(u^*) = \sigma_0 \tag{2.15}$$

5. *Boundary Conditions at* $\pm\infty$
By uniform continuity of u^*, (2.10), and (2.15)

$$\lim_{x\to+\infty} W(u^*(x)) = 0 \tag{2.16}$$

If $\lim_{x\to+\infty} u^*(x)$ does not exist, then

$$\exists \ \{x_i^1\} \ , \ \{x_i^2\} \ \text{ s.t. } \ u^*(x_i^1) \to a_1 \ , \ u^*(x_i^2) \to a_2$$

Hence $\exists \ \{\hat{x}_i\}$ s.t. $W(u^*(\hat{x}_i))$ does not converge to zero, a consequence of the continuity of $u(x)$ and the discreteness of A, in contradiction to (2.16),
Thus

$$\lim_{x\to\pm\infty} u^*(x) = \begin{cases} a \\ a^+ \end{cases} \tag{2.17}$$

$a \ , \ a^+ \in A.$

Claim 1: $u^*(-\infty) = a^-$
Suppose for the sake of contradiction that $a \neq a^-$. Hence $\exists\ x_j \to -\infty$ s.t.

$$|u^*(x_j) - a| = \varepsilon_j \ , \ \ \varepsilon_j \to 0 \tag{2.18}$$

On the other hand $u_k(x) \to a^-$ as $x \to -\infty$.

By a diagonal argument we can choose subsequence of $\{u_j\}$ which we denote again by $\{u_j\}$ s.t.

$$|u_j(x_j) - a| \leq \varepsilon_j \ , \ j = 1, 2, ... \tag{2.19}$$

Since $u_j(x) \to a^-$ as $x \to -\infty$, it follows that

$$\int_{-\infty}^{x_j} \left(\frac{1}{2}|u_j'|^2 + W(u_j) \right) dx \geq \sigma_0 - \varepsilon_j \tag{2.20}$$

This is a consequence of $J(u_j) \geq \sigma_0$ and Exercise 1 below.

On the other hand from $W(u_j(0)) = W_0$ and the equicontinuity of u_j by (2.10), for $\delta > 0$ small, fixed

$$\int_{-\delta}^{\delta} W(u_j)dx \geq \delta W_0 \tag{2.21}$$

Hence by (2.19), (2.20)

$$\int_{-\delta}^{\delta} \left(\frac{1}{2}|u_j'|^2 + W(u_j) \right) dx \geq \sigma_0 - \varepsilon_j + \delta W_0 \tag{2.22}$$

contradicting that $\{u_j\}$ is a minimizing sequence.
Claim 2: $u^*(+\infty) = a^+ \neq a^-$
Once more we proceed by contradiction with a similar argument. So suppose $u^*(+\infty) = a^-$. As in (2.19), $\exists$ sequence $x_j \to +\infty \ , \ u_j(x_j) \to a^-$. From $W(u_j(0)) = W_0$ and the equicontinuity of u

$$\int_{-\delta}^{\delta} W(u_j)dx \geq \delta W_0 \tag{2.23}$$

for $\delta > 0$ small fixed.

Since $u_j \in \mathscr{A} \Rightarrow u_j(x) \to a \neq a^-$ as $x \to +\infty$.

Hence

$$\int_{x_j}^{+\infty} \left(\frac{1}{2}|u_j'|^2 + W(u_j) \right) dx \geq \sigma_0 - C_W\delta^2, \tag{2.24}$$

by Exercise 1 below.

Thus (2.23), (2.24) give

$$\int_{-\delta}^{+\infty}\left(\frac{1}{2}|u_j'|^2+W(u_j)\right)dx \geq \sigma_0 - C_W\delta^2+\delta W_0,$$

contradicting that $\{u_j\}$ is a minimizing sequence.

The proof of Theorem 3 is complete. □

Exercise 1 Let a_i , $a_j \in \{W=0\}$, $s_+ > s_-$. Let $v:(s_-,s_+)\to\mathbb{R}^m$ minimize

$$J_{(s_-,s_+)}(v)=\int_{s_-}^{s_+}\left(\frac{1}{2}|v'|^2+W(v)\right)dx$$

subject to $|v(s_-)-a_i|=|v(s_+)-a_j|=\delta$

Then

$$J_{(s^-,s^+)}(v)\geq\sigma_{ij}-C_W\delta^2 \quad , \; C_W \text{ a positive constant determined by } (\mathbf{h1}).$$

3 The Variational Maximum Principle

3.1 Hypotheses

$$(H_{VMP})\begin{cases} W:\mathbb{R}^m\to\mathbb{R}\ ,\ \text{non-negative}\ ,\ W\in C(\mathbb{R}^m;\mathbb{R})\\ a\in\mathbb{R}^m\quad ,\ W(a)=0\\ (0,r_0]\ni r\to W(a+r\xi)\ \text{non decreasing for every }\xi\in\mathbb{R}^m\\ \qquad\text{with }|\xi|=1\ ,\ W(a+r_0\xi)>0\end{cases} \tag{3.1}$$

$A\subset\mathbb{R}^n$, open, bounded, ∂A Lipschitz.

Theorem 4 *Let $u\in W^{1,2}_{loc}(A;\mathbb{R}^m)\cap L^\infty(A;\mathbb{R}^m)$ be a **minimizer** of*

$$J(u,A)=\int_A\left(\frac{1}{2}|\nabla u|^2+W(u)\right)dx \tag{3.2}$$

***with respect to its Dirichlet conditions on** ∂A:*

$$J(u+v,A)\geq J(u,A)\ ,\ \forall\, v\in C_0^1(A;\mathbb{R}^m). \tag{3.3}$$

Assume that

$$|u(x)-a|\leq r\ \ on\ \partial A,\ \ 0<2r\leq r_0 \tag{3.4}$$

Then

$$|u(x) - a| \leq r \;\; on \;\; A. \tag{3.5}$$

Theorem 4 was established in [6]. Our presentation follows [11].
Comments (Difference with the usual maximum principle) This is a purely variational result while the usual maximum principle is a calculus fact that is based on the equation and holds for all solutions.

Exercise 2 Consider for $W(u) = \frac{1}{2}(u^2 - 1)^2$,

$$\varepsilon^2 u'' - W'(u) = 0$$

For $\varepsilon > 0$ small there are (periodic) solutions which clearly violate (3.5) (phase plane).

Exercise 3 For W convex (3.5) holds by the classical maximum principle:

$$0 = u'' - W'(u) = u'' - \frac{W'(u)}{u} u$$

Remark 4 Equation (3.5) does not hold for local minimizers via the second variation.

Before giving the proof we introduce the *polar representation* for a map $u(x)$.

$$\begin{aligned} u(x) &= a + |u(x) - a| \frac{u(x) - a}{|u(x) - a|} =: a + \rho(x)\vec{n}(x) \\ \rho(x) &:= |u(x) - a| \;,\; \vec{n}(x) = \begin{cases} \frac{u(x)-a}{|u(x)-a|} \;, & u(x) \neq a \\ 0 \;, & u(x) = a \end{cases} \end{aligned} \tag{3.6}$$

Formally

$$|\nabla u|^2 = |\nabla \rho|^2 + \rho^2(x)|\nabla \vec{n}(x)|^2 \tag{3.7}$$

leading to the polar form of the energy

$$J_A(u) = \int_A \left\{ \frac{1}{2} \left(|\nabla \rho|^2 + \rho^2 |\nabla \vec{n}(x)|^2 \right) + W(a + \rho \vec{n}) \right\} dx \tag{3.8}$$

The point is that we will consider perturbations of $u(x)$ only in the radial part, keeping $\vec{n}$ fixed:

For $u(\cdot) \in W^{1,2}(A; \mathbb{R}^m) \cap L^\infty(A; \mathbb{R}^m)$, and $f : \mathbb{R} \to \mathbb{R}$ locally Lipschitz, $f(0) = 0$ we consider the perturbation

$$\tilde{u}(x) = a + f(\rho(x))\vec{n}(x) \tag{3.9}$$

Then

$$\int_A |\nabla \tilde{u}|^2 dx = \int_A (|f'(\rho)\nabla\rho|^2 + f^2(\rho)|\nabla\vec{n}|^2)dx \tag{3.10}$$

It is a calculus fact for Sobolev functions that $\tilde{u}(\cdot) \in W^{1,2}(A;\mathbb{R}^m) \cap L^\infty(A;\mathbb{R}^m)$ and that (3.10), (3.8) hold rigorously.

Exercise 4 ([11]) For variations of the special form

$$f(s) = sg(s) \quad , \quad g : \mathbb{R} \to \mathbb{R} \text{ Lipschitz}$$

we can bypass $\vec{n}(x)$ and express $|\nabla\tilde{u}(x)|$ as follows:

$$|\nabla\tilde{u}|^2 = (f'(\rho))^2|\nabla\rho|^2 + (|\nabla u|^2 - |\nabla\rho|^2)(\frac{f(\rho)}{\rho})^2. \tag{3.11}$$

Notice that $|\nabla u|^2 \geq |\nabla\rho|^2$, and that moreover if

$$|f'| \leq 1 \quad , \quad |g| \leq 1 \tag{3.12}$$

then

$$|\nabla\tilde{u}(x)|^2 \leq |\nabla u(x)|^2 \tag{3.13}$$

3.2 Proofs

Theorem 4 will follow from the following replacement result:
The cut-off lemma

Lemma 1 *Let W satisfy the hypotheses $(\boldsymbol{H_{VMP}})$ and let $A \subset \mathbb{R}^n$, open bounded, with Lipschitz boundary. Suppose that*

$$u(\cdot) \in W^{1,2}(A;\mathbb{R}^m) \cap L^\infty(A;\mathbb{R}^m)$$

If the following two conditions hold

(I) $|u(x) - a| \leq r$ *on* ∂A , $0 < 2r \leq r_0$,
(II) $\mathcal{L}^n(A \cap \{|u(x) - a| > r\}) > 0$,

then $\exists\ \tilde{u}(\cdot) \in W^{1,2}(A;\mathbb{R}^m) \cap L^\infty(A;\mathbb{R}^m)$ *such that*

$$\begin{cases} \tilde{u} = u \ \ on \ \ \partial A \\ |\tilde{u}(x) - a| \leq r \ \ , \ on \ \ A \\ J_A(\tilde{u}) < J_A(u) \end{cases}$$

Proof We will assume without loss of generally that A is connected. Let $\tilde{u}$ as in (3.9)

Case 1(easy): $\rho(x) \leq r_0$ *a.e. in A*

Let

$$f(s) = \frac{\min\{s, r\}}{s} s = g(s)s \tag{3.14}$$

By (3.13) ($|f'| \leq 1$, $|g| \leq 1$)

$$\int_A |\nabla \tilde{u}(x)|^2 dx \leq \int_A |\nabla u(x)|^2 dx \tag{3.15}$$

Note that in the case of equality in (3.15), via (3.11) we obtain

$$\begin{aligned} 0 &= \int_A |\nabla \tilde{u}|^2 dx - \int_A |\nabla u|^2 dx \\ &= \int_A |\nabla \rho|^2 ((f'(\rho))^2 - 1) dx + \int_A (|\nabla u|^2 - |\nabla \rho|^2)(g^2(\rho) - 1) dx \\ &\leq - \int_{A \cap \{\rho \geq r\}} |\nabla \rho|^2 dx. \end{aligned} \tag{3.16}$$

Thus

$$\nabla \rho = 0 \ \ , \ \text{a.e. on} \ \ A \cap \{\rho \geq r\}. \tag{3.17}$$

Hence

$$\nabla(\tilde{\rho} - \rho) = 0 \ \ , \ \text{a.e. on} \ \ A \ \ (\tilde{\rho} = f(\rho)). \tag{3.18}$$

Hence by connectedness

$$\tilde{\rho}(x) - \rho(x) = \text{const.} \quad \text{a.e. in} \ \ A \tag{3.19}$$

and since

$$\begin{aligned} &\tilde{\rho}(x) - \rho(x) = 0 \ \ , \ \text{on} \ \partial A \ \text{in the trace sense} \\ &\Rightarrow \tilde{\rho}(x) - \rho(x) = 0 \ , \ \text{a.e. in} \ \ A \end{aligned} \tag{3.20}$$

in contradiction to (II).

Thus we have strict inequality in (3.15).

On the other hand

$$\begin{aligned} \int_A W(\tilde{u}(x))dx &= \int_A W(a + f(\rho(x))\vec{n}(x))dx \\ &\leq \int_A W(a + \rho(x)\vec{n}(x))dx = \int_A W(u(x))dx \end{aligned} \tag{3.21}$$

Case 1 is settled. Notice that this argument has not utilized W. In contrast in what follows W plays the main role.

Case 2: $\mathcal{L}^n(A \cap \{\rho > r_0\}) > 0$

Consider the cut-off functions:

$$\alpha(s) = \begin{cases} 1 & , \ s \leq r \\ \frac{2r-s}{r} & , \ r \leq s \leq 2r \\ 0 & , \ s \geq 2r \end{cases}$$

$$f(s) = \min\{s, r\}\alpha(s) \ , \ g(s) = \frac{f(s)}{s},$$

(Reflection along $|u - a| = r$).

Define

$$\tilde{u}(x) = a + g(\rho(x))(u(x) - a) \tag{3.22}$$

By (3.13)

$$|\nabla \tilde{u}(x)|^2 \leq |\nabla u(x)|^2 \tag{3.23}$$

In order to proceed further with the proof of Case 2 we will need the following lemma which is also of independent interest. □

Lemma 2 ("Continuity" of Sobolev Functions) *Let $A \subset \mathbb{R}^n$ open, bounded, connected with Lipschitz boundary, $f \in W^{1,2}(A; \mathbb{R})$ satisfies*

$$\begin{cases} f \leq \hat{r} \ \text{on} \ \partial A \ \text{(trace sense)} \\ \mathcal{L}^n(A \cap \{\hat{s} < f\}) > 0 \ , \ \text{some} \ \hat{s} > \hat{r} \end{cases} \tag{3.24}$$

Then

$$\mathcal{L}^n(A \cap \{\hat{r} < f < \hat{s}) > 0 \tag{3.25}$$

Proof of Lemma 2 Let

$$E_1 = A \cap \{f \le \hat{r}\}\ ,\ E_2 = A \cap \{\hat{r} < f \le \hat{s}\}\ ,\ E_3 = A \cap \{\hat{s} < f\} \tag{3.26}$$

Define

$$\begin{aligned} \sigma(x) &= \min\{f(x), \hat{s}\} = \begin{cases} f(x)\ , & x \in E_1 \cup E_2 \\ \hat{s}\ , & x \in E_3 \end{cases} \\ \tau(x) &= \max\{\sigma(x), \hat{r}\} = \begin{cases} \hat{r}\ , & x \in E_1 \\ \hat{s}\ , & x \in E_3 \\ f(x)\ , & x \in E_2 \end{cases} \end{aligned} \tag{3.27}$$

Suppose for the sake of contradiction that

$$\mathcal{L}^n(A \cap \{\hat{r} < f \le \hat{s}) = 0. \tag{3.28}$$

Then

$$\mathcal{L}^n(E_2) = 0 \quad \text{and} \quad \tau(x) = \begin{cases} \hat{r}\ , & x \in E_1 \\ \hat{s}\ , & x \in E_3. \end{cases}$$

Hence

$$\nabla\tau(x) = 0 \quad \text{a.e. in } A \tag{3.29}$$

On the other hand the min and max of Sobolev functions produce Sobolev functions [28]. Hence τ is Sobolev and the connectedness of A together with (3.29) implies that $\tau \equiv$ constant [27, Exercise 11, p.307]. Hence necessarily by (3.24)

$$\tau \equiv \hat{s}\ ,\ \mathcal{L}^n(E_3) > 0\ ,$$

and also $\mathcal{L}^n(E_1) = 0$. Thus by the trace theorem $f > \hat{s}$ a.e. in A. Thus $f \ge \hat{s}$ on ∂A, which contradicts $(3.24)_{(1)}$.

The proof of Lemma 2 is complete.

Conclusion. Let $\varepsilon > 0$ such that $W > 0$ on $r_0 \le |u - a| \le r_0 + \varepsilon$. We define the sets $E_1 := A \cap \{\rho \le r_0\}$, $E_2 := A \cap \{r_0 < \rho \le r_0 + \varepsilon\}$, $E_3 := A \cap \{\rho > r_0 + \varepsilon\}$. From $\mathcal{L}^n(A \cap \{\rho > r_0\}) > 0$ we obtain that if $\mathcal{L}^n(E_2) = 0$, then necessarily $\mathcal{L}^n(E_3) > 0$. But $\rho \le r < r_0 + \varepsilon$ on ∂A. Hence by Lemma 2 $\mathcal{L}^n(E_2) > 0$. Therefore, $\mathcal{L}^n(E_2) > 0$ under any circumstances. On $A \cap \{r \le \rho \le 2r\}$ we have: $W(\tilde{u}(x)) = W(a + r\alpha(\rho(x))\vec{n}(x)) \le W(a + r\vec{n}(x)) \le W(a + \rho(x)\vec{n}(x)) = W(u(x))$. On the other hand on $A \cap \{\rho > 2r\}$ we have $0 = W(\tilde{u}(x)) \le W(u(x))$, while on E_2 $0 = W(\tilde{u}(x)) < W(u(x))$. Therefore $J_A(\tilde{u}) < J_A(u)$ and the proof of the cut-off lemma is complete. □

4 The (Vector) Caffarelli-Córdoba Density Estimate

In this lecture we are interested in entire solutions of

$$\Delta u - W_u(u) = 0 \quad , \quad u : \mathbb{R}^n \to \mathbb{R}^m \tag{4.1}$$

which are also minimizers, in the sense of (1.24), that satisfy the estimates

$$|u(x) - a| < M \ , \ |\nabla u(x)| < M \ , \ W(a) = 0.$$

We now introduce the density estimate by considering first the motivation behind it that comes from minimal partitions. Our argument below is formal.

Consider a minimal surface $\Sigma^{n-1} \subset \mathbb{R}^n$, separating D from D^c. Let $x \in \Sigma^{n-1}$ and consider $B_r(x)$ which is partitioned by Σ^{n-1} in two parts D_r and D_r^c. Assume that $\mathcal{L}^n(D_\delta) > 0$ for all small $\delta \in (0, \delta_0)$.

Let

$$V(r) = \mathcal{L}^n(D_r), \quad A(r) = \mathcal{H}^{n-1}(\Sigma^{n-1} \cap B_r) \tag{4.2}$$

where S_r stands for the spherical cup bounding D_r, $\mathcal{H}^{n-1}$ for the $(n-1)-$ Hausdorff measure.

Consider the following computation.

$$\begin{aligned}
c_0(V(r))^{\frac{n-1}{n}} &\leq [\mathcal{H}^{n-1}(\Sigma^{n-1} \cap B_r) + \mathcal{H}^{n-1}(S_r)], \quad \text{(by the isometric inequality)},\\
c_0(V(r))^{\frac{n-1}{n}} &\leq [2\mathcal{H}^{n-1}(S_r)] \quad \text{by minimality of } \Sigma^{n-1},\\
c_0(V(r))^{\frac{n-1}{n}} &\leq [2V'(r)] \quad \text{by the coarea formula or Fubini}
\end{aligned} \tag{4.3}$$

Hence

$$V'(r) \geq \frac{c_0}{2}(V(r))^{\frac{n-1}{n}}.$$

Integrating

$$(V(r))^{\frac{1}{n}} - (V(\delta))^{\frac{1}{n}} \geq \frac{1}{n}\frac{c_0}{2}(r - \delta) \ , \ r \geq \delta.$$

Since $\delta > 0$ arbitrary we obtain

$$V(r) \geq (\frac{c_0}{2n} r)^n \ , \ r > 0. \tag{4.4}$$

This estimate together with the corresponding one for D_r^c expresses that both D and D^c have uniform positive density at each x for all $r \in [0, \infty]$:

$$0 < \lambda_1 \leq \frac{\mathcal{L}^n(D \cap B_r(x))}{\mathcal{L}^n(B_r(x))} \leq \lambda_2 < 1. \tag{4.5}$$

Our interest in (4.5) lies at $r = \infty$, that relates to Bernstein/Liouville type theorems.

The analogy with the diffuse interface is via the identification

$$A(r) = \int_{B_r \cap \{|u-a| \leq \lambda\}} W(u)dx, \quad V(r) = \mathcal{L}^n(B_r \cap \{|u - a| > \lambda\}) \tag{4.6}$$

where $W(a) = 0$ and $\lambda > 0$ any number such that

$$d_0 = \text{dist}(a, \{W = 0\} \setminus \{a\}) \geq \lambda. \tag{4.7}$$

Hypotheses

$$(\boldsymbol{H_d}) \begin{cases} W \in C^2(\mathbb{R}^m; [0, \infty)\ ,\ W(a) = 0, \\ W \text{ is } C^2 \text{ in a neighborhood of } a, \\ C_2|\xi|^2 \geq \xi^T W_{uu}(a_i)\xi \geq C_1|\xi|^2\ ,\ C_1\ ,\ C_2 > 0. \end{cases}$$

Theorem 5 *Assume W satisfies ($\boldsymbol{H_d}$), Ω open, possibly coinciding with $\mathbb{R}^n$ $n \geq 1$, $u : \Omega \subset \mathbb{R}^n \to \mathbb{R}^m$ a De Giorgi minimizer satisfying the bounds below* (4.1). *Then for any $\mu_0 > 0$ and any $\lambda \in (0, d_0)$ the condition*

$$\mathcal{L}^n(B_{r_0}(x_0) \cap \{|u - a| > \lambda\} \geq \mu_0 \tag{4.8}$$

implies

$$\mathcal{L}^n(B_r(x_0) \cap \{|u - a| > \lambda\}) \geq Cr^n, \quad r \geq r_0 \tag{4.9}$$

as long as $B_r(x_0) \subset \Omega$, $C = C(W, \mu_0, \lambda, r_0, M)$.

We note that the definition of minimizing solution is a slight modification of (1.24), (1.23), with V open, bounded in Ω, instead of $\mathbb{R}^n$.

For the proof we refer the reader to [7, Th 5.2]. Although the arguments there are quite intricate, the idea behind it is (4.3). The proof modulo the polar form and certain observations is quite close, to the scalar case [17]. The theorem for the vector case is from [5].

For minimizers the following estimate holds ([7, Lemma 5.1]),

$$J(u, B_r(x_0)) \leq C_0 r^{n-1}\ ,\ \text{some } C_0 > 0\ ,\ r > 0 \tag{4.10}$$

which together with the density estimate gives control on the regions Ω_i where u is close to a_i , $W(a_i) = 0$.

Exercise 5 Utilizing (4.10) show that the validity of the theorem for one value of $\lambda \in (0, d_0)$ implies its validity for all $\lambda' \in (0, d_0)$.

Exercise 6 Assume $\{W = 0\} = \{a_1, a_2\}$, and assume that (H_d) holds. Then show that given $0 < \theta < |a_1 - a_2|$, the condition

$$\mathcal{L}^n(B_1(x_0) \cap \{|u - a_1| \leq \theta\}) \geq \mu_0 > 0 \tag{4.11}$$

implies the estimate:

$$\mathcal{L}^n(B_r(x_0) \cap \{|u - a_1| \leq \theta\}) \geq Cr^n \tag{4.12}$$

for $r \geq 1$ as long as $B_r \subset \Omega$.

5 The Proof of Theorem 1: Outline and Details for the Main Estimate

The most obvious difficulty that one faces as soon as the symmetry hypothesis is removed is the lack of an priori center for the junction.

Our strategy is as follows: we start with a Dirichlet problem for the rescaled problem on the unit disc B_1, by assigning smooth Dirichlet date g_ε,

$$g_\varepsilon = \begin{cases} a_1 \ , \\ a_2 \ , & \text{except at } O(\varepsilon) \text{ transition layers (see (5.8))} \\ a_3 \ , \end{cases} \tag{5.1}$$

$$u_\varepsilon \ : \ \min_{u=g_\varepsilon} \int_{B_1(0)} \left(\frac{\varepsilon}{2} |\nabla u|^2 + \frac{1}{\varepsilon} W(u) \right) dz \tag{5.2}$$

The objective is to rescale back u_ε *with respect to a carefully chosen point* $z_0(\varepsilon)$, and capture the entire solution as a limit along a subsequence $\varepsilon_n \to 0$,

$$\begin{aligned} u_\varepsilon(\eta) = u_\varepsilon(z) \ , \ \eta = \frac{z - z_0(\varepsilon)}{\varepsilon} \\ u(\eta) = \lim_{\varepsilon \to 0} u_\varepsilon(\eta) \end{aligned} \tag{5.3}$$

Here is the outline of the proof:

Step 1: Lower Bound
This together with a sharp but relatively easy upper bound gives a 1st estimate of the center $z_0(\varepsilon)$.

Step 2: Localization of the diffuse interface

$$I_{\varepsilon,\gamma} = \{z \in B_1 : |u_\varepsilon - a_i| > \gamma \ , \ \text{all } i = 1, 2, 3\} \tag{5.4}$$

in an $\varepsilon^{1/4}$ neighborhood of $\partial\mathcal{P}$, the part of the singular minimal cone in B_1.

This is accomplished by taking into account the way the lower bound is obtained and utilizing the sharp upper bound together with the variational maximum principle.

Step 3: The $\varepsilon-$width of the transition layers. Here the tool is the vector Caffarelli-Cordoba estimate.

Step 4: Identification of points P_ε , Q_ε , R_ε such that

$$\begin{aligned} u_\varepsilon(P_\varepsilon) \sim a_1 \ , \ u_\varepsilon(Q_\varepsilon) \sim a_2 \ , \ u_\varepsilon(R_\varepsilon) \sim a_3 \\ d(P_\varepsilon, Q_\varepsilon) \ , \ d(R_\varepsilon, Q_\varepsilon) \leq \varepsilon \end{aligned} \tag{5.5}$$

and thus by rescaling and a simple compactness argument obtain a nontrivial $u(x)$ defined on $\mathbb{R}^2$.

Step 5: Extraction of a sequence $t_k \to +\infty$ such that $u(t_k x) \to u_0(x) = \sum_{j=1}^{3} a_j \chi_{D_j}(x)$, where $\{D_1, D_2, D_3\}$ the minimal partition corresponding to $\partial\mathcal{P}$. We follow the proof of Modica [40, Theorem 3] with the necessary modifications to fit our setting which is minimal partitions as opposed to minimal surfaces. This requires some basic geometric measure theory, compactness and the characterization of the minimal cones on the plane.

This is the outline of the proof in [8]. For completing the proof and establishing that u_0 does not depend on the sequence,

$$\lim_{t \to +\infty} u(tx) = u_0(x) \tag{5.6}$$

is highly nontrivial and required the paper [35]. The proof of (6) utilized analogs of step 1 and step 2 above with the emphasis on the lower bound, and also two tools, from [47] which we mention briefly. These are the *asymptotic monotonicity* of

$$\tilde{W}(r) = \frac{1}{R} \int_{B_1} W(u) dx$$

and the *asymptotic equipartition*, which together can be put in the form

$$\begin{aligned} &\tilde{W}(R_2) - \tilde{W}(R_1) \\ &\geq -\sqrt{\frac{c_1}{2} \frac{R_2^{1/2}}{R_1}} \left\{ \int_{B_{R_2} \setminus B_{R_1}} \left(\sqrt{W(u)} - \frac{1}{2} |\nabla u| \right)^{1/2} dx \right\}^{1/2} , 0 < R_1 < R_2 \end{aligned}$$

and justified by the key estimate

$$\int_{B_R} \left(\sqrt{W(u)} - \frac{1}{\sqrt{2}} |\nabla u| \right)^2 dx < C_2 R^{1-\alpha}, \, some \, \alpha \in (0, 1)$$

These are PDE facts (as opposed to our purely variational approach) and two dimensional, as can be seen from Pohozaev's identity below for $n = 2$

$$\int_{B_R} \left(\frac{n-2}{2} |\nabla u|^2 + n W(u) \right) dx = R \int_{\partial B_R} \left(\frac{1}{2} |\nabla u|^2 + W(u) - \left| \frac{\partial u}{\partial \nu} \right|^2 \right) dS.$$

In the rest of this lecture we focus on the proof of the lower bound. For simplicity we take $\sigma_{12} = \sigma_{13} = \sigma_{23} = \sigma$, and the angles α_i equal to $\frac{2\pi}{3}$. We do not introduce any symmetry hypothesis on the solution.

Lemma 3 (Lower/Upper Bound) *$\exists$ constants $C = C(W)$, $\varepsilon_0 = \varepsilon_0(W)$, such that for $\varepsilon \leq \varepsilon_0$ it holds:*

$$3\sigma - C\varepsilon^{1/3} \leq \int_{B_1} \left(\frac{\varepsilon}{2} |\nabla u_\varepsilon|^2 + \frac{1}{\varepsilon} W(u_\varepsilon) \right) dz \leq 3\sigma + C\varepsilon \tag{5.7}$$

Proof of the Lower Bound
Firstly we note that the power $1/3$ is not important. Any fractional power would be sufficient for the proof of the theorem.

Take a $c_0 > 0$, and fix it. We require $u_\varepsilon(e^{i\theta}) = g_\varepsilon(\theta)$ on ∂B_1,

$$g_\varepsilon(\theta) := \begin{cases} a_2 + g_0(\frac{\theta - (\frac{\pi}{2} - c_0\varepsilon)}{2c_0\varepsilon})(a_1 - a_2), & \theta \in [\frac{\pi}{2} - c_0\varepsilon, \frac{\pi}{2} + c_0\varepsilon) \\ a_1, & \theta \in [\frac{\pi}{2} + c_0\varepsilon, \frac{7\pi}{6} - c_0\varepsilon), \\ a_1 + g_0(\frac{\theta - (\frac{7\pi}{6} - c_0\varepsilon)}{2c_0\varepsilon})(a_3 - a_1), & \theta \in [\frac{7\pi}{6} - c_0\varepsilon, \frac{7\pi}{6} + c_0\varepsilon) \\ a_3, & \theta \in [\frac{7\pi}{6} + c_0\varepsilon, \frac{11\pi}{6} - c_0\varepsilon), \\ a_3 + g_0(\frac{\theta - (\frac{11\pi}{6} - c_0\varepsilon)}{2c_0\varepsilon})(a_2 - a_3), & \theta \in [\frac{11\pi}{6} - c_0\varepsilon, \frac{11\pi}{6} + c_0\varepsilon), \\ a_2, & \theta \in [\frac{11\pi}{6} + c_0\varepsilon, 2\pi) \end{cases} \tag{5.8}$$

where $g_0 : [0, 1] \to [0, 1]$ is a strictly increasing smooth function, with $g_0(0) = 0$, $g_0(1) = 1$, $|g_0'(x)| \leq 2$.

Notice that the 2nd and 3rd leg of the tripod (Fig. 5) are at $(-\frac{\sqrt{3}}{2}, -\frac{1}{2})$, $(\frac{\sqrt{3}}{2}, -\frac{1}{2})$. Set the family of the horizontal line segments γ_y , $y \in [-\frac{1}{2}, 1)$ as

$$\gamma_y := \{(x, y) \,:\, x \in \mathbb{R}\} \cap B_1 , \tag{5.9}$$

and

$$\lambda_i(y) := \mathcal{L}^1(\gamma_y \cap \{|u(x, y) - a_i| < \varepsilon^{1/6}\}) \, , \, i = 1, 2, 3 \; (u_\varepsilon \equiv u). \tag{5.10}$$

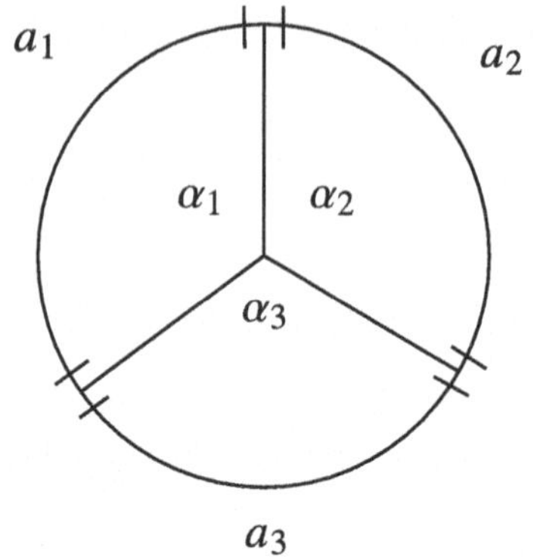

Fig. 5 The Dirichlet problem on the unit disc. The sectors D_1, D_2, D_3 are as in Fig. 4(b) in the Introduction

where $\mathcal{L}^1$ stands for the 1− dimensional Lebesgue measure. Note that by the boundary condition $\lambda_1(y) > 0$, $\lambda_2(y) > 0$ for $y \in [-\frac{1}{2} + c_0\varepsilon, 1 - c_0\varepsilon)$.

Let y^* be defined by

$$y^* := \min\{y \in [-\frac{1}{2} + c_0\varepsilon, 1] \ : \ \lambda_1(y) + \lambda_2(y) \geq \mathcal{L}^1(\gamma_y) - \varepsilon^{1/3}\}. \tag{5.11}$$

and note that for $1 - y^2 < \frac{\varepsilon^{2/3}}{4}$, $\mathcal{L}^1(\gamma_y) - \varepsilon^{1/3} < 0$, and therefore the set in which we minimize above is not empty. We define

$$\Omega_1 = B_1 \cap \{y \geq y^*\} \ , \ \Omega_2 = B_1 \cap \{y \leq y^*\} \tag{5.12}$$

(see Fig. 6 below).

Notice that the definition of y^* gives a boundary condition on the upper boundary of Ω_2, that is u there takes mostly values near a_1 or a_2:

$$\lambda_1(y^*) + \lambda_2(y^*) = \mathcal{L}^1(\gamma_y) - \varepsilon^{1/3}. \tag{5.13}$$

Therefore in Ω_1 on the basis of the boundary conditions we do not expect a y−gradient, but only an x−gradient involving the transition $a_1 \to a_2$. On the other hand in Ω_2, on the basis of the definition of y^*, we expect u on most γ_y sections, at some point, to get close to a_3. This suggests two transitions $a_1 \to a_3 \to a_2$ in the x− direction. In the y− direction in Ω_2 on the basis of (5.13) we expect one transition either $a_1 \to a_3$ or $a_2 \to a_3$. Thus in Ω_2 we expect a full gradient.

Estimate in Ω_1

Utilizing the boundary conditions and Exercise 1 we see that integrating the energy density on γ_y we obtain

$$\int_{-\sqrt{1-y^2}}^{\sqrt{1-y^2}} \left(\frac{\varepsilon}{2}|\partial_x u|^2 + \frac{1}{\varepsilon}W(u)\right) dx \geq \sigma \tag{5.14}$$

Thus

$$\iint_{\Omega_1}\left(\frac{\varepsilon}{2}|\nabla u|^2+\frac{1}{\varepsilon}W(u)\right)dx \geq \int_{y^*}^{1-c_0\varepsilon}\int_{-\sqrt{1-y^2}}^{\sqrt{1-y^2}}\left(\frac{\varepsilon}{2}|\partial_x u|^2+\frac{1}{\varepsilon}W(u)\right)dx$$
$$\geq \max\{\sigma(1-c_0\varepsilon-y^*),0\} \tag{5.15}$$

Estimate in Ω_2

First we verify that in the most γ_y- sections, $y < y^*$ the solution u at some point gets close to a_3. More precisely we claim that

$$\mathcal{L}^1(\{y\ :\ -\frac{1}{2}+c_0\varepsilon < y < y^*\ ,\ \lambda_3(y)=0\}) < C\varepsilon^{1/3}. \tag{5.16}$$

For this purpose we utilize the following (easy) upper bound

$$\int_{B_1}\left(\frac{\varepsilon}{2}|\nabla u|^2+\frac{1}{\varepsilon}W(u)\right)dz < \hat{C}\ ,\quad \text{some constant } \hat{C},\quad \text{that follows from (4.10).} \tag{5.17}$$

Note that for (5.16) to be meaningful and nontrivial we need that $y^* \geq -\frac{1}{2}+c_0\varepsilon+C\varepsilon^{1/3}$, for some constant. We will see shortly after that for $y^* \in [-\frac{1}{2}, -\frac{1}{2}+c_0\varepsilon+C\varepsilon^{1/3}]$ the lower bound we seek is easy to establish.

We now verify the claim for (5.16).

Let

$$S := \{y\ :\ -\frac{1}{2}+c_0\varepsilon < y < y^*\ ,\ \ \lambda_3(y)=0\}, \tag{5.18}$$

and notice that for any $y \in S$ the definition of y^* and S imply that

$$\lambda_1(y)+\lambda_2(y)+\lambda_3(y) < \mathcal{L}^1(\gamma_y)-\varepsilon^{1/3}\ ,\quad \text{that is}$$
$$\mathcal{L}^1(\{x\in[-\sqrt{1-y^2},\sqrt{1-y^2}]\ :\ |u(x,y)-a_i| > \varepsilon^{1/6}\ ,\ \forall i\}) > \varepsilon^{1/3}.$$

Then by our assumption on the potential W in **(H1)** in Sect. 1

$$W(u(x,y)) \geq C_1\varepsilon^{1/3} \text{ when } |u(x,y)-a_i| > \varepsilon^{1/6}\ ,\ \forall i$$

and via (5.17)

$$\hat{C} > \frac{1}{\varepsilon}\int_S\int_{\gamma_y} W(u)dxdy \geq \frac{1}{\varepsilon}C_1\varepsilon^{1/3}\mathcal{L}^1(S)$$

from which we obtain

$$\mathcal{L}^1(S) \leq C\varepsilon^{1/3}\ , \quad \text{for some constant } C \text{ depending on } W.$$

and so the claim is established.

Next we dismiss the case $y^* \in [-\frac{1}{2}, -\frac{1}{2} + c_0\varepsilon + C\varepsilon^{1/3}]$.

Set

$$K_0 := \{x \in [-\frac{\sqrt{3}}{2} + c_0\varepsilon, \frac{\sqrt{3}}{2} - c_0\varepsilon] \ : \ |u(x, y^*) - a_i| < \varepsilon^{1/6}\ , \ i = 1 \text{ or } 2\}$$

By (5.13), $\mathcal{L}^1(K_0) \geq \sqrt{3} - 2c_0\varepsilon - \varepsilon^{1/3}$. Thus utilizing Exercise 1

$$\begin{aligned}\iint_{\Omega_2} \left(\frac{\varepsilon}{2}|\nabla u|^2 + \frac{1}{\varepsilon}W(u)\right) dxdy &\geq \int_{K_0} \int_{-\sqrt{1-x^2}}^{y^*} \left(\frac{\varepsilon}{2}|\partial_y u|^2 + \frac{1}{\varepsilon}W(u)\right) dydx \\ &\geq (\sqrt{3} - 2c_0\varepsilon - \varepsilon^{1/3})(\sigma - C_W\varepsilon^{1/3}).\end{aligned} \tag{5.19}$$

Combining this with (5.15) and taking into account the restriction on y^* we obtain

$$\begin{aligned}&\iint_{B_1} \left(\frac{\varepsilon}{2}|\nabla u|^2 + \frac{1}{\varepsilon}W(u)\right) dxdy \\ &\geq (\frac{3}{2} + \sqrt{3} - 3c_0\varepsilon - (C+1)\varepsilon^{1/3})(\sigma - C_W\varepsilon^{1/3}),\end{aligned} \tag{5.20}$$

which clearly satisfies the lower bound in (5.7), for $\varepsilon > 0$ small.

Now we consider the case $y^* \geq -\frac{1}{2} + c_0\varepsilon + C\varepsilon^{1/3}$. For any $x \in [-\frac{\sqrt{3}}{2} + c_0\varepsilon, \frac{\sqrt{3}}{2} - c_0\varepsilon]$ we set

$$\zeta(x) := \min\{y^*, \sqrt{1-x^2}\}$$

We also introduce the sets

$$\begin{aligned}K &:= \{x \in [-\frac{\sqrt{3}}{2} + c_0\varepsilon, \frac{\sqrt{3}}{2} - c_0\varepsilon] \ : \ |u(x, \zeta(x)) - a_i| < \varepsilon^{1/6}\ , \ i = 1 \text{ or } 2\} \\ L &:= \{y \in [-\frac{1}{2} + c_0\varepsilon, y^*] \ : \ \lambda_3(y) > 0\}\end{aligned} \tag{5.21}$$

We note that

$$\mathcal{L}^1(K) \geq \sqrt{3} - 2c_0\varepsilon - \varepsilon^{1/3}\ , \ \mathcal{L}^1(L) \geq y^* + \frac{1}{2} - C\varepsilon^{1/3} - c_0\varepsilon \tag{5.22}$$

Moreover, we introduce a parameter $\theta \in (0, \frac{\pi}{2})$ and split the potential W into two parts, $W = (sin^2\theta)W + (cos^2\theta)W$ and compute the energy in the vertical and the horizontal direction respectively.

For any $x \in K$

$$|u(x, \zeta(x)) - a_i| < \varepsilon^{1/6}\ , \ \text{ for } i = 1, \text{ or } 2, \ u(x, -\sqrt{1-x^2}) = a_3$$

Vertical Direction

$$\begin{aligned}
&\int_{-\sqrt{1-x^2}}^{\zeta(x)} \left(\frac{\varepsilon}{2}|\partial_y u|^2 + \frac{sin^2\theta}{\varepsilon} W(u)\right) dy \\
&= sin\theta \int_{-\sqrt{1-x^2}}^{\zeta(x)} \left(\frac{\varepsilon}{2sin\theta}|\partial_y u|^2 + \frac{sin\theta}{\varepsilon} W(u)\right) dy \qquad (5.23)\\
&\geq sin\theta(\sigma - C_W \varepsilon^{1/3})
\end{aligned}$$

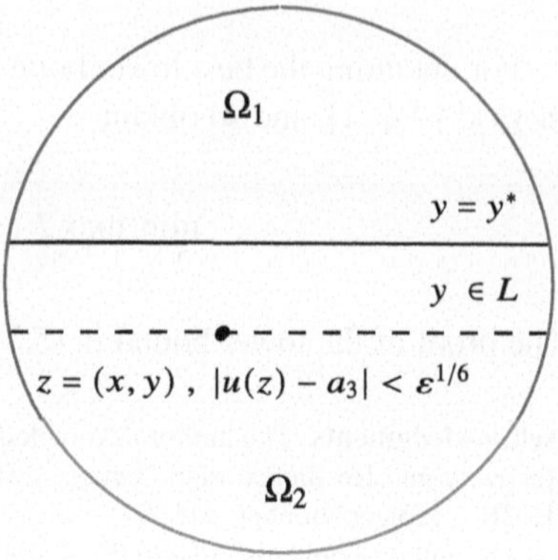

Fig. 6 The slicing of Ω: shaded region Ω_1, *nonshaded region* Ω_2

Horizontal Direction
Similarly for any $y \in L$ (see Fig. 6 above)

$$\begin{aligned}
&u(-\sqrt{1-y^2}, y) = a_1\ ,\ u(\sqrt{1-y^2}, y) = a_2\ ,\ \exists\, (x_0, y) \in \gamma_y\ , \\
&|u(x_0, y) - a_3| < \varepsilon^{1/6} \qquad (5.24)
\end{aligned}$$

$$\begin{aligned}
&\int_{\gamma_y} \left(\frac{\varepsilon}{2}|\partial_x u|^2 + \frac{cos^2\theta}{\varepsilon} W(u)\right) dx \\
&= cos\theta \left\{\int_{-\sqrt{1-y^2}}^{x_0} + \int_{x_0}^{\sqrt{1-y^2}}\right\} \left(\frac{\varepsilon}{2cos\theta}|\partial_x u|^2 + \frac{cos\theta}{\varepsilon} W(u)\right) dx \qquad (5.25)\\
&\geq 2cos\theta(\sigma - C_W \varepsilon^{1/3}).
\end{aligned}$$

Thus (5.23), (5.25) give

$$\iint_{\Omega_2}\left(\frac{\varepsilon}{2}|\nabla u|^2+\frac{1}{\varepsilon}W(u)\right)dxdy \geq \left(sin\theta\mathcal{L}^1(K)+2cos\theta\mathcal{L}^1(L)\right)(\sigma - C_W\varepsilon^{1/3})$$

Also

$$\iint_{\Omega_1}\left(\frac{\varepsilon}{2}|\nabla u|^2+\frac{1}{\varepsilon}W(u)\right)dxdy \geq \sigma(1-c_0\varepsilon - y^*).$$

Therefore

$$\begin{aligned}\int_{B_1}\left(\frac{\varepsilon}{2}|\nabla u|^2+\frac{1}{\varepsilon}W(u)\right)dz &\geq \left[2y^*(cos\theta-\frac{1}{2})+\sqrt{3}sin\theta+cos\theta+1\right]\sigma - C\varepsilon^{1/3}\\ &= F(y^*,\theta)\sigma - C\varepsilon^{1/3} \quad , \ \varepsilon<\varepsilon_0(W).\end{aligned}$$

For obtaining the best lower bound we maximize first in θ, and then we minimize in $y^* \in [-\frac{1}{2}, 1]$, and so obtain

$$\min_{y^*}(\max_{\theta} F(y^*,\theta)) \geq F(0,\frac{\pi}{3}) = 3 \tag{5.26}$$

The proof of the lower bound in (5.7) is complete. □

Acknowledgments The author acknowledges the "Basic research Financing" under the National Recovery and Resilience Plan "Greece 2.0" funded by the European Union-NextGeneration EU (H.F.R.I. Project Number: 016097).

The author would also like to thank the anonymous referees for their valuable suggestions and corrections, which improved the previous version of the paper.

References

1. G. Alberti, L. Ambrosio, X. Cabre, On a long-standing conjecture of E. De Giorgi: symmetry in 3D for general nonlinearities and a local minimality property. Acta Appl. Math. **65**, 9–33 (2001)
2. S. Allen, J. Cahn, A microscopic theory for antiphase motion and its application to antiphase domain coarsening. Acta Metail. **27**, 1084–1095 (1979)
3. N.D. Alikakos, Some basic facts on the system $\Delta u - W_u(u) = 0$. Proc. Am. Math. Soc. **139**, 153–162 (2011)
4. N.D. Alikakos, F. Giorgio, On the connection problem for potentials with several global minima. Indiana Univ. Math. J. **57**(4), 1871–1906 (2008)
5. N.D. Alikakos, F. Giorgio, Density estimates for vector minimizers and applications. Discrete Continuous Dyn. Syst. **35**(12), 5631–5663 (2015)
6. N.D. Alikakos, F. Giorgio, A maximum principle for systems with variational structure and an application to standing waves. J. Eur. Math. Soc. **17**(7), 1547–1567 (2015)

7. N.D. Alikakos, F. Giorgio, *Panayotis Smyrnelis: Elliptic Systems of Phase Transition Type*. Progress in Nonlinear Differential Equations and Their Applications (Birkhauser, Basel, 2018)
8. N.D. Alikakos, Z. Geng, On the triple junction problem without symmetry hypotheses. Arch. Ration. Mech. Anal. **248**, 24 (2024)
9. N.D. Alikakos, S. Betelú, X. Chen, Explicit stationary solutions in multiple well dynamics and non-uniqueness of interfacial energy densities. Eur. J. Appl. Math. **17**(5), 525–556 (2006)
10. L. Ambrosio, X. Cabre, Entire solution of semilinear elliptic equations in $\mathbb{R}^3$ and a conjecture of De Giorgi. J. Am. Math. Soc. **13**, 725–739 (2000)
11. P. Antonopoulos, P. Smyrnelis, A maximum principle for the system $\Delta u - \nabla W(u) = 0$. Comptes Rendus. Math. **354**(6), 595–600 (2016)
12. P. Antonopoulos, P. Smyrnelis, *On Minimizers of the Hamiltonian System $u'' = \nabla W(u)$ and on the Existence of Heteroclinic, Homoclinic and Periodic Orbits*. Indiana Univ. Math. J. **65**(5), 1503–1524 (2016)
13. M.T. Barlow, R.F. Bass, C. Gui, The Liouville property, and a conjecture of De Giorgi. Commun. Pure Appl. Math. **53**, 1007–1038 (2000)
14. L. Bronsard, R.V. Kohn, Motion by mean curvature as the signular limit of Ginzburg-Landau dynamics. J. Differ. Equ. **90**(2), 211-237 (1991)
15. L. Bronsard, F. Reitich, On three-phase boundary motion and the singular limit of a vector-valued Ginzburg-Landau equation. Arch. Ration. Mech. Anal. **124**, 355–379 (1993)
16. L. Bronsard, C. Gui, M. Schatzman, A three-layered minimizer in $\mathbb{R}^2$ for a variational problem with a symmetric three-well potential. Commun. Pure Appl. Math. **49**(7), 677–715 (1996)
17. L. Caffarelli, A. Cordoba, Uniform convergence of a singular perturbation problem. Commun. Pure Appl. Math. **48**, 1–12 (1995)
18. L. Caffarelli, N. Garofalo, F. Segala, A gradient bound for entire solutions of quasi-linear equations and its consequences. Commun. Pure Appl. Math. **47**(11), 1457–1473 (1994)
19. J.W. Cahn, J.E. Hilliard, Free energy of a nonuniform system I. Interfacial free energy. J. Chem. Phys. **2**, 258–267 (1958)
20. J. Carr, R.L. Pego, Metastable patterns in solutions of $u_t = \varepsilon^2 u_{xx} - f(u)$. Commun. Pure Appl. Math. **42**(5), 523–576 (1989)
21. X. Chen, Generation and propagation of interfaces for reaction-diffusion equations. J. Differ. Equ. **96**(1), 116–141 (1992)
22. O. Chodosh, *Lecture Notes on Geometric Features of the Allen-Cahn equation* (Summer School, Princeton, 2019)
23. E. De Giorgi, Una estensione del teorema di Bernstein. Ann. Scuola Norm. Sup. Pisa (3), 19 (1965)
24. E. De Giorgi, Convergence problems for functionals and operators. *Proceedings of the International Meeting on Recent Methods in Nonlinear Analysis (Rome, 1978)* (Pitagora, Bologna, 1979), pp. 131–188
25. P. de Mottoni, M. Schatzman, Geometrical evolution of developed interface. Math. Soc. **347**(5), 1533–1589 (1995)
26. M. del Pino, M. Kowalczyk, J. Wei, On De Giorgi's conjecture in dimension $N \geq 9$. Ann. Math. (2) **174**(3), 1485–1569 (2011)
27. L.C. Evans, *Partial Differential Equations*. Graduate Studies in Mathematics, vol. 19, 2nd edn. (American Mathematical Society, Providence, 2010)
28. L.C. Evans, R.F. Gariepy, *Measure Theory and Fine Properties of Functions*, Revised edn. (1st edn.) (Chapman and Hall/CRC, Boca Raton, 2015)
29. L.C. Evans, H.M. Soner, P.E. Souganidis, Phase transitions and generalized motion mean curvature. Commun. Pure Appl. Math. **45**(9), 1097–1123 (1992)
30. A. Farina, E. Valdinoci, The state of the art for a conjecture of De Giorgi and related problems. *Recent Progress Reaction-Diffusion Systems and Viscosity Solutions*. Gianni Gilardi Pietro Poggi-Corradini eds, (World Scientific, 2009), pp. 74–96
31. J. Fischer, A. Marveggio, Quantitative convergence of the vectorial Allen-Cahn equation towards multiphase mean curvature flow. Ann. Inst. H. Poincaré **41**(5), 1117–1178 (2024)

32. G. Fusco, Minimizing under relaxed symmetry constraints: Triple and N-junctions. Annali Scuola Normale Superiore - Classe Di Scienze **25**(2), 884–940 (2024)
33. G. Fusco, J.K. Hale, Slow-motion manifolds, dormant instability, and singular perturbations. J. Dyn. Differ. Equ. **1**, 75–94 (1989)
34. G. Fusco, G.F. Gronchi, M. Novaga, On the existence of connecting orbits for critical values of the energy. J. Differ. Equ. **263**(12), 8848–8872 (2017)
35. Z. Geng, Uniqueness of the blow-down limit for a triple junction problem. Commun. Pure Appl. Math. **78**, 500–534 (2025)
36. N. Ghoussoub, C. Gui, On a conjecture of De Giorgi and some related problems. Math. Ann. **311**, 481–491 (1998)
37. D. Gilbarg, N.S. Trudinger, *Elliptic Partial Differential Equations of Second Order*, 2nd edn. (Springer, Berlin, 1998)
38. T. Ilmanen, Convergence of the Allen-Cahn equation to Brakke's motion by mean curvature. J. Differ. Geom. **38**(2), 417–461 (1993)
39. K. Kawasaki, T. Ohta, Kinetic drumhead model of interface I. Progr. Theor. Phys. **67**, 147–163 (1982)
40. L. Modica, Convergence to minimal surfaces problem and global solution of $\Delta u = 2(u^3 - u)$, in Proceedings of the International Meeting on Recent Methods in Nonlinear Analysis (Rome, 1978) (Pitagora, Bologna, 1979)
41. L. Modica, S. Mortola, Un esempio di Γ convergenza. (Italian) Boll. Un. Mat. Ital. B (5) **14**(1), 285–299 (1977)
42. A. Monteil, F. Santambrogio, Metric methods for heteroclinic connections. Math. Methods Appl. Sci. **41**(3), 1019–1024 (2016)
43. W.W. Mullins, Two-dimensional motion of idealized grain boundaries. J. Appl. Phys. **27**, 900–904 (1956)
44. P.H. Rabinowitz, Homoclinic and heteroclinic orbits for a class of Hamiltonian systems. Calc. Var **1**, 1–36 (1993)
45. O. Savin, Regularity of flat level sets in phase transitions. Ann. Math. (2) **169**(1), 41–78 (2009)
46. O. Savin, *Minimal Surfaces and Minimizers of the Ginzburg-Landau Energy*. Contemporary Mathematics, vol. 528 (Americal Mathematical Society, Providence, 2009)
47. É. Sandier, P. Sternberg, Allen–Cahn solutions with triple junction structure at infinity. Commun. Pure Appl. Math. **77**(11) 4163–4211 (2024)
48. P. Smyrnelis, Connecting orbits in Hilbert spaces and applications to P.D.E. Commun. Pure Appl. Anal. **19**(5), 2797–2818 (2020)
49. P. Sternberg, Vector-valued local minimizers of nonconvex variational problems. Rocky Mt. J. Math. **21**(2), 799–807 (1991)
50. A. Zuniga, P. Sternberg, On the heteroclinic connection problem for multi-well gradient systems. J. Differ. Equ. **261**(7), 3987–4007 (2016)

Part II
Analysis Tools

Introduction to Quantitative De Giorgi Methods

Giovanni Brigati and Clément Mouhot

1 Introduction

This section presents the motivation and scientific context. The new theory we will present in the last section can be called **regularity theory of kinetic equations with rough coefficients**. But it is a new extension of a classical one. It has three independent origins which we briefly summarise in the three next subsections.

1.1 The Solution to Hilbert's 19th Problem

At the ICM (International Congress of Mathematicians) of 1900, [46], D. Hilbert posed twenty-three mathematical problems to the mathematical community for the incoming century. The 19th of them asked *which are the regular variational problems*.

Hilbert defined a **regular variational problem** as a variational problem whose Euler-Lagrange equation is an elliptic partial differential equation with analytic coefficients. Therefore, concretely Hilbert's 19th problem asks whether, for the class of nonlinear partial differential equations arising as Euler-Lagrange equations of

G. Brigati
Institute of Science and Technology Austria (ISTA), Klosterneuburg, Austria
e-mail: giovanni.brigati@ist.ac.at

C. Mouhot (✉)
Department of Pure Mathematics and Mathematical Statistics, University of Cambridge, Cambridge, UK
e-mail: c.mouhot@dpmms.cam.ac.uk

© The Author(s), under exclusive license to Springer Nature Switzerland AG 2026
N. D. Alikakos, C. Villani (eds.), *Analysis Techniques for Mathematical Physics and Geometry – and Music*, Lecture Notes in Mathematics 2390,
https://doi.org/10.1007/978-3-032-11396-2_3

such variational problem, solutions are analytic; this implies that the local minimisers of the original variational problem exists and are analytic. The underlying motivation is that Laplace's equation admits analytic solutions only, and, at the same time, it is the Euler–Lagrange equation associated with the minimisation of the Dirichlet energy over a bounded domain. Hilbert's 19th problem consists therefore in determining how general this observation is.

Hilbert's question can be reduced to investigating the smoothness of local minimisers, as analyticity follows by studying the convergence of the Taylor series, as observed by Bernstein [13]. Let $\mathcal{U}$ be a bounded domain of $\mathbb{R}^n$. Let $\mathrm{L} : \mathbb{R}^n \to \mathbb{R}$ be a **Lagrangian** function, assumed to be convex. We then consider the variational problem

$$\min_u \mathcal{E}(u) := \int_{\mathcal{U}} \mathrm{L}(\nabla_x u)\,\mathrm{d}x, \tag{1}$$

where $u : \mathcal{U} \to \mathbb{R}$ varies in an appropriate functional space (which can and often does include suitable boundary conditions). If the Lagrangian is given by

$$\mathrm{L}(p) = \frac{1}{2}\,|p|^2, \tag{2}$$

then $\mathcal{E}$ is the Dirichlet energy, and solutions to (1) satisfy the linear Laplace equation

$$\Delta u = 0 \text{ on } \mathcal{U}$$

with appropriate boundary conditions. If

$$\mathrm{L}(p) = \sqrt{1 + |p|^2} \tag{3}$$

is the area element of the surface defined by the height function u, then $\mathcal{E}$ is the area functional, and the variational problem corresponds to finding **minimal surfaces**. The nonlinear Euler-Lagrange equation is then

$$\nabla_x \cdot \left(\frac{\nabla_x u}{\sqrt{1 + |\nabla_x u|^2}} \right) = 0 \text{ on } \mathcal{U} \tag{4}$$

with, again, appropriate boundary conditions.

In general, we derive the Euler-Lagrange equations associated with (1). Let u be a solution to (1), then for all $w \in \mathrm{C}_c^\infty(\mathcal{U})$ we have

$$\forall \varepsilon > 0, \int_{\mathcal{U}} \mathrm{L}\left(\nabla_x u + \varepsilon \nabla_x w\right)\,\mathrm{d}x \geq \int_{\mathcal{U}} \mathrm{L}(\nabla_x u)\,\mathrm{d}x.$$

By Taylor-expanding the left-hand side of the last display, and dividing by ε, we get

$$\int_{\mathcal{U}} \left[\nabla L(\nabla_x u) \cdot \nabla_x w \right] \mathrm{d}x = 0,$$

which is the weak formulation for (the notation ":" stands for contracting tensors)

$$\nabla_x \cdot \left[\nabla L(\nabla_x u) \right] = \left[\nabla^2 L(\nabla_x u) \right] : \nabla_x^2 u = 0 \tag{5}$$

in the space of distributions $\mathcal{D}'(\mathcal{U})$. This is an elliptic PDE in non-divergence form in general. Let us reason a priori and consider a partial derivative $f = \partial_{x_k} u$ for $k = 1, \ldots, n$. Then, if u solves (5), the new unknown f solves (repeated indices are summed)

$$\partial_{x_i} \cdot (a_{ij}(x)\, \partial_{x_j} f) = 0, \quad \text{with} \quad a_{ij}(x) = \left[\nabla^2 L(\nabla_x u(x)) \right]_{ij}, \tag{6}$$

which is a second-order partial differential equation in divergence form. Note that when the matrix of coefficients $A = (a_{ij})$ is definite positive—which corresponds to the strict convexity of L—then u satisfies the maximum principle and Bernstein's method allows estimating the gradient of u in a pointwise manner (see for instance [61]). Therefore, given suitable boundary conditions, we can assume $\nabla u \in L^\infty$.

In these lecture notes, we focus on the **interior regularity** of solutions to equations of the form (6), and of their kinetic counterparts. This means working away from the boundary, thus we will not discuss further boundary conditions. The regularity at the boundary is an interesting problem, still currently mostly open in kinetic theory. Solving the 19th Hilbert problem consists in establishing the analytic regularity of solutions to (5)–(6). The difficulty is that the matrix of coefficients $A = (a_{i,j})$ depends on the gradient of the solution ∇u, resulting in a non-linear bootstrap.

We make the following hypotheses on L.

$$L \in C^\infty(\mathbb{R}^n), \tag{H1}$$

$$\exists\, \Lambda > 0 : \forall\, p \in \mathbb{R}^n,\ \Lambda^{-1} \le \nabla^2 L(p) \le \Lambda. \tag{H2}$$

The first hypothesis can be relaxed appropriately when one is interested in establishing only finite regularity of the solution u. The second hypothesis is a uniform coercivity/ellipticity condition on the matrix A, combined with an upper bound. In terms of L, it corresponds to a uniform convexity combined with a subquadratic control on the growth, together with a pointwise bound on the second derivative.

Going back to our two examples, the Dirichlet Lagrangian (2) yields $A = \nabla^2 L = \mathrm{Id}$ which satisfies our assumption, and the minimal surface Lagrangian (3) yields

$$\nabla_p^2 L(p) = \frac{\mathrm{Id} + \left(|p|^2 \mathrm{Id} - p \otimes p \right)}{\left(1 + |p|^2 \right)^{\frac{3}{2}}}$$

which satisfies a modified form of the hypothesis when we restrict p to a bounded region (this turns out to be sufficient for the local regularity theory).

Here are chronologically the main steps in the resolution of Hilbert's 19th problem:

1904: Bernstein [13] proves in dimension $n = 2$ that analyticity follows from C^3 regularity for solutions to (1)–(5), under our assumption on L, see also [85] for general dimensions.

1934: Schauder [86] proves that if $A \in C^{\alpha}_{\mathrm{loc}}(\mathcal{U})$ for $\alpha \in (0, 1)$ (Hölder regularity), then the solution u to the non-divergence elliptic equation (5) is $C^{2+\alpha}_{\mathrm{loc}}(\mathcal{U})$. A bootstrap argument can then be implemented by differentiating the equation (the Schauder theory survives when lower order terms are added). It shows that solutions u are C^{∞}. Since we know already $\nabla_x u \in L^{\infty}$, the missing link is the Hölder regularity of $\nabla_x u$ (implying that of A), which triggers Schauder's estimates.

1938: Morrey [69] gives a solution which is specific to the two-dimensional case $n = 2$.

1957: De Giorgi and Nash independently [22, 80] solve the problem. De Giorgi shows that weak L^2 solutions to

$$\begin{cases} \nabla_x \cdot (A\, \nabla_x f) = 0 \text{ on } \mathcal{U}, \\ A \text{ symmetric measurable and } \Lambda^{-1} \le A \le \Lambda \text{ for some } \Lambda > 0, \end{cases} \tag{7}$$

are such that $f \in C^{\alpha}_{\mathrm{loc}}(\mathcal{U})$ for some $\alpha > 0$. Such an elliptic PDE is said to have **rough coefficients**. Nash proves a similar theorem for linear parabolic equations with rough coefficients.

1960: Moser [70, 71, 73] introduces an alternative approach closer to "energy estimates" in PDE analysis, which contains many influencing ideas. The related Kruzhkov contribution [58] simplifies this approach. In particular, these papers contain the first proof of the **Harnack inequality** for elliptic and parabolic equations with rough coefficients.

Together, the last steps of this program are referred to as the **De Giorgi–Nash–Moser theory** (DGNM). The viewpoint adopted in this theory, which had a huge influence in PDE analysis, is that *the nonlinearity is traded for the understanding of a linear problem with **low regularity coefficients***.

Nash performed integral estimates on the fundamental solution to (7) and its logarithm, while De Giorgi studied the level sets, see for instance [21, 23] for modern presentations. De Giorgi's proof is elliptic, while Nash's proof is parabolic, but both approaches can treat both cases. They also both adapt thoroughly to the following *hypoelliptic equations of type I* (see the next subsection)

$$\sum_{i,j=1}^{m} D_i \left(a_{ij} D_j f\right) = 0$$

where $(D_i)_{i=1}^m$ is a collection of first-order differential operators such that they generate a Lie algebra of full rank

$$\mathrm{Lie}\,(D_1, \cdots, D_m) = \mathrm{span}\left(\partial_{x_1}, \cdots, \partial_{x_n}\right),$$

where the Lie bracket of the Lie algebra is the commutator of vector fields. The hypotheses on the matrix $A = (a_{i,j})_{1\leq i,j\leq m}$ are the same as in (7). This case is important in sub-Riemannian geometry, see for instance [1], but will not be discussed in these notes.

The classical DGNM theory had two important restrictions. The first one is that it deals with divergence-form PDEs. This is not merely a technical restriction, as all approaches—by De Giorgi, Nash or Moser—crucially rely on **energy estimates**. These energy estimates are also called **Caccioppoli inequalities**: see (20) in the elliptic case and (40) in the parabolic case and (60) in the kinetic (hypoelliptic) case.

A similar result of Hölder regularity was later obtained by Krylov and Safonov [62] for solutions to non-divergence equations of the form

$$a_{ij}(x)\,\partial^2_{x_i x_j} f = 0,$$

with a symmetric matrix $A = (a_{i,j})_{i,j}$ of rough coefficients that satisfies $\Lambda^{-1} \leq A \leq \Lambda$ for some $\Lambda > 0$. We also refer to [3, 60, 64] for important previous contributions. The methods are however different in the non-divergence case.

The second restriction of the classical DGNM theory is that it does not cover the *hypoelliptic case of type II* (see the next subsection) which is ubiquitous in physics, and such an extension is the object of these lecture notes.

We however also focus in these notes on obtaining constructive proofs with quantitative estimates on the constants and exponents. This leads us to revisit certain results even in the classical elliptic and parabolic cases.

There are two motivations behind this. The first motivation is rooted in mathematical physics. The classical DGNM theory was developed to solve Hilbert's 19th problem about the regularity of energy minimisers. And the extension of this classical theory to the hypoelliptic equations appearing in kinetic theory was developed to obtain the conditional regularity of solutions to the Landau equation (and to the Boltzmann equation for long-range interactions, in the non-local case). In both problems, a mathematical solution is only truly satisfying if the constants of regularity obtained are of orders compatible with the physics itself. And it is difficult to ensure this without developing quantitative methods of proof. The second motivation is internal to mathematics. The search for quantitative methods often deepens the understanding of the mathematical structures and suggests new connexions. In our case, it has lead us to connecting the control of integral oscillation (through inequalities of Poincaré-type) to that of constructing particular trajectories solving a control problem.

1.2 Hörmander's Hypoellipticity Theory

The DGNM theory revolutionised the study of elliptic and parabolic equations. It was however not covering a fundamental class of related equations, whose importance has been recognised at least since Kolmogorov [56]. In this note, Kolmogorov investigates a model equation driven by a first-order differential operator combined with diffusion only in certain directions. Let x, v be the variables respectively in $\mathbb{R}^d$. Then, **Kolmogorov's equation** reads, in its simplest version:

$$\begin{cases} \partial_t f + v \cdot \nabla_x f = \Delta_v f, \quad f = f(t,x,v), \\ f(0,\cdot,\cdot) = f_0 \text{ which is a probability density on } \mathbb{R}^{2d}. \end{cases} \tag{8}$$

This equation corresponds to the evolution of the law of the random process

$$\begin{cases} \mathrm{d}X_t = V_t \, \mathrm{d}t, \\ \mathrm{d}V_t = \mathrm{d}W_t, \end{cases} \tag{9}$$

where $(W_t)_t$ is a standard $d-$dimensional Brownian motion. Indeed, (9) is a time-integrated Brownian motion, which gives its name to the note [56].

Kolmogorov computes an explicit formula for the fundamental solution to (8), by combining Fourier analysis and Duhamel's principle. Given the initial data $f_0 = \delta_{x_0,v_0}$ in (8), one gets

$$G(t,x,v) = \left(\frac{\sqrt{3}}{2\pi t^2} \right)^d \exp\left(-3 \frac{|x - x_0 - t(v+v_0)/2|^2}{t^3} - \frac{|v - v_0|^2}{4t} \right). \tag{10}$$

The fundamental solution (10) is smooth in all variables for $t > 0$, and therefore all solutions to (8) are smooth as well for $t > 0$. However, ellipticity in x fails for (8). Intuitively, the noise in v is "transferred" to the x variable by the transport drift operator $v \cdot \nabla_x$, see Fig. 1.

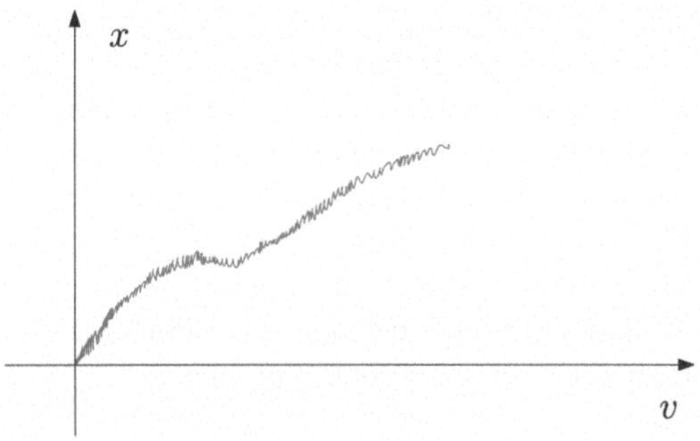

Fig. 1 Noise in v propagates to the x variable thanks to the transport drift $v \cdot \nabla_x$

After some seminal works in specific cases such as [83], a complete understanding of this phenomenon was developed by Hörmander [48] in 1967. His starting point was Kolmogorov's calculation showing that for second-order operators with smooth coefficients, while ellipticity is a sufficient condition for the smoothness of the fundamental solution (he called the latter property "hypoellipticity"), it is *not* necessary. For the Kolmogorov equation, regularisation is produced by the interaction between the vector fields ∇_v and $v \cdot \nabla_x$.

More precisely, we say that a linear operator L is **hypoelliptic** if

$$Lu \in C^\infty_{\text{loc}}(\mathcal{U}) \implies u \in C^\infty_{\text{loc}}(\mathcal{U}),$$

in an open subset $\mathcal{U} \subset \mathbb{R}^n$ (again, we do not address the issue of the boundaries). When restricting to second-order operators with smooth coefficients, the standard **ellipticity**, i.e. the coercivity of the principal symbol, implies this condition. It is however not necessary. Consider, for $m \in \mathbb{N}^*$, a family $(X_i)_{i=1}^m$ of smooth vector fields and the operator

$$L = X_0 + \sum_{i=1}^m X_i^2. \tag{11}$$

The main theorem of [48] is a necessary and sufficient condition for operators of the form (11) to be hypoelliptic: the Lie algebra of $X_0, \dots, X_m$ must satisfy

$$\dim \operatorname{Lie}(X_0, \cdots, X_m) = n.$$

When $X_0 = 0$ (no first-order term) such operators write as a sum of squares and we call them **hypoelliptic operators of type I**; otherwise we call them **hypoelliptic operators of type II**. Hypoelliptic operators of type II have also been called **ultraparabolic** in the literature.

The Kolmogorov operator is the archetypical example of the type II. Let

$$\mathcal{K} := \partial_t + v \cdot \nabla_x + \sum_{i=1}^d \partial_{v_i}^\star \partial_{v_i}, \tag{12}$$

be the operator driving (8), with $t \in \mathbb{R}$, $x, v \in \mathbb{R}^d$. This has the form (11) with

$$n = 2d + 1, \quad X_0 = \partial_t + v \cdot \nabla_x, \text{ and } X_i = \partial_{v_i} \text{ for } i = 1, \cdots, d.$$

The fundamental computation is

$$[X_0, X_i] = -\partial_{x_i}, \quad \forall\, i = 1, \cdots, d$$

which allows recovering all ∂_{x_i}'s, and finally ∂_t is recovered by linear combination. Hence,

$$\dim \mathrm{Lie}(X_0, X_1, \cdots, X_d) = n$$

at every point and $\mathcal{K}$ is hypoelliptic. By inspecting the (x, v)-Fourier transform of the Kolmogorov fundamental solution, one sees that hypoelliptic operators have in general *Gevrey* rather than *analytic* regularisation (with exponent $2/3$ in the case of the Kolmogorov operator, but this exponent depends on the Lie algebra).

1.3 Mathematical Kinetic Theory

The bedrock of kinetic theory is Newton's classical mechanics. Along Newton's theory, decisive progresses were later made for electric-magnetic forces (Ampère, Faraday, Maxwell…), large velocities and large scales (Lorentz, Poincaré, Einstein, Minskowski…), small scales and quantum physics (Planck, Einstein, Bohr, Heisenberg, Schrödinger, Dirac, Bose…). However, all these works concern the study of a single object (a planet, an electron,…) or a small number of those, such as the planets in the solar system or the electrons in an atom. Many natural phenomena pertain to systems made of a very large number of objects (or particles, or agents…). The observable universe contains 10^9–10^{12} galaxies, each of them made of billions of stars (about 10^{10} in the Milky Way). A gas is composed of about 10^{23} molecules per mole (the Avogadro constant). Crowds, flocks of birds, shoals of fishes, are made up of thousands of individuals. In such situations, it is often impossible to track each particle/molecule/individual.

In the 1867 paper [67] (and in subsequent papers), Maxwell laid the mathematical foundations of statistical mechanics. The statistical approach was developed to understand these large systems through their statistical distributions in a way that is rigorously connected to the microscopic dynamics. This is in contrast with the previously invented fluid mechanics, which combines conservation laws and adhoc closure laws to derive evolution equations at the macroscopic level. The theory invented by Maxwell is called *kinetic* due to the presence of a "kinetic" variable, the microscopic velocity, which is inaccessible to observation. We have therefore three scales of description of complex systems, ordered by increasing level of detail:

(1) a **macroscopic** scale (fluid dynamics, observable);
(2) a **mesoscopic** scale (statistical mechanics, where the behaviour of the unknown in some variables is only observable on average);
(3) a **microscopic** scale (Newton's mechanics, fully not observable).

At the mesoscopic scale, the fundamental equation—and also the oldest one—is the **(Maxwell-)Boltzmann equation**, invented in [15, 67]:

$$\partial_t f + v \cdot \nabla_x f = Q(f, f). \tag{13}$$

The unknown is a non-negative time-dependent probability distribution function $f = f(t, x, v)$, with $x \in \Omega \subset \mathbb{R}^d$, $v \in \mathbb{R}^d$, and Ω a spatial domain with suitable boundary conditions. The **collision operator** Q is given by

$$Q(f, f)(t, x, v) =$$
$$\int_{v_\star \in \mathbb{R}^d} \int_{\sigma \in \mathbb{S}^{d-1}} \mathrm{B}(v - v_\star, \omega) \left[f(t, x, v') f(t, x, v'_\star) - f(t, x, v) f(t, x, v_\star) \right] \mathrm{d}\omega \, \mathrm{d}v_\star,$$

where the variables $v_\star$, v', $v'_\star$ are related to v as follows (given $\sigma \in \mathbb{S}^{d-1}$)

$$v' = \frac{v + v_\star}{2} + \frac{|v - v_\star|}{2}\sigma, \qquad v'_\star = \frac{v + v_\star}{2} - \frac{|v - v_\star|}{2}\sigma.$$

This is a parametrisation of the degrees of freedom left over once the conservation laws of the elastic binary collision between particles with identical mass—see Fig. 2—have been taken into account:

$$\begin{cases} v + v_\star = v' + v'_\star, & \text{(conservation of momentum)} \\ |v|^2 + |v_\star|^2 = |v'|^2 + |v'_\star|^2, & \text{(conservation of kinetic energy).} \end{cases}$$

Observe that $Q(f, f)$ statistically accounts for collisions at time t, at a point x, between two particles with pre-collisional velocities v', $v'_\star$, and post-collisional velocities v, $v_\star$, see Fig. 2. The collision kernel B encodes the frequencies at which collisions with different pairs of velocities occur, and $Q(f, f)$ collects the weighted balance-sheet of all possible collisions in the velocity space. Q is a bilinear nonlocal operator acting only in the kinetic variable v.

By invariance (see [14]), the collision kernel B is a function of $|v - v_\star|$ and $\cos\theta$, so that $\mathrm{B} = \mathrm{B}(|v - v_\star|, \cos\theta)$, where θ is the angle between $v - v_\star$ and $v' - v'_\star$, i.e. between $v - v_\star$ and σ. The computation of Maxwell shows that $\mathrm{B} = |v - v_\star|$ for hard spheres interacting by elastic collisions. When particles interact through a

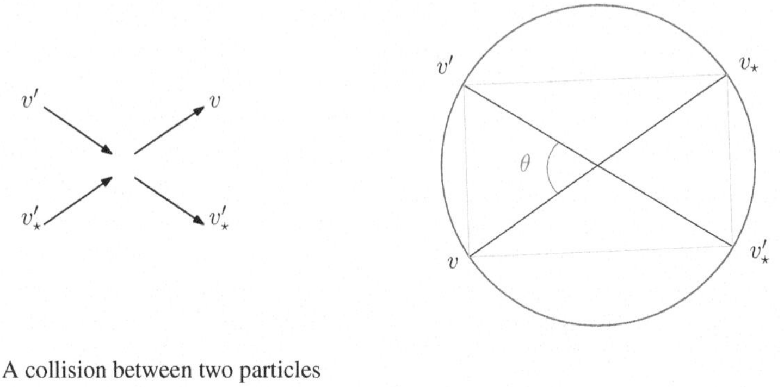

Fig. 2 A collision between two particles

repulsive force proportional to $r^{-\alpha}$ with r the distance between particles and $\alpha > 2$, we have in dimension $d = 3$

$$B = |v - v_\star|^\gamma \, b(\cos\theta), \quad b(\cos\theta) \sim_{\theta\to 0} \theta^{-2-2s},$$

with

$$\gamma = \frac{\alpha - 5}{\alpha - 1}, \qquad s = \frac{1}{\alpha - 1}.$$

Except in the hard-sphere case, the kernel B is therefore *not* integrable on $\mathbb{S}^{d-1}$:

$$\begin{aligned}\int_{\mathbb{S}^2} B &= |v - v_\star|^\gamma \int_{\mathbb{S}^2} b(\cos\theta)\, \mathrm{d}\sigma \\ &= 2\pi\, |v - v_\star|^\gamma \int_0^\pi b(\cos\theta) \sin\theta\, \mathrm{d}\theta \\ &\approx 2\pi\, |v - v_\star|^\gamma \int_0^\pi \theta^{-1-2s}\, \mathrm{d}\theta = +\infty.\end{aligned}$$

This singularity in θ implies that Q manifests non-local fractional ellipticity. Artificially removing such singularity is called **Grad's cutoff**, and accordingly the non-modified regime is called **non-cutoff long-range interaction**. The challenge of extending the De Giorgi–Nash–Moser theory to the Boltzmann equation with non-cutoff long-range interaction was, to our knowledge, first raised in [91].

It helps intuition if one draws an analogy with local differential operators. The dependence of B on $|v - v_\star|$ is analogous to the growth or decay of the coefficients of a local differential operator, while the singularity of order $2s$ at $\theta \sim 0$ is reminiscent of the order of differentiability for this operator.

It means heuristically that one expects the collision operator Q to share similar features with the following rough approximation

$$Q \approx \nabla_v^s \cdot \left(A\, \nabla_v^s\right) + \text{lower order terms},$$

with A growing like $|v|^\gamma$ at large v. This analogy is backed up by the regularity estimates [4, 6] and by the spectral gap estimates [38, 78]. Moreover, for $\gamma = 0$, Fourier methods apply just like they do for constant coefficients PDEs. And when $\alpha \to 2$ and $s \to 1$ the Boltzmann kernel indeed resembles more and more a local diffusion operator (with diffusion coefficients depending on the solution in a non-local manner). The limit case $\alpha = 2$ corresponds to the **Coulomb repulsive interaction**; the Boltzmann operator is ill-defined in this case, and must be modified, as we discuss below.

Another fundamental observation is that Boltzmann's equation (13) "contains" the compressible hydrodynamic equations on the macroscopic fields (local density, local momentum and local temperatures):

$$\rho = \int_{\mathbb{R}^d} f \, dv, \qquad \rho u = \int_{\mathbb{R}^d} f \, v \, dv, \qquad \rho T = \frac{1}{d} \int_{\mathbb{R}^d} f \, |v - u|^2 \, dv.$$

More precisely: such dynamics can, at least formally, be derived from the Boltzmann equation in certain scaling limits. Viscous incompressible hydrodynamics can also be obtained by a scaling limit. Hence, the Cauchy problem for (13) is expected to be at least difficult as the Cauchy problem for the Euler and Navier-Stokes systems.

Going back to the limit case $\alpha = 2$, one sees that the singular integral does not make sense anymore, and the Boltzmann operator must be modified. Instead, one must use the **Landau equation** (invented by Landau [63, 88] in 1936)

$$\partial_t f + v \cdot \nabla_x f = Q_L(f, f), \tag{14}$$

with the **Landau operator** (since it does not act on the variables t, x, these variables are omitted from the formula for clarity)

$$Q_L(f, f)(v) = \nabla_v \cdot \left(\int_{v_\star \in \mathbb{R}^d} |v - v_\star|^{-1} \, \Pi_{(v-v^\star)^\perp} \Big[f(v_\star) \, \nabla_v f(v) - f(v) \, \nabla_{v_\star} f(v_\star) \Big] \, dv_\star \right). \tag{15}$$

On can rewrite Q_L as a non-linear drift-diffusion operator with coefficients that depend in a non-local manner on the solution:

$$Q_L(f, f) = \nabla_v \cdot \left(A[f] \, \nabla_v f + B[f] \, f \right). \tag{16}$$

The global Cauchy theory for the Landau–Coulomb equation is open in general, but has recently been solved in the spatially homogeneous case [43], following many previous partial progresses [5, 24, 25, 30, 31, 33, 87, 92, 94] (this non-exaustive list is restricted to works dealing with the spatially homogeneous Landau equation). We give two open problems hereby motivating a regularity theory for nonlinear hypoelliptic PDEs.

We denote by **hydrodynamic fields** the functions

$$(t, x) \quad \mapsto \quad \int_{\mathbb{R}^d} f \, dv, \quad \int_{\mathbb{R}^d} v f \, dv, \quad \int_{\mathbb{R}^d} |v|^2 f \, dv, \quad \int_{\mathbb{R}^d} f \ln f \, dv.$$

Problem 1 (Conditional Regularity for the Boltzmann Equation) *Show that a solution on* $[0, T]$ *to* (13) *with non-cutoff long-range interactions, whose hydrodynamic fields satisfy suitable a priori pointwise bounds on* $[0, T]$, *is smooth on* $(0, T]$.

This problem is now solved for $\alpha > 3$:

- L^∞ bounds hold on f conditionally to bounds on the hydrodynamic fields by [87],

- polynomial decay estimates hold on f conditionally to bounds on the hydrodynamic fields by [51],
- finally weak Harnack inequality, Hölder regularity, Schauder estimates, and smoothness are proved, conditionally to bounds on the hydrodynamic fields, in the series of papers [52–54].

For $\alpha \in (2, 3)$ the problem is open.

Problem 2 (Conditional Regularity for the Landau Equation) *Show that a solution on* $[0, T]$ *to* (14)*, which satisfies suitable a priori pointwise bounds on the hydrodynamic fields, is smooth on* $(0, T]$.

This second problem in roughly the limit case $\alpha = 2$ of the first problem. It is still open, but a partial answer has been obtained for less singular, mollified versions, of the Landau operator. Consider for $\gamma \geq -3$, a modification of the operator (15) where the singular term $|v - v_\star|^{-1}$ in the integrand is replaced by $|v - v_\star|^{\gamma+2}$. The original case of Coulombian interactions corresponds to $\gamma = -3$. For $\gamma \geq -2$, we have the following results:

- L^∞ bounds hold on f conditionally to bounds on the hydrodynamic fields by [87],
- Harnack inequality and Hölder regularity hold on f conditionally to bounds on the hydrodynamic fields by [34],
- Smoothness is proved conditionally to bounds on the hydrodynamic fields by [45, 50].

For $\gamma \in [-3, 2)$ the problem is open.

To finish this brief overview, let us add that many other problems in mathematical physics would certainly benefit from a robust and quantitative theory combining the approaches of De Giorgi–Nash–Moser and Hörmander. We will present some steps towards such theory, but much remains to be explored; in particular the boundary regularity and non-divergence equations.

1.4 Notation

We denote by n or d the dimension. We denote by $\lesssim$ and $\gtrsim$ inequalities with a constant independent of the parameters at hand, or $\lesssim_k$ for an inequality with a constant depending on a parameter k. Given $\mathcal{U} \subset \mathbb{R}^n$ open set, we write $\tilde{\mathcal{U}} \subset\subset \mathcal{U}$ when $\tilde{\mathcal{U}}$ is an open subset of $\mathcal{U}$ so that $\tilde{\mathcal{U}}$ has compact closure included in $\mathcal{U}$. We finally denote $f_+ := \max(f, 0)$ for a scalar function f, and $B_r := B(0, r)$ the ball of $\mathbb{R}^n$ with radius $r > 0$.

2 The Elliptic Case

This section is partly inspired by the lecture notes [68, 90]. This is a concise self-contained account of the regularity theory of De Giorgi in [22], together with more recent quantitative alternative arguments for the control of oscillations. We also include brief presentations of the alternative approaches of Moser and Kruzhkov [58, 70], highlighting the similarities and differences with that of De Giorgi.

2.1 The Result of De Giorgi

Let us go back to Hilbert's 19th problem. We consider $\mathcal{U}$ an open subset of $\mathbb{R}^n$, a Lagrangian $L : \mathbb{R}^n \to \mathbb{R}$ and the minimisation problem (1). When $L(p) = |p|^2/2$, then $\mathcal{E}$ is the standard Dirichlet form and the Euler-Lagrange equation of the minimisation problem is simply the Laplace equation. Weak solutions to the Laplace equation are harmonic in $\mathcal{U}$, from which one deduces the smoothness and real analyticity in $\mathcal{U}$ (interior regularity). Hilbert's question asked whether this interior regularity hold for more general Lagrangians L that satisfy (H1)–(H2). The original motivation of De Giorgi was to answer this question.

Lemma 3 *The following two steps reduce Hilbert's* 19*th problem to the regularity theory of elliptic PDEs with rough coefficients:*

(i) Let $\mathrm{L} \in C^2(\mathbb{R}^n)$ *so that* $\nabla^2 \mathrm{L} \le \Lambda$ *on* $\mathbb{R}^n$ *for some* $\Lambda > 0$. *Then all minimisers of* (1) *in* $\mathrm{H}^1(\mathcal{U})$ *are also weak solutions to the Euler–Lagrange equation* (5).
(ii) Let $\mathrm{L} \in C^2(\mathbb{R}^n)$ *so that* $\Lambda^{-1} \le \nabla^2 \mathrm{L} \le \Lambda$ *on* $\mathbb{R}^n$ *for some* $\Lambda > 0$. *Let* $u \in \mathrm{H}^1(\mathcal{U})$ *be a weak solution to* (5), $\tilde{\mathcal{U}} \subset\subset \mathcal{U}$, *and* $i = 1, \cdots, n$. *Then* $u \in H^2(\tilde{\mathcal{U}})$ *and* $f := \partial_{x_i} u \in H^1(\tilde{\mathcal{U}})$ *is a weak solution to* (6) *on* $\tilde{\mathcal{U}}$.

Proof

(i) Given $\varepsilon > 0$ and $w \in C_c^\infty(\mathcal{U})$, we have

$$\int_{\mathcal{U}} \mathrm{L}(\nabla_x u \pm \varepsilon \nabla_x w)\, \mathrm{d}x \ge \int_{\mathcal{U}} \mathrm{L}(\nabla_x u)\, \mathrm{d}x.$$

Note that integrands are integrable because $\nabla_x u$, $\nabla_x w \in L^2(\mathcal{U})$ and L is sub-quadratic. Then, for all $x \in \mathcal{U}$, we have by Taylor expansion

$$\mathrm{L}(\nabla_x u(x) \pm \varepsilon\, \nabla_x w(x)) \le \mathrm{L}(\nabla_x u(x)) \pm \varepsilon \nabla \mathrm{L}(\nabla_x u(x)) \cdot \nabla_x w(x) + \Lambda \frac{\varepsilon^2}{2} |\nabla_x w(x)|^2.$$

The last two equations yield

$$\pm \int_{\mathcal{U}} \nabla \mathrm{L}(\nabla_x u) \cdot \nabla_x w \, \mathrm{d}x \geq -\Lambda \frac{\varepsilon}{2} \int_{\mathcal{U}} |\nabla_x w|^2 \, \mathrm{d}x.$$

By letting $\varepsilon \to 0$, we deduce

$$\forall\, w \in C_c^\infty(\mathcal{U}), \quad \int_{\mathcal{U}} \nabla \mathrm{L}(\nabla_x u) \cdot \nabla_x w \, \mathrm{d}x = 0,$$

which is the weak form of (5).

(ii) By symmetry, a covering argument, translation and dilation, it is enough to prove the claim when $\mathcal{U} = B_3, \tilde{\mathcal{U}} = B_1, i = 1$. For $-1 \leq h \leq 1$, consider

$$\tau_h u(x) := \frac{u(x + h\, e_1) - u(x)}{h},$$

and η a smooth function such that

$$\eta \equiv 1 \text{ on } B_1, \quad \eta \equiv 0 \text{ on } B_2^c, \quad \text{and } |\nabla_x \eta| \leq 2.$$

The operators $\eta\, \tau_h$, and $\tau_h\, (\eta\, \cdot)$ map $L^2(B_3)$ to itself and map $H^1(B_3)$ to $H_0^1(B_3)$. We then test (5) against $\phi := \tau_{-h}\left(\eta^2 \tau_h u\right)$. Note that squaring the localisation function η is standard in energy estimates and related to the L^2 structure. We obtain:

$$
\begin{aligned}
0 = &\int_{B_3} \nabla_x \phi \cdot \nabla L(\nabla_x u)\, \mathrm{d}x = \\
&- \int_{B_3} \eta\, \nabla_x(\tau_h u) \cdot \eta\, \tau_h(\nabla L(\nabla_x u))\, \mathrm{d}x - 2 \int_{B_3} (\nabla_x \eta)\, (\tau_h u)\, \eta \tau_h(\nabla L(\nabla_x u))\, \mathrm{d}x = \\
&- \int_{B_3} \eta\, \nabla_x(\tau_h u) \left(\left(\int_0^1 \nabla^2 L(\nabla_x u + \theta\, (\nabla_x u(x + h\, e_i) \right.\right. \\
&\left.\left. - \nabla_x u(x)))\, d\theta \right)\, \eta \nabla_x(\tau_h u)\, \mathrm{d}x + \\
&- 2 \int_{B_3} (\nabla_x \eta)(\tau_h u) \left(\left(\int_0^1 \nabla^2 L(\nabla_x u + \theta\, (\nabla_x u(x + h\, e_i) \right.\right. \\
&\left.\left. - \nabla_x u(x)))\, d\theta \right)\, \eta\, \nabla_x(\tau_h u)\, \mathrm{d}x.
\end{aligned}
$$

Using $\Lambda^{-1} \leq \nabla^2 L \leq \Lambda$, we deduce

$$
\begin{aligned}
\frac{1}{\Lambda} \int_{B_3} \eta^2 |\nabla_x \tau_h u|^2 \, \mathrm{d}x \leq 2\Lambda &\left(\int_{B_3} |\nabla_x \eta|^2 \, |\tau_h u|^2 \, \mathrm{d}x \right)^{1/2} \\
&\left(\int_{B_3} \eta^2 \, |\nabla_x \tau_h u|^2 \, \mathrm{d}x \right)^{1/2},
\end{aligned}
$$

which shows

$$\int_{B_3} \eta^2 |\nabla_x \tau_h u|^2 \, \mathrm{d}x \lesssim \Lambda^4 \int_{B_3} |\nabla_x \eta|^2 \, |\tau_h u|^2 \, \mathrm{d}x \leq \Lambda^4 \, \|u\|^2_{H^1(B_3)}.$$

By letting $h \to 0$, we deduce $f = \partial_i u \in H^1(B_1)$ and

$$\|f\|_{H^1(B_1)} \lesssim \Lambda^2 \|u\|_{H^1(B_3)}$$

(see [27] for results on generalised derivatives). Then given $\phi \in H^1_0(B_1)$, we have $\tau_{-h}\phi \in H^1_0(B_2)$ and

$$0 = -\int_{B_2} \tau_h \phi \, \nabla_x \cdot (\nabla L(\nabla_x u)) \, \mathrm{d}x = \int_{B_2} \nabla_x \phi \cdot \tau_h (\nabla L(\nabla_x u)) \, \mathrm{d}x.$$

Finally, by letting $h \to 0$, we deduce

$$\begin{aligned} 0 = \int_{B_1} \nabla_x \phi \cdot \partial_{x_1} \nabla L(\nabla_x u) \, \mathrm{d}x &= \int_{B_1} \nabla_x \phi \cdot \nabla^2 L(\nabla_x u) \nabla_x f \, \mathrm{d}x \\ &- \int_{B_1} \nabla_x \phi \cdot A \nabla_x f \, \mathrm{d}x, \end{aligned}$$

which is the weak form of (18) on B_1 for f.

□

As we have discussed above, the previous works of Bernstein, Schauder and others have showed that if $u \in C^{1+\alpha}_{\mathrm{loc}}(\mathcal{U})$ solves (5), then u is smooth and real analytic in $\mathcal{U}$. The solution to Hilbert's 19th problem is therefore reduced to proving $\nabla_x u \in C^\alpha_{\mathrm{loc}}(\mathcal{U})$ for some $\alpha \in (0,1)$, i.e. that any weak H^1 solution f to (6) is Hölder continuous.

Let us recall the definition of Hölder spaces. For any open set $\mathcal{U} \subset \mathbb{R}^n$, we define the **Hölder norm** $C^\alpha(\mathcal{U})$ as follows:

$$\|f\|_{C^\alpha(\mathcal{U})} := \|f\|_{L^\infty(\mathcal{U})} + [f]_{C^\alpha(\mathcal{U})}, \qquad [f]_{C^\alpha(\mathcal{U})} := \sup_{x,y \in \mathcal{U}} \frac{|f(x) - f(y)|}{|x-y|^\alpha}, \tag{17}$$

where $[f]_{C^\alpha(\mathcal{U})}$ is the **Hölder seminorm** and $\alpha \in (0,1)$. The standard notation $C^\alpha_{\mathrm{loc}}(\mathcal{U})$ denotes the set of functions that are in $C^\alpha_{\mathrm{loc}}(\tilde{\mathcal{U}})$ for every $\tilde{\mathcal{U}} \subset\subset \mathcal{U}$.

Theorem 4 (Elliptic De Giorgi–Nash–Moser) *Given $\Lambda > 0$, there is $\alpha \in (0,1)$, depending only $\Lambda > 0$ and the dimension n, and there is $C > 0$, depending only on Λ and the dimension n, so that the following holds:*

Let $f \in H^1(\mathcal{U})$ be a weak solution on the open set $\mathcal{U} \subset \mathbb{R}^n$ to the equation

$$\nabla_x \cdot (A \nabla_x f) = 0, \tag{18}$$

with $A : \mathcal{U} \to \mathbb{R}^{2n}$ *a symmetric matrix with measurable entries so that*

$$\Lambda^{-1} \leq A(x) \leq \Lambda \quad \textit{for almost every } x \in \mathcal{U}.$$

Then, we have $f \in C^{\alpha}_{\mathrm{loc}}(\mathcal{U})$.

Moreover, given $\delta > 0$, *there is* $C > 0$, *depending only on* Λ, δ *and the dimension* n, *so that for any* $\tilde{\mathcal{U}} \subset\subset \mathcal{U}$ *so that* $\mathrm{dist}(\tilde{\mathcal{U}}, \mathcal{U}) \geq \delta$, *we have*

$$\|f\|_{C^{\alpha}(\tilde{\mathcal{U}})} \leq C\|f\|_{\mathrm{L}^2(\mathcal{U})}.$$

Remarks 5

(1) Notice that the strict convexity $\nabla^2 \mathrm{L} \geq \Lambda^{-1}$ is necessary. In dimension $n = 1$, let L be a double-well shaped function

$$\mathrm{L}(p) = |p|^2 - 2|p|.$$

Then, one can check that $u(x) = |x|$ is a minimiser of (1) in $H^1((-2, 2))$, but u' is not Hölder continuous.

(2) Nash proved a similar result for parabolic equations with rough coefficients, which includes this theorem as a particular case. We discuss the parabolic case in the next section. Moser and Kruzhkov later also recovered this theorem by other proofs.

(3) We emphasize that this is **not** a perturbation result around the case of constant coefficients: the equation may not be locally close to an elliptic PDE with constant coefficients, whichever the scale one zooms in. The entries of A being merely measurable, they may oscillate fast at **every** scale.

2.2 The First De Giorgi Lemma

The argument of De Giorgi is divided into two parts, corresponding roughly first to the control of the L^{∞} part of the Hölder norm at every scale (first De Giorgi Lemma) and second to the control of the oscillation and thus the Hölder semi-norm (second De Giorgi Lemma). The first lemma plays a key role in the proof of the second lemma.

Lemma 6 (First De Giorgi Lemma: Gain of Integrability) *Given* $\Lambda > 0$, *there is* $\delta > 0$, *depending only on* Λ *and the dimension* n, *so that the following holds:*

Let $f \in H^1(B_1)$ *be a weak subsolution to*

$$\nabla_x \cdot (A \nabla_x f) \geq 0$$

with A symmetric measurable so that

$$\forall x \in \mathcal{U}, \quad \Lambda^{-1} \leq A(x) \leq \Lambda.$$

Then, we have

$$\|f_+\|_{L^2(B_1)} \leq \delta \quad \Longrightarrow \quad \|f_+\|_{L^\infty(B_{1/2})} \leq \frac{1}{2}.$$

Remarks 7

(1) This lemma is scale-invariant provided that the integrals are normalised:

$$\forall r \in (0,1], \qquad \left(\int_{B_r} f_+^2 \frac{dx}{|B_r|} \right)^{1/2} \leq \delta \Longrightarrow \|f_+\|_{L^\infty(B_{r/2})} \leq \frac{1}{2}.$$

(2) By covering, translation and dilation, this result implies that, when the Lagrangian L satisfies (H1)–(H2), any minimiser u of (1) in $H^1(\mathcal{U})$ is in $W^{1,\infty}_{\mathrm{loc}}(\mathcal{U})$.

(3) Lemma 6 implies that $f \in H^1(\mathcal{U})$ weak solution to (18) is $L^\infty_{\mathrm{loc}}(\mathcal{U})$. As soon as $n \geq 2$, this is not implied by Sobolev's embeddings. Let us also emphasize that even in situations where the L^∞ bound on f can be obtained by maximum principle arguments, this lemma proves a stronger result because of the scale-invariant estimates.

(4) The super-exponential convergence of the De Giorgi–Moser iteration below is reminiscent of Newton's iterative method, used in the KAM Theorem [8, 57, 72] and in the so-called Nash–Moser implicit function theorem [74, 75, 79].

Proof of Lemma 6 (De Giorgi's Version) The proof combines an iterative scheme, a localised energy estimate, the use of the Sobolev embedding and the Chebyshev inequality.

A. The iterative scheme For all $k \geq 0$, we consider balls with radii

$$r_k = \frac{1}{2} + \frac{1}{2^{k+1}} \quad \text{so that } r_0 = 1 \text{ and } r_k \downarrow \frac{1}{2}$$

(see Fig. 3), and energy levels

$$e_k := \frac{1}{2} - 2^{-k-1} \quad \text{so that } e_0 = 0 \text{ and } e_k \uparrow \frac{1}{2}.$$

The goal is to establish estimates by induction on

$$\forall k \geq 0, \qquad E_k := \int_{B_{r_k}} (f - e_k)_+^2 \, dx.$$

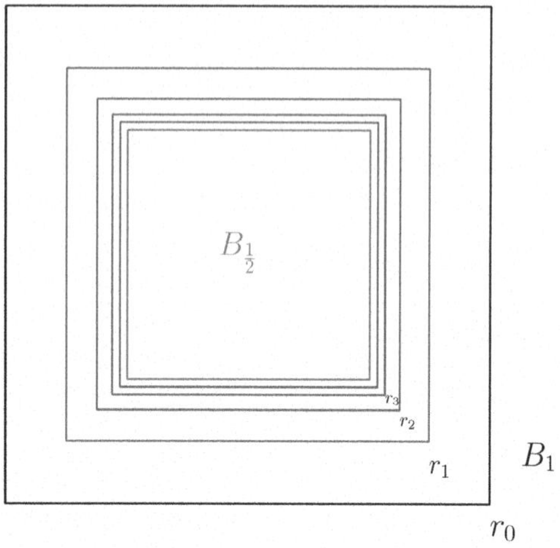

Fig. 3 The sequence of balls in the De Giorgi–Moser iteration

Indeed $E_0 = \|f\|^2_{L^2(B_1)}$, and if we prove that there are $C > 0$ and $\beta > 1$ such that

$$\forall k \geq 1, \quad E_k \leq C^k \, E_{k-1}^{\beta} \tag{19}$$

then we deduce by induction

$$E_k \leq C^{k+\beta(k-1)+\beta^2(k-2)+\dots+\beta^{k-1}} \, E_0^{\beta^k} \leq \left(C' \, E_0\right)^{\beta^k}.$$

where we have used

$$k + \beta(k-1) + \dots + \beta^{k-1} = \frac{k+1}{1-\beta} + \beta^k \, \frac{\beta}{(1-\beta)^2} \lesssim \beta^k.$$

If $E_0 \, C' < 1$, which is ensured by $\|f_+\|_{L^2(B_1)} \leq \delta$ with $\delta < 1/C'$, then

$$\int_{B_{\frac{1}{2}}} \left(f - \frac{1}{2}\right)_+^2 \mathrm{d}x = E_\infty = \lim_{k \to \infty} E_k = 0,$$

which concludes the proof.

The rest of the proof is devoted to establishing (19). This inequality is nonlinear, due to the hidden collision of scalings in the equation which relates the L^2 norms of f and of the gradient of f: this collision of scalings is revealed by the energy estimate, Sobolev's inequality and Chebyshev's inequality.

B. Localised energy estimate The following inequality is a localised version of the by-now standard $L^2 \to H^1$ ellipticity estimate, which still holds for divergence-form elliptic equations with rough coefficients:

$$\forall \phi \in C_c^1(B_1), \quad \int_{B_1} |\nabla_x(\phi f_+)|^2 \, \mathrm{d}x \leq \Lambda^2 \, \|\nabla_x \phi\|^2_{\mathrm{L}^\infty(B_1)} \int_{B_1 \cap \operatorname{supp} \phi} f_+^2 \, \mathrm{d}x. \tag{20}$$

It is also called **Caccioppoli inequality**, in memory of [19] where Caccioppoli introduced such inequalities in his study of elliptic equations. The motivation of Caccioppoli was to provide an L^2 analogue of the Cauchy inequalities for holomorphic functions, and mean value type inequalities for harmonic functions. Holomorphic functions indeed satisfy the Cauchy-Riemann equations, an elliptic system, and the Cauchy integral formula implies a pointwise control on the first derivative of a function on a small ball by the values of the function itself on a larger ball, with a constant that involves the difference in the radii of the two balls. Similar estimates can be derived for harmonic functions thanks to the mean value property. Caccioppoli worked on Hilbert's 19-th problem in the 1930s:[1] he extended the method of Bernstein beyond the two-dimensional case, and reduced the assumption from C^3 to C^2 regularity on the minimiser to (1).
To prove it, we integrate $\nabla_x \cdot (A \nabla_x f) \geq 0$ against $\phi^2 f_+$ to get

$$\begin{aligned}
0 \leq \int_{B_1} \nabla_x \cdot (A \nabla_x f) \, \phi^2 f_+ \, \mathrm{d}x &= - \int_{B_1} (A \nabla_x f_+) \cdot \nabla_x (\phi^2 f_+) \, \mathrm{d}x \\
&= - \int_{B_1} (A \nabla_x f_+) \cdot \nabla_x (\phi f_+) \, \phi \, \mathrm{d}x - \int_{B_1} (A \nabla_x f_+ \cdot \nabla_x \phi) \, (f_+ \phi) \, \mathrm{d}x \\
&= - \int_{B_1} A \nabla_x (\phi f_+) \cdot \nabla_x (\phi f_+) \, \mathrm{d}x \\
&\qquad - \int_{B_1} A \nabla_x \phi \cdot \nabla_x (f_+ \phi) \, f_+ \, \mathrm{d}x - \int_{B_1} (A \nabla_x f_+ \cdot \nabla_x \phi)(f_+ \phi) \, \mathrm{d}x \\
&= - \int_{B_1} A \nabla_x (\phi f_+) \cdot \nabla_x (\phi f_+) \, \mathrm{d}x + \int_{B_1} (A \nabla_x \phi \cdot \nabla_x \phi) f_+^2 \, \mathrm{d}x,
\end{aligned}$$

(the integration by parts have no boundary terms since ϕ being zero on ∂B_1), which implies

$$\int_{B_1} A \nabla_x (\phi f_+) \cdot \nabla_x (\phi f_+) \, \mathrm{d}x \leq \int_{B_1} (A \nabla_x \phi \cdot \nabla_x \phi) f_+^2 \, \mathrm{d}x.$$

We deduce (20) thanks to the inequality $\Lambda^{-1} \leq A \leq \Lambda$ almost everywhere on the matrix A.

C. Sobolev's embedding Let $(\phi_k)_k \subset C_c^\infty(B_1)$ be a family of cutoff functions such that

[1] See the biographical note at https://mathshistory.st-andrews.ac.uk/Biographies/Caccioppoli/.

$$\forall k \geq 1, \quad \mathbf{1}_{B_{r_k}} \leq \phi_k \leq \mathbf{1}_{B_{r_{k+1}}}, \qquad |\nabla_x \phi_k| \leq c_0 2^k,$$

for some c_0 which does not depend on k. We bound from above

$$E_k = \int_{B_{r_k}} (f - e_k)_+^2 \, \mathrm{d}x \leq \int_{B_1} \phi_k^2 \, (f - e_k)_+^2 \, \mathrm{d}x.$$

For all $k \geq 1$, the function $(f - e_k)$ also satisfies (18) by linearity and because constants are solutions. We then apply (20) with $\phi := \phi_k$ to get

$$\int_{B_1} |\nabla_x (\phi_k \, (f - e_k)_+)|^2 \, \mathrm{d}x \leq \Lambda^2 \, \|\nabla_x \phi_k\|_{L^\infty(B_1)}^2 \int_{B_1 \cap \operatorname{supp} \phi_k} (f - e_k)_+^2 \, \mathrm{d}x. \tag{21}$$

Let us define the following integrability exponent

$$p_* := \frac{2d}{d-2} \text{ for } d \geq 3, \quad \text{or any } \; p_* \in (2, \infty) \text{ for } d = 1, 2.$$

Then, Sobolev's inequality [27] applied to the function $\phi_k \, (f - e_k)_+$ over B_1 combined with the localised energy estimate (21) yields

$$\begin{aligned}
\|\phi_k \, (f - e_k)_+\|_{L^{p_*}(B_1)}^2 &\leq C_{\text{Sob}}^2 \int_{B_1} |\nabla_x (\phi_k (f) e_k)_+)|^2 \, \mathrm{d}x \\
&\leq C_{\text{Sob}}^2 \, \Lambda^2 \, c_0^2 \, 2^{2k} \int_{B_1 \cap \operatorname{supp} \phi} (f - e_k)_+^2 \, \mathrm{d}x.
\end{aligned}$$

On $\operatorname{supp} \phi_k \subset B_{r_{k-1}}$, we have $\phi_{k-1} \equiv 1$ so

$$(f - e_k)_+ \leq (f - e_{k-1})_+ \, \phi_{k-1}$$

and we deduce

$$\begin{aligned}
\|\phi_k \, (f - e_k)_+\|_{L^{p_*}(B_1)}^2 &\leq C_{\text{Sob}}^2 \, \Lambda^2 \, c_0^2 \, 2^{2k} \int_{B_1 \cap \operatorname{supp} \phi} (f - e_k)_+^2 \, \mathrm{d}x \\
&\leq C_{\text{Sob}}^2 \, \Lambda^2 \, c_0^2 \, 2^{2k} \, E_{k-1}.
\end{aligned}$$

D. Chebyshev's inequality We apply Hölder's inequality with p_* and its conjugate exponent. (To fix ideas, we write things for $d \geq 3$ and $p_* = 2d/(d-2)$, the case of $d = 1$ or 2 is an easy variation.)

$$\begin{aligned}
\forall k \geq 1, \quad E_k &= \int_{B_1} (\phi_k \, (f - e_k)_+)^2 \, \mathrm{d}x \\
&\leq \|\phi_k (f - e_k)_+\|_{L^{p_*}(B_1)}^2 \, |\{\phi_k (f - e_k)_+ > 0\}|^{\frac{2}{n}}.
\end{aligned} \tag{22}$$

Equation (22) is the key-step of the proof: together with Chebychev's inequality, it will allow us to exploit the nonlinearity hidden in the relation between f and $\nabla_x f$ (Caccioppoli's inequality) and Sobolev's inequality.
We observe that

$$\{\phi_k(f - e_k)_+ > 0\} \subset \left\{\phi_{k-1}\,(f - e_{k-1})_+ \geq 2^{-k-1}\right\}$$

since $\phi_{k-1} \equiv 1$ on the support of ϕ_k, and

$$f \geq e_k = 1 - 2^{-k-1} \implies f - e_{k-1} = f - \left(\frac{1}{2} - 2^{-k}\right) \geq 2^{-k} - 2^{-k-1} = 2^{-k-1}.$$

Therefore, by Chebyshev's inequality:

$$\begin{aligned} |\{\phi_k(f - e_k)_+ > 0\}| &\leq \left|\left\{\phi_{k-1}\,(f - e_{k-1})_+ \geq 2^{-k-1}\right\}\right| \\ &\leq 2^{2(k+1)} \int_{B_1} (\phi_{k-1}(f - e_{k-1})_+)^2 \, dx \\ &= 2^{2(k+1)}\, E_{k-1}. \end{aligned}$$

The last display, combined with (22), gives

$$E_k \leq C_{\text{Sob}}^2\, c_0^2\, \Lambda^2\, 2^{2k+2(k+1)\frac{2}{d}}\, E_{k-1}^{1+\frac{2}{d}},$$

which proves (19) with $\beta \geq 1 + \frac{2}{d} > 1$.

□

Proof of Lemma 6 (Moser's Version) The proof by Moser [70] of the first De Giorgi lemma is also based on an iterative scheme. The proofs of De Giorgi and Moser are conceptually close, so that they are often collectively close the "De Giorgi–Moser iteration". The main idea of Moser is to fully exploit the fact that the energy estimate and the Sobolev embedding both continue to yield the same inequality for weak subsolutions, not just for weak solutions, i.e. for functions f so that

$$\nabla_x \cdot (A\nabla_x f) \geq 0. \tag{23}$$

Moser's iteration uses the same sequence of radii, but instead of considering energy levels to measure integrability, it considers a *sequence of Lebesgue exponents*

$$p_k := 2\left(\frac{p_*}{2}\right)^k$$

with

$$p_* := 2 + \frac{4}{d-2} \text{ for } d \geq 3 \text{ or } p_* \in (2, +\infty) \text{ in } d = 1, 2.$$

So we have $p_0 = 2$ and $p_k \uparrow \infty$. Instead of E_k, the proof establishes estimates by induction on the following Lebesgue norms:

$$\forall k \geq 0, \quad F_k := \|f_+\|_{\mathrm{L}^{p_k}(B_{r_k})}.$$

Note that $F_0 = \|f_+\|_{L^2(B_1)}$ and $F_\infty = \|f_+\|_{L^\infty(B_{1/2})}$.

The key remark is that if f is a subsolution (23) and $\Phi : \mathbb{R} \to \mathbb{R}$ is C^2 with $\Phi', \Phi'' \geq 0$, then $\Phi(f)$ is also a subsolution:

$$\nabla_x \cdot (A\nabla_x \Phi(f)) = \Phi''(f) A\nabla_x f \cdot \nabla_x f + \Phi'(f) \nabla_x \cdot (A\nabla_x f) \geq 0.$$

By approximation it implies in particular that f_+ is a subsolution, and then that any $f_+^{p_k/2}$ is also a subsolution, since

$$\Phi(z) := z^{\frac{p_k}{2}}$$

satisfies the assumptions above for $p_k/2 > 1$. Applying (20)—which is valid for subsolutions—to $f_+^{p_k/2}$, we find

$$\forall k \geq 1, \quad \left\| \nabla_x \left(\phi_k \, f_+^{\frac{p_k}{2}} \right) \right\|_{\mathrm{L}^2(B_1)} \lesssim \Lambda^2 2^k \, F_k^{\frac{p_k}{2}},$$

where $(\phi_k)_k$ is the same cutoff family as in the proof of De Giorgi. Sobolev's inequality then yields

$$\|f_+\|^{\frac{p_k}{2}}_{\mathrm{L}^{p_k\left(1+\frac{2}{d-2}\right)}(B_{r_{k+1}})} \lesssim \Lambda^2 2^k \, F_k^{\frac{p_k}{2}}.$$

This shows

$$F_{k+1} \leq \Lambda^2 C^{\frac{2}{p_k}} \, 2^{\frac{2}{p_k}} \, F_k,$$

for an explicit $C = C(n, \Lambda)$. Since the Cauchy product $\prod_{k\geq 0} C^{\frac{2}{p_k}} 2^{\frac{2}{p_k}}$ is converging, we deduce $F_\infty \lesssim \Lambda^2 F_0$, which concludes the proof. Note that this approach makes it clear that $\delta \sim \Lambda^{-2}$ in Lemma 6. □

2.3 *The Improved First De Giorgi Lemma*

In fact, the first De Giorgi lemma can be improved so that the initial integrability assumption can be a positive power as low as wanted. This only requires one more iteration argument.

Lemma 8 (Improved First De Giorgi Lemma) *Given $\Lambda > 0$ and $\zeta_0 \in (0, 2]$, there is $C > 0$, depending only on Λ, ζ_0 and the dimension n, so that the following holds:*

Let f be a weak subsolution in $H^1(B_R)$ to

$$\nabla_x \cdot (A \nabla_x f) \geq 0$$

with A symmetric measurable so that

$$\Lambda^{-1} \leq A(x) \leq \Lambda \text{ for almost every } x \in B_R.$$

Then, for any $\zeta \in (\zeta_0, 2]$ and $0 < r < R$, we have

$$\|f\|_{L^\infty(B_r)} \leq C(R-r)^{-\frac{n}{\zeta}} \left(\frac{1}{|B_R|} \int_{B_R} f^\zeta \, \mathrm{d}x \right)^{\frac{1}{\zeta}}.$$

Proof By using the scaling of the equation in the first De Giorgi lemma, one can prove the following inequality between two general balls B_r and B_R:

$$\|f\|_{L^\infty(B_r)} \lesssim (R-r)^{-\frac{n}{2}} \left(\frac{1}{|B_R|} \int_{B_R} f^2 \, \mathrm{d}x \right)^{\frac{1}{2}}.$$

Together, the Young inequality and the inequality

$$\left(\frac{1}{|B_R|} \int_{B_R} f^2 \, \mathrm{d}x \right)^{\frac{1}{2}} \lesssim \left(\frac{1}{|B_R|} \int_{B_R} f^\zeta \, \mathrm{d}x \right)^{\frac{1}{2}} \|f\|_{L^\infty(Q_R)}^{1-\frac{\zeta}{2}},$$

imply that the function $\mathcal{N}(r) := \|f\|_{L^\infty(Q_r)}$ verifies the following inequality for $0 < r < r' < R$ and some $C > 0$:

$$\mathcal{N}(r) \leq \frac{1}{2}\mathcal{N}(r') + C(r'-r)^{-\frac{n}{\zeta}} \left(\frac{1}{|B_R|} \int_{B_R} f^\zeta \, \mathrm{d}x \right)^{\frac{1}{\zeta}}.$$

We define the increasing sequence of radii:

$$r_{k+1} := r_k + \varepsilon \left(1 - \alpha^{-1} \right) \alpha^{-k}$$

for $k \geq 0$, some $\alpha > 1$, $\varepsilon := (R - r)$ and $r_0 := r$. The sequence converges to $r_\infty = R$. Note that the scaling ratio $\alpha > 1$ between the successive balls is not 2 as in the proof of Lemma 6; it will be chosen later so as to make the geometric growth slow enough.

Then, the previous display implies

$$\mathcal{N}(r_k) \leq \frac{1}{2}\mathcal{N}(r_{k+1}) + C\varepsilon^{-\frac{n}{\zeta}}(\alpha - 1)^{-\frac{n}{\zeta}}\alpha^{\frac{kn}{\zeta}} \left(\frac{1}{|B_R|}\int_{B_R} f^\zeta \,\mathrm{d}x\right)^{\frac{1}{\zeta}},$$

so that for any $\ell \geq 1$,

$$\mathcal{N}(r) \leq \frac{\mathcal{N}(r_{\ell+1})}{2^\ell} + C\varepsilon^{-\frac{n}{\zeta}}(\alpha - 1)^{-\frac{n}{\zeta}} \left(\sum_{k=1}^{\ell} 2^{-k}\alpha^{\frac{kn}{\zeta}}\right) \left(\frac{1}{|B_R|}\int_{B_R} f^\zeta \,\mathrm{d}x\right)^{\frac{1}{\zeta}}.$$

The series in the right-hand side converges for the choice $\alpha := (3/2)^{\zeta/n}$:

$$\sum_{k=1}^{\ell} 2^{-k}\alpha^{\frac{kn}{\zeta}} \leq \frac{2}{2 - \alpha^{n/\zeta}} = 4,$$

and we finally deduce by letting $\ell \to \infty$:

$$\mathcal{N}(r) \lesssim \varepsilon^{-\frac{n}{\zeta}}(\alpha - 1)^{-\frac{n}{\zeta}} \left(\frac{1}{|B_R|}\int_{B_R} f^\zeta \,\mathrm{d}x\right)^{\frac{1}{\zeta}},$$

from which the conclusion follows (using that $\zeta \geq \zeta_0 > 0$). □

Remarks 9 On can track the dependency in ζ of the constant in this proof: at first order the constant diverges as $\zeta \to 0$ like $\zeta \exp[-n(\ln \zeta)/\zeta]$.

2.4 *The Second De Giorgi Lemma*

This second lemma concerns regularity, not just integrability. Since the coefficients of equation (18) have no regularity, it is not possible to access the regularity of f by differentiating the equation. The approach of De Giorgi is to study the *pointwise oscillation* of the solution. The oscillation compares locally the supremum of f with its infimum. De Giorgi proves that such oscillation decreases geometrically with the scaling when one zooms in on a point.

The overall structure of the proof of the second De Giorgi lemma is as follows:

$$\boxed{\begin{array}{l}\text{Intermediate value Lemma} \Longrightarrow \text{Decrease of supremum Lemma} \\ \Longrightarrow \text{Reduction of oscillation Lemma} \Longrightarrow \text{Local Hölder regularity.}\end{array}} \tag{24}$$

Lemma 10 (Second De Giorgi Lemma: Reduction of Oscillation) *Given $\Lambda > 0$, there is $\nu \in (0,1)$, depending only on Λ and the dimension n, so that the following holds:*

Let f be a weak solution in $H^1(B_3)$ to

$$\nabla_x \cdot (A \nabla_x f) = 0$$

where A is symmetric measurable and so that

$$\Lambda^{-1} \le A(x) \le \Lambda \text{ for almost every } x \in B_3.$$

Then, we have

$$\operatorname{osc}_{B_{\frac{1}{2}}} f \le \nu \operatorname{osc}_{B_2} f, \tag{25}$$

where we denote $\operatorname{osc}_B f := \sup_B f - \inf_B f$.

Note that this lemma requires f to be **both** a subsolution and a supersolution, i.e. to be a solution. Intuitively, being a subsolution and taking values far below the supremum imposes decrease of the supremum when zooming in on a point (see Lemma 12), while being a supersolution and taking values far above the infimum imposes increase of the infimum when zooming in on a point (which is Lemma 12 applied to the opposite of the supersolution). Both are combined to obtain the reduction of oscillation.

Let us show that Lemma 10 implies the Hölder regularity, which thus concludes the proof of Theorem 4.

Corollary 11 *Given $\Lambda > 0$, there is $\alpha \in (0,1)$, depending only on Λ and the dimension n, so that the following holds:*

Let f be a weak solution in $H^1(\mathcal{U})$ with $\mathcal{U} \subset \mathbb{R}^n$ open set, to

$$\nabla_x \cdot (A \nabla_x f) = 0$$

where A is symmetry measurable so that

$$\Lambda^{-1} \le A(x) \le \Lambda \text{ for almost every } x \in \mathcal{U}.$$

Then, we have $f \in C^\alpha_{\mathrm{loc}}(\mathcal{U})$.

Moreover, given any $\delta > 0$, there is $C > 0$, depending on Λ, δ and the dimension n, so that for any $\tilde{\mathcal{U}} \subset\subset \mathcal{U}$ with $\mathrm{dist}(\tilde{\mathcal{U}}, \partial\mathcal{U}) \ge \delta$, we have

$$\|f\|_{C^\alpha(\tilde{\mathcal{U}})} \le C \|f\|_{L^2(\mathcal{U})}. \tag{26}$$

Proof We already have that $f \in L^\infty_{loc}(\mathcal{U})$ by Lemma 6 with the estimate

$$\forall \hat{\mathcal{U}} \subset\subset \mathcal{U}, \quad \|f\|_{L^\infty(\hat{\mathcal{U}})} \le C \|f\|_{L^2(\mathcal{U})} \tag{27}$$

for a constant $C > 0$ depending on Λ, $\mathrm{dist}(\hat{\mathcal{U}}, \partial\mathcal{U})$ and the dimension n.
Let $\tilde{\mathcal{U}} \subset\subset \hat{\mathcal{U}} \subset\subset \mathcal{U}$ so that

$$d_0 := \mathrm{dist}(\tilde{\mathcal{U}}, \partial\mathcal{U}) \le \frac{1}{2}\mathrm{dist}(\tilde{\mathcal{U}}, \partial\hat{\mathcal{U}}). \tag{28}$$

Given $x_0 \in \tilde{\mathcal{U}}$, we define the sequence of centred rescaled functions

$$\forall k \ge 1, \ \forall y \in B_1, \quad \tilde{g}_k(y) := g_{k-1}\left(\frac{y}{4}\right)$$
$$\forall y \in B_1, \quad g_0(y) := f\left(x_0 + \frac{d_0 y}{4}\right).$$

Note that each function g_k is a rescaling of f and solves

$$\nabla_x \cdot [A_k \nabla_x g_k] = 0,$$

in B_2, with the diffusion matrix

$$A_k(y) := A\left(x_0 + d_0 2^{-2k} y\right).$$

This new matrix is still measurable and satisfies

$$\Lambda^{-1} \le A_k(x) \le \Lambda \ \text{ for almost every } x \in B_2.$$

Therefore, we can apply Lemma 10 to each g_k and deduce

$$\forall k \ge 1, \quad \mathrm{osc}_{B_{\frac{1}{2}}} g_k \le \nu \, \mathrm{osc}_{B_2} g_k.$$

Combined with the observation

$$\forall k \ge 1, \quad \mathrm{osc}_{B_2} g_{k+1} = \mathrm{osc}_{B_{\frac{1}{2}}} g_k,$$

we obtain by induction

$$\mathrm{osc}_{B_{4^{-k}}} g_1 \le \nu^{k-1} \mathrm{osc}_{B_2} g_1 \le 2\, \nu^{k-1} \|f\|_{L^\infty(\hat{\mathcal{U}})}.$$

Then, given $x \in B(x_0, 1/4)$ we pick up $k \geq 0$ so that

$$|x - x_0| \in \left[4^{-k-2}, 4^{-k-1}\right]$$

and write

$$|f(x) - f(x_0)| \leq 2\,\nu^{k-1}\,\|f\|_{L^\infty(\hat{\mathcal{U}})},$$

so that (using $|x - x_0| \geq 4^{-k-2}$)

$$|f(x) - f(x_0)| \lesssim \left(4^\alpha \nu\right)^k\,|x - x_0|^\alpha\,\|f\|_{L^\infty(\hat{\mathcal{U}})}.$$

We then choose

$$\alpha := -\frac{\ln \nu}{2\ln 2}$$

so that $4^\alpha \nu = 1$, and deduce f is C^α at x_0. The α does *not* depend on x_0, but only on the constant ν of Lemma 10.

We have therefore proved

$$\|f\|_{C^\alpha(\tilde{\mathcal{U}})} \leq C\|f\|_{L^\infty(\hat{\mathcal{U}})}$$

with a constant depending only on ν and d_0. Combined with (27) and (28), it implies (26) and concludes the proof. □

We shall show that Lemma 10 is implied by the following lemma.

Lemma 12 (Decrease of Supremum—Measure-to-Pointwise Upper Bound) *Given $\Lambda > 0$ and $\mu > 0$, there is $\lambda \in (0,1)$, depending only on Λ, μ and the dimension n such that the following holds:*

Let $f \leq 1$ be a weak subsolution in $H^1(B_2)$ to

$$\nabla_x \cdot [A\nabla_x f] \geq 0,$$

where A is symmetric measurable so that

$$\Lambda^{-1} \leq A(x) \leq \Lambda \text{ for almost every } x \in \mathcal{U}.$$

We assume that f satisfies the following condition in measure:

$$|B_1 \cap \{f \leq 0\}| \geq \mu |B_1|.$$

Then, we have

$$\sup_{B_{\frac{1}{2}}} f \leq 1 - \lambda.$$

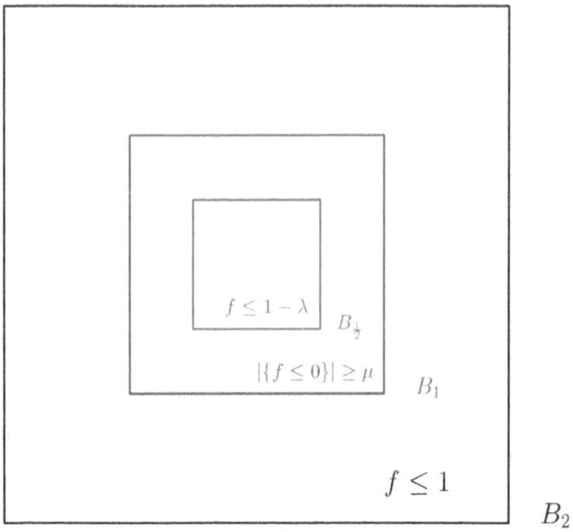

Fig. 4 The setting of the decrease of maximum lemma

Remarks 13 The setting is represented in Fig. 4. Note that the infima and suprema are in fact essential infima and suprema. A posteriori however, once the Hölder regularity is established, those would be bounds on a continuous function which is defined everywhere.

Proof of Lemma 10 When Assuming Lemma 12 Take f a solution to (18) in $H^1(B_3)$. It is then bounded on B_2 by the first De Giorgi lemma. Assume $\operatorname{osc}_{B_2} f > 0$, otherwise the statement is trivial. Then, we define

$$g(x) := \frac{2}{\operatorname{osc}_{B_2} f}\left[f(x) - \frac{\sup_{B_2} f + \inf_{B_2} f}{2}\right]. \tag{29}$$

This new unknown g is a solution to (18), is valued in $[-1, 1]$, and satisfies

$$\operatorname{osc}_{B_2} g = 2.$$

Then, certainly,

$$\text{either } |B_1 \cap \{g \leq 0\}| \geq \frac{1}{2}|B_1| \quad \text{or} \quad |B_1 \cap \{g \geq 0\}| \geq \frac{1}{2}|B_1|.$$

In the first case, we apply Lemma 12 to the subsolution g, with $\mu := 1/2$:

$$\operatorname{osc}_{B_{\frac{1}{2}}} g \leq (2 - \lambda) \leq \left(1 - \frac{\lambda}{2}\right) \operatorname{osc}_{B_2} g$$

for some $\lambda \in (0, 1)$. Going back to f, this means

$$\operatorname{osc}_{B_{\frac{1}{2}}} f \leq \left(1 - \frac{\lambda}{2}\right) \operatorname{osc}_{B_2} f,$$

yielding Lemma 10 with $\nu = 1 - \lambda/2$. In the second case, we perform the same argument on the subsolution $-g$. Note that we are therefore using that both f and $-f$ are subsolutions, i.e. that f is a solution. □

To show Lemma 12, the main idea in De Giorgi's approach is to establish an H^1 version of the intermediate value lemma. Indeed, if f were *nearly everywhere non-positive* in the sense

$$|B_1 \cap \{f \le 0\}| \ge |B_1| - \delta^2,$$

for some small $\delta > 0$, we would have

$$\|f_+\|_{L^2(B_1)} = \left(\int_{B_1} f_+^2 \,dx\right)^{\frac{1}{2}} \le \left(\int_{B_1 \cap \{f>0\}} 1 \,dx\right)^{\frac{1}{2}} \le \delta,$$

where we have used $f \le 1$. Then, provided δ is small enough to apply Lemma 6, we would conclude

$$\sup_{B_{1/2}} f \le \frac{1}{2}$$

as required. Our assumption is however merely

$$|B_1 \cap \{f \le 0\}| \ge \mu |B_1|.$$

The following H^1 intermediate value lemma of De Giorgi allows bridging the gap between these two assumptions by a finite induction.

Lemma 14 (De Giorgi Intermediate Value Lemma) *There is $C > 0$, depending only on the dimension n, such that the following holds:*

Let g be a measurable function on B_1 so that $g_+ \in H^1(B_1)$.

Then, we have

$$|B_1 \cap \{0 < g < 1/2\}| \ge \frac{C}{\|\nabla_x g_+\|^2_{L^2(B_1)}} |B_1 \cap \{g \ge 1/2\}|^2 \, |B_1 \cap \{g \le 0\}|^{2-\frac{2}{n}}. \tag{30}$$

Remarks 15

(1) The case $\|\nabla_x g_+\|_{L^2(B_1)} = 0$ is trivial and implicitly omitted.
(2) Inequalities of this form would be straightforward to prove if g were regular and the constant were allowed to depend on a C^1 norm.
(3) A weaker statement where $1/2$ is replaced by some $\theta \in (0, 1)$ close to 1 would be sufficient for the sequel, see later and [34].
(4) Unlike its counterpart in the kinetic setting (see later), this result does not rely on the PDE (18), but only on the H^1 regularity of g_+. The non-constructive

proof below shows in fact that the result still holds when replacing the bound on g_+ in $H^1(B_1)$ by the weaker fractional Sobolev norm $H^s(B_1)$ with $s \geq 1/2$. However, the statement becomes false with the fractional Sobolev norm $H^s(B_1)$ for $s \in (0, 1/2)$, since indicator functions belong to this Sobolev space.

(5) Nevertheless, importantly, the De Giorgi intermediate value lemma still holds, once properly formulated, for subsolutions to nonlocal elliptic equations with rough coefficients, even for orders of fractional ellipticity $2s$ with $s \in (0, 1/2)$ (which corresponds to the H^s regularity in the energy estimate). This is because the non-local version of the energy estimate includes an extra "cross-term" specific to non-local operators, see [20]. A constructive method based on trajectories is also possible in this case, and it extends to the parabolic and kinetic cases, see [7, 66].

(6) This lemma is also sometimes called the "*isoperimetric lemma of De Giorgi*" because it gives a control of the measure of interior values $\{0 < \tilde{g} < 1/2\}$ by the measure of boundary values $\{\tilde{g} = 0\}$ and $\{\tilde{g} = 1/2\}$.

Proof of Lemma 14 We present different methods of proof.

A. Non-constructive method by contradiction It is possible to give a short proof by contradiction at the cost of constructive estimates. Consider a contradiction sequence g_k so that

$$|B_1 \cap \{0 < g_k < 1/2\}| \xrightarrow{k\to\infty} 0,$$

and

$$|B_1 \cap \{g_k \geq 1/2\}| \geq a_1,$$

$$|B_1 \cap \{g_k \leq 0\}| \geq a_2,$$

for some $a_1, a_2 > 0$ independent of k. Let the sequence $\tilde{g}_k$ be defined by

$$\tilde{g}_k(x) = \begin{cases} 0, & \text{for } g_k(x) \leq 0, \\ g_k(x), & \text{for } 0 < g_k(x) < 1/2, \\ 1/2, & \text{for } g_k(x) \geq 1/2. \end{cases}$$

We can assume by compactness that a subsequence of the sequence $\tilde{g}_k$ is converging to some g_∞ strongly in $L^2(B_1)$, weakly in $H^1(B_1)$, and almost everywhere. These convergences imply

$$g_\infty \in H^1(B_1)$$

and

$$|B_1 \cap \{0 < g_\infty < 1/2\}| = 0,$$
$$|B_1 \cap \{g_\infty \geq 1/2\}| \geq a_1 > 0$$
$$|B_1 \cap \{g_\infty \leq 0\}| \geq a_2 > 0.$$

Therefore $2g_\infty = \mathbf{1}_\mathcal{B}$ is the indicator function of a Borel set $\mathcal{B} \subset B_1$ whose measure is positive and strictly less than $|B_1|$.
Let us prove that $\mathbf{1}_\mathcal{B}$ is not in $H^1(B_1)$. A function $g \in H^1(B_1)$ satisfies, for any $\epsilon \in (0, 1)$, the following inequality

$$\forall h \in B_\epsilon, \quad \int_{B_{1-\epsilon}} |\mathbf{1}_\mathcal{B}(x+h) - \mathbf{1}_\mathcal{B}(x)|^2 \, \mathrm{d}x \lesssim \|g\|_{H^1(B_1)} |h|^2.$$

When we consider $g = \mathbf{1}_\mathcal{B}$, the integrand $|\mathbf{1}_\mathcal{B}(x+h) - \mathbf{1}_\mathcal{B}(x)|^2$ is valued in $\{0, 1\}$, so the exponent 2 can be changed to 1 indifferently. Given $m \in \mathbb{N}^*$ and some fixed $h_0 \in B_\epsilon$, we have then

$$\begin{aligned}
\int_{B_{1-\epsilon}} |\mathbf{1}_\mathcal{B}(x+h_0) - \mathbf{1}_\mathcal{B}(x)|^2 \, \mathrm{d}x &= \int_{B_{1-\epsilon}} |\mathbf{1}_\mathcal{B}(x+h_0) - \mathbf{1}_\mathcal{B}(x)| \, \mathrm{d}x \\
&\leq \sum_{j=0}^{m-1} \int_{B_{1-\epsilon}} |\mathbf{1}_{\mathcal{B}-jh_0/m}(x) - \mathbf{1}_{\mathcal{B}-(j+1)h_0/m}(x)| \, \mathrm{d}x \\
&\leq \sum_{j=0}^{m-1} \int_{B_{1-\epsilon}+jh_0/m} |\mathbf{1}_\mathcal{B}(x) - \mathbf{1}_{\mathcal{B}-h_0/m}(x)| \, \mathrm{d}x \\
&\leq \sum_{j=0}^{m-1} \int_{B_{1-\epsilon}+jh_0/m} |\mathbf{1}_\mathcal{B}(x) - \mathbf{1}_{\mathcal{B}-h_0/m}(x)|^2 \, \mathrm{d}x \\
&\lesssim \sum_{j=0}^{m-1} \left|\frac{h_0}{m}\right|^2 \lesssim m^{-1} |h_0|,
\end{aligned}$$

where in the last line, we have used the H^1 bound on $\mathbf{1}_\mathcal{B}$. Therefore, by letting $m \to \infty$, we deduce

$$\int_{B_{1-\epsilon}} |\mathbf{1}_\mathcal{B}(x+h_0) - \mathbf{1}_\mathcal{B}(x)|^2 \, \mathrm{d}x = 0.$$

This implies that $\mathbf{1}_\mathcal{B}$ is constant in $B_{1-\epsilon}$, and by letting $\epsilon \to 0$, we deduce that $\mathbf{1}_\mathcal{B}$ is constant in B_1, which contradicts the fact that the measure of $\mathcal{B}$ is non-zero and strictly less than that of B_1.
Note that this argument can be easily generalised to assuming only $H^s(B_1)$ for $s \in (1/2, 1)$, rather than $H^1(B_1)$. The fractional Sobolev space $H^s(B_1)$ can indeed by defined by requiring that the following seminorm is finite:

$$\int_{x,y\in B_1} \frac{|g(x)-g(y)|^2}{|x-y|^{n+2s}}\,\mathrm{d}x\,\mathrm{d}y < \infty$$

for $g \in H^s(B_1)$. When we consider $g = \mathbf{1}_{\mathcal{B}}$, again the integrand $|\mathbf{1}_{\mathcal{B}}(y)-\mathbf{1}_{\mathcal{B}}(x)|^2$ is valued in $\{0,1\}$, so the exponent 2 can be changed to 1 indifferently. Given $m \in \mathbb{N}^*$, we have then

$$\begin{aligned}
&\int_{x,y\in B_1} \frac{|\mathbf{1}_{\mathcal{B}}(y)-\mathbf{1}_{\mathcal{B}}(x)|^2}{|y-x|^{n+2s}}\,\mathrm{d}y\,\mathrm{d}x\\
&= \int_{x\in B_1}\int_{h\in\mathbb{R}^n} \mathbf{1}_{x+h\in B_1} \frac{|\mathbf{1}_{\mathcal{B}}(x+h)-\mathbf{1}_{\mathcal{B}}(x)|^2}{|h|^{n+2s}}\,\mathrm{d}h\,\mathrm{d}x\\
&= \int_{x\in B_1}\int_{h\in\mathbb{R}^n} \mathbf{1}_{x+h\in B_1} \frac{|\mathbf{1}_{\mathcal{B}}(x+h)-\mathbf{1}_{\mathcal{B}}(x)|}{|h|^{n+2s}}\,\mathrm{d}h\,\mathrm{d}x\\
&\le \sum_{j=0}^{m-1}\int_{x\in B_1}\int_{h\in\mathbb{R}^n} \mathbf{1}_{x+h\in B_1} \frac{|\mathbf{1}_{\mathcal{B}-jh/m}(x)-\mathbf{1}_{\mathcal{B}-(j+1)h/m}(x)|}{|h|^{n+2s}}\,\mathrm{d}h\,\mathrm{d}x\\
&\le \sum_{j=0}^{m-1}\int_{x\in B_1+jh/m}\int_{h\in\mathbb{R}^n} \mathbf{1}_{x+\frac{(m-j)h}{m}\in B_1} \frac{|\mathbf{1}_{\mathcal{B}}(x)-\mathbf{1}_{\mathcal{B}-h/m}(x)|}{|h|^{n+2s}}\,\mathrm{d}h\,\mathrm{d}x\\
&\le m^{-2s}\sum_{j=0}^{m-1}\int_{x\in B_1+jh/m}\int_{h\in\mathbb{R}^n} \mathbf{1}_{x+\frac{(m-j)h}{m}\in B_1} \frac{|\mathbf{1}_{\mathcal{B}}(x)-\mathbf{1}_{\mathcal{B}-h/m}(x)|^2}{|h/m|^{n+2s}}\,\mathrm{d}\left(\frac{h}{m}\right)\mathrm{d}x\\
&\lesssim m^{1-2s},
\end{aligned}$$

where in the last line, we have used the $H^s(B_1)$ bound on $\mathbf{1}_{\mathcal{B}}$. Therefore, by letting $m \to \infty$, we deduce

$$\int_{x,y\in B_1} \frac{|\mathbf{1}_{\mathcal{B}}(x)-\mathbf{1}_{\mathcal{B}}(y)|^2}{|x-y|^{n+2s}}\,\mathrm{d}x\,\mathrm{d}y = 0.$$

This implies that $\mathbf{1}_{\mathcal{B}}$ is constant in B_1, which again contradicts the fact that the measure of $\mathcal{B}$ is non-zero and strictly less than that of B_1.
In the limit case $s = 1/2$, it is still possible to prove that $\mathbf{1}_{\mathcal{B}}$ is not in $H^{1/2}(B_1)$; the indicator function is however in $H^s(B_1)$ for $s \in [0,1/2)$. We refer for instance to [29] for these results.

B. Constructive method based on trajectories This proof was given in [90]. The idea is to connect the points where $g > 1/2$ with those where $g < 0$ via trajectories driven by the vector fields ∂_{x_i} inducing the diffusion; in this elliptic setting these trajectories are simply straight lines, see Fig. 5.

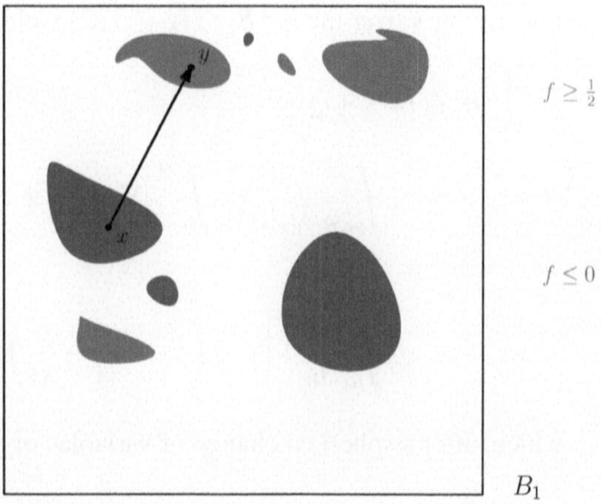

Fig. 5 Trajectories in the elliptic case

Let $\tilde{g}$ be defined by

$$\tilde{g}(x) = \begin{cases} 0, & \text{for } g(x) \leq 0, \\ g(x), & \text{for } 0 < g(x) < 1/2, \\ 1/2, & \text{for } g(x) \geq 1/2. \end{cases}$$

This modified unknown satisfies $|\nabla_x \tilde{g}| \leq |\nabla_x g_+|$ on B_1. Given then any

$$x \in B_1 \cap \{g \leq 0\},$$

$$y \in B_1 \cap \{g \geq 1/2\},$$

we Taylor-expand between the two points:

$$\frac{1}{2} = \tilde{g}(y) - \tilde{g}(x) = \int_0^1 (y - x) \cdot \nabla_x \tilde{g}(x + t(y - x))\, \mathrm{d}t$$

(the manipulations can be a posteriori justified by density of smooth functions). Hence, we have

$$\begin{aligned} \frac{1}{2} &\leq \int_0^{|y-x|} \left| \nabla_x \tilde{g}\left(x + s\frac{y - x}{|y - x|}\right) \right| \mathrm{d}s \\ &\leq \int_0^{\infty} \left| \nabla_x \tilde{g}\left(x + s\frac{y - x}{|y - x|}\right) \right| \mathbf{1}_{x + s\frac{y-x}{|y-x|} \in B_1}\, \mathrm{d}s. \end{aligned}$$

Integrating y over the set $B_1 \cap \{g \geq 1/2\}$ yields

$$\frac{1}{2} |B_1 \cap \{g \geq 1/2\}| \leq \int_{y \in B_1 \cap \{g \geq 1/2\}} \int_0^\infty \left| \nabla_x \tilde{g} \left(x + s \frac{y-x}{|y-x|} \right) \right| \mathbf{1}_{x+s\frac{y-x}{|y-x|} \in B_1} \, \mathrm{d}s \, \mathrm{d}y$$

$$\leq \int_{B_1} \int_0^\infty \left| \nabla_x \tilde{g} \left(x + s \frac{y-x}{|y-x|} \right) \right| \mathbf{1}_{x+s\frac{y-x}{|y-x|} \in B_1} \, \mathrm{d}s \, \mathrm{d}y,$$

which, after a spherical change of variables of y around x gives

$$\frac{1}{2} |B_1 \cap \{g \geq 1/2\}| \leq \left(\int_0^2 r^{n-1} \, \mathrm{d}r \right) \int_{\sigma \in \mathbb{S}^{d-1}} \int_0^\infty |\nabla_x \tilde{g}\,(x + s\,\sigma)| \, \mathbf{1}_{x+s\sigma \in B_1} \, \mathrm{d}s \, \mathrm{d}\sigma$$

$$\leq \frac{2^n}{n} \int_{\mathbb{R}^n} \frac{|\nabla_x \tilde{g}(z)|}{|x-z|^{n-1}} \mathbf{1}_{z \in B_1} \, \mathrm{d}z$$

$$\leq \frac{2^n}{n} \int_{B_1} \frac{|\nabla_x \tilde{g}(z)|}{|x-z|^{n-1}} \, \mathrm{d}z,$$

thanks to another spherical change of variables $(s, \sigma) \mapsto z = x + s\,\sigma$. Integrating x over $B_1 \cap \{g \leq 0\}$ gives

$$\frac{1}{2} |B_1 \cap \{g \leq 0\}| \, |B_1 \cap \{g \geq 1/2\}| \leq \frac{2^n}{n} \int_{B_1 \cap \{g \leq 0\}} \int_{B_1} \frac{|\nabla_x \tilde{g}(z)|}{|x-z|^{n-1}} \, \mathrm{d}z \, \mathrm{d}x.$$

We now use the standard rearrangement inequality

$$\int_{B_1 \cap \{g \leq 0\}} \frac{\mathrm{d}x}{|z-x|^{n-1}} \leq \int_{B_{r_0}} \frac{\mathrm{d}x}{|x|^{n-1}}$$

$$\leq r_0$$

$$\lesssim |B_1 \cap \{g \leq 0\}|^{\frac{1}{n}},$$

with $r_0 > 0$ is chosen so that

$$|B_{r_0}| = |B_1 \cap \{g \leq 0\}|.$$

Altogether, we obtain

$$\begin{aligned}
&\frac{1}{2}|B_1 \cap \{g \le 0\}|\,|B_1 \cap \{g \ge 1/2\}| \\
&\le \frac{2^n}{n}\left(\int_{B_1} |\nabla_x \tilde{g}|\right) |B_1 \cap \{g \le 0\}|^{\frac{1}{n}} \\
&\le \frac{2^n}{n}\left(\int_{B_1 \cap \{0<g<1/2\}} |\nabla_x \tilde{g}|\right) |B_1 \cap \{g \le 0\}|^{\frac{1}{n}} \\
&\le \frac{2^n}{n}\|\nabla_x g_+\|_{L^2(B_1)}\, |B_1 \cap \{0 < g < 1/2\}|^{\frac{1}{2}}\, |B_1 \cap \{g \le 0\}|^{\frac{1}{n}}
\end{aligned}$$

which concludes the proof with $C = n^2 2^{-2-2n}$.

C. Less sharp but simpler constructive method based on the Poincaré inequality This argument was proposed in [39, 40]. It is also—in another respect—close to the method we will deploy in the hypoelliptic setting. We apply the Poincaré inequality in the L^1 setting to $\tilde{g}$ in B_1:

$$\int_{B_1} \left|\tilde{g} - \langle \tilde{g} \rangle_{B_1}\right| \mathrm{d}x \lesssim \int_{B_1} |\nabla_x \tilde{g}| \,\mathrm{d}x \tag{31}$$

where $\langle \tilde{g} \rangle_{B_1}$ is the average of $\tilde{g}$ on B_1.

Then, we observe

$$\frac{|B_1 \cap \{g \ge 1/2\}|}{|B_1|} \le \frac{2}{|B_1|}\int_{B_1 \cap \{g \ge 1/2\}} \tilde{g} \,\mathrm{d}x \le 2\langle \tilde{g} \rangle_{B_1}$$

so that

$$\begin{aligned}
|B_1 \cap \{g \le 0\}||B_1 \cap \{g \ge 1/2\}| &\le 2|B_1| \int_{B_1 \cap \{g \le 0\}} \langle \tilde{g} \rangle_{B_1} \,\mathrm{d}x \\
&\le 2|B_1| \int_{B_1 \cap \{g \le 0\}} \left|\tilde{g}(x) - \langle \tilde{g} \rangle_{B_1}\right| \mathrm{d}x \\
&\le 2|B_1| \int_{B_1} \left|\tilde{g}(x) - \langle \tilde{g} \rangle_{B_1}\right| \mathrm{d}x.
\end{aligned}$$

We have added $g(x)$ in the second integrand because it is zero on the domain of integration.

We then plug the Poincaré inequality (31) in this last estimate:

$$|B_1 \cap \{g \le 0\}||B_1 \cap \{g \ge 1/2\}| \lesssim \int_{B_1} |\nabla_x \tilde{g}| \,\mathrm{d}x.$$

In this last integrand, we observe that $\nabla_x \tilde{g} = 0$ almost everywhere when restricted to the sets of points where $\tilde{g} = 0$ or $\tilde{g} = 1/2$. These sets are respectively made up of local minima and maxima of $\tilde{g}$: the gradient would be zero for $g \in C^1$ by standard differential calculus, and the proof in H^1 is done by approximation, see [28, Section 4.2.2, Theorem 4-(iii), p. 129]. Therefore

$$\begin{aligned}\int_{B_1} |\nabla_x \tilde{g}| \, \mathrm{d}x &= \int_{B_1\cap\{0<g<1/2\}} |\nabla_x \tilde{g}| \, \mathrm{d}x \\ &\lesssim |B_1 \cap \{0 < g < 1/2\}|^{\frac{1}{2}} \, \|\nabla_x \tilde{g}\|_{L^2(B_1)} \\ &\lesssim |B_1 \cap \{0 < g < 1/2\}|^{\frac{1}{2}} \, \|\nabla_x g_+\|_{L^2(B_1)}\end{aligned}$$

by Cauchy-Schwarz' inequality and the comparison $|\nabla_x \tilde{g}| \leq |\nabla_x g_+|$. Altogether we deduce the inequality

$$|B_1 \cap \{0 < g < 1/2\}| \geq \frac{C}{\|\nabla_x g_+\|^2_{L^2(B_1)}} \, |B_1 \cap \{g \geq 1/2\}|^2 \, |B_1 \cap \{g \leq 0\}|^2$$

for some constant $C > 0$. This alternative argument proves a slightly weaker inequality where the last exponent is 2 instead of $2 - 2/n$, which results in a worse lower bound for sets with small measure.

□

Proof of the Decrease of Supremum Lemma 12 from the Intermediate Value Lemma 14

Let $f \leq 1$ subsolution in $H^1(B_2)$ to

$$\nabla_x \cdot (A \nabla_x f) \geq 0$$

with A symmetric measurable so that $\Lambda^{-1} \leq A \leq \Lambda$ almost everywhere. We assume that f satisfies the measure condition

$$|B_1 \cap \{f \leq 0\}| \geq \mu |B_1|$$

for some $\mu > 0$.

Let us define the following sequence of new unknowns on B_2:

$$\forall k \geq 0, \quad g_k := 2^k \left[f - \left(1 - 2^{-k} \right) \right].$$

They satisfy the inductive relation

$$\forall k \geq 0, \quad g_{k+1} = 2 g_k - 1$$

and the property $g_k \leq 1$ for all $k \geq 0$ (see Fig. 6 for the representation of the successive zooms on values of f close to 1).

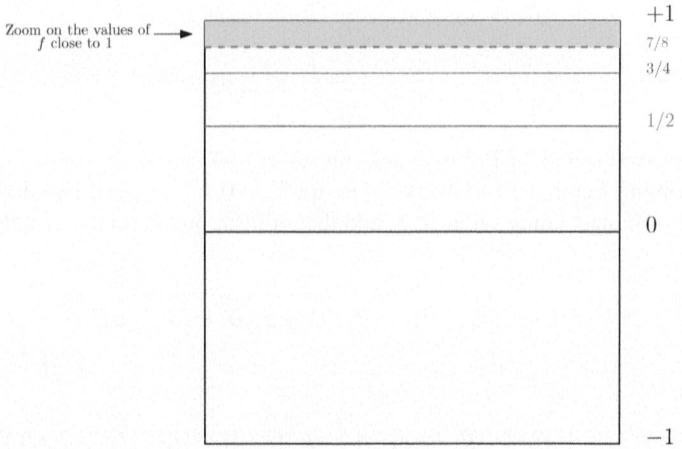

Fig. 6 Successive zooms and dilations on the values of f close to 1

Thus, $(g_k)_+$ is a subsolution on B_2 and is always valued in $[0, 1]$. By relating the $H^1(B_1)$ norm of $(g_k)_+$ to the $L^2(B_2)$ norm of $(g_k)_+$, the Caccioppoli inequality (20) (local energy estimate) therefore shows that $(g_k)_+ \in H^1(B_1)$ with a bound independent of $k \geq 0$.

We now want to apply the intermediate value lemma to each $(g_k)_+$. Since $f(x) \leq 0$ implies $g_k(x) \leq 0$ for all $k \geq 0$, we have

$$|B_1 \cap \{g_k \leq 0\}| \geq |B_1 \cap \{f \leq 0\}| \geq \mu|B_1|.$$

If the inequality

$$\int_{B_1} |(g_0)_+|^2 = \int_{B_1} |f_+|^2 \leq \delta^2$$

holds with δ as in the first De Giorgi Lemma 6, then this Lemma 6 indeed applies and implies $\sup_{B_{1/2}} f \leq 1/2$, so we are done. Suppose then, instead, that

$$\int_{B_1} (g_0)_+^2 \, \mathrm{d}x > \delta^2,$$

and consider any $k_0 \geq 1$ such that

$$\forall\, 0 \leq k \leq k_0, \quad \int_{B_1} (g_k)_+^2 \, \mathrm{d}x > \delta^2. \tag{32}$$

Then, for $k = 0, \cdots, k_0 - 1$, we have that

$$|B_1 \cap \{g_k \geq 1/2\}| = |B_1 \cap \{g_{k+1} \geq 0\}|$$
$$\geq \int_{B_1} g_{k+1}^2 \, dx > \delta^2,$$

where we have used $g_k = 1/2 + g_{k+1}/2$ and $g_{k+1} \leq 1$.

By applying Lemma 14 to every $(g_k)_+$ for $k = 0, \cdots, k_0 - 1$, we deduce that there is $\alpha > 0$, depending on n, μ, δ, and the uniform bound on the $H^1(B_1)$ norm, such that

$$\forall k = 0, \cdots, k_0 - 1, \quad |B_1 \cap \{g_k \in (0, 1/2)\} \geq \alpha |B_1|.$$

Then,

$$|B_1 \cap \{g_{k_0} \leq 0\}| \geq |B_1 \cap \{g_{k_0-1} \in (0, 1/2)\}| + |B_1 \cap \{g_{k_0-1} \leq 0\}|$$
$$\geq \alpha |B_1| + |B_1 \cap \{g_{k_0-1} \leq 0\}|$$
$$\geq \cdots$$
$$\geq k_0 \alpha |B_1| + |B_1 \cap \{g_0 \leq 0\}| \geq (\mu + \alpha\, k_0)|B_1|.$$

Therefore $\mu + k_0\alpha \leq 1$ which shows that k_0 cannot exceed the value

$$K_0 := \frac{(1 - \mu)}{\alpha}.$$

We can therefore define $k_0 \geq 1$ as the largest integer so that (32) holds. Then, $k_0 + 1 \leq K_0 + 1$, and we have

$$\int g_{k_0+1}^2 \, dx \leq \delta^2.$$

This implies, by Lemma 6, that

$$\sup_{B_{1/2}} g_{k_0+1} \leq 1/2$$

which translates into the bound

$$\sup_{B_{1/2}} f \leq 1 - 2^{-k_0-2}$$

on f. This concludes the proof with $\lambda := 2^{-k_0-2} \geq 2^{-(1-\mu)/\alpha-2}$. □

Remarks 16 By tracking the behaviour of the constant α in terms of δ, and then the behaviour of δ in terms of Λ, it is possible to obtain the more explicit formula

$$\lambda \sim Ce^{-C'\frac{(1-\mu)}{\mu^{2-2/n}}\Lambda^4}$$

for some constant $C, C' > 0$ depending only on n. The constant μ is $1/2$ in practice when applying the intermediate value lemma to derive the decrease of maximum, so we deduce

$$\lambda \sim Ce^{-C''\Lambda^4}$$

for some new absolute constant $C'' > 0$. We can also track the Hölder regularity exponent in terms of Λ, with the help of explicit formulae $\alpha = -(\ln \nu)/(2 \ln 2)$ and $\nu = 1 - \lambda/2$. This leads finally to $\alpha \sim [C/(4 \ln 2)]e^{-C''\Lambda^4}$. These estimates on the constants are obviously not optimal, but the exponential dependency of α in terms of a power of the ellipticity constant Λ is likely to be structural.

2.5 *The Method of Moser for Controlling the Oscillations*

The approach of Moser [70, 71] regarding the second De Giorgi lemma differs from that of De Giorgi in two crucial aspects:

(1) Moser measures oscillations by *integrals* rather than in a *pointwise* manner, specifically through a **Poincaré inequality**.
(2) Influenced by Nash's paper [80] and in particular by the "G bound" in [80],[2] Moser derives energy estimates on the logarithm of the solution and exploits the collision of scales at that level.

Moser first "pushes" further the first De Giorgi lemma. He denotes

$$\Phi(p,r) := \left(\frac{1}{|B_r|} \int_{B_r} f_+^p \, \mathrm{d}x \right)^{\frac{1}{p}}$$

which converges to $\sup_{B_r} f_+$ for $p \to +\infty$ and to $\min_{B_r} f_+$ for $p \to -\infty$. He then shows that for f subsolution then

$$\forall\, p > 0, \quad \Phi(+\infty, 1) \lesssim_p \Phi(p, 2),$$

and for f supersolution then

$$\forall\, p > 0, \quad \Phi(-\infty, 1) \gtrsim_p \Phi(-p, 2).$$

[2] See also "*The masterpieces of John Forbes Nash Jr*" by De Lellis in [47] for a recent detailed rewriting of the proof of Nash.

These two estimates are proved by the localised energy estimate, the Sobolev inequality and the De Giorgi–Moser iteration.

He then invokes Poincaré's inequality to prove a universal bound on the mean oscillation of the logarithm of the solution (see (33) below), and applies a result by John and Nirenberg [55] about functions with bounded mean oscillations (applied to the logarithm of the solution) to deduce

$$\Phi(p, 2) \lesssim \Phi(-p, 2)$$

for $p > 0$ small. To satisfy the assumptions of the John-Nirenberg result and prove the latter inequality, Moser must prove that if f is a positive supersolution then $F = \ln f$ satisfies

$$\int_{B_r(x_0)} \left(F - \langle F \rangle_{B_r(x_0)}\right)^2 \, dx \lesssim |B_r(x_0)| \tag{33}$$

for any ball $B_r(x_0)$ included in the domain, where $\langle F \rangle_{B_r(x_0)}$ is the average on the ball. Thanks to the Poincaré inequality, the inequality (33) is implied by

$$\int_{B_r(x_0)} |\nabla_x F|^2 \, dx \lesssim r^{-2} |B_r(x_0)|. \tag{34}$$

This universal bound (34) follows from the localised energy estimate on the subequation

$$\nabla_x \cdot (A \nabla_x F) + A \nabla_x F \cdot \nabla_x F \leq 0$$

satisfied by $F = \ln f$. The second non-negative term $A \nabla_x F \cdot \nabla_x F$ is produced by having taken logarithm of the solution, and is the key to this universal estimate.

The chain of inequalities then implies $\Phi(+\infty, 1) \lesssim \Phi(-\infty, 1)$ which finally proves the following **Harnack inequality**: there is $C_{\mathrm{H}} > 0$, depending only on n and Λ, so that for any $f \geq 0$ non-negative solution to (18) on B_2, then

$$\sup_{B_1} f \leq C_{\mathrm{H}} \min_{B_1} f.$$

Let us show how this powerful inequality implies the reduction of oscillation (25). By scaling the Harnack inequality also holds in $B_{1/2}$. The solution f to (18) on B_2 is bounded on B_1 by the first De Giorgi lemma. So we can perform a rescaling similar to (29) and define the new unknown

$$g(x) := \frac{2}{\operatorname{osc}_{B_1} f} \left[f(x) - \frac{\sup_{B_1} f + \inf_{B_1} f}{2} \right]$$

which solves (18) in B_2.

This g satisfies

$$\min_{B_1} g = -1 \text{ and } \max_{B_1} g = +1,$$

and we denote

$$m_1 := \max_{B_{\frac{1}{2}}} g \text{ and } m_0 := \min_{B_{\frac{1}{2}}} g.$$

We then apply the Harnack inequality in $B_{1/2}$ to the following two non-negative solutions to (18) in B_1: $1 - g \geq 0$ and $g + 1 \geq 0$. This yields

$$1 - m_0 = \max_{B_{1/2}}(1 - g) \leq C_{\mathrm{H}} \min_{B_{1/2}}(1 - g) = C_{\mathrm{H}}(1 - m_1),$$

$$1 + m_1 = \max_{B_{1/2}}(g + 1) \leq C_{\mathrm{H}} \min_{B_{1/2}}(g + 1) = C_{\mathrm{H}}(1 + m_0).$$

Summing the two inequalities we get

$$m_1 - m_0 \leq \frac{C_{\mathrm{H}} - 1}{C_{\mathrm{H}} + 1} 2$$

which is exactly the reduction of oscillation with constant

$$\nu := \frac{C_{\mathrm{H}} - 1}{C_{\mathrm{H}} + 1} \in (0, 1).$$

We note that the Moser approach seems to yield better constants in the Harnack inequality and Hölder regularity in terms of the elliptic constant Λ: we refer to [26] for a discussion.

2.6 The Method of Kruzhkov for Controlling the Oscillations

Kruzhkov's idea [58] is to push even further the study of the logarithm of the solution that was crucial in both Nash and Moser's arguments. He extracts more information from the "collision of scales" on the logarithm of the solution (see below) and simplifies the proof by directly proving the decrease of maximum lemma. (In fact, he proves the equivalent symmetric form: the *increase of the minimum* for supersolutions.)

Let us consider $f \geq 0$ a non-negative supersolution in $H^1(B_2)$ to

$$\nabla_x \cdot (A \nabla_x f) \leq 0$$

with the usual assumptions on A, and so that

$$|B_1 \cap \{f \geq 1\}| \geq \mu |B_1| \tag{35}$$

for some $\mu > 0$. The following argument of Kruzhkov shows that there is $\lambda \in (0, 1)$ so that $f \geq \lambda$ on $B_{1/2}$. This proves Lemma 12 on $1 - f$.

Let us consider the function

$$\forall z > 0, \quad G(z) = (-\ln z + z - 1)\mathbf{1}_{z \leq 1}.$$

It is twice differentiable, convex, and satisfies

$$\forall z \in \mathbb{R}_+^*, \qquad \begin{cases} G'(z) = \dfrac{z-1}{z} \mathbf{1}_{z \leq 1} \leq 0, \\ G''(z) = \dfrac{1}{z^2} \mathbf{1}_{z \leq 1} \geq 0 \\ G''(z) \geq G'(z)^2. \end{cases}$$

With the help of this function, we define the new unknown

$$g := G(f_\delta) \qquad \text{with the notation} \qquad f_\delta := \frac{f + \delta}{1 + \delta},$$

and for $\delta > 0$ to be chosen later. By translation and dilation, f_δ is also a supersolution, and g is a subsolution which satisfies

$$\begin{aligned} 0 \leq A\nabla_x g \cdot \nabla_x g &= G''(f_\delta) A\nabla_x f_\delta \cdot \nabla_x f_\delta \\ &\leq G''(f_\delta) A\nabla_x f_\delta \cdot \nabla_x f_\delta + \underbrace{G'(f_\delta)\nabla_x \cdot (A\nabla_x f_\delta)}_{\geq 0} = \nabla_x \cdot (A\nabla_x g). \end{aligned}$$

Note that the term $G'(f_\delta)\nabla_x \cdot (A\nabla_x f_\delta)$ is non-negative because $G' \leq 0$ and f is a supersolution, so $\nabla_x \cdot (A\nabla_x f_\delta) \leq 0$.

Integrating this inequation against a smooth localisation function ϕ so that

$$\mathbf{1}_{B_1} \leq \phi \leq \mathbf{1}_{B_2},$$

we deduce

$$\int_{\mathbb{R}^n} (A\nabla_x g \cdot \nabla_x g)\, \phi^2 \, \mathrm{d}x \leq \int_{\mathbb{R}^n} [\nabla_x \cdot (A\nabla_x g)]\, \phi^2 \, \mathrm{d}x.$$

By integration by parts, Cauchy-Schwarz' inequality, and the bounds $\Lambda^{-1} \leq A \leq \Lambda$ on the matrix A, we deduce from the last estimate:

$$\int_{\mathbb{R}^n} |\nabla_x g|^2 \phi^2 \, \mathrm{d}x \le \Lambda^2 \left(\int_{\mathbb{R}^n} |\nabla_x g|^2 \phi^2 \, \mathrm{d}x \right)^{\frac{1}{2}} \left(\int_{\mathbb{R}^n} |\nabla_x \phi|^2 \, \mathrm{d}x \right)^{\frac{1}{2}}.$$

We end up with the universal bound

$$\|\nabla_x g\|^2_{\mathrm{L}^2(B_1)} \lesssim \Lambda^4 \int_{B_2} |\nabla_x \phi|^2 \lesssim \Lambda^4. \tag{36}$$

We then want to apply the Poincaré inequality to g in B_1 in order to control the L^2 norm. This requires to control the average. This is where we use the assumption (35) to control the average by the L^2 norm. Indeed $f \ge 1$ is equivalent to $f_\delta \ge 1$ which is itself equivalent to $g = G(f_\delta) = 0$. Therefore

$$\langle g \rangle_{B_1} = \frac{1}{|B_1|} \int_{\{f<1\} \cap B_1} g \, \mathrm{d}x \le \sqrt{1-\mu} \left(\frac{1}{|B_1|} \int_{B_1} g^2 \, \mathrm{d}x \right)^{\frac{1}{2}}.$$

Since by orthogonality we have

$$\frac{1}{|B_1|} \int_{B_1} g^2 \, \mathrm{d}x \le \frac{1}{|B_1|} \int_{B_1} \left[g - \langle g \rangle_{B_1} \right]^2 \, \mathrm{d}x + \langle g \rangle^2_{B_1},$$

we deduce from the two last inequalities

$$\frac{1}{|B_1|} \int_{B_1} g^2 \, \mathrm{d}x \le \frac{1}{\mu |B_1|} \int_{B_1} \left[g - \langle g \rangle_{B_1} \right]^2 \, \mathrm{d}x.$$

We then apply the Poincaré inequality to g on B_1:

$$\frac{1}{|B_1|} \int_{B_1} g^2 \, \mathrm{d}x \le \frac{1}{\mu |B_1|} \int_{B_1} \left[g - \langle g \rangle_{B_1} \right]^2 \, \mathrm{d}x \lesssim \frac{1}{\mu} \|\nabla_x g\|^2_{\mathrm{L}^2(B_1)} \lesssim \frac{\Lambda^4}{\mu}.$$

We finally use the first De Giorgi Lemma 6 on the non-negative subsolution g:

$$\|g\|_{L^\infty(B_{1/2})} \lesssim \|g\|_{L^2(B_1)} \lesssim \frac{\Lambda^2}{\sqrt{\mu}}.$$

But the supremum of g on $B_{1/2}$ can be expressed in terms of the infimum of f since G is decreasing and we obtain

$$\left| \ln \left(\frac{\inf_{Q_{1/2}} f + \delta}{1+\delta} \right) \right| \lesssim \frac{\Lambda^2}{\sqrt{\mu}}.$$

This proves the lower bound

$$\inf_{Q_{1/2}} f \geq \lambda \quad \text{with} \quad \lambda = Ce^{-C' \frac{\Lambda^2}{\sqrt{\mu}}} > 0$$

for some constant $C, C' > 0$. This concludes the proof.

Remarks 17

(1) In fact, Kruzhkov's approach also leads to an alternative proof of the Harnack inequality. The increase of the minimum is combined with an estimate of propagation (spreading) of the lower bound to obtain a "weak Harnack inequality" which, combined with the first De Giorgi lemma (gain of integrability), implies the Harnack inequality (see the next remark). We refer to [41, 52, 59] for more details and for the extension to the kinetic case.
(2) The Harnack inequality readily follows from the so-called **weak Harnack inequality** combined with the first De Giorgi lemma. The weak Harnack inequality states as follows: there are $\zeta > 0$ and $C_{\mathrm{WH}} > 0$, depending only on n and Λ, so that for any $f \geq 0$ non-negative solution to (18) on B_2, then

$$\int_{B_{\frac{1}{2}}} f^{\zeta} \, \mathrm{d}x \leq C_{\mathrm{WH}} \inf_{B_1} f.$$

(3) An alternative route to the weak Harnack inequality is to simply deduce it from the measure-to-pointwise upper bound (decrease of supremum lemma) by a covering argument: we refer to [44, Proof of Lemma 5.14] for a non-constructive argument in the elliptic case, and to [42, 66] for a constructive argument that extends to the kinetic case.
(4) The dependency of the constants obtained by the Kruzhkov approach in terms of the elliptic constant Λ has not been explored to our knowledge.

3 The Parabolic Case

This was the setting first considered by Nash [80]. However, the approaches of De Giorgi, Moser and Kruzhkov were also extended to this case. The parabolic second De Giorgi lemma was however only proved by non-constructive methods, until the recent works [39–42]. We only sketch the proofs when they resemble the elliptic case.

We consider $\mathcal{U} \subset \mathbb{R}^n$ open set, $T > 0$, and the equation

$$\partial_t f = \nabla_x \cdot (A \nabla_x f) \quad \text{in } (0, T) \times \mathcal{U} \tag{37}$$

with a matrix-valued function $A = A(t, x)$. We define the Hölder space $C^{\alpha}((0, T) \times \mathcal{U})$ as in (17) by replacing $\mathcal{U}$ by $(0, T) \times \mathcal{U} \subset \mathbb{R} \times \mathbb{R}^n$. The parabolic counterpart of the main Theorem 4 of the elliptic case is the following statement.

Theorem 18 (Parabolic De Giorgi–Nash–Moser) *Given $\Lambda > 0$, there is $\alpha \in (0, 1)$, depending only on Λ and the dimension n, so that the following holds:*

Let f be a weak solution to (37) on $(0, T) \times \mathcal{U}$ with $\mathcal{U} \subset \mathbb{R}^n$ open set, in the functional setting

$$L_t^\infty((0, T); L_x^2(\mathcal{U})) \cap L_t^2(0, T); H_x^1(\mathcal{U})),$$

and with $A : [0, T] \times \mathcal{U} \to \mathbb{R}^{2n}$ a symmetric measurable matrix-valued function so that

$$\Lambda^{-1} \leq A(t, x) \leq \Lambda \text{ for almost every } t \in [0, T] \text{ and } x \in \mathcal{U}.$$

Then $f \in C^\alpha_{\text{loc}}((0, T) \times \mathcal{U})$.

Moreover, given $\delta > 0$ and $0 < \delta < T_0 < T_1 < T - \delta$, there is $C > 0$, depending only δ, Λ and the dimension n, so that for any $\tilde{\mathcal{U}} \subset\subset \mathcal{U}$ with $\operatorname{dist}(\tilde{\mathcal{U}}, \partial\mathcal{U}) \geq \delta$, we have

$$\|f\|_{C^\alpha((T_0, T_1)\times\tilde{\mathcal{U}})} \leq C\|f\|_{L^2((0,T)\times\mathcal{U})}.$$

3.1 Modification of the Scaling and Cylinders

This modification is imposed by the scaling of the equation. Because one derivative in time is equated with two derivatives in space, the invariant scaling is different on both variables, namely

$$(t, x) \mapsto (r^2 t, rx).$$

The different powers on the two variables t and x reflect the orders of the partial derivatives in t and x in the equation.

Combined with the fact that parabolic regularisation is forward in time, the base **parabolic cylinders** that replace the balls are

$$Q_r := (-r^2, 0] \times B_r. \tag{38}$$

3.2 The Parabolic First De Giorgi Lemma

The statement of the parabolic version of the first De Giorgi lemma remains quite similar to the elliptic case. Note however the additional assumption $\partial_t f \in L_t^2((-1, 0); H_x^{-1}(B_1))$ on the weak subsolutions. This ensures that one can easily performs approximation procedures and a priori estimates. Such regularity on $\partial_t f$ is always satisfied by weak solutions by combining the equality $\partial_t f = \nabla_x \cdot (A\nabla_x f)$

in the sense of distributions with the estimate $A\nabla_x f \in L^2$ provided by the parabolic Caccioppoli inequality (40) (the energy estimate) and the pointwise bounds on A. It is however not always satisfied by subsolutions due to the presence of a defect measure. Such a subsolution f indeed merely satisfies

$$\partial_t f = \nabla_x \cdot (A\nabla_x f) - m$$

for a non-negative defect measure m. In general, this measure m will not belong to $L^2_t((-1,0); H^{-1}_x(B_1))$.

Lemma 19 (Parabolic First De Giorgi Lemma) *Given $\Lambda > 0$, there is $\delta > 0$, depending only on Λ and the dimension n, so that the following holds:*

Let f be a weak subsolution in Q_1 to

$$\partial_t f \le \nabla_x \cdot (A\nabla_x f) \tag{39}$$

in the functional setting

$$f \in L^\infty_t((-1,0); L^2_x(B_1)) \cap L^2_t((-1,0); H^1_x(B_1)) \text{ and}$$
$$\partial_t f \in L^2_t((-1,0); H^{-1}_x(B_1)),$$

with A symmetric measurable so that

$$\Lambda^{-1} \le A(t,x) \le \Lambda \text{ for almost every } (t,x) \in Q_1.$$

Then, we have

$$\|f_+\|_{L^2(Q_1)} \le \delta \quad \Longrightarrow \quad \|f_+\|_{L^\infty(Q_{1/2})} \le \frac{1}{2}.$$

This estimate is translation and scaling-invariant, provided one uses the appropriate parabolic scaling $(t,x) \mapsto (r^2 t, rx)$.

The setting of the parabolic first De Giorgi lemma is represented in Fig. 7.

Proof of Lemma 19 The De Giorgi–Moser iterative scheme can be easily adapted. For instance, in the De Giorgi viewpoint, the sequences of cylinders and energy levels are

$$\begin{cases} T_k := -\frac{1}{4} - \frac{1}{2^{k+1}}, \quad T_0 = -1, \quad T_k \uparrow -\frac{1}{4}, \\ r_k = \frac{1}{2} + \frac{1}{2^{k+1}}, \quad r_0 = 1, \quad r_k \downarrow \frac{1}{2}, \\ e_k = \frac{1}{2} - \frac{1}{2^{k+1}}, \quad e_0 = 0, \quad e_k \uparrow \frac{1}{2}, \\ \tilde{Q}_k := (T_k, 0) \times B_{r_k}, \\ E_k := \displaystyle\int_{\tilde{Q}_k} (f - e_k)_+^2 \,\mathrm{d}t\,\mathrm{d}x. \end{cases}$$

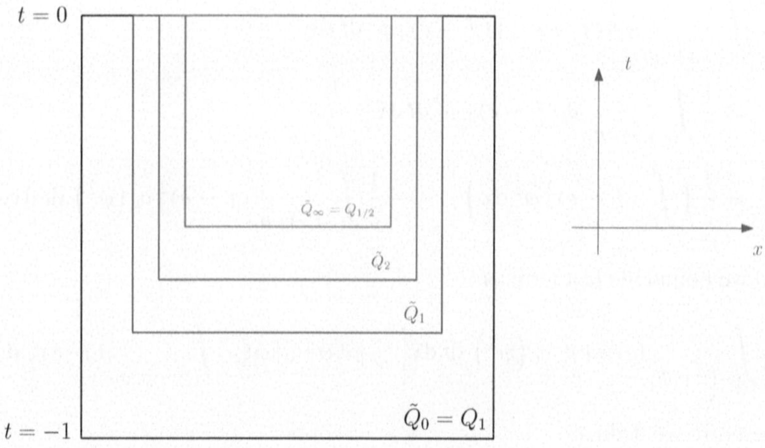

Fig. 7 Setting of the first De Giorgi lemma in the parabolic case

Note that the cylinders $\tilde{Q}_k$ are not exactly of the form (38); instead they are constructed so as to converge to $Q_{1/2}$.

One establishes again a nonlinear inequality between successive E_k terms, and the proof is based on a localised energy estimate, Sobolev's embedding and Chebyshev's inequality, as in the elliptic case. Only the localised energy estimate and the gain of integrability $L^2 \to L^{p>2}$ in all variables differ from the elliptic proof. We focus only on these two substeps.

A. Localised energy estimate We write the energy estimate in a way that is suitable for both the De Giorgi and Moser approaches to the iteration. Consider an energy level $e \geq 0$; it is used in De Giorgi's iteration, but fixed to zero in Moser's iteration). Consider a smooth localisation function ϕ with support in $(-1, 0] \times B_1$: note that ϕ is zero at initial time but not necessarily at final time. We integrate (39) against $(f - e)_+\phi^2$ on $(-1, \tau] \times B_1$ for $\tau \in (-1, 0]$:

$$\int_{(-1,\tau]\times B_1} (\partial_t (f - e)_+) (f - e)_+ \phi^2 \, \mathrm{d}t \, \mathrm{d}x$$

$$= - \int_{(-1,\tau]\times B_1} (A\nabla_x (f - e)_+ \cdot \nabla_x (f - e)_+) \, \phi^2 \, \mathrm{d}t \, \mathrm{d}x$$

$$- 2 \int_{(-1,\tau]\times B_1} (A\nabla_x (f - e)_+ \cdot \nabla_x \phi) \, \phi (f - e)_+ \, \mathrm{d}t \, \mathrm{d}x.$$

The left-hand side writes

$$
\begin{aligned}
&\int_{(-1,\tau]\times B_1} (\partial_t (f-e)_+)\,(f-e)_+\phi^2 \,\mathrm{d}t\,\mathrm{d}x \\
&= \frac{1}{2}\int_{(-1,\tau]\times B_1} \partial_t (f-e)_+^2\phi^2 \,\mathrm{d}t\,\mathrm{d}x \\
&= \frac{1}{2}\left(\int_{B_1} (f-e)_+^2\phi^2\,\mathrm{d}x\right)\Big|_{t=\tau} - \frac{1}{2}\int_{(-1,\tau]\times B_1} (f-e)_+^2\partial_t\left(\phi^2\right)\mathrm{d}t\,\mathrm{d}x,
\end{aligned}
$$

and we bound the last term as

$$
\left|\frac{1}{2}\int_{(-1,\tau]\times B_1} (f-e)_+^2\partial_t\left(\phi^2\right)\mathrm{d}t\,\mathrm{d}x\right| \lesssim \|\partial_t\phi\|_\infty\|\phi\|_\infty \int_{Q_1\cap \operatorname{supp}\phi} (f-e)_+^2\,\mathrm{d}t\,\mathrm{d}x.
$$

Therefore we deduce

$$
\begin{aligned}
&\frac{1}{2}\left(\int_{B_1} (f-e)_+^2\phi^2\,\mathrm{d}x\right)\Big|_{t=\tau} + \int_{(-1,\tau]\times B_1} |\nabla_x (f-e)_+|^2\,\phi^2\,\mathrm{d}t\,\mathrm{d}x \\
&\lesssim \Lambda\|\partial_t\phi\|_\infty\|\phi\|_\infty \int_{Q_1\cap \operatorname{supp}\phi} (f-e)_+^2\,\mathrm{d}t\,\mathrm{d}x \\
&\qquad + \Lambda\int_{(-1,\tau]\times B_1} |A\nabla_x(f-e)_+\cdot\nabla_x\phi|\,\phi(f-e)_+\,\mathrm{d}t\,\mathrm{d}x \\
&\lesssim \Lambda\|\partial_t\phi\|_\infty\|\phi\|_\infty \int_{Q_1\cap \operatorname{supp}\phi} (f-e)_+^2\,\mathrm{d}t\,\mathrm{d}x \\
&\qquad + \Lambda^2\left(\int_{(-1,\tau]\times B_1} |\nabla_x(f-e)_+|^2\,\phi\,\mathrm{d}t\,\mathrm{d}x\right)^{\frac{1}{2}} \\
&\left(\int_{(-1,\tau]\times B_1} |\nabla_x\phi|^2\,(f-e)_+^2\,\mathrm{d}t\,\mathrm{d}x\right)^{\frac{1}{2}}.
\end{aligned}
$$

By splitting the last term into two squares, we get

$$
\begin{aligned}
&\left(\int_{B_1} (f-e)_+^2\phi^2\,\mathrm{d}x\right)\Big|_{t=\tau} + \int_{(-1,\tau]\times B_1} |\nabla_x (f-e)_+|^2\,\phi^2\,\mathrm{d}t\,\mathrm{d}x \\
&\qquad \lesssim \Lambda^4\left(1+\|\nabla_x\phi\|_\infty + \|\partial_t\phi\|_\infty\right) \int_{Q_1\cap \operatorname{supp}\phi} (f-e)_+^2\,\mathrm{d}t\,\mathrm{d}x.
\end{aligned}
$$

By varying the final time $\tau \in (-1, 0]$, we finally obtain the following **parabolic Caccioppoli inequality**, which is the parabolic counterpart to (20):

$$\frac{1}{2}\sup_{-1<\tau\le 0}\left(\int_{B_1}(f-e)_+^2\phi^2\,\mathrm{d}x\right)\Big|_{t=\tau}+\int_{Q_1}|\nabla_x(f-e)_+|^2\phi^2\,\mathrm{d}t\,\mathrm{d}x$$

$$\lesssim \Lambda^4\left(1+\|\nabla_x\phi\|_\infty+\|\partial_t\phi\|_\infty\right)\int_{Q_1\cap\operatorname{supp}\phi}(f-e)_+^2\,\mathrm{d}t\,\mathrm{d}x. \tag{40}$$

Given ϕ_k a sequence of smooth localisation functions such that

$$\mathbf{1}_{\tilde{Q}_{k+1}}\le\phi_k\le\mathbf{1}_{\tilde{Q}_k}\quad\text{and}\quad|\nabla_{t,x}\varphi_k|\lesssim 2^k,$$

the sequence of localised energy estimate is therefore

$$\sup_{T_{k+1}\le\tau\le 0}\left(\int_{B_1}\phi_{k+1}^2(f-e_{k+1})_+^2\,\mathrm{d}x\right)\Big|_{t=\tau}+\int_{Q_1}|\nabla_x(\phi_k(f-e_{k+1})_+)|^2\,\mathrm{d}t\,\mathrm{d}x$$

$$\lesssim\Lambda^4 2^k E_k.$$

B. Parabolic Sobolev embedding The second term of the left-hand side of the last equation yields some Sobolev regularity in x, which leads to a gain of integrability in this variable by the standard Sobolev embedding in x with t fixed. The gain of integrability in t is ensured by the first term on the left-hand side. By combining the two estimates on the $L^2_t L^{p>2}_x$ norm and the $L^\infty_t L^2_x$ norm, one controls the solution in $L^{p_*}_{t,x}$ with $p_*>2$. □

3.3 The Improved Parabolic First De Giorgi Lemma

Similarly to the elliptic case, we can improve the first De Giorgi lemma as follows.

Lemma 20 (Improved Parabolic First De Giorgi Lemma) *Given $\Lambda>0$ and $\zeta_0\in(0,2]$, there is $C>0$, depending only on Λ, ζ_0 and the dimension n, so that the following holds:*

Let f be a weak subsolution in the parabolic cylinder Q_R to

$$\partial_t f\le\nabla_x\cdot(A\nabla_x f)$$

in the functional setting

$$f\in L^\infty_t((-R^2,0);L^2_x(B_R))\cap L^2_t((-R^2,0);H^1_x(B_R))\ \textit{and}$$
$$\partial_t f\in L^2_t((-R^2,0);H^{-1}_x(B_R)),$$

with A symmetric measurable so that

$$\Lambda^{-1}\le A(t,x)\le\Lambda\ \textit{for almost every}\ (t,x)\in Q_R.$$

Then for any $\zeta \in (\zeta_0, 2]$ and $0 < r < R$, we have

$$\|f\|_{L^\infty(Q_r)} \leq C(R-r)^{-\frac{2+n}{\zeta}} \left(\frac{1}{|Q_R|} \int_{Q_R} f^\zeta \, \mathrm{d}t \, \mathrm{d}x \right)^{\frac{1}{\zeta}}.$$

Proof By using the scaling of the equation in the parabolic first De Giorgi lemma, one can prove the following inequality between two general cylinders Q_r and Q_R:

$$\|f\|_{L^\infty(B_r)} \lesssim (R-r)^{-1-\frac{n}{2}} \left(\frac{1}{|Q_R|} \int_{Q_R} f^2 \, \mathrm{d}t \, \mathrm{d}x \right)^{\frac{1}{2}}.$$

Together, the Young inequality and the inequality

$$\left(\frac{1}{|Q_R|} \int_{Q_R} f^2 \, \mathrm{d}t \, \mathrm{d}x \right)^{\frac{1}{2}} \lesssim \left(\frac{1}{|Q_R|} \int_{Q_R} f^\zeta \, \mathrm{d}t \, \mathrm{d}x \right)^{\frac{1}{2}} \|f\|_{L^\infty(Q_R)}^{1-\frac{\zeta}{2}}$$

imply that the function $\mathcal{N}(r) := \|f\|_{L^\infty(Q_r)}$ verifies the following inequality for some $C > 0$ and any $0 < r < r' < R$:

$$\mathcal{N}(r) \leq \frac{1}{2}\mathcal{N}(r') + C(r'-r)^{-\frac{2+n}{\zeta}} \left(\frac{1}{|Q_R|} \int_{Q_R} f^\zeta \, \mathrm{d}t \, \mathrm{d}x \right)^{\frac{1}{\zeta}}.$$

By the same iteration along the sequence of radii

$$r_{k+1} = r_k + \varepsilon \left(1 - \alpha^{-1}\right) \alpha^{-k}$$

for $k \geq 1$, some $\alpha > 1$, $\varepsilon := R - r$ and $r_0 := r$, we deduce as in the elliptic case that

$$\mathcal{N}(r) \lesssim_{\zeta_0} \varepsilon^{-\frac{2+n}{\zeta}} \left(\frac{1}{|Q_R|} \int_{Q_R} f^\zeta \, \mathrm{d}t \, \mathrm{d}x \right)^{\frac{1}{\zeta}},$$

from which the conclusion follows. □

3.4 The Parabolic Second De Giorgi Lemma

The parabolic second De Giorgi lemma (reduction of oscillation) resembles the elliptic case, and it implies the Theorem 18 by an argument also similar to the elliptic case, but accounting for the slightly different geometry of the parabolic cylinders.

More generally, the overall structure of the proof of the parabolic second De Giorgi lemma is similar to the elliptic case (24), with parabolic versions of each step.

Lemma 21 (Parabolic Second De Giorgi Lemma: Reduction of Oscillation) *Given $\Lambda > 0$, there is $\nu \in (0, 1)$, depending only on Λ and the dimension n, so that the following holds:*

Let f be a weak solution in Q_3 to

$$\partial_t f = \nabla_x \cdot (A \nabla_x f) \tag{41}$$

in the functional setting $L^2((-9, 0]; H^1(B_3))$, where A is symmetric measurable and so that

$$\Lambda^{-1} \leq A(t, x) \leq \Lambda \text{ for almost every } (t, x) \in Q_3.$$

Then, we have

$$\operatorname{osc}_{Q_{\frac{1}{2}}} f \leq \nu \operatorname{osc}_{Q_2} f, \tag{42}$$

where we denote $\operatorname{osc}_B f := \sup_B f - \inf_B f$.

Note again that this lemma requires f to be **both** a subsolution and a supersolution, i.e. to be a solution. Let us show that Lemma 21 implies the Hölder regularity, which thus concludes the proof of Theorem 18.

Corollary 22 *Given $\Lambda > 0$, there is $\alpha \in (0, 1)$, depending only on Λ and the dimension n, so that the following holds:*

Let f be a weak solution in $L^2((0, T); H^1(\mathcal{U}))$, with $\mathcal{U} \subset \mathbb{R}^n$ an open set, to

$$\partial_t f = \nabla_x \cdot (A \nabla_x f),$$

where A is symmetric measurable and so that

$$\Lambda^{-1} \leq A(t, x) \leq \Lambda \text{ for almost every } (t, x) \in (0, T) \times \mathcal{U}.$$

Then, we have $f \in C^{\alpha}_{\text{loc}}((0, T) \times \mathcal{U})$.

Moreover, given any $\delta \in (0, T/2)$, there is $C > 0$, depending on Λ, δ and the dimension n, so that for any $\tilde{\mathcal{U}} \subset\subset \mathcal{U}$ with $\operatorname{dist}(\tilde{\mathcal{U}}, \partial\mathcal{U}) \geq \delta$ and $\delta < T_0 < T_1 < T - \delta$, we have

$$\|f\|_{C^{\alpha}((T_0, T_1) \times \tilde{\mathcal{U}})} \leq C \|f\|_{L^2(\mathcal{U})}. \tag{43}$$

Proof We already have that $f \in \mathrm{L}^{\infty}_{\text{loc}}(\mathcal{U})$ by Lemma 19 with the estimate

$$\forall \hat{\mathcal{U}} \subset\subset \mathcal{U}, \quad \|f\|_{L^{\infty}((\delta/2, T-\delta/2) \times \hat{\mathcal{U}})} \leq C \|f\|_{L^2((0,T) \times \mathcal{U})} \tag{44}$$

for a constant $C > 0$ depending on Λ, $\operatorname{dist}(\hat{\mathcal{U}}, \partial\mathcal{U})$, δ and the dimension n.

Let $\tilde{\mathcal{U}} \subset\subset \hat{\mathcal{U}} \subset\subset \mathcal{U}$ so that

$$\text{dist}(\tilde{\mathcal{U}}, \partial \mathcal{U}) \leq \frac{1}{2} \text{dist}(\tilde{\mathcal{U}}, \partial \hat{\mathcal{U}}), \quad \text{dist}(\tilde{\mathcal{U}}, \partial \mathcal{U}) \in \left[\frac{\delta}{4}, \frac{\delta}{2}\right], \quad \text{dist}(\tilde{\mathcal{U}}, \partial \hat{\mathcal{U}}) \in \left[\frac{\delta}{2}, \delta\right]. \tag{45}$$

Given $(t_0, x_0) \in (T_0, T_1) \times \tilde{\mathcal{U}}$, we define the following sequence of centred rescaled functions

$$\forall k \geq 1, \ \forall (s, y) \in Q_1, \quad \tilde{g}_k(s, y) := g_{k-1}\left(\frac{s}{16}, \frac{y}{4}\right)$$

$$\forall (s, y) \in Q_1, \quad g_0(s, y) := f\left(t_0 + \frac{\delta}{32} + \frac{\delta s}{16}, x_0 + \frac{\delta y}{4}\right).$$

Note that each function g_k is a rescaling of f and solves

$$\partial_t g_k = \nabla_x \cdot [A_k \nabla_x g_k]$$

in Q_2, with the diffusion matrix

$$A_k(s, y) := A\left(t_0 + \frac{\delta}{32} + \frac{\delta s}{2^{4k}}, x_0 + \frac{\delta y}{2^{2k}}\right).$$

This new diffusion matrix is still measurable and satisfies

$$\Lambda^{-1} \leq A_k(s, y) \leq \Lambda \quad \text{for almost every } (s, y) \in Q_2.$$

Therefore, we can apply Lemma 10 to each g_k and deduce

$$\forall k \geq 1, \quad \text{osc}_{Q_{\frac{1}{2}}} g_k \leq \nu \, \text{osc}_{Q_2} g_k.$$

The scaling relation between successive terms in the sequence g_k yields

$$\forall k \geq 1, \quad \text{osc}_{Q_2} g_{k+1} = \text{osc}_{Q_{\frac{1}{2}}} g_k,$$

so we obtain by induction

$$\text{osc}_{Q_{4^{-k}}} g_1 \leq \nu^{k-1} \text{osc}_{Q_2} g_1 \leq 2 \, \nu^{k-1} \, \|f\|_{L^\infty((T_0/2) \times \hat{\mathcal{U}})}.$$

Then, given a point

$$(t, x) \in \left(t_0 - \frac{\delta}{16}, t_0 + \frac{\delta}{16}\right) \times B\left(x_0, \frac{\delta}{4}\right),$$

we pick up $k \geq 0$ so that

$$|t - t_0| \in \left[\delta 16^{-k-2}, \delta 16^{-k-1}\right], \quad |x - x_0| \in \left[\delta 4^{-k-2}, \delta 4^{-k-1}\right],$$

and write

$$|f(t,x) - f(t_0,x_0)| \le 2\,\nu^{k-1}\,\|f\|_{L^\infty(\hat{\mathcal{U}})},$$

so that (using $|t - t_0| \ge \delta 16^{-k-2}$ and $|x - x_0| \ge \delta 4^{-k-2}$)

$$|f(t,x) - f(t_0,x_0)| \lesssim \Big[(16^\alpha \nu)^k\, |t - t_0|^\alpha + \left(4^{2\alpha}\nu\right)^k\, |x - x_0|^{2\alpha} \Big]\, \|f\|_{L^\infty((\delta/2, T-\delta/2)\times\hat{\mathcal{U}})}.$$

We then choose

$$\alpha := -\frac{\ln \nu}{4 \ln 2}$$

so that $16^\alpha \nu = 4^{2\alpha}\nu = 1$, and deduce f is C^α at (t_0, x_0). (In fact, note that the Hölder regularity exponent is α in t and 2α in x, which reflects the order of the partial derivatives in each coordinate in the equation). The α does *not* depend on (t_0, x_0), but only on the constant ν of Lemma 10.

We have therefore proved

$$\|f\|_{C^\alpha((\delta, T-\delta)\times\hat{\mathcal{U}})} \le C\|f\|_{L^\infty((\delta/2, T-\delta/2)\times\hat{\mathcal{U}})}$$

with a constant depending only on ν and d_0. Combined with (44) and (45), it implies (43) and concludes the proof. □

We shall show that Lemma 21 is implied by the following lemma.

Lemma 23 (Parabolic Decrease of Supremum Lemma) *Given $\Lambda > 0$, there is $\lambda \in (0,1)$, depending only on Λ and the dimension n, so that the following holds:*

Let f be a weak subsolution in Q_2 to

$$\partial_t f \le \nabla_x \cdot (A \nabla_x f)$$

in the functional setting

$$f \in L^\infty((-4,0); L^2(B_2)) \cap L^2((-4,0); H^1(B_2))$$
$$\partial_t f \in L^2((-4,0); H^{-1}(B_2)),$$

with A symmetric measurable so that

$$\Lambda^{-1} \le A(t,x) \le \Lambda \ \textit{for almost every } t \in [-4,0] \textit{ and } x \in B_2,$$

and such that $f \leq 1$ *and, for some* $\mu > 0$,

$$|\tilde{Q} \cap \{f \leq 0\}| \geq \mu |\tilde{Q}| \quad \textit{with} \quad \tilde{Q} := (-3/2, -1) \times B_1.$$

Then, we have

$$f \leq 1 - \lambda \;\; \textit{on} \;\; Q_{1/2} = (-1/2, 0) \times B_{1/2}.$$

Proof of Lemma 21 When Assuming Lemma 23 Take f a weak solution to (41) in Q_3. Assume $\operatorname{osc}_{B_2} f > 0$, otherwise the statement is trivial. Then, can perform a rescaling similar to (29) and define the new unknown

$$g(x) := \frac{2}{\operatorname{osc}_{B_2} f} \left[f(x) - \frac{\sup_{B_2} f + \inf_{B_2} f}{2} \right].$$

This new unknown g is a solution to (41), is valued in $[-1, 1]$, and satisfies

$$\operatorname{osc}_{B_2} g = 2.$$

Then, certainly,

$$\text{either } |Q_1 \cap \{g \leq 0\}| \geq \frac{1}{2} |Q_1| \quad \text{or} \quad |Q_1 \cap \{g \geq 0\}| \geq \frac{1}{2} |Q_1|.$$

In the first case, we apply Lemma 23 to the subsolution g, with $\mu := 1/2$:

$$\operatorname{osc}_{B_{\frac{1}{2}}} g \leq (2 - \lambda) \leq \left(1 - \frac{\lambda}{2}\right) \operatorname{osc}_{B_2} g$$

for some $\lambda \in (0, 1)$. Going back to f, this means

$$\operatorname{osc}_{B_{\frac{1}{2}}} f \leq \left(1 - \frac{\lambda}{2}\right) \operatorname{osc}_{B_2} f,$$

yielding Lemma 21 with $\nu = 1 - \lambda/2$. In the second case, we perform the same argument on the subsolution $-g$. Note that we are therefore using that both f and $-f$ are subsolutions, i.e. that f is a solution. □

In turn, the parabolic decrease of supremum Lemma 23 is itself implied by a parabolic version of the De Giorgi intermediate value lemma, which we now state. The proof that Lemma 24 implies Lemma 23 is exactly similar to the elliptic case and is not repeated.

Lemma 24 (Parabolic Intermediate Value Lemma) *Given* $\Lambda > 0$, $\mu > 0$ *and* $\delta > 0$, *there is* $\alpha > 0$, *depending only on* Λ, μ, δ *and the dimension* n, *so that the following holds:*

Let f be a weak subsolution in Q_2 to

$$\partial_t f \leq \nabla_x \cdot (A \nabla_x f)$$

in the functional setting

$$f \in L^\infty((-4,0); L^2(B_2)) \cap L^2((-4,0); H^1(B_2))$$
$$\partial_t f \in L^2((-4,0); H^{-1}(B_2)),$$

with A symmetric measurable so that

$$\Lambda^{-1} \leq A(t,x) \leq \Lambda \text{ for almost every } t \in [-4,0] \text{ and } x \in B_2,$$

and such that $f \leq 1$ and, for some $\mu, \delta > 0$,

$$|\{f \geq 1/2\} \cap Q_1| \geq \delta |Q_1|,$$
$$|\{f \leq 0\} \cap \tilde{Q}| \geq \mu |\tilde{Q}|$$
$$\text{where } \tilde{Q} := (-3/2, -1) \times B_1.$$

Then, we have

$$|\{0 < f < 1/2\} \cap (Q_1 \cup \tilde{Q})| \geq \alpha |Q_1 \cup \tilde{Q}|.$$

The setting of the parabolic decrease of maximum lemma and the parabolic intermediate value lemma is represented in Fig. 8.

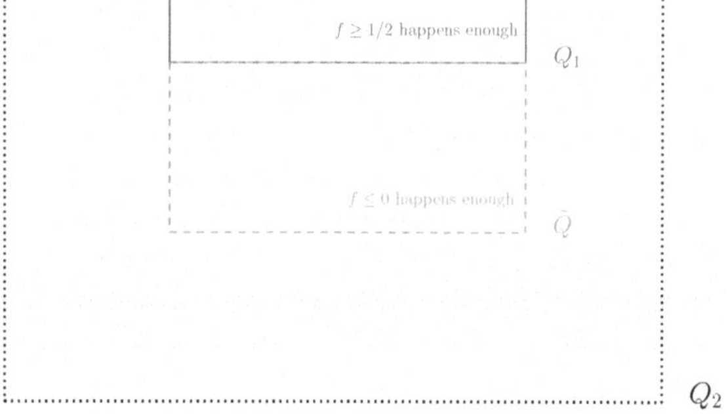

Fig. 8 Setting of the second De Giorgi lemma in the parabolic case

We first briefly sketch the extension of the non-constructive argument by compactness and contradiction to the parabolic setting, and then turn to the constructive argument based on trajectories.

Proof of Lemma 24 by Non-constructive Compactness Method Consider a sequence f_k a of weak subsolutions in Q_2 so that $f_k \leq 1$ and

$$|\{f_k \geq 1/2\} \cap Q_1| \geq \delta |Q_1| > 0,$$

$$|\{f_k \leq 0\} \cap (-3/2, -1) \times B_1| \geq \mu |(-3/2, -1) \times B_1|,$$

$$|\{0 < f_k < 1/2\} \cap (Q_1 \cup \tilde{Q})| \to 0.$$

Then $(f_k)_+$ is valued in $[0, 1]$ so uniformly bounded in $L^2(Q_2)$, and the energy estimate implies it is uniformly bounded in $L^2_t H^1_x(Q_1 \cup \tilde{Q})$.

Moreover by integrating the inequation against a localisation function (*without* squaring the unknown) one can establish a uniform bound of $\partial_t (f_k)_+$ in

$$M^1(Q_1 \cup \tilde{Q}) \cap L^\infty_t W^{-1,1}_x(Q_1 \cup \tilde{Q}),$$

where M^1 is the space of non-negative measure normed by the total variation. These bounds are sufficient to apply the Aubin-Lions lemma [10, 65] (or equivalent compactness results) and deduce that, for possibly a subsequence that we still denote as the original sequence, $(f_k)_+ \to g$ strongly in $L^2_{t,x}(Q_1 \cup \tilde{Q})$, almost everywhere, and weakly in $L^2_t H^1_x(Q_1 \cup \tilde{Q})$.

These convergences are now strong enough to deduce that the limit satisfies

$$|\{0 < g < 1/2\} \cap (Q_1 \cup \tilde{Q})| = 0, \tag{46}$$

$$|\{g \geq 1/2\} \cap Q_1| \geq \delta |Q_1| > 0, \tag{47}$$

$$|\{g \leq 0\} \cap \tilde{Q}| \geq \mu |\tilde{Q}|. \tag{48}$$

And the weak convergence in $L^2_t H^1_x$ implies that for almost every $t \in (-3/2, 0)$:

$$\nabla_x g(t, \cdot) \in L^2(B_1).$$

The $H^1(B_1)$ bound in x for almost every t, combined with (46), provide us with the assumptions to apply the elliptic intermediate value Lemma 14 for almost every t. It implies, for almost every t, that either $g(t, \cdot) = 0$ almost everywhere on B_1, or $g(t, \cdot) \geq 1/2$ almost everywhere on B_1. But the lower bound (48) implies on the one hand that there are $t \in (-3/2, -1)$ such that $g(t, \cdot) = 0$ almost everywhere on B_1. On the other hand, the energy estimate shows that the function $t \mapsto \|g(t, \cdot)\|_{L^2(B_1)}$ cannot jump upwards on $(-3/2, 0)$. Therefore, $g(t, x) = 0$ almost everywhere on $(0, 1) \times B_1$ which contradicts (47). □

Remarks 25

(1) Note that the parabolic intermediate value lemma now assumes f to be a subsolution to the PDE, not simply to be in a certain Sobolev space. It could in fact be traded for the infinite family of energy estimates, which would be sufficient to prove the parabolic first De Giorgi lemma, the parabolic decrease of supremum lemma and the parabolic intermediate value lemma. Functions satisfying such infinite family of energy estimates are said to belong to a **De Giorgi class**. In the kinetic case, no equivalent of the De Giorgi classes has been identified yet, and the kinetic counterparts of these results are proved only for subsolutions to the PDE itself.

(2) As in the elliptic setting, the cutoff value $1/2$ in the intermediate value lemma could be replaced by $\theta \in (0, 1)$ close to 1, resulting in a slightly weaker statement that is nevertheless sufficient for proving the reduction of oscillation.

(3) This contradiction argument was extended to the hypoelliptic case in [34], as we discuss in the next section.

(4) Note the time delay between $\tilde{Q}$ and $Q_{1/2}$ in the parabolic decrease of supremum lemma, but the absence of a time delay between Q_1 and $\tilde{Q}$ in the parabolic intermediate value lemma. In the kinetic setting, a gap is needed **also** in the intermediate value lemma, as first noted in [40]. Intuitively, this is because when the diffusive term vanishes, the trajectories are simply vertical in the (t, x) representation in the parabolic case, whereas they can be tilted in many ways in the kinetic case: we give a counter-example in the next section.

(5) In both the elliptic and parabolic settings (as well as in the kinetic setting in the next section), variants of the previous results hold in the presence of lower order terms, provided their coefficients have sufficient integrability. We do not address this issue to keep these notes short, but lower order terms are treated in the references [39, 40] in the parabolic case for instance.

(6) The functional setting of the weak subsolutions we consider could be relaxed to

$$f \in L^1_t((0, T); L^2_x(\mathcal{U})) \cap L^2_t(0, T); H^1_x(\mathcal{U}))$$

by careful approximation procedures, see [11, 26].

Proof of Lemma 24 by Constructive Method Based on Trajectories Lemma 24 is in fact implied by the following "parabolic Poincaré inequality": □

Theorem 26 (Parabolic Poincaré Inequality) *Given $\Lambda > 0$, there is $C > 0$, depending only on Λ and the dimension d, so that the following holds:*

Let $r = 1/20$ and $f \geq 0$ be a weak subsolution in Q_1 to

$$\partial_t f \leq \nabla_x \cdot (A \nabla_x f)$$

with A symmetric measurable so that

$$A(t, x) \leq \Lambda \text{ for almost every } (t, x) \in Q_1.$$

Then f satisfies the following inequality

$$\int_{Q_r^+} \left(f - \langle f \rangle_{Q_r^-} \right)_+ \mathrm{d}t\, \mathrm{d}x \le C \int_{Q_1} |\nabla_x f| \, \mathrm{d}t\, \mathrm{d}x, \tag{49}$$

where Q_r^- is defined by

$$Q_r^- := \left\{ -3r^2 < t < -2r^2, \quad |x| < r \right\},$$

and the average $\langle f \rangle_{Q_r^-}$ is defined by

$$\langle f \rangle_{Q_r^-} := \frac{1}{|Q_r^-|} \int_{Q_r^-} f \, \mathrm{d}t\, \mathrm{d}x.$$

Remarks 27

(1) This parabolic Poincaré inequality is inspired by the kinetic Poincaré inequality in Theorem 41 in the next section, to which we refer for further discussion. A partial version of this parabolic Poincaré inequality was first obtained in [39, 40].

(2) Intuitively this theorem measures how much (in L^1) the solution f can grow above its past average over Q_r^-, in terms of a bound on $\nabla_x f$ in L^1 in a bigger cube *in the present*, provided that f is a subsolution on such a bigger cube. This is therefore a **time-oriented integral control on the oscillation**. For instance if $\nabla_x f \equiv 0$, then f depends only on t and solves $\partial_t f \le 0$ so is non-increasing, and the inequality (49) is clear.

(3) This theorem only requires the bound from above on A, not the coercivity. However, the coercivity of A is needed to derive the parabolic intermediate value lemma from the parabolic Poincaré inequality.

Proof of Theorem 26 (the Parabolic Poincaré Inequality) The full, and more powerful, argument will be presented in the next section in the kinetic case. We only stress the modifications and simplifications in the parabolic case with respect to the kinetic case. The setting with the three cylinders Q_r^-, Q_r and Q_r^0 between Q_r^- and Q_r is similar to the kinetic case and is explained in the next section (see Fig. 11). The intermediate cylinder is useful for introducing a smooth localisation function.

A. Construction of the trajectories We uses the letter "z" to denote the joint variable $z = (t, x)$.

Given three points $z_+ \in Q^+$, $z_0 \in Q^0$ and $z_- \in Q^-$, we want to construct two paths $s \to \Gamma_+(s)$ and $s \to \Gamma_-(s)$ for $s \in [0, 1]$ such that

$$\Gamma_+(s) = (T_+(s), X_+(s)), \qquad \Gamma_+(0) = z_+, \qquad \Gamma_+(1) = z_0,$$

$$\Gamma_-(s) = (T_-(s), X_-(s)), \qquad \Gamma_-(0) = z_-, \qquad \Gamma_-(1) = z_0,$$

where

$$\begin{cases} \frac{\mathrm{d}}{\mathrm{d}s} X_\pm(s) = \mathsf{m}_\pm s^{-\frac{1}{2}} \\ \frac{\mathrm{d}}{\mathrm{d}s} T_\pm(s) = \delta_\pm, \end{cases} \tag{50}$$

where $\delta_\pm := t_0 - t_\pm$ and $\mathsf{m}_\pm \in \mathbb{R}^d$. Solving the differential equations yields

$$\begin{cases} X_\pm(s) = x_\pm + 2\delta_\pm s^{\frac{1}{2}} \mathsf{m}_\pm, \\ T_\pm(s) = s t_0 + (1-s) t_\pm. \end{cases} \tag{51}$$

The boundary conditions $\Gamma_\pm(1) = z_0$ impose $\mathsf{m}_\pm = (2\delta_\pm)^{-1}(x_0 - x_\pm)$. We deduce $X_\pm(s) = \Phi^s_\pm(x_0)$ with

$$\nabla_{x_0} \left(\Phi^s_\pm\right)^{-1} \sim O\left(s^{-\frac{1}{2}}\right)$$

which remains integrable. Observe also that our choice of control function is such that its derivative

$$X'_\pm(s) \sim O\left(s^{-\frac{1}{2}}\right)$$

is integrable on $s \in [0, 1]$, which implies that the trajectories are bounded (with integrable tangent vector field).

B. Proof of the parabolic Poincaré inequality by following the trajectories
We refer again to the next section for the details. Thanks to the estimates in the last two displays, we can argue exactly as in the proof detailed in the kinetic case, to obtain

$$\int_{Q^+} \left(f(z_+) - \langle f \rangle_{Q^-}\right)_+ \mathrm{d}z_+ \lesssim \int_{Q^+} \int_0^1 \int_{Q^0} |(\nabla_x f)(\Gamma_+(s))| \, s^{-\frac{1}{2}} \, \mathrm{d}s \, \mathrm{d}z_0 \, \mathrm{d}z_+$$
$$+ \int_{Q^-} \int_0^1 \int_{Q^0} |(\nabla_x f)(\Gamma_-(s))| \, s^{-\frac{1}{2}} \, \mathrm{d}s \, \mathrm{d}z_0 \, \mathrm{d}z_-,$$

and the end of the proof is similar: we are now integrating in both Q^0 and $Q_\pm$, so we are now in a position to use a change of variable not only with the intermediate variable z_0, but also with the future/past variables $z_\pm$. Note that it was not possible to use the integration in $Q_\pm$ before because of the positive value around the Q^0 integral. Using these changes of variables, we deduce

$$\int_{Q^\pm}\int_0^1\int_{Q^0} |(\nabla_x f)(\Gamma_\pm(s))|\, s^{-\frac{1}{2}}\, \mathrm{d}s\, \mathrm{d}z_0\, \mathrm{d}z_\pm \lesssim \int_{Q_1} |\nabla_x f|\, \mathrm{d}t\, \mathrm{d}x,$$

which concludes the proof of the parabolic Poincaré inequality (49).
The parabolic Poincaré inequality (49) then implies the parabolic intermediate value Lemma 24. The proof is exactly similar to the argument in the kinetic case in the next section, i.e. the proof of the kinetic intermediate value Lemma 39 from the kinetic Poincaré inequality (90). One can apply the kinetic proof to a function independent of the space variable to recover the parabolic proof. (In fact, the argument does not depend on the equation but only on the Poincaré-type inequality and the Caccioppoli inequality.) We therefore refer the reader to the next section for the proof.

□

3.5 *Parabolic Versions of the Moser and Kruzhkov Methods*

The Moser approach was extended to the parabolic case in [73, 76, 77]. The first paper in this series had a mistake (which was corrected a few years later), which illustrates that the John-Nirenberg result is not a convenient tool in the parabolic case. Instead, in the later paper [77], Moser improved and simplified the argument: in order to "bridge the gap" between negative and positive exponents and prove

$$\Phi(p,2) \lesssim \Phi(-p,2)$$

for $p > 0$ small, Moser combines a universal bound on the L^2 norm of the gradient of the logarithm of the solution with a lemma by Bombieri and Giusti [16]. This method provides a sharper dependency of the constant in the Harnack inequality in terms of the ellipticity constant Λ. We refer to the recent contribution [82] for a re-interpretation of this Moser-Bombieri-Giusti approach based on trajectories in the parabolic case, and a discussion of the dependency of the Harnack constant in terms of Λ. We also refer to [9, 89] for the extension of Moser's method to some quasilinear parabolic equations.

The Kruzhkov approach was extended to the parabolic case in [59]. The argument is quite close to the elliptic case. The main difference is that the right-hand side of (36), in the parabolic case, also includes the L^∞ norm of g, albeit with a power one:

$$\|\nabla_x g\|^2_{L^2(Q_1)} \lesssim \Lambda^4 \left(\int_{Q_2} |\nabla_x \phi|^2 + \int_{Q_2} g^2\, |\partial_t \phi|^2\right) \lesssim \Lambda^4 \left(1 + \|g\|_{L^\infty(Q_1)}\right).$$

This is due to the error terms produced by the time derivative operator.

The difference in powers between the left-hand and right-hand sides still allows the argument to work. One gets finally

$$\|g\|^2_{L^\infty(Q_{1/2})} \lesssim_\Lambda 1 + \|g\|_{L^\infty(Q_1)}$$

and we then express the norm on the left-hand side in terms of the infimum on the solution f on $Q_{1/2}$, and we bound the norm on the right-hand side by the trivial upper bound on $G(f_\delta)$.

This leads to the inequality

$$\left|\ln\left(\frac{\inf_{Q_{1/2}} f + \delta}{1+\delta}\right)\right|^2 \lesssim_\Lambda 1 + \left|\ln\left(\frac{\delta}{1+\delta}\right)\right|$$

which implies a strictly positive lower bound on $\inf_{Q_{1/2}} f$.

4 The Kinetic Case

We focus in this section on the simplest class of local hypoelliptic equations of type II (in the terminology of Hörmander): the **Kolmogorov equation**. The underlying motivation is the study of the Landau equation (14)–(16), which, we recall, has the following form

$$\partial_t f + v \cdot \nabla_x f = \nabla_v \cdot (A[f]\,\nabla_v f) + B[f] \cdot \nabla_v f + (\nabla_v \cdot B[f])\, f$$

for an unknown $f = f(t, x, v) \geq 0$, for $t \in \mathbb{R}$, $(x, v) \in \mathcal{U} \subset \mathbb{R}^d \times \mathbb{R}^d$, where $A[f]$ and $B[f]$ are linear integral operators on the v variable applied to the unknown f.

Following the same approach as De Giorgi and Nash in their solution to Hilbert's 19th problem, i.e. trading off nonlinearity for the study of linear equations with rough coefficients, we consider instead the following linear equation:

$$\partial_t f + v \cdot \nabla_x f = \nabla_v \cdot [A \nabla_v f] + B \cdot \nabla_v f + S, \tag{52}$$

on $[0, T]$, with $A = A(t, x, v)$ a measurable symmetric matrix-valued function, $B = B(t, x, v)$ a measurable vector-valued function, $S = S(t, x, v)$ a measurable scalar function. We assume that the coefficients and source term satisfy

$$\left\{\begin{array}{c} \Lambda^{-1} \leq A(t, x, v) \leq \Lambda \\ |B(t, x, v)| \leq \Lambda \\ |S(t, x, v)| \leq C_S \end{array}\right\} \quad \text{for almost every } t \in [0, T] \text{ and } (x, v) \in \mathcal{U}, \tag{53}$$

for some $\Lambda > 0$ and $C_S > 0$.

The main result reviewed this section is the local Hölder regularity of the weak solutions to (52), under the assumptions (53). This, in turn, implies the Hölder regularity of weak a priori solutions to the Landau equation (14), provided we assume these solutions to be bounded in L^∞ in all variables and to satisfy uniform pointwise bounds on their hydrodynamic fields. The details of this corollary are included in [34]. This is one of the steps in the solution to the open problem of the introduction (conditional regularity for the Landau equation). Because of the structure of the Landau equation, this corollary requires the presence of the lower order terms $B \cdot \nabla_v f$ and S in (52).

In these notes, we however focus only on the simplest form of (52) when $B = 0$ and $S = 0$. This simplifies the setting, the calculations and makes the core ideas more visible. The treatment of these lower order terms merely adds technical complication and is not essential to the proofs. Again, the full details are discussed in [34], and we also refer to the recent review [32] about the work [34]. We however emphasize that the L^∞ hypothesis on the a priori solutions to the Landau is made because of the assumptions we require on the source term S in the main Hölder regularity result.

The model problem and main statement of the section are therefore as follows. We consider $\mathcal{U} \subset \mathbb{R}^d \times \mathbb{R}^d$ an open set, $T > 0$, and the equation

$$\partial_t f + v \cdot \nabla_v f = \nabla_v \cdot (A \nabla_v f) \quad \text{in } (0, T) \times \mathcal{U} \tag{54}$$

with $A = A(t, x, v)$. We define the Hölder space $C^\alpha((0, T) \times \mathcal{U})$ as in (17) by replacing $\mathcal{U}$ by $(0, T) \times \mathcal{U} \subset \mathbb{R} \times \mathbb{R}^d \times \mathbb{R}^d$. The kinetic counterpart of the main Hölder regularity result is the following statement.

Theorem 28 (Kinetic De Giorgi–Nash–Moser [7, 26, 34, 41, 42, 81, 93]) *Given $\Lambda > 0$, there is $\alpha \in (0, 1)$, depending only on Λ and the dimension d, so that the following holds:*

Let f be a weak solution to (54) *on $[0, T] \times \mathcal{U}$, with $\mathcal{U} \subset \mathbb{R}^{2d}$ an open set, in the functional setting*

$$L^\infty_t((0, T); L^2_{x,v}(\mathcal{U})) \cap L^2_t(0, T); L^2_x H^1_v(\mathcal{U})),$$

with $A : [0, T] \times \mathcal{U} \to \mathbb{R}^{n \times n}$ a symmetric measurable matrix-valued function so that

$$\Lambda^{-1} \le A(t, x, v) \le \Lambda \text{ for almost every } t \in [0, T] \text{ and } (x, v) \in \mathcal{U}. \tag{55}$$

Then, we have $f \in C^\alpha_{\mathrm{loc}}((0, T) \times \mathcal{U})$.

Moreover, given $\delta > 0$ and $\delta < T_0 < T_1 < T - \delta$, there is $C > 0$, depending only on Λ, δ and the dimension d, so that for any $\tilde{\mathcal{U}} \subset\subset \mathcal{U}$ with $\mathrm{dist}(\tilde{\mathcal{U}}, \partial\mathcal{U}) \ge \delta$, *we have*

$$\|f\|_{C^\alpha((T_0,T_1)\times\tilde{\mathcal{U}})} \le C\|f\|_{L^2((0,T)\times\mathcal{U})}.$$

4.1 Modification of the Scaling and Cylinders

In the sequel, we write $z := (t, x, v) \in \mathbb{R}^{1+2d}$ for the full set of variables, and $\mathcal{T} := \partial_t + v \cdot \nabla_x$ for the transport operator.

The invariant transformations of the class of equations (54) that satisfy (55) are slightly more intricate than in the parabolic case. The scaling now must take into account the balance between the three differential operators in the equation, which yields the **kinetic scaling**

$$(t, x, v) \mapsto (r^2 t, r^3 x, rv) \quad \text{for } r > 0. \tag{56}$$

Besides we still have invariance by translation in time and in space, and we have an additional invariance involving the velocity variable, albeit modified by the inertia principle. This results in the **invariance by Galilean translations**

$$(t, x, v) \mapsto (t_0 + t, x_0 + x + tv_0, v_0 + v) \quad \text{for } (t_0, x_0, v_0) \in \mathbb{R}^{1+2d}. \tag{57}$$

The scaling invariance naturally leads to defining the following base **kinetic cylinders** for $r > 0$

$$Q_r := rQ_1 := (-r^2, 0] \times B_{r^3} \times B_r, \tag{58}$$

where the left multiplication rQ_1 corresponds to the scaling law. The invariance by Galilean transformation naturally leads to the following more general translated kinetic cylinders

$$\begin{aligned} Q_r(z_0) &= z_0 \circ [rQ_1] \\ &= \Big\{(t, x, r) \, : \, -r^2 < t - t_0 \le 0, \ |x - x_0 \\ &\qquad -(t - t_0)v_0| < r^3, \ |v - v_0| < r\Big\}, \end{aligned}$$

where the $\circ$ left operation corresponds to the (non-commutative) Galilean translation.

4.2 The Kinetic First De Giorgi Lemma

The main difficulty in adapting the first De Giorgi lemma, by comparison with the parabolic case, lies in the degeneracy of (54), which lacks ellipticity in the x variable. We sketch two arguments for the core part of the proof, the gain of integrability from L^2 to L^{p_*} with some $p_* > 2$ in all variables. We first present the approach of [34], based on the so-called "averaging lemma", and, second, the strategy introduced in [84] and used in [42, 93]. The second strategy leads to sharper results. Its drawback however is that it relies on the fundamental solution of the Kolmogorov equation. So it might not be as robust as the first strategy for future applications.

Overall, the statement remains quite similar to the parabolic case. Note the additional assumption $\mathcal{T}f \in L^2_t((-1,0); L^2_x H^{-1}_v(B_1 \times B_1))$ made on the weak subsolutions. Just like the assumption made on $\partial_t f$ in the parabolic case, it ensures that we can easily perform approximation procedures and a priori estimates. Such regularity on $\mathcal{T}f$ is always satisfied by weak solutions by combining the equality $\mathcal{T}f = \nabla_v \cdot (A\nabla_v f)$ in the sense of distributions with the estimate $A\nabla_v f \in L^2$ provided by the kinetic Caccioppoli inequality (60) (the energy estimate) and the pointwise bounds on A. It is however not always satisfied by subsolutions due to the presence of a defect measure, as in the parabolic case.

Lemma 29 (Kinetic First De Giorgi Lemma) *Given $\Lambda > 0$, there is $\delta > 0$, depending only on Λ and the dimension d, so that the following holds:*

Let f be a weak subsolution in Q_1 to

$$\partial_t f + v \cdot \nabla_x f \le \nabla_v \cdot (A\nabla_v f) \tag{59}$$

in the functional setting

$$f \in L^\infty_t((-1,0); L^2_{x,v}(B_1 \times B_1)) \cap L^2_t((-1,0); L^2_x H^1_v(B_1 \times B_1))$$
$$\mathcal{T}f \in L^2_t((-1,0); L^2_x H^{-1}_v(B_1 \times B_1)),$$

with A symmetric measurable so that

$$\Lambda^{-1} \le A(t,x,v) \le \Lambda \text{ for almost every } (t,x,v) \in Q_1.$$

Then, we have

$$\|f_+\|_{L^2(Q_1)} \le \delta \quad \Longrightarrow \quad \|f_+\|_{L^\infty(Q_{1/2})} \le \frac{1}{2}.$$

This estimate is translation-invariant under the Galilean transformations (57)*, and scaling-invariant under the kinetic scaling* (56).

Proof of Lemma 29 The setting of the kinetic first De Giorgi lemma is represented in Fig. 9: it is schematically very similar to the parabolic case, provided the new definition (58) of cylinders is adopted. For instance, in the De Giorgi viewpoint, the sequence of cylinders and energy levels in the iteration is

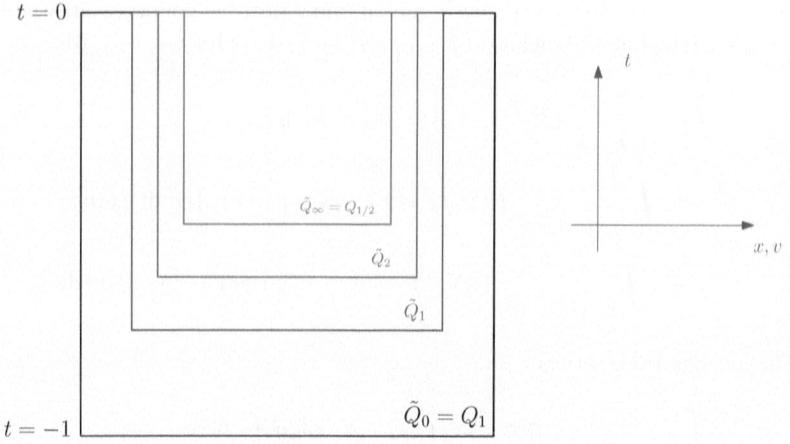

Fig. 9 Setting of the first De Giorgi lemma in the kinetic case

$$\begin{cases} T_k := -\frac{1}{4} - \frac{1}{2^{k+1}}, & T_0 = -1, \quad T_k \uparrow -\frac{1}{4}, \\ r_k^x = \frac{1}{8} + \frac{1}{2^{k+1}}, & r_0 = 1, \quad r_k \downarrow \frac{1}{8}, \\ r_k^v = \frac{1}{2} + \frac{1}{2^{k+1}}, & r_0 = 1, \quad r_k \downarrow \frac{1}{2}, \\ e_k = \frac{1}{2} - \frac{1}{2^{k+1}}, & e_0 = 0, \quad e_k \uparrow \frac{1}{2}, \\ \tilde{Q}_k := (T_k, 0) \times B_{r_k^x} \times B_{r_k^v}, \\ E_k := \int_{\tilde{Q}_k} (f - e_k)_+^2 \, \mathrm{d}t \, \mathrm{d}x \, \mathrm{d}v. \end{cases}$$

Note again that the cylinders $\tilde{Q}_k$ are not exactly of the form (58); instead they are constructed so as to converge to $Q_{1/2}$.

One establishes again a nonlinear inequality between successive E_k terms. The proof is based on a **localised energy estimate**, a small **gain of integrability in all variables** (playing the role of a "kinetic Sobolev embedding") and Chebyshev's inequality. Only the localised energy estimate and the gain of integrability from L^2 to L^{p_*} for some $p_* > 2$ in all variables—the "kinetic Sobolev embedding"—differ from the previous proofs. We only discuss these two substeps. The other steps in the De Giorgi–Moser iteration are similar to the elliptic and parabolic cases.

A. Localised energy estimate We write the energy estimate in a way that is suitable for both the De Giorgi and Moser approaches to the iteration, as in the previous proofs.

Consider an energy level $e \geq 0$; it is used in De Giorgi's iteration, but fixed to zero in Moser's iteration. Consider a smooth localisation function ϕ with support in $(-1, 0] \times B_1 \times B_1$: note that ϕ is zero at initial time but not necessarily at final time. We integrate (59) against $(f-e)_+\phi^2$ on $[-1, \tau]$ for $\tau \in (-1, 0]$:

$$\begin{aligned}
&\int_{(-1,\tau]\times B_1\times B_1} [\mathcal{T}(f-e)_+]\,(f-e)_+\phi^2 \,\mathrm{d}t\,\mathrm{d}x\,\mathrm{d}v \\
&= -\int_{(-1,\tau]\times B_1\times B_1} [A\nabla_v(f-e)_+ \cdot \nabla_v(f-e)_+]\,\phi^2 \,\mathrm{d}t\,\mathrm{d}x\,\mathrm{d}v \\
&\quad - 2\int_{(-1,\tau]\times B_1\times B_1} [A\nabla_v(f-e)_+ \cdot \nabla_v\phi]\,\phi(f-e)_+ \,\mathrm{d}t\,\mathrm{d}x\,\mathrm{d}v.
\end{aligned}$$

The left-hand side writes

$$\begin{aligned}
&\int_{(-1,\tau]\times B_1\times B_1} [\mathcal{T}(f-e)_+]\,(f-e)_+\phi^2 \,\mathrm{d}t\,\mathrm{d}x\,\mathrm{d}v \\
&= \frac{1}{2}\int_{(-1,\tau]\times B_1\times B_1} \left[\partial_t(f-e)_+^2 + v\cdot\nabla_x(f-e)_+^2\right]\phi^2 \,\mathrm{d}t\,\mathrm{d}x\,\mathrm{d}v \\
&= \frac{1}{2}\left(\int_{B_1\times B_1} (f-e)_+^2\phi^2 \,\mathrm{d}x\,\mathrm{d}v\right)\Big|_{t=0} \\
&\quad - \frac{1}{2}\int_{(-1,\tau]\times B_1\times B_1} (f-e)_+^2\,\mathcal{T}\left(\phi^2\right) \,\mathrm{d}t\,\mathrm{d}x\,\mathrm{d}v,
\end{aligned}$$

and we bound the last term as

$$\begin{aligned}
&\left|\frac{1}{2}\int_{(-1,\tau]\times B_1\times B_1} (f-e)_+^2\,\mathcal{T}\left(\phi^2\right) \,\mathrm{d}t\,\mathrm{d}x\,\mathrm{d}v\right| \\
&\qquad\qquad \lesssim \|\mathcal{T}\phi\|_\infty\|\phi\|_\infty \int_{Q_1\cap\operatorname{supp}\phi} (f-e)_+^2 \,\mathrm{d}t\,\mathrm{d}x\,\mathrm{d}v.
\end{aligned}$$

Therefore we deduce

$$\begin{aligned}
&\frac{1}{2}\left(\int_{B_1\times B_1} (f-e)_+^2\phi^2 \,\mathrm{d}x\,\mathrm{d}v\right)\Big|_{t=\tau} + \int_{(-1,\tau]\times B_1\times B_1} |\nabla_v(f-e)_+|^2\,\phi^2 \,\mathrm{d}t\,\mathrm{d}x\,\mathrm{d}v \\
&\lesssim \Lambda\|\mathcal{T}\phi\|_\infty\|\phi\|_\infty \int_{Q_1\cap\operatorname{supp}\phi} (f-e)_+^2 \,\mathrm{d}t\,\mathrm{d}x\,\mathrm{d}v \\
&\quad + \Lambda\int_{(-1,\tau]\times B_1\times B_1} |A\nabla_v(f-e)_+ \cdot \nabla_v\phi|\,\phi(f-e)_+ \,\mathrm{d}t\,\mathrm{d}x\,\mathrm{d}v \\
&\lesssim \Lambda\|\mathcal{T}\phi\|_\infty\|\phi\|_\infty \int_{Q_1\cap\operatorname{supp}\phi} (f-e)_+^2 \,\mathrm{d}t\,\mathrm{d}x\,\mathrm{d}v
\end{aligned}$$

$$+\Lambda^2\left(\int_{(-1,\tau]\times B_1\times B_1}|\nabla_v(f-e)_+|^2\,\phi\,\mathrm{d}t\,\mathrm{d}x\,\mathrm{d}v\right)^{\frac{1}{2}}\times$$
$$\left(\int_{(-1,\tau]\times B_1\times B_1}|\nabla_v\phi|^2\,(f-e)_+^2\,\mathrm{d}t\,\mathrm{d}x\,\mathrm{d}v\right)^{\frac{1}{2}}.$$

By splitting the last term into two squares, we get

$$\left(\int_{B_1\times B_1}(f-e)_+^2\phi^2\,\mathrm{d}x\,\mathrm{d}v\right)\Bigg|_{t=\tau}+\int_{(-1,\tau]\times B_1\times B_1}|\nabla_v(f-e)_+|^2\,\phi^2\,\mathrm{d}t\,\mathrm{d}x\,\mathrm{d}v$$
$$\lesssim\Lambda^4\left(1+\|\nabla_v\phi\|_\infty+\|\mathcal{T}\phi\|_\infty\right)\int_{Q_1\cap\,\mathrm{supp}\,\phi}(f-e)_+^2\,\mathrm{d}t\,\mathrm{d}x\,\mathrm{d}v.$$

Finally, we vary the final time $\tau\in(-1,0]$ to deduce the following **kinetic Caccioppoli inequality**, which is the kinetic counterpart to (40):

$$\sup_{-1<\tau\le 0}\left(\int_{B_1\times B_1}(f-e)_+^2\phi^2\,\mathrm{d}x\,\mathrm{d}v\right)\Bigg|_{t=\tau}+\int_{Q_1}|\nabla_v(f-e)_+|^2\,\phi^2\,\mathrm{d}t\,\mathrm{d}x\,\mathrm{d}v$$
$$\lesssim\Lambda^4\left(1+\|\nabla_v\phi\|_\infty+\|\mathcal{T}\phi\|_\infty\right)\int_{Q_1\cap\,\mathrm{supp}\,\phi}(f-e)_+^2\,\mathrm{d}t\,\mathrm{d}x\,\mathrm{d}v. \tag{60}$$

Given ϕ_k a sequence of smooth localisation functions such that

$$\mathbf{1}_{\tilde{Q}_{k+1}}\le\phi_k\le\mathbf{1}_{\tilde{Q}_k}\quad\text{and}\quad|\nabla_{t,x,v}\phi_k|\lesssim 2^k,$$

the sequence of localised energy estimate is therefore

$$\sup_{T_{k+1}\le\tau\le 0}\left(\int_{B_1\times B_1}\phi_{k+1}^2(f-e_{k+1})_+^2\,\mathrm{d}x\,\mathrm{d}v\right)\Bigg|_{t=\tau}$$
$$+\int_{Q_1}|\nabla_x(\phi_k(f-e_{k+1})_+)|^2\,\mathrm{d}t\,\mathrm{d}x\,\mathrm{d}v\lesssim\Lambda^4 2^k E_k.$$

Note on the one hand that the energy estimate only provides control on the Sobolev regularity in the v variable. It misses part of the variables (the space variable x), in contrast to the parabolic case. On the other hand, just like in the parabolic case, the gain of integrability in t is ensured by the first term on the left hand side.

B. Gain of integrability (proof by averaging lemma) The goal of this part of the proof is to recover a gain of integrability in all variables. Once this gain of integrability is obtained, with a different (smaller) exponent $p_*>2$ as

compared to the elliptic and the parabolic cases, the structure of the iteration is exactly similar. Let us consider a smooth localisation function ϕ so that $\mathbf{1}_{\tilde{Q}_{k+1}} \le \phi \le \mathbf{1}_{\tilde{Q}_k}$, which satisfies $|\nabla_{t,x,v}\phi| \lesssim 2^k$.
Then, by applying the standard Sobolev inequality in the v variable to (60) (dropping the first term on the left hand side), we obtain

$$\|(f-e)_+\phi\|_{L^2_{t,x}L^p_v(\tilde{Q}_{k+1})} \lesssim \Lambda^4 2^k \|(f-e)_+\|_{L^2(Q_1\cap \operatorname{supp}\phi)}$$

for some $p > 2$.

To recover the gain of integrability in x we would like to apply the following standard result in kinetic theory: □

Lemma 30 (Averaging Lemma [17, 35, 36]) *Given $f = f(t,x,v) \in L^2(\mathbb{R}^{1+2d})$ which satisfies the equation*

$$\mathcal{T}f = \partial_t f + v\cdot\nabla_x f = \nabla\cdot H_1 + H_0 \quad \textit{with} \quad H_0, H_1 \in L^2(\mathbb{R}^{1+2d}),$$

then for any $\varphi = \varphi(v) \in C_c^\infty(\mathbb{R}^d)$, we have

$$\forall s \in \left(0, \frac{1}{3}\right), \quad (t,x) \mapsto \int_{\mathbb{R}^d} \varphi f \,\mathrm{d}v \ \textit{belongs to} \ H^s(\mathbb{R}^{1+d}) \tag{61}$$

with a bound that is controlled linearly by the sum of the $L^2(\mathbb{R}^{1+2d})$ norms of f, H_1 and H_2, and with a constant depending on φ.
We also have (Bouchut's version in [17, Theorem 1.3]):

$$\left\|D_x^{\frac{1}{3}} f\right\|^2_{L^2(\mathbb{R}^{1+2d})} + \left\|D_t^{\frac{1}{3}} f\right\|^2_{L^2(\mathbb{R}^{1+2d})}$$
$$\lesssim \|f\|_{L^2(\mathbb{R}^{1+2d})} + \|\nabla_v f\|^2_{L^2(\mathbb{R}^{1+2d})} + \|H_0\|^2_{L^2(\mathbb{R}^{1+2d})} + \left\|(1+|v|^2)H_1\right\|^2_{L^2(\mathbb{R}^{1+2d})} \tag{62}$$

where $D_\cdot^{1/3} = (-\Delta_\cdot)^{1/3}$ is the standard fractional derivative.

Note that we will review the proof based on (62) (as done in [34]), because it is slightly shorter and cleaner. It is however possible to use (61) instead: for non-negative solutions the estimate gives a control $H^s_{t,x}L^1_v$ provided we localise in velocity, and combined with the energy estimate $L^\infty_t L^2_{x,v} \cap L^2_{t,x}H^1_v$ and some interpolation, this would give the gain of integrability in all variables by a comparison argument as discussed below. The interest of (61), by comparison with (62), is that such estimate remain true for more general transport operators $\partial_t + b(v)\cdot\nabla_x$ where b satisfies weak non-degeneracy assumptions (see [32] for a more detailed discussion).

In order to apply (62), we face a difficulty. It does not apply to subsolutions of our problem because of the lack of $L^2_{t,x} H^{-1}_v(\mathbb{R}^{1+2d})$-regularity of the defect measure. Indeed the subsolutions, say f, we are dealing with, satisfy merely

$$\mathcal{T} f \leq h$$

with $h \in L^2_{t,x} H^{-1}_v(\mathbb{R}^{1+2d})$. Expressed with a defect measure this means

$$\mathcal{T} f = h - m$$

with $h \in L^2_{t,x} H^{-1}_v(\mathbb{R}^{1+2d})$ and m a non-negative measure on $\mathbb{R}^{1+2d}$. We will solve this issue by using a comparison principle, since we are *in fine* only interested in integrability properties of f, which are preserved by comparison.

Let us consider a weak subsolution f in Q_1, and let us assume without loss of generality that it is non-negative. We define the new unknown $\tilde{f} := f\phi_k$ on $\mathbb{R}^{1+2d}$, extending it by zero for $t > 0$. It satisfies the inequation

$$\mathcal{T}\tilde{f} \leq \nabla \cdot (A \nabla \tilde{f}) + \nabla_v \cdot H_1 + H_0$$

on the whole $\mathbb{R}^{1+2d}$, with

$$\begin{cases} H_0 := f\,(\nabla_v \cdot (A \nabla_v \phi_k) - \mathcal{T}\phi_k)\,\mathbf{1}_{t \leq 0}, \\[2ex] H_1 := A \nabla_v f \cdot \nabla_v \phi_k \mathbf{1}_{t \leq 0}. \end{cases}$$

Note that we have used that $\partial_t \mathbf{1}_{t \leq 0} \leq 0$.

The kinetic Caccioppoli inequality (60) (local energy estimate) on f implies

$$\|\nabla_v f\|_{L^2(\tilde{Q}_k)} \lesssim \|f\|_{L^2(\tilde{Q}_{k-1})}. \tag{63}$$

All supports in the equation for $\tilde{f}$ are included in $\tilde{Q}_k$ so we use (63) to deduce

$$\|H_0\|_{L^2(\mathbb{R}^{1+2d})} + \|H_1\|_{L^2(\mathbb{R}^{1+2d})} \lesssim \|f\|_{L^2(\tilde{Q}_k)}.$$

The unknown $\tilde{f}$ is zero at $t = -1$ because of the localisation by ϕ_1. We define g that is zero for $t \leq -1$ and satisfies the equation

$$\mathcal{T} g = \nabla \cdot (A \nabla g) + \nabla_v \cdot H_1 + H_0 \tag{64}$$

on $[-1, +\infty) \times \mathbb{R}^{2d}$.

By comparison we have $0 \leq \tilde{f} \leq g$ on Q_1. Moreover, we integrate (64) on $(-\infty, \tau] \times B_1 \times B_1$ for any $\tau > -1$ to get

$$\sup_{\tau>-1} \|g(\tau,\cdot,\cdot)\|_{L^2(\mathbb{R}^{2d})} + \|\nabla_v g\|_{L^2(\mathbb{R}^{1+2d})}$$
$$\lesssim \Lambda^2 \left(\|H_0\|_{L^2(\mathbb{R}^{1+2d})} + \|H_1\|_{L^2(\mathbb{R}^{1+2d})}\right) \lesssim \|f\|_{L^2(Q_1)}. \tag{65}$$

Therefore, given a smooth time localisation function $\theta = \theta(t)$ so that

$$\mathbf{1}_{(-1,0]} \leq \theta \leq \mathbf{1}_{(-1,2]},$$

the new unknown $\tilde{g}(t,x,v) := \theta(t)g(t,x,v)$ satisfies the following modified version of the equation (64):

$$\mathcal{T}\tilde{g} = \nabla \cdot (A\nabla \tilde{g}) + \nabla_v \cdot H_1 + \tilde{H}_0 \quad \text{where} \quad \tilde{H}_0 = H_0 + \theta'(t)g.$$

And the estimate (65) implies

$$\|\tilde{g}\|_{L^2(\mathbb{R}^{1+2d})} + \|\nabla_v \tilde{g}\|_{L^2(\mathbb{R}^{1+2d})} + \|\tilde{H}_0\|_{L^2(\mathbb{R}^{1+2d})} + \|H_1\|_{L^2(\mathbb{R}^{1+2d})} \lesssim \|f\|_{L^2(Q_1)}.$$

We can therefore apply (62) to $\tilde{g}$ and, together with the last bound, it implies

$$\left\|D_x^{\frac{1}{3}}\tilde{g}\right\|_{L^2(\mathbb{R}^{1+2d})} + \left\|D_t^{\frac{1}{3}}\tilde{g}\right\|_{L^2(\mathbb{R}^{1+2d})} + \|\tilde{g}\|_{L^2(\mathbb{R}^{1+2d})} + \|\nabla_v \tilde{g}\|_{L^2(\mathbb{R}^{1+2d})} \lesssim \|f\|_{L^2(Q_1)}. \tag{66}$$

Therefore by Sobolev's embedding we get $\tilde{g} \in L^{p_*}(\mathbb{R}^{1+2d})$, for some $p_* \in (2,p)$. This local integrability is finally inherited by f on $\tilde{Q}_{k+1}$ since $f = \tilde{f} \leq g = \tilde{g}$ on $\tilde{Q}_{k+1}$. Tracking down all these estimates leads to the inequality

$$\|(f-e)_+\|_{L^{p_*}(\tilde{Q}_{k+1})} \lesssim 2^k \|(f-e)_+\|_{L^2(Q_1 \cap \operatorname{supp}\phi_k)}.$$

Remarks 31 In the final comparison between f and g, the Sobolev regularity in t and x seems lost. However, by integrating the inequation on f:

$$\mathcal{T}f = \partial_t f + v \cdot \nabla_x f = \nabla_v \cdot (A\nabla_x f) - m$$

against a smooth localisation function, and without squaring the unknown, one can deduce a bound on the total variation of the defect measure $m \leq 0$ in terms of the L^2 norm of f. Such a bound in total variation translates into a control on the $W^{-1/2,q}_{t,x,v}$ norm of m, for some $q > 1$ close to 1. Then, more sophisticated forms of averaging lemmas can be invoked to obtain a gain of regularity $W^{s,q}$ in t, x for $s > 0$ small. It means that even subsolutions have some amount of Sobolev regularity, which can be useful for compactness arguments for instance. This low level of Sobolev regularity for subsolutions is also used quantitatively in [42].

C. Gain of integrability (proof by fundamental solutions) This is an alternative proof of the step of "gain of integrability" $L^2 \to L^{p_*}$ with $p_* > 0$ in all variables. This argument was proposed in [84]. The idea is to use the explicit fundamental solutions to the Kolmogorov equation, when A is the identity matrix, and treat the actual operator as a source term. It leads to sharper results but is less robust since it crucially relies on knowning explicitly the fundamental solutions of the equation when the rough coefficients are constant.

We study the following auxiliary problem:

Lemma 32 (Estimates on Kolmogorov's Equation) *Let $f \geq 0$ be a locally integrable function on $\mathbb{R}^{1+2d}$ that satisfies*

$$\mathcal{K}f := (\mathcal{T} - \Delta_v) f = \nabla_v \cdot F_1 + F_2 - m, \tag{67}$$

with $F_1, F_2 \in L^1 \cap L^2(\mathbb{R}^{1+2d})$, $m \geq 0$ a measure on $\mathbb{R}^{1+2d}$. Then, we have

$$\|f\|_{L^{2+\frac{1}{d}}(\mathbb{R}^{1+2d})} \lesssim \|F_1\|_{L^2(\mathbb{R}^{1+2d})} + \|F_2\|_{L^2(\mathbb{R}^{1+2d})}, \tag{68}$$

$$\|f\|_{L^1_{t,v} W_x^{\sigma_1,1}(\mathbb{R}^{1+2d})} \lesssim \|F_1\|_{L^1(\mathbb{R}^{1+2d})} + \|F_2\|_{L^1(\mathbb{R}^{1+2d})} + \|m\|_{\mathrm{TV}(\mathbb{R}^{1+2d})}, \tag{69}$$

$$\|f\|_{L^1_{x,v} W_t^{\sigma_2,1}(\mathbb{R}^{1+2d})} \lesssim \|F_1\|_{L^1(\mathbb{R}^{1+2d})} + \|F_2\|_{L^1(\mathbb{R}^{1+2d})} + \|m\|_{\mathrm{TV}(\mathbb{R}^{1+2d})}, \tag{70}$$

for any $\sigma_1 \in (0, 1/3)$ and $\sigma_2 \in (0, 1/2)$. In (69) and (70), the constants are proportional to $(1/3 - \sigma_1)^{-1}$ and $(1/2 - \sigma_2)^{-1}$, respectively.

Remarks 33 It is an interesting question to know whether the maximal regularity $L^1_{t,v} W_x^{1/3,1}$ is attained in such an L^1 setting, in which the Young inequality for weak Lebesgue spaces (see for instance [37, Theorem 1.4.25, p. 73], and various other standard inequalities, degenerate.

Proof of Proposition 32 We express f in terms of Kolmogorov's fundamental solution:

$$f(t,x,v) = \int_{\mathbb{R}^{1+2d}} G\left(t - t', x - x' - (t - t')v', v - v'\right) \mathcal{K}(f)\left(t', x', v'\right) \mathrm{d}z', \tag{71}$$

where the Galilean translation is used to define the modified convolution, and

$$G(t,x,v) = \begin{cases} \left(\dfrac{3}{4\pi^2 t^4}\right)^{\frac{d}{2}} \exp\left[-\dfrac{3\left|x - \frac{t}{2}v\right|^2}{t^3} - \dfrac{|v|^2}{4t}\right], & t > 0, \\ 0, & t \leq 0, \end{cases} \tag{72}$$

is the fundamental solution to $\mathcal{K}(G) = \delta_0$. The calculation of G by Kolmogorov is based on the Fourier transform in (x, v) and the characteristic method. This classical result is exposed in [49, Chapter VII, Section 7.6, p. 211], and a recent concise re-writing can be found in [18, Appendix A].

Since $f, G, m \geq 0$, we deduce that

$$0 \leq f \leq \int_{\mathbb{R}^{1+2d}} G\left(t-t', x-x'-(t-t')v', v-v'\right)[\nabla_v \cdot F_1+F_2]\left(t', x', v'\right) \mathrm{d}z'.$$

Then, we have, denoting by $L^{p,\infty}$ the weak Lebesgue space,

$$|\nabla_v G(t,x,v)| + t|\nabla_x G(t,x,v)| \leq \frac{1}{t^{2d+\frac{1}{2}}} \exp\left[-\frac{3\left|x-\frac{t}{2}v\right|^2}{t^3} - \frac{|v|^2}{4t}\right]$$

so that

$$G,\ \nabla_v G(t,x,v),\ t\nabla_x G(t,x,v) \in L^{\frac{2d+1}{2d+1/2},\infty}(\mathbb{R}^{1+2d}).$$

By integration by parts and Young's inequality (easily extended to our modified convolution product), we deduce (68).

To prove (69) and (70), the idea is to use, roughly, that $\nabla_v G$ and $t\nabla_x G$ "almost" belong to $L^1_{t,v} W^{1/3,1}_x$ and $L^1_{x,v} W^{1/2,1}_t$, i.e. in the sense of *weak* Lebesgue spaces. We refer to [12, Section 1.3, p. 7] for the definition of weak Lebesgue spaces, as well as the more general family of Lorentz spaces they belong to. We only sketch the argument. Define for $\varepsilon > 0$ the following decomposition:

$$G := G_\varepsilon + G_\varepsilon^\perp, \quad \text{with} \quad \begin{cases} G_\varepsilon(t,x,v) = \chi\left(\frac{t}{\varepsilon}\right) G(t,x,v) \\ G_\varepsilon(t,x,v) = \left(1-\chi\left(\frac{t}{\varepsilon}\right)\right) G(t,x,v) \end{cases}$$

for some smooth cutoff function $\chi \in C^\infty(\mathbb{R}_+)$ such that $\mathbf{1}_{[0,1]} \leq \chi \leq \mathbf{1}_{[0,2]}$. The formula (71) then implies a corresponding decomposition $f = f_\varepsilon + f_\varepsilon^\perp$ on f.

Arguing similarly as above, by integration by parts and Young's inequality, one gets

$$\begin{cases} \|f_\varepsilon\|_{L^1_{t,x,v}} \lesssim \varepsilon^{\frac{1}{2}} \left(\|F_1\|_{L^1} + \|F_2\|_{L^1} + \|m\|_{\mathrm{TV}}\right) a \\ \|f_\varepsilon^\perp\|_{L^1_{t,v} \mathrm{W}^{k,1}_x} \lesssim \varepsilon^{-\frac{1}{2}-\frac{3k}{2}} \left(\|F_1\|_{L^1} + \|F_2\|_{L^1} + \|m\|_{\mathrm{TV}}\right) \\ \|f_\varepsilon^\perp\|_{L^1_{x,v} \mathrm{W}^{k,1}_t} \lesssim \varepsilon^{-\frac{1}{2}-k} \left(\|F_1\|_{L^1} + \|F_2\|_{L^1} + \|m\|_{\mathrm{TV}}\right) \end{cases}$$

for any $k \geq 1$.

The convergence $O(\varepsilon^{1/2})$ to zero in the first line is due to the shrinking domain of integration in time when integrating $|\nabla_v G(t,x,v)| + t|\nabla_x G(t,x,v)| \sim t^{-1/2}$. The blow-up $O(\varepsilon^{-1/2-3k/2})$ in the second line follows from integrating

$$\nabla_x^k \nabla_v G(t,x,v) \text{ and } t\nabla_x^{k+1} G(t,x,v) \quad \sim t^{-\frac{1}{2}-\frac{3k}{2}}$$

away from the singularity $t \geq \varepsilon$ (and using the time integrability ensured by $k \geq 1$). The blow-up $O(\varepsilon^{-1/2-k})$ in the third line follows from integrating

$$\partial_t^k \nabla_v G(t,x,v) \text{ and } \partial_t^k t \nabla_x^k G(t,x,v) \quad \sim t^{-\frac{1}{2}-k}$$

away from the singularity $t \geq \varepsilon$ (and using again the time integrability ensured by $k \geq 1$).

By the Littlewood-Paley decomposition or any equivalent interpolation argument, one then deduces that $f \in L^1_{t,v} W_x^{\sigma_1,1}$ for $\sigma_1 \in [0, 1/3)$, and that $f \in L^1_{x,v} W_t^{\sigma_2,1}$ for $\sigma_2 \in [0, 1/2)$. □

We now deduce the gain of integrability on subsolutions, as well as gain of (low level) Sobolev regularity in x. We consider $f \geq 0$ a sub-solution to (59) on $\tilde{Q}_k$, and we claim:

$$\|f\|_{L^p(\tilde{Q}_{k+1})} + \|f\|_{L^1_{t,v} W_x^{\sigma_1,1}(\tilde{Q}_{k+1}))} + \|f\|_{L^1_{x,v} W_t^{\sigma_1,1}(\tilde{Q}_{k+1}))} \lesssim 2^k \|f\|_{L^2(\tilde{Q}_k)}, \tag{73}$$

with $\sigma_1 \in [0, 1/3)$, $\sigma_2 \in [0, 1/2)$ and a constant degenerating as $\sigma_1 \to 1/3$ or $\sigma_2 \to 1/2$.

Given ϕ_k the usual smooth localisation function such that $\mathbf{1}_{\tilde{Q}_{k+1}} \leq \phi_k \leq \mathbf{1}_{\tilde{Q}_k}$, we define $\tilde{f} := f\phi_k$ for $t \leq 0$ and $\tilde{f} = 0$ for $t > 0$. Then $\tilde{f}$ satisfies the equation

$$\mathcal{K}\tilde{f} = \nabla_v \cdot F_1 + F_0 - \tilde{m} \tag{74}$$

with

$$\begin{aligned}
F_1 &= (A\nabla_v f)\phi_k - (\nabla_v f)\phi_k - f\nabla_v \phi_k, \\
F_0 &= -A\nabla_v f \cdot \nabla_v \phi_k + f\mathcal{T}\phi_k - \nabla_v \phi_k \cdot \nabla_v f, \\
\tilde{m} &= \phi_k m.
\end{aligned}$$

and m the defect measure of the subsolution f.

The local energy estimate (60) and the compact support imply

$$\|F_1\|_{L^1(\mathbb{R}^{1+2d})} + \|F_2\|_{L^1(\mathbb{R}^{1+2d})} + \|F_1\|_{L^2(\mathbb{R}^{1+2d})} + \|F_2\|_{L^2(\mathbb{R}^{1+2d})} \lesssim \|f\|_{L^2(\tilde{Q}_k)}.$$

By integrating (74) against ϕ_k (without squaring the unknown) we also obtain

$$\|\tilde{m}\|_{\mathrm{TV}(\mathbb{R}^{1+2d})} \lesssim \|f\|_{L^2(\tilde{Q}_k)}.$$

Therefore we can apply (68)–(69)–(70) to deduce (73).

4.3 The Improved Kinetic First De Giorgi Lemma

Similarly to the elliptic and parabolic cases, we can improve the first De Giorgi lemma as follows.

Lemma 34 (Improved Kinetic First De Giorgi Lemma) *Given $\Lambda > 0$ and $\zeta_0 \in (0, 2]$, there is $C > 0$, depending only Λ, ζ_0 and the dimension d, so that the following holds:*

Let f be a weak subsolution in a kinetic cylinder Q_R to

$$\partial_t f + v \cdot \nabla_x f \le \nabla_v \cdot (A \nabla_v f)$$

in the functional setting

$$f \in L^\infty_t((-R^2, 0); L^2_{x,v}(B_{R^3} \times B_R)) \cap L^2_t((-R^2, 0); L^2_x H^1_v(B_{R^3} \times B_R))$$
$$\mathcal{T}f \in L^2_t((-R^2, 0); L^2_x H^{-1}_v(B_{R^3} \times B_R)),$$

with A symmetric measurable so that

$$\Lambda^{-1} \le A(t, x) \le \Lambda \textit{ for almost every } (t, x, v) \in Q_R.$$

Then, we have, for any $\zeta \in (\zeta_0, 2]$ and $0 < r < R$,

$$\|f\|_{L^\infty(Q_r)} \le C(R - r)^{-\frac{2+4d}{\zeta}} \left(\frac{1}{|Q_R|} \int_{Q_R} f^\zeta \, \mathrm{d}z \right)^{\frac{1}{\zeta}}.$$

Proof By using the scaling of the equation in the kinetic first De Giorgi lemma, one can prove the following inequality between two general cylinders Q_r and Q_R:

$$\|f\|_{L^\infty(B_r)} \lesssim (R - r)^{-1-2d} \left(\frac{1}{|Q_R|} \int_{Q_R} f^2 \, \mathrm{d}z \right)^{\frac{1}{2}}.$$

The Young's inequality and the inequality

$$\left(\frac{1}{|Q_R|} \int_{Q_R} f^2 \, \mathrm{d}z \right)^{\frac{1}{2}} \lesssim \left(\frac{1}{|Q_R|} \int_{Q_R} f^\zeta \, \mathrm{d}z \right)^{\frac{1}{2}} \|f\|^{1-\frac{\zeta}{2}}_{L^\infty(Q_R)},$$

imply that the function $\mathcal{N}(r) := \|f\|_{L^\infty(Q_r)}$ verifies the following inequality for some $C > 0$ and any $0 < r < r' < R$:

$$\mathcal{N}(r) \le \frac{1}{2}\mathcal{N}(r') + C(r' - r)^{-\frac{2+4d}{\zeta}} \left(\frac{1}{|Q_R|} \int_{B(0,R)} f^\zeta \, \mathrm{d}z \right)^{\frac{1}{\zeta}}.$$

By the same iteration along the sequence of radii

$$r_{k+1} = r_k + \varepsilon \left(1 - \alpha^{-1}\right) \alpha^{-k}$$

for $k \geq 0$, some $\alpha > 1$, $\varepsilon := R - r$ and $r_0 := r$, we deduce as in the elliptic and parabolic cases that

$$\mathcal{N}(r) \lesssim_{\zeta_0} \varepsilon^{-\frac{2+4d}{\zeta}} \left(\frac{1}{|Q_R|} \int_{Q_R} f^\zeta \, \mathrm{d}z \right)^{\frac{1}{\zeta}},$$

from which the conclusion follows. □

4.4 The Kinetic Second De Giorgi Lemma

Inspired by the use of Poincaré's inequality as an *integral measure of the oscillation* in the works of Moser [70, 73], Kruzhkov [58], and later in [93] in a modified fashion, we introduce the strategy of [42], in the form simplified and optimised in [7, 26, 81]. This method does not involve the equation verified by the logarithm of a solution to (59), as in Nash's, Moser's and Kruzhkov's methods.

The kinetic second De Giorgi lemma (reduction of oscillation) resembles the elliptic case, and it implies the Theorem 28 by an argument also similar to the parabolic case, but accounting for the slightly different geometry of the kinetic cylinders.

More generally, the overall structure of the proof of the kinetic second De Giorgi lemma is similar to the elliptic and parabolic cases (24), with kinetic versions of each step.

Lemma 35 (Kinetic Second De Giorgi Lemma: Reduction of Oscillation) *Given $\Lambda > 0$, there is $\nu \in (0, 1)$, depending only on Λ and the dimension n, so that the following holds:*

Let f be a weak solution on Q_3 to

$$\mathcal{T} f = \nabla_v \cdot (A \nabla_v f)$$

in the functional setting $L^2((-9, 0) \times B_{27}; H^1(B_3))$, where A is symmetric measurable and so that

$$\Lambda^{-1} \leq A(t, x, v) \leq \Lambda \text{ for almost every } (t, x, v) \in Q_3.$$

Then, we have

$$\operatorname{osc}_{Q_{\frac{1}{2}}} f \leq \nu \operatorname{osc}_{Q_2} f, \tag{75}$$

where we denote $\operatorname{osc}_B f := \sup_B f - \inf_B f$.

Note again that this lemma requires f to be **both** a subsolution and a supersolution, i.e. to be a solution. Let us show that Lemma 35 implies the Hölder regularity, which thus concludes the proof of Theorem 28.

Corollary 36 *Given* $\Lambda > 0$, *there is* $\alpha \in (0, 1)$, *depending only on* Λ *and the dimension* n, *so that the following holds:*

Let f *be a weak solution in* $(0, T) \times \mathcal{U}$, *with* $\mathcal{U} \subset \mathbb{R}^{2d}$ *and open set, to*

$$\mathcal{T}f = \nabla_v \cdot (A \nabla_v f),$$

where A *is symmetric measurable and so that*

$$\Lambda^{-1} \le A(t, x, v) \le \Lambda \text{ for almost every } (t, x, v) \in (0, T) \times \mathcal{U}.$$

Then, we have $f \in C^{\alpha}_{\text{loc}}((0, T) \times \mathcal{U})$.

Moreover, given any $\delta \in (0, T/2)$, *there is* $C > 0$, *depending on* Λ, δ *and the dimension* n, *so that for any* $\tilde{\mathcal{U}} \subset\subset \mathcal{U}$ *with* $\text{dist}(\tilde{\mathcal{U}}, \partial\mathcal{U}) \ge \delta$ *and* $\delta < T_0 < T_1 < T - \delta$, *we have*

$$\|f\|_{C^{\alpha}((T_0, T_1) \times \tilde{\mathcal{U}})} \le C \|f\|_{L^2(\mathcal{U})}. \tag{76}$$

Proof We already have that $f \in L^{\infty}_{\text{loc}}(\mathcal{U})$ by Lemma 29 with the estimate

$$\forall \hat{\mathcal{U}} \subset\subset \mathcal{U}, \quad \|f\|_{L^{\infty}((T_0/2, T) \times \hat{\mathcal{U}})} \le C \|f\|_{L^2((0,T) \times \mathcal{U})} \tag{77}$$

for a constant $C > 0$ depending on Λ, $\text{dist}(\hat{\mathcal{U}}, \partial\mathcal{U})$, T_0 and the dimension n.

Let $\tilde{\mathcal{U}} \subset\subset \hat{\mathcal{U}} \subset\subset \mathcal{U}$ so that

$$\text{dist}(\tilde{\mathcal{U}}, \partial\mathcal{U}) \le \frac{1}{2}\text{dist}(\tilde{\mathcal{U}}, \partial\hat{\mathcal{U}}), \quad \text{dist}(\tilde{\mathcal{U}}, \partial\mathcal{U}) \in \left[\frac{\delta}{4}, \frac{\delta}{2}\right], \quad \text{dist}(\tilde{\mathcal{U}}, \partial\hat{\mathcal{U}}) \in \left[\frac{\delta}{2}, \delta\right]. \tag{78}$$

Given $(t_0, x_0, v_0) \in (T_0, T_1) \times \tilde{\mathcal{U}}$, we define the sequence of centred rescaled functions

$$\forall k \ge 1,\ \forall (s, y, w) \in Q_1, \quad \tilde{g}_k(s, y, w) := g_{k-1}\left(\frac{s}{16}, \frac{y}{64}, \frac{w}{4}\right)$$

$$\forall (s, y) \in Q_1, \quad g_0(s, y, w) := f\left(t_0 + \frac{\delta}{32} + \frac{\delta s}{16}, x_0 + \frac{\delta y}{64}, v_0 + \frac{\delta w}{4}\right).$$

Note that each function g_k is a rescaling of f and solves

$$\mathcal{T}g_k = \nabla_x \cdot [A_k \nabla_x g_k]$$

in Q_2, with the diffusion matrix

$$A_k(s, y, w) := A\left(t_0 + \frac{\delta}{32} + \frac{\delta s}{2^{4k}}, x_0 + \frac{\delta y}{2^{6k}}, v_0 + \frac{\delta w}{2^{2k}}\right).$$

This new diffusion matrix is still measurable and satisfies

$$\Lambda^{-1} \le A_k(s, y, w) \le \Lambda \text{ for almost every } (s, y) \in Q_2.$$

Therefore, we can apply Lemma 35 to each g_k and deduce

$$\forall k \ge 1, \quad \operatorname{osc}_{Q_{\frac{1}{2}}} g_k \le \nu \operatorname{osc}_{Q_2} g_k.$$

The scaling relation between successive terms in the sequence g_k yields

$$\forall k \ge 1, \quad \operatorname{osc}_{Q_2} g_{k+1} = \operatorname{osc}_{Q_{\frac{1}{2}}} g_k,$$

so we obtain by induction

$$\operatorname{osc}_{Q_{4^{-k}}} g_1 \le \nu^{k-1} \operatorname{osc}_{Q_2} g_1 \le 2\,\nu^{k-1} \|f\|_{L^\infty((\delta/2, T-\delta/2)\times\hat{\mathcal{U}})}.$$

Then, given a point

$$(t, x, v) \in \left(t_0 - \frac{\delta}{16}, t_0 + \frac{\delta}{16}\right) \times B\left(x_0, \frac{\delta}{64}\right) \times B\left(v_0, \frac{\delta}{4}\right),$$

we pick up $k \ge 0$ so that

$$\begin{cases} |t - t_0| \in \left[\delta 16^{-k-2}, \delta 16^{-k-1}\right], \\ |x - x_0| \in \left[\delta 64^{-k-2}, \delta 64^{-k-1}\right], \\ |v - v_0| \in \left[\delta 4^{-k-2}, \delta 4^{-k-1}\right], \end{cases}$$

and write

$$|f(t, x, v) - f(t_0, x_0, v_0)| \le 2\,\nu^{k-1} \|f\|_{L^\infty((\delta/2, T-\delta/2)\times\hat{\mathcal{U}})},$$

so that (using $|t - t_0| \ge \delta 16^{-k-2}$, $|x - x_0| \ge \delta 64^{-k-2}$ and $|v - v_0| \ge \delta 4^{-k-2}$)

$$|f(t, x, v) - f(t_0, x_0, v_0)| \lesssim \left[\left(16^{3\alpha}\nu\right)^k |t - t_0|^{3\alpha} + \left(64^{2\alpha}\nu\right)^k |x - x_0|^{2\alpha} + \left(4^{6\alpha}\nu\right)^k |v - v_0|^{6\alpha}\right] \|f\|_{L^\infty((\delta/2, T-\delta/2)\times\hat{\mathcal{U}})}.$$

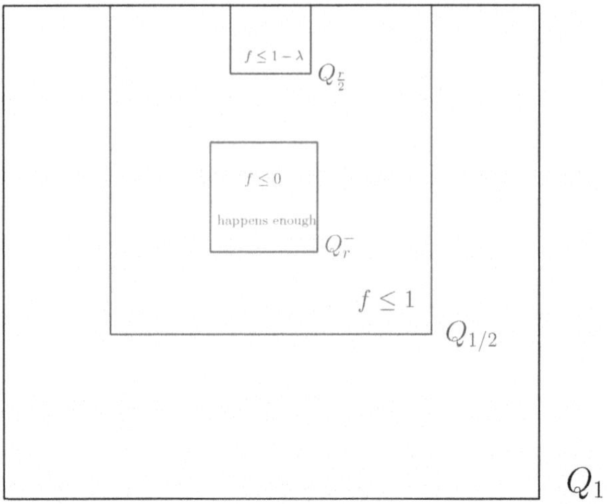

Fig. 10 Setting of the decrease of supremum lemma and intermediate value lemma in the kinetic case

We then choose

$$\alpha := -\frac{\ln \nu}{12 \ln 2}$$

so that $16^{3\alpha}\nu = 64^{2\alpha}\nu = 4^{6\alpha}\nu = 1$, and deduce f is C^α at (t_0, x_0, v_0). (In fact, note that the Hölder regularity exponent is 3α in t, 2α in x and 6α in v, which reflects the scaling of the equation.) The α does *not* depend on (t_0, x_0, v_0), but only on the constant ν of Lemma 35.

We have therefore proved

$$\|f\|_{C^\alpha((\delta,T-\delta)\times\tilde{\mathcal{U}})} \le C\|f\|_{L^\infty((\delta/2,T-\delta/2,\hat{\mathcal{U}})}$$

with a constant depending only on ν and δ. Combined with (77) and (78), it implies (76) and concludes the proof. □

Lemma 35 is implied by the following kinetic decrease of supremum lemma, whose setting is represented in Fig. 10. The proof is exactly similar to the parabolic case and is not repeated.

Lemma 37 (Kinetic Decrease of Supremum Lemma) *Given $\Lambda > 0$ and $\delta \in (0, 1)$, there is $\lambda \in (0, 1)$, depending only Λ, δ and the dimension d, so that the following holds:*

Let $r = 1/20$, and let f be a weak subsolution over Q_1 to

$$\partial_t f + v \cdot \nabla_x f \le \nabla_v \cdot (A \nabla_v f),$$

with A symmetric measurable so that

$$\Lambda^{-1} \le A(t,x) \le \Lambda \text{ for almost every } (t,x,v) \in Q_1,$$

and that satisfies $f \le 1$ on $Q_{1/2}$ and the measure condition

$$|\{f \le 0\} \cap Q_r^-| \ge \delta |Q_r^-|,$$

where we have denoted

$$Q_r^- := \left\{ -3r^2 < t < -2r^2, \quad |x| < r^3, \quad |v| < r \right\}. \tag{79}$$

Then, we have

$$\sup_{Q_{\frac{r}{2}}^-} f \le 1 - \lambda.$$

Remarks 38 By tracking the dependency of the constants in the proof, one finds $\lambda \approx \delta^{2(1+\delta^{-2})}$. This (bad) exponential dependency is due to the finite De Giorgi inductive argument used to deduce this lemma from the intermediate value lemma, where the number of steps potentially grows exponential as $\delta \to 0$. This lemma was first proved in [34] by non-constructive method, and then in [41, 42] by constructive methods.

Finally the kinetic decrease of supremum lemma is itself implied by a kinetic version of De Giorgi intermediate value lemma, whose setting is again represented in Fig. 10. The proof that Lemma 39 implies the decrease of supremum lemma is exactly similar to the elliptic and parabolic cases and is not repeated.

Lemma 39 (Kinetic De Giorgi Intermediate Value Lemma) *Given $\Lambda > 0$ and $\delta_1, \delta_2 \in (0,1)$, there are $\theta, \nu \in (0,1)$, depending only on Λ, δ_1, δ_2 and the dimension d, so that the following holds:*

Let $r = 1/20$, and let f be a subsolution over Q_2 to

$$\partial_t f + v \cdot \nabla_x f \le \nabla_v \cdot (A \nabla_v f),$$

with A symmetric measurable so that

$$\Lambda^{-1} \le A(t,x) \le \Lambda \text{ for almost every } (t,x,v) \in Q_2,$$

and that satisfies $f \le 1$ on Q_2, and the measure conditions

$$|\{f \le 0\} \cap Q_r^-| \ge \delta_1 |Q_r^-|, \tag{80}$$

$$|\{f \ge 1-\theta\} \cap Q_r| \ge \delta_2 |Q_r| \tag{81}$$

where Q_r^- is defined as before, see (79).

Then, we have

$$|\{0 < f < 1 - \theta\} \cap Q_1| \geq \nu |Q_1|.$$

Remarks 40

(1) By tracking the constants in the proof one finds $\theta \approx \delta_1\delta_2$ and $\nu \approx \delta_1\delta_2$ for small δ_1, δ_2.
(2) This lemma was first proved in [34] by non-constructive method, and then in [42] by a constructive method.
(3) The gap in time is required, as shown by the following counter-example. The function $\mathbf{1}_{x+ct<a}$ is a subsolution on Q_1 for any $a \in \mathbb{R}$ and $c > 1$. If Q_r^- and Q_r were too close, a line of discontinuity of the form $x + ct = a$ could cross both and the previous subsolution would contradict the conclusion.

We first give a proof by contradiction that extends the contradiction argument in the elliptic and parabolic cases, and then turn to a method based on trajectories and Poincaré-type inequalities.

Proof of the Kinetic Intermediate Value Lemma 39 by Compactness This argument was first proposed in [34], but is reminiscent of similar compactness arguments in the De Giorgi theory and in homogenisation theory. In this compactness argument, we can prove a slightly stronger statement where $\theta = 1/2$.

A. The setting We argue by contradiction. Let us consider a contradiction sequence $(f_k)_{k\geq 0}$ of subsolutions in Q_2 to

$$\mathcal{T} f_k = \nabla_v \cdot (A_k \nabla_v f_k) - \mu_k \tag{82}$$

for some non-negative measures μ_k, such that $f_k \leq 1$ for all $k \geq 1$ and, for some fixed $\delta_1, \delta_2 > 0$:

$$|\{f_k \geq 1/2\} \cap Q_r| \geq \delta_1 |Q_r|$$
$$|\{f_k \leq 0\} \cap Q_r^-| \geq \delta_2 |Q_r^-|$$
$$|\{0 < f_k < 1/2\} \cap Q_1| \to 0 \quad \text{as} \quad k \to +\infty.$$

The same assumptions still hold for the positive parts $(f_k)_+$ of each function f_k in the sequence.

B. A priori estimates on the contradiction sequence First, by using the local energy estimate (60) (the kinetic Caccioppoli inequality) and the first De Giorgi lemma, we obtain uniform bounds on the $L^\infty_{\text{loc}}(Q_2)$ and $(L^2_t H^1_x)_{\text{loc}}(Q_2)$ norms of the $(f_k)_+$'s. Second, by integrating against a localisation function without squaring the unknown, we obtain a uniform bound on the local total variation of the μ_k's in Q_2.

C. Limit by compactness Using standard compactness theorem, we have (still writing f_k and A_k and μ_k for the subsequences, in an abuse of notation)

$$(f_k)_+ \overset{*}{\rightharpoonup} F \text{ in } L^\infty_{\mathrm{loc}}(Q_2)$$

and

$$\nabla_v (f_k)_+ \rightharpoonup \nabla_v F \quad \text{and} \quad A_k \nabla_v (f_k)_+ \rightharpoonup H \quad \text{in } L^2_{\mathrm{loc}}(Q_2) \tag{83}$$

for some weak limit F and H_1, and

$$\mu_k \rightharpoonup \mu$$

in the space of non-negative measure, thanks to the bound on the total variation. By weak limit, we deduce uniform bounds on the $L^2_t H^1_x(Q_1)$ norm of F, on the $L^2(Q_1)$ norm of H, and on the total variation of μ, and we have the equation

$$\mathcal{T} F = \nabla_v \cdot H - \mu. \tag{84}$$

Thanks to the averaging lemma (61), we then improve the convergence of $(f_k)_+$ (taking if necessary a further subsequence), so that we have the strong convergence

$$f_k^+ \to F \text{ in } L^p_{\mathrm{loc}}(Q_2) \quad \text{for } 1 \le p < +\infty.$$

It implies the convergence in probability and thus the function F satisfies

$$|\{F \ge 1/2\} \cap Q_r| \ge \delta_1 |Q_r| \tag{85}$$

$$|\{F = 0\} \cap \hat{Q}_r^-| \ge \delta_2 |Q_r^-| \tag{86}$$

$$|\{0 < F < 1/2\} \cap Q_1| = 0.$$

We now use the elliptic version of the intermediate value lemma (Lemma 14): for almost every $(t, x) \in (-1, 0] \times B_1$,

$$\begin{cases} \text{either} \quad \text{for almost every } v \in B_1, \quad F(t,x,v) \le 0, \\ \text{or} \quad \text{for almost every } v \in B_1, \quad F(t,x,v) \ge \dfrac{1}{2}. \end{cases}$$

In other words, $F(t, x, v) = F_0(t, x, v) \mathbf{1}_E(t, x)$ with $F_0 \ge 1/2$ and some measurable set $E \subset B_1 \times (-1, 0)$. In view of (85) and (86), this set E satisfies

$$\begin{cases} |E \cap (-r^2, 0] \times B_{r^3}| > 0 \\ |(-3r^2, -2r^2] \times B_{r^3} \setminus E| > 0. \end{cases} \tag{87}$$

D. Reaching a contradiction We thus get from (84) the following inequality

$$\partial_t F + v \cdot \nabla_x F \leq \nabla_v \cdot H$$

in $Q_{3/2}$. Then, choosing $v_0 \in B_1$, and averaging locally around v_0 with a smooth mollifier ζ we get (since F only depends on (t, x))

$$\partial_t F + v_0 \cdot \nabla_x F \leq \int_{\mathbb{R}^d} \Big[|H(x, v, t) \nabla_v \zeta(v - v_0)| \Big] \, \mathrm{d}v \tag{88}$$

in $(t, x) \in B_1 \times (-1, 0)$. Since $H \in L^2_{\mathrm{loc}}(Q_2)$, we deduce, for $v_0 \in B_1$ and $(t, x) \in B_1 \times (-1, 0)$, that

$$\partial_t F + v_0 \cdot \nabla_x F \leq 0 \text{ as long as } F(t, x, \cdot) = 0. \tag{89}$$

This is due to the fact that F cannot jump from 0 to $1/2$ without the left-hand side producing a positive Dirac mass, which is not allowed by the L^2_{loc} upper bound (88).

The rest of the argument is similar to the parabolic case: since $F(t, \cdot, \cdot)$ is almost everywhere non-positive for some times in the past cylinder (with non-zero measure), the previous inequality shows that it cannot become positive, which contradicts F being almost everywhere greater than $1/2$ in a non-zero measure set of the future cylinder. □

We now turn to the constructive method. We will prove that the kinetic intermediate value Lemma 39 is implied, with quantitative constants, by the following "kinetic Poincaré inequality". The setting of the kinetic Poincaré inequality is again given by Fig. 10.

The overall structure of the constructive proof of the kinetic second De Giorgi lemma is therefore as follows:

Kinetic Poincaré inequality $\Longrightarrow$ Kinetic intermediate value lemma
$\Longrightarrow$ Kinetic decrease of the supremum $\Longrightarrow$ Kinetic reduction of oscillation.

Note that the coercivity of the diffusion matrix A is not required for the proof of the Poincaré inequality, but only for the later steps.

Theorem 41 (Kinetic Poincaré Inequality) *Given $\Lambda > 0$, there is $C > 0$, depending only on Λ and the dimension d, and there is $r \in (0, 1)$, so that the following holds:*

Let $f \geq 0$ be a weak subsolution in Q_1 to

$$\partial_t f + v \cdot \nabla_x f \leq \nabla_v \cdot (A \nabla_v f)$$

with A symmetric measurable so that

$$A(t,x,v) \leq \Lambda \text{ for almost every } (t,x,v) \in Q_1.$$

Then f satisfies the following inequality

$$\int_{Q_r^+} \left(f - \langle f \rangle_{Q_r^-}\right)_+ \mathrm{d}z \leq C \int_{Q_1} |\nabla_v f| \,\mathrm{d}z, \tag{90}$$

where Q_r^- is defined as before, see (79), *and the average $\langle f \rangle_{Q_r^-}$ is defined by*

$$\langle f \rangle_{Q_r^-} := \frac{1}{|Q_r^-|} \int_{Q_r^-} f \,\mathrm{d}z.$$

Remarks 42

(1) This kinetic Poincaré inequality was developed in the series of works [7, 26, 42, 81]. The two key inspirations were first a weaker modified form of such inequality proposed before in [93], albeit with a complicated corrector replacing the average $\langle f \rangle_{Q_r^-}$ (this corrector itself solving another hypoelliptic problem), and second the kinetic Poincaré inequality for constant diffusion matrix obtained in [2].
(2) Intuitively this theorem measures how much (in L^1) the solution f can grow above its past average over Q_r^-, in terms of a bound on $\nabla_v f$ in L^1 in a bigger cube *in the present*, provided that f is a subsolution on such a bigger cube. This is therefore a **time-oriented integral control on the oscillation**. For instance if $\nabla_v f \equiv 0$, then f solves $\mathcal{T} f \leq 0$ and decays along the transport lines: it cannot increase its local average as time flows from past to future.
(3) This theorem only requires the bound from above on A, not the coercivity. However, in the sequel we will use the coercivity of A in order to control the L^2 norm of $\nabla_v f$ in the right hand side of the previous display by the L^2 norm of f via the energy estimate.
(4) Notice the gap between the two small cubes. It is required for subsolutions, since the kinetic Poincaré inequality (90) implies the kinetic intermediate value Lemma 39, which fails for subsolutions when no gap is present.
(5) The characterisation of the optimal constant in (90) is an interesting question, open to our knowledge.

Proof of the Intermediate Value Lemma 39 from the Poincaré Inequality (90)
Given f as in the statement of the intermediate value Lemma 39, its positive part f_+ is also a subsolution verifying exactly the same assumptions as f.

We can therefore apply the Poincaré inequality (90) to f_+:

$$\int_{Q_r} \left(f_+ - \langle f_+ \rangle_{Q_r^-}\right)_+ \mathrm{d}z \lesssim \int_{Q_1} |\nabla_v f_+| \,\mathrm{d}z. \tag{91}$$

We now study the average on Q_r^- and use the first measure condition (80):

$$\langle f_+ \rangle_{Q_r^-} = \frac{1}{|Q_r^-|} \int_{Q_r^-} f_+ \,\mathrm{d}z \leq \frac{|\{f > 0\} \cap Q_r^-|}{|Q_r^-|} \leq 1 - \delta_1.$$

We deduce the following lower bound on the left hand side of (91):

$$\begin{aligned} \int_{Q_r} \left(f_+ - \langle f_+ \rangle_{Q_r^-} \right)_+ \mathrm{d}z &\geq \int_{Q_r} (f_+ - (1-\delta_1))_+ \,\mathrm{d}z \\ &\geq \int_{Q_r \cap \{f > 1-\theta\}} (\delta_1 - \theta)\,\mathrm{d}z \\ &\geq |Q_r| \delta_2 (\delta_1 - \theta). \end{aligned}$$

Then, we bound from above the right-hand side of (91):

$$\begin{aligned} \int_{Q_1} |\nabla_v f_+| \,\mathrm{d}z &= \int_{Q_1 \cap \{f \leq 0\}} |\nabla_v f_+| \,\mathrm{d}z \\ &\quad + \int_{Q_1 \cap \{1-\theta > f > 0\}} |\nabla_v f_+| \,\mathrm{d}z + \int_{Q_1 \cap \{f \geq 1-\theta\}} |\nabla_v f_+| \,\mathrm{d}z \\ &=: I_1 + I_2 + I_3. \end{aligned}$$

We now consider the three terms I_1, I_2 and I_3 successively. First, we have $I_1 = 0$, since $\nabla_v f_+ = 0$ almost everywhere over $\{f_+ = 0\}$, see again [28, Section 4.2.2, Theorem 4-(iii), p. 129]. Second, we have

$$\begin{aligned} I_2 &\lesssim |\{0 < f < 1-\theta\} \cap Q_1|^{1/2} \left(\int_{Q_1} |\nabla_v f_+|^2 \,\mathrm{d}z \right)^{1/2} \\ &\lesssim |\{0 < f < 1-\theta\} \cap Q_1|^{1/2} \left(\int_{Q_2} f_+^2 \,\mathrm{d}z \right)^{1/2} \\ &\lesssim |\{0 < f < 1-\theta\} \cap Q_1|^{1/2}, \end{aligned}$$

where we have used the energy estimate (60) (the kinetic Caccioppoli inequality) between the first and second lines, and the bound $0 \leq f_+ \leq 1$ between the second and third lines.

Third and finally, we have

$$\begin{aligned} I_3 &= \int_{Q_1} \left| \nabla_v \Big[(f - (1-\theta))_+ + (1-\theta) \Big] \right| \mathrm{d}z \\ &= \int_{Q_1} |\nabla_v (f - (1-\theta))_+| \,\mathrm{d}z \end{aligned}$$

$$\lesssim \left(\int_{Q_1} |\nabla_v (f - (1-\theta))_+|^2 \, \mathrm{d}z \right)^{1/2}$$

$$\lesssim \left(\int_{Q_2} |(f - (1-\theta))_+|^2 \, \mathrm{d}z \right)^{1/2} \lesssim \theta,$$

where we have used the energy estimate (60) (the kinetic Caccioppoli inequality) between the third and fourth lines on the subsolution $(f - (1-\theta))_+$, and the bound $0 \le (f - (1-\theta))_+ \le \theta$ in the last line.

Altogether, we obtain

$$\delta_2(\delta_1 - \theta) \lesssim |\{0 < f < 1 - \theta\} \cap Q_1|^{1/2} + \theta.$$

Provided we choose $\theta \le \delta_1/2$, we deduce

$$\delta_1 \, \delta_2 \lesssim |\{0 < f < 1 - \theta\} \cap Q_1|^{1/2} + \theta.$$

Provided we furthermore impose $\theta \lesssim \delta_1 \delta_2$, we finally get

$$\delta_1 \, \delta_2 \lesssim |\{0 < f < 1 - \theta\} \cap Q_1|^{1/2},$$

from which the conclusion follows. □

Proof of Theorem 41 (The Kinetic Poincaré Inequality) We present the argument of [7, 26], inspired by [42], the subsequent work [81] and control theory. The idea is to replace the straight lines of the constructive argument in the elliptic case by appropriate trajectories generated by the Hörmander vector fields $\partial_t + v \cdot \nabla_x$ and ∇_v, and use the *upper bound* $(\partial_t + v \cdot \nabla_x) f \le \nabla_v \cdot (A \nabla_v f)$ provided by the assumption that f is a subsolution; the right-hand side is then controlled after integrating by parts but the Jacobian of the parametrisation by the *departing* cylinder must be controlled. We also introduce an intermediate cylinder Q_r^0 between Q_r^- and Q_r^+ and a smooth localisation function in this intermediate cylinder: this variable z_0 is used for the integration by parts (see Fig. 11).

We recall the notation $z = (t, x, v)$.

A. Construction of the trajectories Given three points $z_+ \in Q_r^+$, $z_0 \in Q_r^0$ and $z_- \in Q_r^-$, we want to construct two paths $s \to \Gamma_+(s)$ and $s \to \Gamma_-(s)$ for $s \in [0, 1]$ such that (see Figs. 11 and 12)

$$\Gamma_+(s) = (T_+(s), X_+(s), V_+(s)), \qquad \Gamma_+(0) = z_+, \qquad \Gamma_+(1) = z_0,$$

$$\Gamma_-(s) = (T_-(s), X_-(s), V_-(s)), \qquad \Gamma_-(0) = z_-, \qquad \Gamma_-(1) = z_0,$$

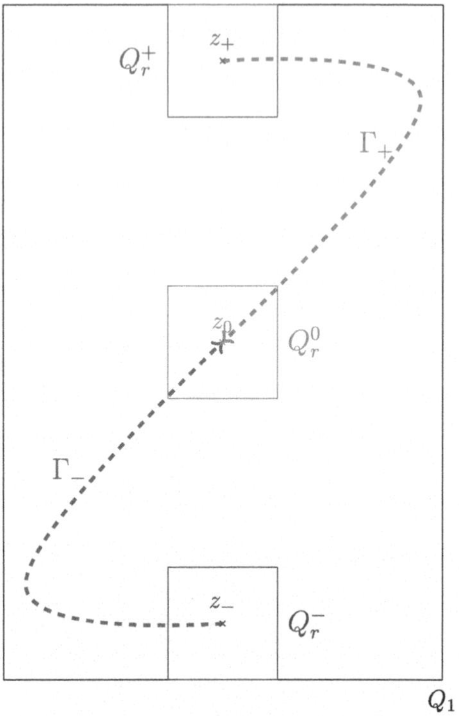

Fig. 11 Construction of the trajectories. The curve Γ_+ connects any point $z_+ \in Q_r^+$ to some intermediate point $z_0 \in Q_r^0$, whereas the curve Γ_- connects any point $z_- \in Q_r^-$ to some intermediate point $z_0 \in Q_r^0$

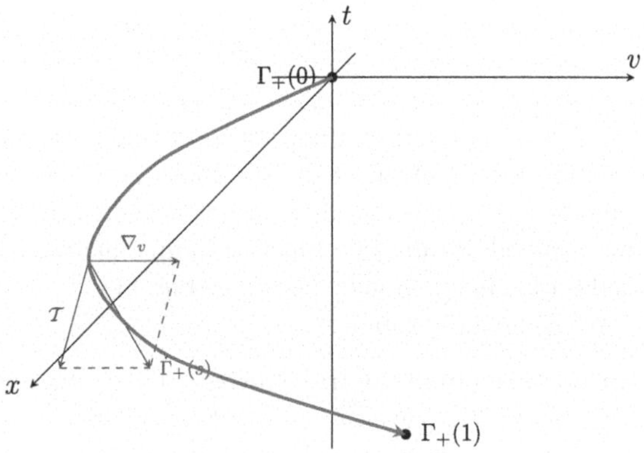

Fig. 12 Construction of the trajectories. The curve Γ_+ connects any point $\Gamma_+(0) = z_+ \in Q_r^+$ to some intermediate point $\Gamma_+(1) = z_0 \in Q_r^0$ along the vector fields $\mathcal{T}$ and ∇_v. The tangent vector $\dot\Gamma_+(s)$ is in the span of $\mathcal{T}$ and ∇_v (the third axis, for the x variable, is orthogonal to the plane of the paper in this schematic representation)

where

$$\begin{cases} \frac{\mathrm{d}}{\mathrm{d}s} V_\pm(s) = \mathsf{m}_\pm^{(0)} g_0''(s) + \mathsf{m}_\pm^{(1)} g_1''(s), \\ \frac{\mathrm{d}}{\mathrm{d}s} X_\pm(s) = \delta_\pm V_\pm(s), \\ \frac{\mathrm{d}}{\mathrm{d}s} T_\pm(s) = \delta_\pm, \end{cases} \tag{92}$$

for $\delta_\pm := t_0 - t_\pm$, some control functions $g_i \in C^2((0,1])$ with $g_i(0) = g_i'(0) = 0$ for $i = 0, 1$, and constant vectors $\mathsf{m}_\pm^{(i)} \in \mathbb{R}^d$ for $i = 0, 1$. Note that the forcing term is independent of the position and velocity. This makes it easy to solve these differential equations.

Solving these differential equations then yields

$$\begin{cases} V_\pm(s) = v_\pm + \mathsf{m}_\pm^{(0)} g_0'(s) + \mathsf{m}_\pm^{(1)} g_1'(s), \\ X_\pm(s) = x_\pm + s\delta_\pm v_\pm + \delta_\pm \left[\mathsf{m}_\pm^{(0)} g_0(s) + \mathsf{m}_\pm^{(1)} g_1(s) \right], \\ T_\pm(s) = st_0 + (1-s)t_\pm. \end{cases} \tag{93}$$

Let us denote $\mathbf{1}_d$ the d-vector of 1's, $\mathbf{0}_d$ the d-vector of 0's, Id_d the $d \times d$ identity matrix, $\mathbb{O}$ a zero block of arbitrary size, and

$$\mathsf{M}_\pm := \begin{pmatrix} \mathsf{m}_\pm^{(0)} \\ \mathsf{m}_\pm^{(1)} \end{pmatrix}, \quad \mathsf{W}(s) := \begin{pmatrix} g_0'(s)\mathrm{Id}_d & g_1'(s)\mathrm{Id}_d \\ g_0(s)\mathrm{Id}_d & g_1(s)\mathrm{Id}_d \end{pmatrix},$$

$$\mathsf{W}_\pm^\delta(s) := \begin{pmatrix} \mathrm{Id}_d & \mathbb{O} \\ \mathbb{O} & \delta_\pm \mathrm{Id}_d \end{pmatrix} \mathsf{W}(s), \quad \mathsf{Y}_\pm = \begin{pmatrix} \mathsf{y}_\pm^{(0)} \\ \mathsf{y}_\pm^{(1)} \end{pmatrix} := \begin{pmatrix} v_0 - v_\pm \\ x_0 - x_\pm - \delta_\pm v_\pm \end{pmatrix}.$$

The boundary conditions $\Gamma_\pm(1) = z_0$ impose

$$\mathsf{W}_\pm^\delta(1)\mathsf{M}_\pm = \mathsf{Y}_\pm \quad \Longrightarrow \quad \mathsf{M}_\pm = \mathsf{W}_\pm^\delta(1)^{-1}\mathsf{Y}_\pm, \tag{94}$$

provided that the Wronskian matrix $\mathsf{W}(1)$ is invertible at $s = 1$. We deduce

$$\begin{pmatrix} V_\pm(s) \\ X_\pm(s) \end{pmatrix} = \mathsf{W}_\pm^\delta(s)\mathsf{M}_\pm + \begin{pmatrix} \mathrm{Id}_d & \mathbb{O} \\ \delta_\pm s\mathrm{Id}_d & \mathrm{Id}_d \end{pmatrix} \begin{pmatrix} v_\pm \\ x_\pm \end{pmatrix}.$$

Using (94) we obtain

$$\begin{aligned} \begin{pmatrix} V_\pm(s) \\ X_\pm(s) \end{pmatrix} &= \mathsf{W}_\pm^\delta(s)\mathsf{W}_\pm^\delta(1)^{-1}\mathsf{Y}_\pm + \begin{pmatrix} \mathrm{Id}_d & \mathbb{O} \\ \delta_\pm s\mathrm{Id}_d & \mathrm{Id}_d \end{pmatrix} \begin{pmatrix} v_\pm \\ x_\pm \end{pmatrix} \\ &= W_\pm^\delta(s)\left[W_\pm^\delta(1)\right]^{-1} \begin{pmatrix} v_0 \\ x_0 \end{pmatrix} \\ &\quad - W_\pm^\delta(s)\left[W_\pm^\delta(1)\right]^{-1} \begin{pmatrix} v_\pm \\ x_\pm + \delta_\pm v_\pm \end{pmatrix} + \begin{pmatrix} \mathrm{Id}_d & \mathbb{O} \\ \delta_\pm s\mathrm{Id}_d & \mathrm{Id}_d \end{pmatrix} \begin{pmatrix} v_\pm \\ x_\pm \end{pmatrix} \\ &= \mathfrak{A}^s \begin{pmatrix} v_0 \\ x_0 \end{pmatrix} + \mathfrak{B}^s =: \Phi_\pm^s(x_0, v_0) \end{aligned} \tag{95}$$

which is an affine function with matrix $\mathfrak{A}^s := \mathsf{W}_\pm^\delta(s)[\mathsf{W}_\pm^\delta(1)]^{-1}$ and a vector $\mathfrak{B}^s$ that depend only on s and $(x_\pm, v_\pm)$.

We still have to prove that the matrix $\mathfrak{A}^s$ is invertible for $s \in (0, 1]$. If so, given $s \in (0, 1]$, the derivative along the second variable $v_\pm(s)$ of the inverse of Φ^s is

$$\nabla_{v_0} \left(\Phi^s_\pm\right)^{-1} = \mathsf{W}^\delta_\pm(1)\left[\mathsf{W}^\delta_\pm(s)\right]^{-1} \begin{pmatrix} \mathbf{1}_d \\ \mathbf{0}_d \end{pmatrix}.$$

We choose, following a new idea in the recent contribution [26],

$$\begin{cases} g_0(s) := s^{\frac{3}{2}} \cos \ln s, \\ g_1(s) := s^{\frac{3}{2}} \sin \ln s. \end{cases}$$

Then the Wronskian matrix is invertible for all $s \neq 0$ with the simple formula

$$\det \mathsf{W}(s) = \left(g_0'(s) g_1(s) - g_0(s) g_1'(s)\right)^d = s^{2d}. \tag{96}$$

Our choice of two linearly control functions g_0 and g_1 is guided by: (1) obtaining the critical exponent $-1/2$ for the forcings g_0'' and g_1'' (scaling condition), and (2) ensuring that the Wronskian matrix is invertible (independence condition). We refer to [26] for a more detailed discussion of these aspects.

Moreover, we have

$$\mathsf{W}^\delta_\pm(s) := \begin{pmatrix} s^{\frac{1}{2}} \left(\frac{3}{2} \cos \ln s - \sin \ln s\right) \mathrm{Id}_d & s^{\frac{1}{2}} \left(\frac{3}{2} \sin \ln s + \cos \ln s\right) \mathrm{Id}_d \\ s^{\frac{3}{2}} \cos \ln s \, \mathrm{Id}_d & s^{\frac{3}{2}} \sin \ln s \, \mathrm{Id}_d \end{pmatrix}$$

which yields

$$\nabla_{v_0} \left(\Phi^s_\pm\right)^{-1} = W^\delta_\pm(1)\left[W^\delta_\pm(s)\right]^{-1} \begin{pmatrix} \mathbf{1}_d \\ \mathbf{0}_d \end{pmatrix} \sim O\left(s^{-\frac{1}{2}}\right). \tag{97}$$

Note that $\nabla_{v_0}(\Phi^s_\pm)^{-1}$ remains integrable on $s \in [0, 1]$ which will be used in the next step.

Observe also that our choice of control functions g_0 and g_1 implies that their derivatives up to order two are integrable on $s \in [0, 1]$, which implies that the trajectories are bounded (with integrable tangent vector field). This allows to choose $r \in (0, 1)$ so that the trajectories we have constructed between points of $Q_r^\pm$ and Q_r^0 stays in the larger base cylinder Q_1, provided that $r \in (0, 1)$ is chosen small enough.

Note finally that the scalings $\nabla_{v_0}(\Phi^s_\pm)^{-1} \sim O(s^{-1/2})$ and $g_i''(s) = O(s^{-1/2})$ are optimal, as discussed in [26]. Trajectories achieving this scaling are called **critical kinetic trajectories** in [26].

B. Proof of the kinetic Poincaré inequality by following the trajectories

We consider the three cylinders $Q^+, Q^-, Q^0 \subset \Omega$ as in Fig. 11, and our subsolution f. Let $\varphi \in C_c^\infty(\mathbb{R}^{2d})$ be a non-negative function of the variables (x, v) (excluding the time component) with compact support in any time-slice of Q^0 and such that

$$\frac{1}{|Q_0|}\int_{Q^0} \varphi \, dz = 1.$$

Then

$$\begin{aligned}
\int_{Q^+} \left(f(z_+) - \langle f \rangle_{Q^-}\right)_+ dz_+ &\leq \frac{1}{|Q^-|}\int_{Q^+}\int_{Q^-} \left(f(z_+) - f(z_-)\right)_+ dz_- \, dz_+ \\
&\leq \frac{1}{|Q^-|}\int_{Q^+}\int_{Q^-} \left(f(z_+) - \langle f\varphi \rangle_{Q^0}\right)_+ dz_- \, dz_+ \\
&\quad + \frac{1}{|Q^-|}\int_{Q^+}\int_{Q^-} \left(\langle f\varphi \rangle_{Q^0} - f(z_-)\right)_+ dz_- \, dz_+,
\end{aligned}$$

which yields the following estimate

$$\begin{aligned}
&\int_{Q^+} \left(f(z_+) - \langle f \rangle_{Q^-}\right)_+ dz_+ \\
&\leq \frac{1}{|Q^0|}\int_{Q^+} \underbrace{\left\{\int_{Q^0} (f(z_+) - f(z_0))\varphi(x_0, v_0)\, dz_0\right\}_+}_{=:\mathcal{I}^+} dz_+ \\
&\quad + \frac{|Q^+|}{|Q^0||Q^-|}\int_{Q^-} \underbrace{\left\{\int_{Q^0} (f(z_0) - f(z_-))\varphi(x_0, v_0)\, dz_0\right\}_+}_{=:\mathcal{I}^-} dz_-, \qquad (98)
\end{aligned}$$

where $z_+ = (t_+ x_+, v_+)$, $z_- = (t_-, x_-, v_-)$ and $z_0 = (t_0, x_0, v_0)$. Note that $t_- < t_0 < t_+$. We now use the previous trajectories to estimate the right hand side.

Using the chain rule, we get

$$\begin{aligned}
\mathcal{I}^\pm &= \pm \int_{Q^0} \Big(f(z_\pm) - f(z_0)\Big)\, \varphi(x_0, v_0)\, dz_0 \\
&= \mp \int_{Q^0}\int_0^1 \frac{d}{ds} f\big(\Gamma_\pm(s)\big)\, \varphi(x_0, v_0)\, ds\, dz_0 \\
&= \mp\delta_\pm \int_{Q^0}\int_0^1 (\mathcal{T}f)\big(\Gamma_\pm(s)\big)\, \varphi(x_0, v_0)\, ds\, dz_0 \\
&\quad \mp \int_{Q^0}\int_0^1 V_\pm'(s)\cdot(\nabla_v f)\,(\Gamma_\pm(s))\, \varphi(x_0, v_0)\, ds\, dz_0,
\end{aligned}$$

which yields the following estimate

$$\mathcal{I}^{\pm} \leq \underbrace{\pm\delta_{\pm} \int_{Q^0} \int_0^1 [\nabla_v (A \nabla_v f)] \left(\Gamma_{\pm}(s)\right) \varphi(x_0, v_0) \, \mathrm{d}s \, \mathrm{d}z_0}_{=: \mathcal{I}_1^{\pm}}$$

$$\underbrace{\mp \int_{Q^0} \int_0^1 \sum_{i=0}^{1} g_i''(s) \mathrm{m}_{\pm}^{(i)} \cdot (\nabla_v f) \left(\Gamma_{\pm}(s)\right) \varphi(x_0, v_0) \, \mathrm{d}s \, \mathrm{d}z_0}_{=: \mathcal{I}_2^{\pm}} . \tag{99}$$

Note that the only difference between the two terms $\mathcal{I}^+$ and $\mathcal{I}^-$ is the role of z_0: in the former case it is the past variable, in the latter it is the future variable.

We now integrate by parts the terms $\mathcal{I}_1^{\pm}$ after a change of variables

$$(x_0, v_0) \mapsto y := \Phi_{\pm}^s(x_0, v_0)$$

for s, t_0 fixed. We recall that $\Phi_{\pm}^s$ is the affine map defined in (95), and that it is invertible for $s \neq 0$.

$$\begin{aligned}
\mathcal{I}_1^{\pm} &= \pm\delta_{\pm} \int_{Q^0} \int_0^1 [\nabla_v \cdot (A \nabla_v f)] \left(\Gamma_{\pm}(s)\right) \varphi(x_0) \, \mathrm{d}s \, \mathrm{d}z_0 \\
&= \pm\delta_{\pm} \int_{Q^0} \int_0^1 [\nabla_v \cdot (A \nabla_v f)] \left(\Phi_{\pm}^s(x_0), s t_0 + (1-s) t_{\pm}\right) \varphi(x_0) \, \mathrm{d}s \, \mathrm{d}t_0 \, \mathrm{d}x_0 \, \mathrm{d}v_0 \\
&= \pm\delta_{\pm} \int_{(\Phi_{\pm}^s \otimes \mathrm{Id})(Q^0)} \int_0^1 [\nabla_v \cdot (A \nabla_v f)] \left(y, s t_0 + (1-s) t_{\pm}\right) \varphi \\
&\quad \left(\left(\Phi_{\pm}^s\right)^{-1}(y)\right) \frac{\mathrm{d}s \, \mathrm{d}t_0 \, \mathrm{d}y}{|\det \mathfrak{A}^s|} \\
&= \pm\delta_{\pm} \int_{(\Phi_{\pm}^s \otimes \mathrm{Id})(Q^0)} \int_0^1 [A \nabla_v f] \left(y, s t_0 + (1-s) t_{\pm}\right) \nabla_v \\
&\quad \left[\varphi\left(\left(\Phi_{\pm}^s\right)^{-1}(y)\right)\right] \frac{\mathrm{d}s \, \mathrm{d}t_0 \, \mathrm{d}y}{|\det \mathfrak{A}^s|} \\
&= \pm\delta_{\pm} \int_{Q^0} \int_0^1 [A \nabla_v f] \left(\Gamma_{\pm}(s)\right) \left\{\nabla_v \left[\varphi\left(\left(\Phi_{\pm}^s\right)^{-1}(y)\right)\right]\right\}_{|y = \Phi_{\pm}^s(x_0)} \mathrm{d}s \, \mathrm{d}t_0 \, \mathrm{d}x_0 \, \mathrm{d}v_0 .
\end{aligned}$$

We then use (97) to bound

$$\begin{aligned}
\left| \left(\nabla_v \left[\varphi\left(\left(\Phi_{\pm}^s\right)^{-1}(y)\right)\right]\right)_{|y = (\Phi_{\pm}^s)^{-1}(x_0)} \right| &\lesssim_{\varphi} \left\| \nabla_{v_0} \left(\Phi_{\pm}^s\right)^{-1} \right\|_{\infty} \\
&\lesssim_{\varphi} s^{-\frac{1}{2}},
\end{aligned}$$

and deduce finally (using the bound on the matrix A)

$$\mathcal{I}_1^{\pm} \lesssim |\delta_{\pm}| \int_{Q^0} \int_0^1 |[\nabla_v f](\Gamma_{\pm}(s))| \, s^{-\frac{1}{2}} \, \mathrm{d}s \, \mathrm{d}t_0 \, \mathrm{d}x_0 \, \mathrm{d}v_0.$$

Now let us turn to $\mathcal{I}_2^{\pm}$ in (99). Using the bound $|g_i''(s)| \lesssim s^{-1/2}$, we deduce

$$\begin{aligned}\mathcal{I}_2^{\pm} &= \mp \int_0^1 \int_{Q^0} \sum_{i=0}^{1} g_i''(s) \mathrm{m}_{\pm}^{(i)} \cdot \nabla_v f(\Gamma_{\pm}(s)) \varphi(x_0, v_0) \, \mathrm{d}s \, \mathrm{d}t_0 \, \mathrm{d}x_0 \, \mathrm{d}v_0 \\ &\lesssim \int_0^1 \int_{Q^0} |(\nabla_v f)(\Gamma_{\pm}(s))| \, s^{-\frac{1}{2}} \, \mathrm{d}s \, \mathrm{d}t_0 \, \mathrm{d}x_0 \, \mathrm{d}v_0.\end{aligned}$$

It follows from (98) and (99) that we are left with estimating

$$\mathcal{J} := \int_{Q^{\pm}} \left\{\mathcal{I}_1^{\pm}\right\}_+ \mathrm{d}z_{\pm} + \int_{Q^{\pm}} \left\{\mathcal{I}_2^{\pm}\right\}_+ \mathrm{d}z_{\pm}.$$

The previous two estimates on $\mathcal{I}_1^{\pm}$ and $\mathcal{I}_2^{\pm}$ imply

$$\mathcal{J} \lesssim \bar{\mathcal{J}} := \int_{Q^{\pm}} \int_{Q^0} \int_0^1 |(\nabla_v f)(\Gamma_{\pm}(s))| \, s^{-\frac{1}{2}} \, \mathrm{d}s \, \mathrm{d}z_{\pm} \, \mathrm{d}z_0.$$

Since we are now integrating in both Q^0 and $Q_{\pm}$, we are now in a position to use a change of variable not only with the intermediate variable z_0, but also with the future/past variables $z_{\pm}$. Note that it was not possible to use the integration in $Q_{\pm}$ before because of the positive value around the Q^0 integral.

We split $\bar{\mathcal{J}}$ as follows, given $s_0 \in (0, 1)$,

$$\begin{aligned}\bar{\mathcal{J}} &= \int_{Q^{\pm}} \int_{Q^0} \int_0^1 \left(\cdots\right) \mathrm{d}s \, \mathrm{d}z_0 \, \mathrm{d}z_{\pm} \\ &= \underbrace{\int_{Q^{\pm}} \int_{Q^0} \int_0^{s_0} \left(\cdots\right) \mathrm{d}s \, \mathrm{d}z_0 \, \mathrm{d}z_{\pm}}_{=:\bar{\mathcal{J}}_1^{\pm}} + \underbrace{\int_{Q^{\pm}} \int_{Q^0} \int_{s_0}^1 \left(\cdots\right) \mathrm{d}s \, \mathrm{d}z_0 \, \mathrm{d}z_{\pm}}_{=:\bar{\mathcal{J}}_2^{\pm}}.\end{aligned}$$

The two changes of variables on each part are represented in Fig. 13.

The part $\bar{\mathcal{J}}_2^{\pm}$ is controlled by using the same change of variables $\Phi_{\pm}^s$ as before:

$$\begin{aligned}\bar{\mathcal{J}}_2^{\pm} &= \int_{Q^0} \int_{s_0}^1 s^{-\frac{1}{2}} \, |(\nabla_v f)(y, st_0 + (1-s)t_{\pm})| \, \frac{\mathrm{d}s \, \mathrm{d}t_0 \, \mathrm{d}y}{|\det \mathfrak{A}^s|} \\ &\lesssim_{s_0} \int_{Q_1} |\nabla_v f(z)| \, \mathrm{d}z, \end{aligned} \tag{100}$$

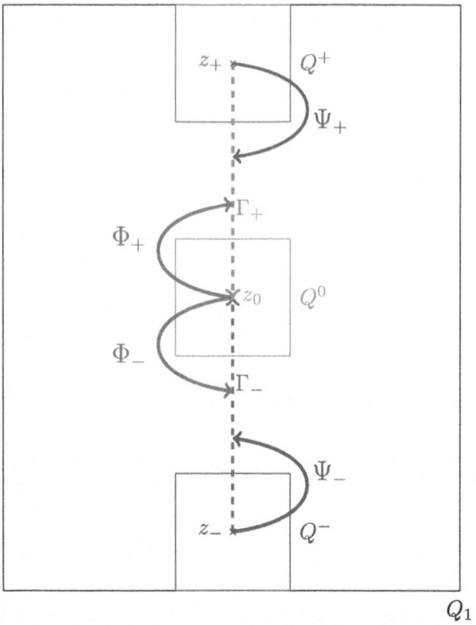

Fig. 13 The change of variables that we use in the proof for some fixed $s \in (0, 1)$. For $s \in (0, s_0)$ we use $\Psi_\pm$ that map $z_\pm$ onto $\Gamma_\pm$, whereas for $s \in (s_0, 1)$ we use $\Phi_\pm$ that map z_0 onto $\Gamma_\pm$

where we have used first that the integral in s avoids the singularity at $s = 0$, and therefore is finite.

To control the part $\bar{\mathcal{J}}_1^{\pm}$ we parametrise $\Gamma_\pm(s)$ by the $z_\pm$ coordinates, for $s \in [0, s_0]$ and $t_0, t_\pm$ all fixed:

$$(t_\pm, x_\pm, v_\pm) \to \Gamma_\pm(s) =: (t_0 s + (1 - s)t_\pm, \Psi^s_\pm(x_\pm)).$$

Since s does not approach 1, we can prove that this change of variables is not singular. The mappings $\Psi^s_\pm$ are determined by solving (95) for $x_\pm$ instead of x_0. It yields

$$\Psi^s_\pm \begin{pmatrix} v_\pm \\ x_\pm \end{pmatrix} = \mathfrak{a}^s \begin{pmatrix} v_\pm \\ x_\pm \end{pmatrix} + \mathfrak{b}^s$$

with the matrix $\mathfrak{a}^s$ and vector $\mathfrak{b}^s$ given by

$$\mathfrak{a}^s_\pm := \left(\mathrm{Id} - \mathsf{W}^\delta_\pm(s)\left[\mathsf{W}^\delta_\pm(1)\right]^{-1}\right) \begin{pmatrix} \mathrm{Id}_d & \mathbb{O} \\ \delta_\pm s \mathrm{Id}_d & \mathrm{Id}_d \end{pmatrix},$$

$$\mathfrak{b}^s_\pm := \mathsf{W}^\delta_\pm(s)\left[\mathsf{W}^\delta_\pm(1)\right]^{-1} \begin{pmatrix} v_0 \\ x_0 \end{pmatrix},$$

which depend only on s and (x_0, v_0).

We have proved that $\mathsf{W}(s) = O(s^{2d})$, which readily implies $\mathsf{W}^{\delta}_{\pm}(s) = O(s^{2d})$. Therefore, $\mathsf{W}^{\delta}_{\pm}(s)$ goes to zero as $s \to 0$. Thus, $\mathfrak{a}^{s}_{\pm}$ is the product of two matrices; the first one is close to identity for $s \in [0, 1]$ close enough to zero and so invertible in this range, while the second matrix is a lower triangular matrix with unit diagonal, which is invertible for all $s \in [0, 1]$. All in all, we deduce that $\mathfrak{a}^{s}_{\pm}$ is invertible with uniformly bounded inverse on $s \in [0, s_0]$ for $s_0 > 0$ small enough.

We apply this change of variables to estimate $\bar{\mathcal{J}}^{\pm}_1$:

$$\bar{\mathcal{J}}^{+}_1 = \int_{Q^0} \int_0^{s_0} s^{-\frac{1}{2}} \left|(\nabla_v f)\,(st_0 + (1-s)t_{\pm}, y)\right| \frac{\mathrm{d}s\,\mathrm{d}t_0\,\mathrm{d}y}{|\det \mathfrak{a}^s|} \lesssim \int_{Q_1} |\nabla_v f(z)|\,\mathrm{d}z. \tag{101}$$

The combination of (100) and (101) yields

$$\int_{Q^+} \left(f(z_+) - \langle f\rangle_{Q^-}\right)_+ \mathrm{d}z_+ \lesssim \mathcal{J} \lesssim \bar{\mathcal{J}} \lesssim \bar{\mathcal{J}}^{+}_1 + \bar{\mathcal{J}}^{-}_1 + \bar{\mathcal{J}}^{+}_2 + \bar{\mathcal{J}}^{-}_2 \lesssim \int_{Q_1} |\nabla_v f(z)|\,\mathrm{d}z.$$

This concludes the proof of (90). □

4.5 *Kinetic Versions of the Moser and Kruzhkov Methods*

A kinetic version of Moser's approach had long seemed out of reach, but the critical kinetic trajectories constructed in [26] have recently made it possible to finally extend this method to the kinetic setting. In this extension, the *critical scaling* of the trajectories is essential in deriving universal bounds on the logarithm of the solution and apply the Bombieri-Giusti result, in the spirit of the paper of Moser [77].

Regarding a kinetic version of Kruzhkov's approach, we refer to [41, 93] for an extension in the kinetic setting. Note, however, that these papers use a weaker form of the kinetic Poincaré inequality, with a complicated corrector, which adds layers of technical complexity to the proofs. These Kruzhkov-type proofs could be simplified by using the kinetic Poincaré inequality (90).

Acknowledgments These lecture notes are an expanded version of the material that the second author taught in various contexts: at Université Paris-Dauphine when he was a "PSL Visiting fellow" in 2023–2024, at the summer school "Festum Pi" in Crete in 2024, and at IHES in March 2025 as a "Cours de l'IHES". The support of these institutions, as well as the feedback from students and colleagues who attended these lectures are gratefully acknowledged. In particular, the first author typed a partial first draft of these notes based on the lectures at University Paris-Dauphine. Helpful extensive comments on a preliminary version of these notes were provided by Amélie Loher, Lukas Niebel, and an anonymous referee. The second author also greatly benefited from the discussions over the years on the De Giorgi theory with Francesca Anceschi, Helge Dietert, François Golse, Jessica Guerand, Cyril Imbert, Amélie Loher, Lukas Niebel, Sergio Polidoro, Annalaura Rebucci, Luis Silvestre, Alexis Vasseur and Rico Zacher. Several parts of the material presented in these lecture notes originate from collaborations with these colleagues. The first author acknowledges the support of a Maria Skłodowska Curie Postdoctoral Fellowship.

References

1. A. Agrachev, D. Barilari, U. Boscain, *A Comprehensive Introduction to Sub-Riemannian Geometry*, vol. 181 (Cambridge University Press, Cambridge, UK, 2019)
2. D. Albritton, S. Armstrong, J.-C. Mourrat, M. Novack, Variational methods for the kinetic Fokker-Planck equation. Anal. PDE **17**, 1953–2010 (2024)
3. A.D. Aleksandrov, Uniqueness conditions and bounds for the solution of the Dirichlet problem. Vestnik Leningrad. Univ. Ser. Mat. Meh. Astronom. **18**, 5–29 (1963)
4. R. Alexandre, L. Desvillettes, C. Villani, B. Wennberg, Entropy dissipation and long-range interactions. Arch. Ration. Mech. Anal. **152**, 327–355 (2000)
5. R. Alexandre, J. Liao, C. Lin, Some a priori estimates for the homogeneous Landau equation with soft potentials. Kinet. Relat. Models **8**, 617–650 (2015)
6. R. Alexandre, Y. Morimoto, S. Ukai, C.-J. Xu, T. Yang, Regularizing effect and local existence for the non-cutoff Boltzmann equation. Arch. Ration. Mech. Anal. **198**, 39–123 (2010)
7. F. Anceschi, H. Dietert, J. Guerand, A. Loher, C. Mouhot, A. Rebucci, *Poincaré inequality and quantitative De Giorgi method for hypoelliptic operators* (2024). Preprint arXiv:2401.12194
8. V.I. Arnold, Proof of a theorem of A. N. Kolmogorov on the preservation of conditionally periodic motions under a small perturbation of the Hamiltonian. Uspehi Mat. Nauk **18**, 13–40 (1963)
9. D.G. Aronson, J. Serrin, Local behavior of solutions of quasilinear parabolic equations. Arch. Ration. Mech. Anal. **25**, 81–122 (1967)
10. J.-P. Aubin, Un théorème de compacité. C. R. Acad. Sci. Paris **256**, 5042–5044 (1963)
11. P. Auscher, C. Imbert, L. Niebel, Weak solutions to Kolmogorov-Fokker-Planck equations: regularity, existence and uniqueness (2024). Preprint arXiv:2403.17464
12. J. Bergh, J. Löfström, *Interpolation Spaces. An Introduction*. Grundlehren Math. Wiss., vol. 223 (Springer, Cham, 1976)
13. S. Bernstein, Sur la nature analytique des solutions des équations aux dérivées partielles du second ordre. Mathematische Annalen **59**, 20–76 (1904)
14. A. Bobylev, The Fourier transform method for the Boltzmann equation for Maxwell molecules. Sov. Phys. Dokl **20**, 820–822 (1976)
15. L. Boltzmann, *Weitere studien über das wärmegleichgewicht unter gasmolekülen* (Kaiserliche-königliche Hof-und Staatsdruckerei, 1872)
16. E. Bombieri, E. Giusti, Harnack's inequality for elliptic differential equations on minimal surfaces. Invent. Math. **15**, 24–46 (1972)
17. F. Bouchut, Hypoelliptic regularity in kinetic equations. J. Math. Pures Appl. (9) **81**, 1135–1159 (2002)
18. E. Bouin, J. Dolbeault, L. Lafleche, C. Schmeiser, Hypocoercivity and sub-exponential local equilibria. Monatsh. Math. **194**, 41–65 (2021)
19. R. Caccioppoli, Limitazioni integrali per le soluzioni di un'equazione lineare ellitica a derivate parziali. Giorn. Mat. Battaglini (4), **4**(80), 186–212 (1951)
20. L. Caffarelli, C.H. Chan, A. Vasseur, Regularity theory for parabolic nonlinear integral operators. J. Am. Math. Soc. **24**, 849–869 (2011)
21. F. Chiarenza, E. Fabes, N. Garofalo, Harnack's inequality for Schrödinger operators and the continuity of solutions. Proc. Am. Math. Soc. **98**, 415–425 (1986)
22. E. De Giorgi, Sulla differenziabilità e l'analiticità delle estremali degli integrali multipli regolari. Mem. Accad. Sci. Torino. Cl. Sci. Fis. Mat. Nat. (3) **3**, 25–43 (1957)
23. C. De Lellis, L. Székelyhidi Jr, *John Nash's Nonlinear Iteration*. Lecture Notes (2016). https://www.math.ias.edu/delellis/sites/math.ias.edu.delellis/files/contr_vol_Nash_11.pdf
24. L. Desvillettes, C. Villani, On the spatially homogeneous Landau equation for hard potentials. I. Existence, uniqueness and smoothness. Commun. Partial Differ. Equ. **25**, 179–259 (2000)
25. L. Desvillettes, C. Villani, On the spatially homogeneous Landau equation for hard potentials. II. H-theorem and applications. Commun. Partial Differ. Equ. **25**, 261–298 (2000)

26. H. Dietert, C. Mouhot, L. Niebel, R. Zacher, *Critical Trajectories in Kinetic Geometry* (2025). In progress
27. L.C. Evans, *Partial Differential Equations*. Graduate Studies in Mathematics, vol. 19, 2nd edn. (American Mathematical Society, Providence, 2010)
28. L.C. Evans, R.F. Gariepy, *Measure Theory and Fine Properties of Functions*. Textbooks in Mathematics, 2nd edn. (CRC Press, Boca Raton, 2025)
29. D. Faraco, K.M. Rogers, The Sobolev norm of characteristic functions with applications to the Calderón inverse problem. Q. J. Math. **64**, 133–147 (2013)
30. N. Fournier, Uniqueness of bounded solutions for the homogeneous Landau equation with a Coulomb potential. Commun. Math. Phys. **299**, 765–782 (2010)
31. N. Fournier, H. Guérin, Well-posedness of the spatially homogeneous Landau equation for soft potentials. J. Funct. Anal. **256**, 2542–2560 (2009)
32. F. Golse, Regularity of solutions of Fokker-Planck equations with rough coefficients. Riv. Mat. Univ. Parma (N.S.) **15**, 143–173 (2024)
33. F. Golse, M.P. Gualdani, C. Imbert, A. Vasseur, Partial regularity in time for the space-homogeneous Landau equation with Coulomb potential. Ann. Sci. Éc. Norm. Supér. (4) **55**, 1575–1611 (2022)
34. F. Golse, C. Imbert, C. Mouhot, A. Vasseur, Harnack inequality for kinetic Fokker-Planck equations with rough coefficients and application to the Landau equation. Ann. Sc. Norm. Super. Pisa, Cl. Sci. (5) **19**, 253–295 (2019)
35. F. Golse, P.-L. Lions, B. Perthame, R. Sentis, Regularity of the moments of the solution of a transport equation. J. Funct. Anal. **76**, 110–125 (1988)
36. F. Golse, B. Perthame, R. Sentis, Un résultat de compacité pour les équations de transport et application au calcul de la limite de la valeur propre principale d'un opérateur de transport. C. R. Acad. Sci. Paris Sér. I Math. **301**, 341–344 (1985)
37. L. Grafakos, *Classical Fourier Analysis*. Graduate Texts in Mathematics, vol. 249, 3rd edn. (Springer, New York, 2014)
38. P.T. Gressman, R.M. Strain, Global classical solutions of the Boltzmann equation without angular cut-off. J. Am. Math. Soc. **24**, 771–847 (2011)
39. J. Guerand, Quantitative parabolic regularity à la De Giorgi, in *Séminaire Laurent Schwartz—Équations aux dérivées partielles et applications*. Année 2018–2019, Ed. Éc. Polytech., Palaiseau, [2018] ©2018, pp. Exp. No. VI, 21
40. J. Guerand, *Quantitative Regularity for Parabolic De Giorgi Classes* (2020). Preprint arXiv:1903.07421
41. J. Guerand, C. Imbert, Log-transform and the weak Harnack inequality for kinetic Fokker-Planck equations. J. Inst. Math. Jussieu **22**, 2749–2774 (2023)
42. J. Guerand, C. Mouhot, Quantitative De Giorgi methods in kinetic theory. J. Éc. polytech. Math. **9**, 1159–1181 (2022)
43. N. Guillen, L. Silvestre, *The Landau Equation Does Not Blow Up* (2025). Preprint arXiv:2311.09420
44. Q. Han, F. Lin, *Elliptic Partial Differential Equations*. Courant Lecture Notes in Mathematics, vol. 1, 2nd edn. (Courant Institute of Mathematical Sciences/American Mathematical Society, New York/Providence, 2011)
45. C. Henderson, S. Snelson, C^∞ smoothing for weak solutions of the inhomogeneous Landau equation. Arch. Ration. Mech. Anal. **236**, 113–143 (2020)
46. D. Hilbert, Mathematische Probleme. Nachrichten von der Gesellschaft der Wissenschaften zu Göttingen, Mathematisch-Physikalische Klasse **1900**, 253–297 (1900)
47. H. Holden, R. Piene (eds.), *The Abel Prize 2013–2017* (Springer, Cham, 2019)
48. L. Hörmander, Hypoelliptic second order differential equations. Acta Math. **119**, 147–171 (1967)
49. L. Hörmander, *The Analysis of Linear Partial Differential Operators. I: Distribution Theory and Fourier Analysis*. Classics in Mathematics, reprint of the 2nd edition 1990 edn. (Springer, Berlin, 2003)

50. C. Imbert, C. Mouhot, The Schauder estimate in kinetic theory with application to a toy nonlinear model. Ann. H. Lebesgue **4**, 369–405 (2021)
51. C. Imbert, C. Mouhot, L. Silvestre, Decay estimates for large velocities in the Boltzmann equation without cutoff. J. Éc. Polytech. Math. **7**, 143–184 (2020)
52. C. Imbert, L. Silvestre, The weak Harnack inequality for the Boltzmann equation without cut-off. J. Eur. Math. Soc. (JEMS) **22**, 507–592 (2020)
53. C. Imbert, L. Silvestre, The Schauder estimate for kinetic integral equations. Anal. PDE **14**, 171–204 (2021)
54. C. Imbert, L.E. Silvestre, Global regularity estimates for the Boltzmann equation without cut-off. J. Am. Math. Soc. **35**, 625–703 (2022)
55. F. John, L. Nirenberg, On functions of bounded mean oscillation. Commun. Pure Appl. Math. **14**, 415–426 (1961)
56. A.N. Kolmogorov, Zufällige Bewegungen (zur Theorie der Brownschen Bewegung). Ann. Math. (2) **35**, 116–117 (1934)
57. A.N. Kolmogorov, On conservation of conditionally periodic motions for a small change in Hamilton's function. Dokl. Akad. Nauk SSSR (N.S.) **98**, 527–530 (1954)
58. S.N. Kruzhkov, A priori bounds for generalized solutions of second-order elliptic and parabolic equations, in *Doklady Akademii Nauk*, vol. 150 (Russian Academy of Sciences, 1963), pp. 748–751
59. S.N. Kruzhkov, A priori bounds and some properties of solutions of elliptic and parabolic equations. Mat. Sb. (N.S.) **65**(107), 522–570 (1964). Translation into English by Zuckerman published by the AMS
60. N.V. Krylov, Sequences of convex functions, and estimates of the maximum of the solution of a parabolic equation. Sibirsk. Mat. Ž. **17**, 290–303, 478 (1976)
61. N.V. Krylov, *Lectures on Elliptic and Parabolic Equations in Hölder Spaces*. Graduate Student in Mathematics, vol. 12 (AMS, American Mathematical Society, Providence, 1996)
62. N.V. Krylov, M.V. Safonov, A certain property of solutions of parabolic equations with measurable coefficients. Izvestiya Rossiiskoi Akademii Nauk. Seriya Matematicheskaya **44**, 161–175 (1980)
63. L.D. Landau, Kinetic equation for the case of coulomb interaction. Phys. Z. Sowjetunion (1936)
64. E.M. Landis, *Second Order Equations of Elliptic and Parabolic Type*. Translations of Mathematical Monographs, vol. 171 (American Mathematical Society, Providence, 1998). Translated from the 1971 Russian original by Tamara Rozhkovskaya
65. J.-L. Lions, Quelques méthodes de résolution des problèmes aux limites non linéaires (Dunod/Gauthier-Villars, Paris, 1969)
66. A. Loher, Quantitative De Giorgi methods in kinetic theory for non-local operators. J. Funct. Anal. **286**, Paper No. 110312, 67 (2024)
67. J.C. Maxwell, On the dynamical theory of gases. Philos. Trans. R. Soc. Lond. 49–88 (1867)
68. C. Mooney, Hilbert's 19th problem revisited. Boll. Unione Mat. Ital. **15**, 483–501 (2022)
69. C.B. Morrey, On the solutions of quasi-linear elliptic partial differential equations. Trans. Am. Math. Soc. **43**, 126–166 (1938)
70. J. Moser, A new proof of De Giorgi's theorem concerning the regularity problem for elliptic differential equations. Commun. Pure Appl. Math. **13**, 457–468 (1960)
71. J. Moser, On Harnack's theorem for elliptic differential equations. Commun. Pure Appl. Math. **14**, 577–591 (1961)
72. J. Moser, On invariant curves of area-preserving mappings of an annulus. Nachr. Akad. Wiss. Göttingen Math.-Phys. Kl. II, **1962**, 1–20 (1962)
73. J. Moser, A Harnack inequality for parabolic differential equations. Commun. Pure Appl. Math. **17**, 101–134 (1964)
74. J. Moser, A rapidly convergent iteration method and non-linear differential equations. II. Ann. Scuola Norm. Sup. Pisa Cl. Sci. (3) **20**, 499–535 (1966)
75. J. Moser, A rapidly convergent iteration method and non-linear partial differential equations. I. Ann. Scuola Norm. Sup. Pisa Cl. Sci. (3) **20**, 265–315 (1966)

76. J. Moser, Correction to: "A Harnack inequality for parabolic differential equations". Commun. Pure Appl. Math. **20**, 231–236 (1967)
77. J. Moser, On a pointwise estimate for parabolic differential equations. Commun. Pure Appl. Math. **24**, 727–740–236 (1971)
78. C. Mouhot, R.M. Strain, Spectral gap and coercivity estimates for linearized Boltzmann collision operators without angular cutoff. J. Math. Pures Appl. (9) **87**, 515–535 (2007)
79. J. Nash, The imbedding problem for Riemannian manifolds. Ann. Math. (2) **63**, 20–63 (1956)
80. J. Nash, Continuity of solutions of parabolic and elliptic equations. Am. J. Math. **80**, 931–954 (1958)
81. L. Niebel, R. Zacher, On a kinetic Poincaré inequality and beyond. J. Funct. Anal. **289**, Paper No. 110899, 18 (2025)
82. L. Niebel, R. Zacher, A trajectorial interpretation of Moser's proof of the Harnack inequality, To appear in Annali della Scuola Normale Superiore di Pisa. Classe di Scienze. Serie V, (2025)
83. L. Nirenberg, F. Treves, Solvability of a first order linear partial differential equation. Commun. Pure Appl. Math. **16**, 331–351 (1963)
84. A. Pascucci, S. Polidoro, The Moser's iterative method for a class of ultraparabolic equations. Commun. Contemp. Math. **6**, 395–417 (2004)
85. I.G. Petrovsky, Sur l'analyticité des solutions des systèmes d'équations différentielles. Rec. Math. N.S. [Mat. Sbornik] **5**(47), 3–70 (1939)
86. J. Schauder, Über lineare elliptische Differentialgleichungen zweiter Ordnung. Mathematische Zeitschrift **38**, 257–282 (1934)
87. L. Silvestre, Upper bounds for parabolic equations and the Landau equation. J. Differ. Equ. **262**, 3034–3055 (2017)
88. D. Ter Haar, *Collected Papers of L.D. Landau* (Science Publishers, Inc., Gordon and Breach, 1965)
89. N.S. Trudinger, Pointwise estimates and quasilinear parabolic equations. Commun. Pure Appl. Math. **21**, 205–226 (1968)
90. A.F. Vasseur, The De Giorgi method for elliptic and parabolic equations and some applications. Lect. Anal. Nonlinear Partial Differ. Equ. **4**, 385–404 (2016)
91. C. Villani, *Propriétés qualitatives des solutions de l'équation de Boltzmann*. Cours Peccot du Collège de France (2003)
92. C. Villani, On the spatially homogeneous Landau equation for Maxwellian molecules. Math. Models Methods Appl. Sci. **8**, 957–983 (1998)
93. W. Wang, L. Zhang, The C^α regularity of a class of non-homogeneous ultraparabolic equations. Sci. China Ser. A: Math. **52**, 1589–1606 (2009)
94. K.-C. Wu, Global in time estimates for the spatially homogeneous Landau equation with soft potentials. J. Funct. Anal. **266**, 3134–3155 (2014)

Optimal Transport Methods for Parabolic Diffusion Equations: The JKO Scheme

Filippo Santambrogio

1 Introduction

Several evolution equations have a gradient structure, meaning that the evolution goes in the direction of the steepest descent of a given global quantity, and exploiting their variational structure provides deep insight into their behavior, allows for general proofs of existence, gives natural procedures to approximate their solutions, and suggests efficient points of view in order to analyze them. Speaking of evolutions of probability densities, the use of optimal transport and the study of those PDEs which have a variational structure w.r.t. the Wasserstein distance W_2 has been popularized in the last decades after the seminal paper by Jordan et al. [37], who introduced a discrete-in-time variational scheme to approximate the Fokker-Planck equation. These lecture notes aim to describe the scheme they introduced, nowadays called JKO scheme, to generalize it to the case where other transport costs instead of the quadratic one are used (thus obtaining different equations at the limit), and to use more recent techniques to study the discrete sequence produced by the scheme.

Section 2 will be devoted to a general introduction to the topic of gradient flows, starting from the Euclidean case and then passing to the abstract metric theory. The whole theory is extended to the case of *nonlinear gradient flows*, which correspond to the equation $x'(t) = -\nabla h^*(\nabla F(x(t)))$ instead of $x'(t) = -\nabla F(x(t))$.

Starting from Sect. 3 we want to focus on the part of the gradient flow theory most related to optimal transport, and we introduce the Monge-Kantorovich optimal transport problem, the Wasserstein spaces, and then present in details the JKO

F. Santambrogio (✉)
Universite Claude Bernard Lyon 1, ICJ UMR5208, CNRS, Ecole Centrale de Lyon, INSA Lyon, Université Jean Monnet, Villeurbanne, France
e-mail: santambrogio@math.univ-lyon1.fr

© The Author(s), under exclusive license to Springer Nature Switzerland AG 2026
N. D. Alikakos, C. Villani (eds.), *Analysis Techniques for Mathematical Physics and Geometry – and Music*, Lecture Notes in Mathematics 2390,
https://doi.org/10.1007/978-3-032-11396-2_4

scheme and the resulting PDEs, which include the Fokker-Planck and porous medium equations.

Section 4 is devoted to different techniques which can be used in order to prove the convergence of the JKO Scheme: the first requires to choose suitable interpolations of the sequence obtained in the JKO and study the relevant compactness properties in order to pass to the limit of terms, the second exploits an integral characterization of the PDE and is bases on lower semicontinuity arguments.

In Sect. 5 we introduce a new inequality involving the gradients of the transported densities and of their Kantorovich potentials, which is of special use when applied to the solution of one step of the JKO scheme. This inequality allows in this case to prove higher-order bounds on these solutions. BV bounds for the porous medium equation are proven, as well as bounds on generalized Fisher informations for some particular variants of the Fokker-Planck equations. Some of these estimates recover well-known bounds from the continuous PDE, some are more original in their general form.

Section 6 provides applications to functional inequalities, focusing on the log-Sobolev inequalities. The idea is to perform a discrete counterpart of the well-known Bakry-Emery theory, using the JKO scheme instead of a continuous flow to differentiate the relevant quantities and compare them. For simplicity, only the most classical log-Sobolev inequality is considered.

Finally, Sect. 7 uses an improvement of the inequality presented in Sect. 5 in order to prove that some second-order integral quantities pass to the limit as the time step $\tau \to 0$. This allows to prove in some regime the strong $L^2_t H^2_x$ convergence of the solution of the JKO scheme for the Fokker-Planck equation, strongly improving previous results.

The content of these notes has been presented in July 2024 in the summer school Festum Pi in Chania (Crete) and, partially, in a mini-course in the Vito Volterra Meeting in June 2024 in Rome.

The whole presentation is not fully detailed, is sometimes sketchy, points to the relevant papers on which these notes are based (see in particular [20, 21, 26, 27, 49, 50, 52] for more details), and very often regularity issues which could be important for a fully rigorous analysis are left to the reader.

2 Gradient Flows and Variants

We present here the main ideas about the general theory of gradient flows, in the Euclidan space and then in metric spaces, also considering what we call *nonlinear gradient flows*.

2.1 The Euclidean Case

We start here from the easiest framework where we can consider gradient flows, i.e. from what happens in the Euclidean space $\mathbb{R}^n$. Most of what we will say stays true in an arbitrary Hilbert space, but we will stick to the finite-dimensional case for simplicity.

Here, given a function $F : \mathbb{R}^n \to \mathbb{R}$, smooth enough, and a point $x_0 \in \mathbb{R}^n$, a gradient flow is just defined as a curve $x(t)$, with starting point at $t = 0$ given by x_0, which moves by choosing at each instant of time the direction which makes the function F decrease as much as possible. More precisely, we consider the solution of the *Cauchy Problem*

$$\begin{cases} x'(t) = -\nabla F(x(t)) & \text{for } t > 0, \\ x(0) = x_0. \end{cases} \tag{1}$$

This is a standard Cauchy problem which has a unique solution if ∇F is Lipschitz continuous, i.e. if $F \in C^{1,1}$. In some cases one could be interested in functions F which are not differentiable, and an easy example is that of convex functions. If F is convex, it could be non-differentiable, but we can replace the gradient with the subdifferential. More precisely, we can consider instead of (1), the following differential inclusion: we look for an absolutely continuous curve $x : [0, T] \to \mathbb{R}^n$ such that

$$\begin{cases} x'(t) \in -\partial F(x(t)) & \text{for a.e. } t > 0, \\ x(0) = x_0, \end{cases} \tag{2}$$

where $\partial F(x) = \{p \in \mathbb{R}^n : F(y) \geq F(x) + p \cdot (y - x) \text{ for all } y \in \mathbb{R}^n\}$. We refer to [47] for all the definitions and notions from convex analysis that one could need here and in the rest of the chapter.

A very interesting feature of these particular Cauchy problems which are gradient flows is their discretization in time. Actually, one can fix a small time step parameter $\tau > 0$ and look for a sequence of points $(x_k^\tau)_k$ defined through the iterated scheme, called *Minimizing Movement Scheme* (see [2, 25]),

$$x_{k+1}^\tau \in \operatorname{argmin}_x F(x) + \frac{|x - x_k^\tau|^2}{2\tau}. \tag{3}$$

Very mild assumptions on F (l.s.c. and some lower bounds, for instance $F(x) \geq C_1 - C_2|x|^2$) are sufficient to guarantee that these problems admit a solution for small τ. When F is convex the operator which associates with the previous point x_k^τ the new point ix_{k+1}^τ, which is unique as soon as $x \mapsto F(x) + \frac{|x - x_k^\tau|^2}{2\tau}$ is strictly convex (which is the case, for instance, if F is convex, or if $D^2 F$ is bounded from below and τ is small), is called in convex optimization *proximal operator*.

We can interpret this sequence of points as the values of the curve $x(t)$ at times $t = 0, \tau, 2\tau, \dots, k\tau, \dots$. It happens that the optimality conditions of the recursive minimization exactly give a connection between these minimization problems and the equation, since we have

$$x_{k+1}^{\tau} \in \operatorname{argmin} F(x) + \frac{|x - x_k^{\tau}|^2}{2\tau} \quad \Rightarrow \quad \nabla F(x_{k+1}^{\tau}) + \frac{x_{k+1}^{\tau} - x_k^{\tau}}{\tau} = 0,$$

i.e.

$$\frac{x_{k+1}^{\tau} - x_k^{\tau}}{\tau} = -\nabla F(x_{k+1}^{\tau}).$$

This expression is exactly the discrete-time *implicit Euler scheme* for $x' = -\nabla F(x)$! (note that in the convex non-smooth case this becomes $\frac{x_{k+1}^{\tau} - x_k^{\tau}}{\tau} \in -\partial F(x_{k+1}^{\tau})$).

Before going on, one can observe that it could be natural to replace the quadratic penalization with other penalizations, for instance another power of the distance. More generally, if $h : \mathbb{R}^n \to \mathbb{R}$ is a convex function, one could consider

$$x_{k+1}^{\tau} \in \operatorname{argmin}_x F(x) + \tau h\left(\frac{x - x_k^{\tau}}{\tau}\right),$$

which reduces to (3) when $h(z) = \frac{1}{2}|z|^2$. In this case the optimality conditions would read as

$$\nabla F(x_{k+1}^{\tau}) + \nabla h\left(\frac{x_{k+1}^{\tau} - x_k^{\tau}}{\tau}\right) = 0, \quad \text{i.e.} \quad \frac{x_{k+1}^{\tau} - x_k^{\tau}}{\tau} = \nabla h^*\left(-\nabla F(x_{k+1}^{\tau})\right)$$

(where we used the Legendre transform h^* defined via $h^*(y) = \sup_x x \cdot y - h(x)$, which is such that $\nabla h^* = (\nabla h)^{-1}$). In this case, we would obtain a discretization of the Cauchy problem

$$\begin{cases} x'(t) = \nabla h^*(-\nabla F(x(t))) & \text{for } t > 0, \\ x(0) = x_0. \end{cases} \tag{4}$$

For simplicity we will always assume that h and h^* are even (and actually, they will most often be radially symmetric—or powers of the norm—so that we can move the minus sign outside ∇h^*). This evolution equation can be called *nonlinear gradient flow*. In the case where $h(z) = \frac{1}{p}|z|^p$ it reads $x'(t) = -\nabla F(x(t)))^{q-1}$, where q is the conjugate exponent of p (i.e. when $\frac{1}{p} + \frac{1}{q} = 1$) and when we write $\mathbf{v}^{\alpha}$ for a vector $\mathbf{v}$ and an exponent $\alpha > 0$ we mean $|\mathbf{v}|^{\alpha-1}\mathbf{v}$ (if $\mathbf{v} \neq 0$, otherwise we take the value 0).

In all these cases, under some regularity assumptions on F, it is possible to prove that, for $\tau \to 0$, the sequence we found, suitably interpolated, converges to the solution of the corresponding evolution equation (see [50]).

2.2 *The Metric Case*

The iterated minimization scheme that we introduced above has another interesting feature: it even suggests how to define solutions for functions F which are only l.s.c., with no gradient at all!

Even more, a huge advantage of this discretized formulation is also that it can easily be adapted to metric spaces. Actually, if one has a metric space (X, d) and a l.s.c. function $F : X \to \mathbb{R} \cup \{+\infty\}$ (under suitable compactness assumptions to guarantee existence of the minimum), one can define (see [2, 25])

$$x_{k+1}^\tau \in \operatorname{argmin}_x F(x) + \frac{d(x, x_k^\tau)^2}{2\tau} \tag{5}$$

and study the limit as $\tau \to 0$. More generally, one can consider, for $p > 1$,

$$x_{k+1}^\tau \in \operatorname{argmin}_x F(x) + \frac{d(x, x_k^\tau)^p}{p\tau^{p-1}}, \tag{6}$$

which corresponds to the case $h(z) = \frac{1}{p}|z|^p$ previously described. We will not deal with the case of general penalizations h, since in the metric case we are only allowed to consider functions of the distance (which roughly corresponds to radial functions h). For simplicity, we will stick to the case of powers of the distance, but a similar analysis could be performed using $\tau h(d(x, x_k^\tau)/\tau)$.

Then, we use the piecewise constant interpolation

$$x^\tau(t) := x_k^\tau \quad \text{for every } t \in ((k-1)\tau, k\tau] \tag{7}$$

and study the limit of x^τ as $\tau \to 0$.

A first important point is to prove that a limit actually exists, i.e. the compactness of the curves $(x^\tau)_\tau$. We start from the easy estimate

$$F(x_{k+1}^\tau) + \frac{d(x_{k+1}^\tau, x_k^\tau)^p}{p\tau^{p-1}} \le F(x_k^\tau), \tag{8}$$

and

$$\sum_{k=0}^{l} d(x_{k+1}^\tau, x_k^\tau)^p \le p\tau^{p-1}\left(F(x_0^\tau) - F(x_{l+1}^\tau)\right) \le C\tau^{p-1}. \tag{9}$$

The Cauchy-Schwartz inequality gives, for $t < s$, $t \in [k\tau, (k+1)\tau[$ and $s \in [l\tau, (l+1)\tau[$ (hence $|l-k| \le \frac{|t-s|}{\tau} + 1$),

$$d(x^\tau(t), x^\tau(s)) \le \sum_{j=k}^{l} d(x^\tau_{j+1}, x^\tau_j) \le \left(\sum_{j=k}^{l} d(x^\tau_{k+1}, x^\tau_k)^p\right)^{1/p} (l+1-k)^{1/q}.$$

We then observe that we have $l+1-k \le \frac{|t-s|}{\tau}+1$ and this allows, together with (9), obtain

$$d(x^\tau(t), x^\tau(s)) \le \left(\sum_{j=k}^{l} d(x^\tau_{k+1}, x^\tau_k)^p\right)^{1/p} \left(\frac{|t-s|}{\tau}+1\right)^{1/q} \le C\left(|t-s|^{1/q}+\tau^{1/q}\right).$$

This provides a Hölder bound on the curves x^τ (with a negligible error of order $\tau^{1/q}$ which disappears at the limit $\tau \to 0$), and allows to extract a uniformly converging subsequence.

It is easier to see what happens if we add some structure to the metric space (X, d), in particular if we assume that (X, d) is a *geodesic space*. This requires a short discussion about curves and geodesics in metric spaces.

Curves and Geodesics in Metric Spaces

We recall that a curve ω is a continuous function defined on a interval, say $[0, 1]$, and valued in a metric space (X, d). As it is a map between metric spaces, it is meaningful to say whether it is Lipschitz or not, but its speed $\omega'(t)$ has no meaning, unless X is a vector space. Surprisingly, it is possible to give a meaning to the modulus of the velocity, $|\omega'|(t)$.

Definition 1 If $\omega : [0, 1] \to X$ is a curve valued in the metric space (X, d) we define the metric derivative of ω at time t, denoted by $|\omega'|(t)$ through

$$|\omega'|(t) := \lim_{h\to 0} \frac{d(\omega(t+h), \omega(t))}{|h|},$$

provided this limit exists.

In the spirit of Rademacher Theorem, it is possible to prove (see [3] or [51]) that, if $\omega : [0, 1] \to X$ is Lipschitz continuous, then the metric derivative $|\omega'|(t)$ exists for a.e. t. Moreover we have, for $t_0 < t_1$,

$$d(\omega(t_0), \omega(t_1)) \le \int_{t_0}^{t_1} |\omega'|(s)\, ds.$$

The same is also true for more general curves, not only Lipschitz continuous.

Definition 2 A curve $\omega : [0, 1] \to X$ is said to be *absolutely continuous* whenever there exists $g \in L^1([0, 1])$ such that $d(\omega(t_0), \omega(t_1)) \leq \int_{t_0}^{t_1} g(s)\, ds$ for every $t_0 < t_1$. The set of absolutely continuous curves defined on $[0, 1]$ and valued in X is denoted by $\mathrm{AC}(X)$.

It is well-known that every absolutely continuous curve can be reparametrized in time (through a monotone-increasing reparametrization) and become Lipschitz continuous. The existence of the metric derivative for a.e. t is also true for $\omega \in \mathrm{AC}(X)$, via this reparametrization.

Some particular curves deserve special attention.

Definition 3 A curve $\omega : [a, b] \to X$ is said to be a constant-speed geodesic between x_0 and $x_1 \in X$ if $\omega(a) = x_0$, $\omega(b) = x_1$ and $|\omega'|(t) = d(x_0, x_1)/(b-a)$ for a.e. $t \in [a, b]$.. A space (X, d) is said to be a *geodesic space* if for every $x_0, x_1 \in X$ there exists a constant-speed geodesic between x_0 and x_1.

We can now come back to the interpolation of the points obtained through the Minimizing Movement scheme (5) and note that, if (X, d) is a geodesic space, then it is possible to replace the piecewise constant interpolation with a piecewise geodesic interpolation. This means defining a curve $x^\tau : [0, T] \to X$ such that $x^\tau(k\tau) = x_k^\tau$ and such that x^τ restricted to any interval $[k\tau, (k+1)\tau]$ is a constant-speed geodesic with speed equal to $d(x_k^\tau, x_{k+1}^\tau)/\tau$. Then, the bounds we proved above read as

$$\int_0^T |x'|(t)^p \mathrm{d}t \leq p(F(x_0) - F(x(T)) \leq C.$$

This proves that the curves x^τ are bounded in $W^{1,p}([0, T]; X)$ and implies equicontinuity since, exactly as in the Euclidean case, one has $W^{1,p} \subset C^{0,\alpha}$ with $\alpha = 1 - 1/p = 1/q$.

The next question is how to characterize the limit curve obtained when $\tau \to 0$, and in particular how to express the fact that it is a gradient flow of the function F. Of course, one cannot try to prove the equality $x' = -\nabla F(x)$, just because neither the left-hand side nor the right-hand side have a meaning in a metric space!

If the space X, the distance d, and the functional F are explicitly known, in some cases it is possible to pass to the limit the optimality conditions of each optimization problem in the discretized scheme, and characterize the limit curves (or the limit curve) $x(t)$. It will be possible to do so in the framework of probability measures, as it will be discussed in Sects. 3 and 4, but not in general. Indeed, without a little bit of (differential) structure on the space X, it is essentially impossible to do so. Hence, if we want to develop a general theory for gradient flows in metric spaces, finer tools are needed. In particular, we need to characterize the solutions of $x' = -\nabla F(x)$ (or $x' \in -\partial F(x)$) by only using metric quantities (in particular, avoiding derivatives, gradients, and more generally vectors). The book by Ambrosio-Gigli-Savaré [4], and in particular its first part (the second being devoted to the space of probability measures) exactly aims at doing so.

The idea is to present alternative characterizations of gradient flows in the smooth Euclidean case, which can be used as a definition of gradient flow in the metric case, since all the quantities which are involved have their metric counterpart.

The first observation is the following: thanks to the definition of Legendre transform h^*, the condition $x'(t) = \nabla h^*(-\nabla F(x(t))$ is equivalent to $h(x'(t)) + h^*(-\nabla F(x(t)) = -x'(t) \cdot \nabla F(x(t))$. Since the inequality $h(x) + h^*(y) \geq x \cdot y$ is always true, we can also say that it is equivalent to the inequality

$$h(x'(t)) + h^*(-\nabla F(x(t))) \leq -x'(t) \cdot \nabla F(x(t)) = -\frac{\mathrm{d}}{\mathrm{d}t} F(x(t)),$$

and this inequality for a.e. t is also equivalent to its integral version

$$\int_0^T h(x'(t))\mathrm{d}t + \int_0^T h^*(-\nabla F(x(t)))dt + F(x(T)) \leq F(x_0).$$

In the case where $h(z) = \frac{1}{p}|z|^p$ is a power this can be written as

$$\frac{1}{p}\int_0^T |x'(t)|^p dt + \frac{1}{q}\int_0^T |\nabla F(x(t)|^q dt + F(x(T)) \leq F(x_0).$$

This condition, called EDI (*Energy Dissipation Inequality*) characterizes gradient flows (or their nonlinear counterpart) in the smooth Euclidean case. When h and h^* are radial, it can also be taken as a definition in the metric case, as all its terms have a meaning in the metric setting. We already saw how to define $|x'|$, let us see how to define the norm of the gradient.

Slope and Modulus of the Gradient

Many definitions of the modulus of the gradient of a function F defined over a metric space are possible. The easiest possible choice is the *local Lipschitz constant*

$$|\nabla F|(x) := \limsup_{y \to x} \frac{|F(x) - F(y)|}{d(x, y)}; \tag{10}$$

another is the *descending slope* (we will often say just *slope*), which is a notion more adapted to the minimization of a function than to its maximization, and hence reasonable for lower semi-continuous functions:

$$|\nabla^- F|(x) := \limsup_{y \to x} \frac{[F(x) - F(y)]_+}{d(x, y)}$$

(note that the slope vanishes at every local minimum point).

In the general theory of Gradient Flows in metric spaces ([4]), another characterization, different from the EDI, is proposed in order to cope with uniqueness and stability results. This is only done in the case $p = 2$ (i.e. for linear gradient flows) and it is based on the following observation: if $F : \mathbb{R}^d \to \mathbb{R}$ is convex, then the inequality

$$F(y) \geq F(x) + p \cdot (y - x) \quad \text{for all } y \in \mathbb{R}^d$$

characterizes (by definition) the vectors $p \in \partial F(x)$ and, if $F \in C^1$, it is only satisfied for $p = \nabla F(x)$. We can pick a curve $x(t)$ and a point y and compute

$$\frac{\mathrm{d}}{\mathrm{d}t}\frac{1}{2}|x(t) - y|^2 = (y - x(t)) \cdot (-x'(t)).$$

Consequently, imposing

$$\frac{\mathrm{d}}{\mathrm{d}t}\frac{1}{2}|x(t) - y|^2 \leq F(y) - F(x(t)$$

(for all y) will be equivalent to $-x'(t) \in -\partial F(x(t))$. This will provide a second characterization (called EVI, *Evolution Variational Inequality*) of gradient flows in a metric environment. Again, all the terms appearing in the above inequality have a metric counterpart (only squared distances and derivatives w.r.t. time appear).

If in the smooth Euclidean case the two notions (EDI and EVI) finally coincide, it is now well-known that in general they are quite different. In particular, it is quite easy (and we will give some hints on how to do) to prove existence of a gradient flow in the EDI sense, but in general uniqueness fails. An example can be found, for instance, in [48]. On the other hand, the EVI notion allows quite easily to prove uniqueness, but the existence of an EVI gradient flow is harder to obtain and, by the way, only holds when F satisfies some convexity assumption (more precisely: it can be proven that whenever F is such that an EVI gradient flow exists for any starting point x_0, then F is *geodesically convex*, a notion on which we wil come back later, see [24]).

2.3 Existence of a Gradient Flow

We can easily understand that, even if the estimate (8) is enough to provide compactness, it will never be enough to characterize the limit curve (indeed, it is satisfied by any discrete evolution where x^τ_{k+1} gives a better value than x^τ_k, without any need for optimality). Hence, we will never obtain either of the two formulations—EDI or EVI—of metric gradient flows.

In order to improve the result, we should exploit how much x^τ_{k+1} is better than x^τ_k. An idea due to De Giorgi allows to obtain the desired result, via a "variational interpolation" between the points x^τ_k and x^τ_{k+1}. In order to do so, once we fix x^τ_k, for every $\theta \in]0, 1]$, we consider the problem

$$m(\theta) := \min_x \quad F(x) + \frac{d(x, x_k^\tau)^p}{p(\theta\tau)^{p-1}}$$

and call $x(\theta)$ any minimizer for this problem, and $m(\theta)$ the minimal value. It is clear that, for $\theta \to 0^+$, we have $x(\theta) \to x_k^\tau$ and $m(\theta) \to F(x_k^\tau)$, and that, for $\theta = 1$, we get back to the original problem with minimizer x_{k+1}^τ. Moreover, the function m is non-increasing and hence a.e. differentiable (actually, we can even prove that it is locally semiconcave). Its derivative $m'(\theta)$ is given by the derivative of the function $\theta \mapsto F(x) + \frac{d^p(x, x_k^\tau)}{p(\theta\tau)^{p-1}}$, computed at the optimal point $x = x(\theta)$ (the existence of $m'(\theta)$ implies that this derivative is the same at every minimal point $x(\theta)$). Hence, a quick computation shows that we have

$$m'(\theta) = -\frac{d^p(x(\theta), x_k^\tau)}{q\theta^p\tau^{p-1}}.$$

Moreover, the optimality condition for the minimization problem with $\theta > 0$ easily show that we have

$$|\nabla^- F|(x(\theta)) \leq \left(\frac{d(x(\theta), x_k^\tau)}{\theta\tau}\right)^{p-1},$$

which wan also be written as $|\nabla^- F|(x(\theta))^q \leq \left(\frac{d(x(\theta), x_k^\tau)}{\theta\tau}\right)^p$.

We now come back to the function m and use

$$m(0) - m(1) = -\int_0^1 m'(\theta)\, d\theta$$

together with

$$-m'(\theta) = \frac{d(x(\theta), x_k^\tau)^p}{q\theta^p\tau^{p-1}} \geq \frac{\tau}{q}|\nabla^- F(x(\theta))|^q. \tag{11}$$

Hence, we get an improved version of (8):

$$F(x_{k+1}^\tau) + \frac{d(x_{k+1}^\tau, x_k^\tau)^p}{p\tau^{p-1}} \leq F(x_k^\tau) - \frac{\tau}{q}\int_0^1 |\nabla^- F(x(\theta))|^q d\theta.$$

If we sum up for $k = 0, 1, 2, \ldots$ and then take the limit $\tau \to 0$, we can prove, for the limit curve, the inequality

$$F(x(T)) + \frac{1}{p}\int_0^T |x'|(t)^p dt + \frac{1}{q}\int_0^T |\nabla^- F(x(t))|^q dt \leq F(x(0)), \tag{12}$$

under some suitable assumptions that we must select. In particular, we need lower-semicontinuity of F in order to handle the term $F(x^{\tau}_{k+1})$ (which will become $F(x(t))$ at the limit), but we also need lower-semicontinuity of the slope $|\nabla^- F|$ in order to handle the corresponding term.

3 Introduction to OT, the Wasserstein Space, and the PDE of Wasserstein Gradient Flows

This section will contain a gentle introduction to the topic of optimal transport and the distances it induces on the space of probability measures, moving forward the applications of these notions to those PDEs which can be expressed as a gradient flow in such a space.

3.1 Monge and Kantorovich Problems

The whole theory of optimal transport was born with the following problem proposed by Monge in 1781 ([45]), that we will express in modern mathematical language (see [49, 54, 55]). Given two probability measures μ, ν on a space X, find a map $T = X \to X$ such that $T_{\#}\mu = \nu$ which minimizes a certain cost. The condition $T_{\#}\mu = \nu$ means that the image measure of μ through T is ν, where $T_{\#}\mu$ is the measure characterized by

$$(T_{\#}\mu)(A) = \mu(T^{-1}(A)) \quad \text{for every measurable set } A,$$

$$\text{or} \quad \int_Y \phi \, \mathrm{d}(T_{\#}\mu) = \int_X \phi \circ T \, \mathrm{d}\mu \quad \text{for every measurable function } \phi.$$

The cost of a transport map T is evaluated in terms of a transport cost $c : X \times X \to \mathbb{R}$ and we minimize the quantity

$$\int_X c(x, T(x)) \, \mathrm{d}\mu(x)$$

among all the maps satisfying $T_{\#}\mu = \nu$. This means that we have a collection of particles, distributed according to μ, that have to be moved, so that they arrange according to a new distribution ν. The cost $c(x, y)$ represents the cost to move a unit mass from x to y. The map T describes the movement, and $T(x)$ represents the destination of the particle originally located at x.

The problem of Monge has stayed with no solution (does a minimizer exist? how to characterize it?...) till the progress made in the 1940s with the work by Kantorovich [38]. For simplicity of the exposition, we will suppose that c is continuous and symmetric: $c(x, y) = c(y, x)$ and X to be a compact metric space.

The formulation proposed by Kantorovich of the problem raised by Monge is the following: consider the problem

$$(KP) \quad \min\left\{\int_{X\times X} c \, \mathrm{d}\gamma \ : \ \gamma \in \Pi(\mu, \nu)\right\}, \tag{13}$$

where $\Pi(\mu, \nu)$ is the set of the so-called *transport plans*, i.e.

$$\Pi(\mu, \nu) = \{\gamma \in \mathcal{P}(X \times X) : (\pi_0)_{\#}\gamma = \mu, \ (\pi_1)_{\#}\gamma = \nu, \}$$

where π_0 and π_1 are the two projections of $X \times X$ onto its factors. These probability measures over $X \times X$ are an alternative way to describe the displacement of the particles of μ: instead of saying, for each x, which is the destination $T(x)$ of the particle originally located at x, we say for each pair (x, y) how many particles go from x to y. It is clear that this description allows for more general movements, since from a single point x particles can a priori move to different destinations y. If multiple destinations really occur, then this movement cannot be described through a map T. It can be easily checked that if $(id, T)_{\#}\mu$ belongs to $\Pi(\mu, \nu)$ then T pushes μ onto ν (i.e. $T_{\#}\mu = \nu$) and the functional takes the form $\int c(x, T(x))\mathrm{d}\mu(x)$, thus generalizing Monge's problem.

The minimizers for this problem are called *optimal transport plans* between μ and ν. Should γ be of the form $(id, T)_{\#}\mu$ for a measurable map $T : X \to X$ (i.e. when no splitting of the mass occurs), the map T would be called *optimal transport map* from μ to ν.

This generalized problem by Kantorovich is much easier to handle than the original one proposed by Monge: for instance in the Monge case we would need existence of at least a map T satisfying the constraints. This is not verified when $\mu = \delta_0$, if ν is not a single Dirac mass. On the contrary, there always exists at least a transport plan in $\Pi(\mu, \nu)$ (for instance we always have $\mu\otimes\nu \in \Pi(\mu, \nu)$). Moreover, one can state that (KP) is the relaxation of the original problem by Monge: if one considers the problem in the same setting, where the competitors are transport plans, but sets the functional at $+\infty$ on all the plans that are not of the form $(id, T)_{\#}\mu$, then one has a functional on $\Pi(\mu, \nu)$ whose relaxation (in the sense of the largest lower-semicontinuous functional smaller than the given one) is the functional in (KP) (see for instance Section 1.5 in [49]).

Anyway, it is important to note that an easy use of the direct method of the Calculus of Variations (i.e. taking a minimizing sequence, saying that it is compact in some topology—here it is the weak convergence of probability measures—finding a limit, and proving semicontinuity, or continuity, of the functional we minimize, so that the limit is a minimizer) proves that a minimum does exist. As a consequence, if one is interested in the problem of Monge, the question may become "does this minimizer come from a transport map T?" (note, on the contrary, that directly attacking by compactness and semicontinuity Monge's formulation is out of reach, because of the non-linearity of the constraint $T_{\#}\mu = \nu$, which is not closed under weak convergence).

Since the problem (KP) is a linear optimization under linear constraints, an important tool will be duality theory, which is typically used for convex problems. We will find a dual problem (DP) for (KP) and exploit the relations between dual and primal. A formal procedure to find the dual problem passes through an inf-sup exchange.

First express the constraint $\gamma \in \Pi(\mu, \nu)$ in the following way : notice that, if γ is a non-negative measure on $X \times X$, then we have

$$\sup_{\varphi, \psi} \int \varphi \, d\mu + \int \psi \, d\nu - \int (\varphi(x) + \psi(y)) \, d\gamma = \begin{cases} 0 & \text{if } \gamma \in \Pi(\mu, \nu) \\ +\infty & \text{otherwise} \end{cases}.$$

Hence, one can remove the constraints on γ by adding the previous sup, since if they are satisfied nothing has been added and if they are not one gets $+\infty$ and this will be avoided by the minimization. We may look at the problem we get and interchange the inf in γ and the sup in φ, ψ: for simplicity we will write $\varphi \oplus \psi$ for the function on $X \times X$ defined via $(x, y) \mapsto \varphi(x) + \psi(y)$ and

$$\min_{\gamma \geq 0} \int c \, d\gamma + \sup_{\varphi, \psi} \left(\int \varphi \, d\mu + \int \psi \, d\nu - \int \varphi \oplus \psi \, d\gamma \right)$$

becomes

$$\sup_{\varphi, \psi} \int \varphi \, d\mu + \int \psi \, d\nu + \inf_{\gamma \geq 0} \int (c - \varphi \oplus \psi) \, d\gamma.$$

Obviously it is not always possible to exchange inf and sup, and the main tools to do this come from convex analysis. We refer to [49], Section 1.6.3 for a simple proof of this fact, or to [54], where the proof is based on Flenchel-Rockafellar duality (see, for instance, [30] or [51]). Anyway, we insist that in this case it is true that $\inf \sup = \sup \inf$.

Afterwards, one can re-write the inf in γ as a constraint on φ and ψ, since one has

$$\inf_{\gamma \geq 0} \int (c(x, y) - (\varphi(x) + \psi(y))) \, d\gamma = \begin{cases} 0 & \text{if } \varphi \oplus \psi \leq c \text{ on } X \times X \\ -\infty & \text{otherwise} \end{cases}.$$

This leads to the following dual optimization problem: given the two probabilities μ and ν and the cost function $c : X \times X \to [0, +\infty]$ we consider the problem

$$(DP) \quad \max \left\{ \int_X \varphi \, d\mu + \int_X \psi \, d\nu \ : \ \varphi, \psi \in C(X), \ \varphi \oplus \psi \leq c \text{ on } X \times X \right\}. \tag{14}$$

This problem does not admit a straightforward existence result, since the class of admissible functions lacks compactness. Yet, it is possible to see that we can restrict the maximization to functions which are $c-$concave. Indeed, we can define a notion of $c-$transform similar to that of Legendre transform and use it as follows

Definition 4 Given a function $\chi : X \to \overline{\mathbb{R}}$ we define its *$c-$transform* (or $c-$conjugate function) by

$$\chi^c(y) = \inf_{x \in X} c(x, y) - \chi(x).$$

Moreover, we say that a function ψ is *$c-$concave* if there exists χ such that $\psi = \chi^c$.

It is quite easy to realize that, given a pair (ϕ, ψ) in the maximization problem (DP), one can always replace it with (ϕ, ϕ^c), and then with (ϕ^{cc}, ϕ^c), and the constraints are preserved and the integrals increased. Moreover, any $c-$concave function shares the same modulus of continuity of the cost c. Hence, if c is uniformly continuous (which is always the case whenever c is continuous and X is compact), one can get a uniform modulus of continuity for suitable minimizing sequences and then prove existence for (DP), by applying Ascoli-Arzelà's Theorem. The functions (φ, ψ) realizing the maximum in (DP) are called *Kantorovich potentials* (we say that φ is the Kantorovich potential for the transport from μ to ν and that ψ is the one for the transport from ν to μ).

We look at two interesting cases. When $c(x, y)$ is equal to the distance $d(x, y)$ on the metric space X, then we can easily see that $c-$concave functions coincide with 1-Lipschitz functions and

$$\varphi \in \mathrm{Lip}_1 \Rightarrow \varphi^c = -\varphi. \tag{15}$$

Another interesting case is the case where $X = \Omega \subset \mathbb{R}^d$ and $c(x, y) = \frac{1}{2}|x - y|^2$. In this case if a function φ is $c-$concave then $x \mapsto \frac{x^2}{2} - \varphi(x)$ is a convex function (and this is an equivalence if $X = \mathbb{R}^d$).

A consequence of (15) is that, in the case where $c = d$, the duality result may be re-written as

$$\min{(KP)} = \max{(DP)} = \max_{\varphi \in \mathrm{Lip}_1} \int_X \varphi \, \mathrm{d}(\mu - \nu). \tag{16}$$

We now concentrate on the case when X is a domain $\Omega \subset \mathbb{R}^d$, and look at the existence of optimal transport maps T. We will use costs c of the form $c(x, y) = h(x - y)$ for a strictly convex function h.

The main tool is the duality result. If we have equality between the minimum of (KP) and the maximum of (DP) and both extremal values are realized, one can consider an optimal transport plan γ and a Kantorovich potential φ and write

$$\varphi(x) + \varphi^c(y) \leq c(x, y) \text{ on } X \times X \ \text{ and } \ \varphi(x) + \varphi^c(y) = c(x, y) \text{ on } \operatorname{spt} \gamma.$$

The equality on $\operatorname{spt}\gamma$ is a consequence of the inequality which is valid everywhere and of

$$\int c \, d\gamma = \int \varphi \, d\mu + \int \varphi^c \, d\nu = \int (\varphi(x) + \varphi^c(y)) \, d\gamma,$$

which implies equality γ–a.e. These functions being continuous, the equality passes to the support of the measure. Once we have that, let us fix a point $(x_0, y_0) \in \operatorname{spt}\gamma$. One may deduce from the previous computations that

$$x \mapsto \varphi(x) - h(x - y_0) \quad \text{is maximal at } x = x_0$$

and, if φ is differentiable at x_0, one gets $\nabla\varphi(x_0) = \nabla h(x_0 - y_0)$ i.e. $y_0 = x_0 - \nabla h^*(\nabla\varphi(x_0))$. This shows that only one unique point y_0 can be such that $(x_0, y_0) \in \operatorname{spt}\gamma$, which means that γ is concentrated on a graph.

Theorem 1 *Assume $c(x, y) = h(x - y)$ for a strictly convex function h. Given μ and ν probability measures on a domain $\Omega \subset \mathbb{R}^d$ there exists an optimal transport plan γ; it is unique and of the form $(id, T)_{\#}\mu$, provided μ is absolutely continuous. Moreover there exists also at least a Kantorovich potential φ, and the gradient $\nabla\varphi$ is uniquely determined μ–a.e. The optimal transport map T and the potential φ are linked by $T(x) = x - \nabla h^*(\nabla\varphi(x))$.*

In the case $h(z) = \frac{1}{2}|z|^2$ we have $T(x) = x - \nabla\varphi(x) = \nabla u(x)$, where $u(x) := \frac{x^2}{2} - \varphi(x)$ is a convex function.

The fact that the optimal map in the quadratic case is the gradient of a convex function is a well-known theorem, due to Brenier ([9, 10], see also [32–34, 42]). The only technical point above is the μ-a.e. differentiability of the potential φ, but φ has the same regularity of c, and Lipschitz functions are differentiable Lebesgue-a.e. (which explains the assumption that μ should be absolutely continuous).

In Theorem 1 only the part concerning the optimal map T is not symmetric in μ and ν: hence the uniqueness (up to additive constants) of the Kantorovich potential is true even if it ν (and not μ) has positive density a.e. (since one can retrieve φ from φ^c and viceversa).

3.2 *The Wasserstein Distances*

Starting from the values of the problem (KP) we can define a set of distances over $\mathcal{P}(X)$.

We mainly consider costs of the form $c(x, y) = |x - y|^p$ in $X = \Omega \subset \mathbb{R}^d$, but the analysis can be adapted to a power of the distance in a more general metric space X. The exponent p will always be taken in $[1, +\infty)$ (we will not discuss the case $p = \infty$) in order to take advantage of the properties of the L^p norms. When Ω is unbounded we need to restrict our analysis to the following set of probabilities

$$\mathcal{P}_p(\Omega) := \left\{ \mu \in \mathcal{P}(\Omega) \, : \, \int_\Omega |x|^p \, \mathrm{d}\mu(x) < +\infty \right\}.$$

The distances that we want to consider are defined in the following way: for any $p \in [1, +\infty)$ set

$$W_p(\mu, \nu) = \big(\min{(KP)} \text{ with } c(x, y) = |x - y|^p\big)^{1/p}.$$

The quantities that we obtain in this way are called Wasserstein distances (this is the standard name nowadays, even if this choice is highly debated). They are very important in many fields of applications and they seem a natural way to describe distances between equal amounts of mass distributed on a same space.

We summarize here some properties of these distances. Most proofs can be found in [49], chapter 5, or in [54] or [4].

Theorem 2 *The quantity W_p defined above is a distance over $\mathcal{P}_p(\Omega)$.*

If X is compact, for any $p \geq 1$ the function W_p is a distance over $\mathcal{P}(X)$ and the convergence with respect to this distance is equivalent to the weak convergence of probability measures.

Without assuming compactness of X, for any $p \geq 1$ the function W_p is a distance over $\mathcal{P}_p(X)$ and, given a measure μ and a sequence $(\mu_n)_n$ in $\mathbb{W}_p(X)$, the following are equivalent:

- *$\mu_n \to \mu$ according to W_p;*
- *$\mu_n \rightharpoonup \mu$ and $\int |x|^p \mathrm{d}\mu_n \to \int |x|^p \mathrm{d}\mu$;*
- *$\int_X \phi \, \mathrm{d}\mu_n \to \int_X \phi \, \mathrm{d}\mu$ for any $\phi \in C^0(X)$ whose growth is at most of order p (i.e. there exist constants A and B depending on ϕ such that $|\phi(x)| \leq A + Bd(x, x_0)^p$ for any x).*

To prove that the convergence according to W_p is equivalent to weak convergence (in the compact case) one first establish this result for $p = 1$, through the use of the duality formula in the form (16). Then it is possible to use the inequalities between the distances W_p to extend the result to a general p. Indeed, as a consequence of Hölder (or Jensen) inequalities, the Wasserstein distances are always ordered, i.e. $W_{p_1} \leq W_{p_2}$ if $p_1 \leq p_2$. Reversed inequalities are possible only if Ω is bounded, and in this case we have, if set $D = \mathrm{diam}(\Omega)$, for $p_1 \leq p_2$, we have

$$W_{p_1} \leq W_{p_2} \leq D^{1-p_1/p_2} W_{p_1}^{p_1/p_2}.$$

After this short introduction to the metric space $\mathbb{W}_p := (\mathcal{P}_p(X), W_p)$ and its topology, since we focus on the Euclidean case where the metric space X is a domain $\Omega \subset \mathbb{R}^d$, we want to study the curves valued in $\mathbb{W}_p(\Omega)$ in connections with PDEs.

The main point is to identify the absolutely continuous curves in the space $\mathbb{W}_p(\Omega)$ with solutions of the continuity equation $\partial_t \varrho_t + \nabla \cdot (\mathbf{v}_t \varrho_t) = 0$ with L^p vector fields $\mathbf{v}_t$. Moreover, we want to connect the L^p norm of $\mathbf{v}_t$ with the metric derivative $|\varrho'|(t)$.

We recall that standard considerations from fluid mechanics tell us that the continuity equation above may be interpreted as the equation ruling the evolution of the density ϱ_t of a family of particles initially distributed according to μ_0 and each of which follows the flow

$$\begin{cases} y_x'(t) = \mathbf{v}_t(y_x(t)) \\ y_x(0) = x. \end{cases}$$

The main theorem in this setting (originally proven in [4]) relates absolutely continuous curves in $\mathbb{W}_p$ with solutions of the continuity equation:

Theorem 3 *Let $(\varrho_t)_{t\in[0,1]}$ be an absolutely continuous curve in $\mathbb{W}_p(\Omega)$ (for $p > 1$ and $\Omega \subset \mathbb{R}^d$ an open domain). Then for a.e. $t \in [0, 1]$ there exists a vector field $\mathbf{v}_t \in L^p(\varrho_t; \mathbb{R}^d)$ such that*

- *the continuity equation $\partial_t \varrho_t + \nabla \cdot (\mathbf{v}_t \varrho_t) = 0$ is satisfied in the sense of distributions,*
- *for a.e. t we have $||\mathbf{v}_t||_{L^p(\varrho_t)} \leq |\varrho'|(t)$ (where $|\varrho'|(t)$ denotes the metric derivative at time t of the curve $t \mapsto \varrho_t$, w.r.t. the distance W_p);*

Conversely, if $(\varrho_t)_{t\in[0,1]}$ is a family of measures in $\mathcal{P}_p(\Omega)$ and for each t we have a vector field $\mathbf{v}_t \in L^p(\varrho_t; \mathbb{R}^d)$ with $\int_0^1 ||\mathbf{v}_t||_{L^p(\varrho_t)}\, dt < +\infty$ solving $\partial_t \varrho_t + \nabla \cdot (\mathbf{v}_t \varrho_t) = 0$, then $(\varrho_t)_t$ is absolutely continuous in $\mathbb{W}_p(\Omega)$ and for a.e. t we have $|\varrho'|(t) \leq ||\mathbf{v}_t||_{L^p(\varrho_t)}$.

Note that, as a consequence of the second part of the statement, the vector field $\mathbf{v}_t$ introduced in the first part must a posteriori satisfy $||\mathbf{v}_t||_{L^p(\varrho_t)} = |\varrho'|(t)$.

We will not give the proof of this theorem, which is quite involved. The main reference is [4] (but the reader can also find alternative proofs in Chapter 5 of [49], in the case where Ω is compact). Yet, if the reader wants an idea of the reason for this theorem to be true, it is possible to start from the case of two time steps: there are two measures ϱ_t and ϱ_{t+h} and there are several ways for moving the particles so as to reconstruct the latter from the former. It is exactly as when we look for a transport. One of these transports is optimal in the sense that it minimizes $\int |T(x)-x|^p d\varrho_t(x)$ and the value of this integral equals $W_p^p(\varrho_t, \varrho_{t+h})$. If we call $\mathbf{v}_t(x)$ the "discrete velocity of the particle located at x at time t, i.e. $\mathbf{v}_t(x) = (T(x) - x)/h$, one has $||\mathbf{v}_t||_{L^p(\varrho_t)} = \frac{1}{h} W_p(\varrho_t, \varrho_{t+h})$. We can easily guess that, at least formally, the result of the previous theorem can be obtained as a limit as $h \to 0$.

Because of the role played by the L^p norm of the velocity field in the continuity equation it turns out that the following functional is extremely useful in optimal transport theory:

$$(\varrho, \mathbf{v}) \mapsto \int_0^1 \int_\Omega |\mathbf{v}_t|^p d\varrho_t$$

to be computed on pairs $(\varrho, \mathbf{v})$ such that $\partial_t \varrho_t + \nabla \cdot (\mathbf{v}_t \varrho_t) = 0$. In particular, we have, for $p > 1$

$$W_p^p(\mu, \nu) = \min \left\{ \int_0^1 \int_\Omega |\mathbf{v}_t|^p \mathrm{d}\varrho_t \, \mathrm{d}t \ : \ \partial_t \varrho_t + \nabla \cdot (\mathbf{v}_t \varrho_t) = 0, \ \varrho_0 = \mu, \ \varrho_1 = \nu \right\}. \tag{17}$$

This is what is usually called *Benamou-Brenier formula* ([8]).

On the other hand, this minimization problem in the variables $(\varrho_t, \mathbf{v}_t)$ has non-linear constraints (due to the product $\mathbf{v}_t \varrho_t$) and the functional is non-convex (since $(s, z) \mapsto s|z|^p$ is not convex). Yet, it is possible to transform it into a convex problem. For this, it is sufficient to switch variables, from $(\varrho_t, \mathbf{v}_t)$ into (ϱ_t, E_t) where $E_t = \mathbf{v}_t \varrho_t$. We then define the functional

$$\mathcal{B}_p(\varrho, E) := \begin{cases} \int_0^1 \int_\Omega |\mathbf{v}_t|^p \mathrm{d}\varrho_t \, \mathrm{d}t & \text{if } E = \varrho \mathbf{v}, \\ +\infty & \text{if not.} \end{cases}$$

We will not provide details here but it is possible to prove that $\mathcal{B}_p$ is convex and lower-semicontinuous for the weak convergence as measures of its variables. Its convexity is the key point of the numerical methods (as it was first done in [8]).

Once we know about curves in their generality, it is interesting to think about geodesics. The following result is a characterization of geodesics in $W_p(\Omega)$ when Ω is a convex domain in $\mathbb{R}^d$. This procedure is also known as *McCann's displacement interpolation* (see [43]).

Theorem 4 *If $\Omega \subset \mathbb{R}^d$ is convex, then all the spaces $\mathbb{W}_p(\Omega)$ are length spaces and if μ and ν belong to $\mathbb{W}_p(\Omega)$, and γ is an optimal transport plan from μ to ν for the cost $c_p(x, y) = |x - y|^p$, then the curve*

$$\mu^\gamma(t) = (\pi_t)_\# \gamma$$

where $\pi_t : \Omega \times \Omega \to \Omega$ is given by $\pi_t(x, y) = (1 - t)x + ty$, is a constant-speed geodesic from μ to ν. In the case $p > 1$ all the constant-speed geodesics are of this form, and, if μ is absolutely continuous, then there is only one geodesic and it has the form

$$\mu_t = [T_t]_\# \mu, \qquad \text{where } T_t := (1 - t)id + tT$$

where T is the optimal transport map from μ to ν. In this case, the velocity field $\mathbf{v}_t$ of the geodesic μ_t is given by $\mathbf{v}_t = (T - id) \circ (T_t)^{-1}$. In particular, for $t = 0$ we have $\mathbf{v}_0 = -\nabla h^*(\nabla \varphi)$ and for $t = 1$ we have $\mathbf{v}_1 = \nabla h^*(\nabla \psi)$, where $h(z) = \frac{1}{p}|z|^p$ and (φ, ψ) are the Kantorovich potentials in the transport from μ to ν for the cost $c(x, y) = h(x - y)$.

3.3 *Minimizing Movement Schemes in the Wasserstein Space and Evolution PDEs*

Thanks to all the theory which has been described so far, it is natural to study (linear or non-linear) gradient flows in the space $\mathbb{W}_p(\Omega)$ and to connect them to PDEs of the form of a continuity equation. The most convenient way to study this is to start from the time-discretized problem, i.e. to consider a sequence of iterated minimization problems:

$$\varrho_{k+1}^{\tau} = \boldsymbol{\Pi}[\varrho_k^{\tau}], \quad \boldsymbol{\Pi}[g] := \operatorname{argmin}_{\varrho} F(\varrho) + \frac{W_p^p(\varrho, g)}{p\tau^{p-1}}. \tag{18}$$

This iterated minimization scheme is called *JKO scheme* (after Jordan, Kinderleher and Otto, [37]). Actually, this name was originally only used for the case $p = 2$ and we will refer to this scheme as W_p–JKO.

Note that we denote now the measures on Ω by the letter ϱ instead of μ or ν because we expect them to be absolutely continuous measures with nice (smooth) densities, and we want to study the PDE they solve. Note that a priori one could use the p–th power of arbitrary distances, and hence use W_q^p instead of W_p^p but in the Wasserstein space $\mathbb{W}_p$ the distance is defined as the power $1/p$ of a transport cost; only in the case $p = q$ the exponent goes away and we are lead to a minimization problem involving $F(\varrho)$ and a transport cost of the form

$$\mathcal{T}_c(\varrho, \nu) := \min \left\{ \int c(x, y)\,\mathrm{d}\gamma \ : \ \gamma \in \Pi(\varrho, \nu) \right\},$$

for $\nu = \varrho_k^{\tau}$. By the way, if we want to be more general, we can also replace the minimization problem in (18) with

$$\boldsymbol{\Pi}[g] := \operatorname{argmin}_{\varrho} F(\varrho) + \mathcal{T}_c(\varrho, g),$$

where $c(x, y) = \tau h(\frac{x-y}{\tau})$ for a strictly convex function h (we come back to (18) when $h(z) = \frac{1}{p}|z|^p$).

In the particular case of the space $\mathbb{W}_p(\Omega)$, which has some additional structure, if compared to arbitrary metric spaces, we would like to give a PDE description of the curves that we obtain as gradient flows, and this will pass through the optimality conditions of the minimization problem (18). In order to study these optimality conditions, we introduce the notion of first variation of a functional. This will be done in a very sketchy and formal way (we refer to Chapter 7 in [49] for more details).

Given a functional $G : \mathcal{P}(\Omega) \to \mathbb{R}$ we call $\frac{\delta G}{\delta \varrho}[\varrho]$, if it exists, the unique (up to additive constants) function such that $\frac{d}{d\varepsilon} G(\varrho + \varepsilon\chi)_{|\varepsilon=0} = \int \frac{\delta G}{\delta \varrho}[\varrho] d\chi$ for every perturbation χ such that, at least for $\varepsilon \in [0, \varepsilon_0]$, the measure $\varrho + \varepsilon\chi$ belongs to

$\mathcal{P}(\Omega)$. The function $\frac{\delta G}{\delta \varrho}[\varrho]$ is called *first variation* of the functional G at ϱ. In order to understand this notion, the easiest possibility is to analyze some examples.

The main classes of examples are the following functionals[1]

$$F(\varrho) = \int f(x, \varrho(x))\mathrm{d}x,$$

where $f : \Omega \times \mathbb{R} \to \mathbb{R}$ is a function which is convex and superlinear in its second variable (and the functional F is set to $+\infty$ if ϱ is not absolutely continuous w.r.t. the Lebesgue measure. In this case it is quite easy to realize that we have

$$\frac{\delta F}{\delta \varrho}[\varrho] = f'(x, \varrho),$$

where the derivative f' is also taken w.r.t. the second variable. A classical example is to consider $f(x, s) = s^m + V(x)s$ or $f(x, \varrho) = s \log s + V(x)s$, where $V : \Omega \to \mathbb{R}$ is regular enough.

It is clear that the first variation of a functional is a crucial tool to write optimality conditions for variational problems involving such a functional. In order to study the problem (18), we need to complete the picture by understanding the first variation of functionals of the form $\varrho \mapsto \mathcal{T}_c(\varrho, \nu)$. The result is the following:

Proposition 1 *Let* $c : \Omega \times \Omega \to \mathbb{R}$ *be a continuous cost function. Then the functional* $\varrho \mapsto \mathcal{T}_c(\varrho, \nu)$ *is convex, and its subdifferential at* ϱ_0 *coincides with the set of Kantorovich potentials* $\{\varphi \in C^0(\Omega) : \int \varphi \, \mathrm{d}\varrho_0 + \int \varphi^c \, \mathrm{d}\nu = \mathcal{T}_c(\varrho, \nu)\}$. *Moreover, if there is a unique c-concave Kantorovich potential* φ *from* ϱ_0 *to* ν *up to additive constants, then we also have* $\frac{\delta \mathcal{T}_c(\cdot, \nu)}{\delta \varrho}[\varrho_0] = \varphi$.

Even if a complete proof of the above proposition is not totally trivial (and Chapter 7 in [49] only provides it in the case where Ω is compact), one can guess why this is true from the following considerations. Using the duality formula we have

$$\mathcal{T}_c(\varrho, \nu) = \max_{\varphi, \psi : \varphi \oplus \psi \le c} \int_\Omega \varphi \, \mathrm{d}\varrho + \int_\Omega \psi \, \mathrm{d}\nu.$$

This expresses $\mathcal{T}_c$ as a supremum of linear functionals in ϱ and shows convexity. Standard considerations from convex analysis allow to identify the subdifferential as the set of functions φ attaining the maximum. An alternative point of view is to consider the functional $\varrho \mapsto \int \varphi \, \mathrm{d}\varrho + \int \psi \, \mathrm{d}\nu$ for fixed (φ, ψ), in which case the first variation is of course φ; then it is easy to see that the first variation of the supremum may be obtained (in case of uniqueness) just by selecting the optimal pair (φ, ψ).

[1] Note that in some cases the functionals that we use are actually valued in $\mathbb{R} \cup \{+\infty\}$, and we restrict to a suitable class of perturbations χ which make the corresponding functional finite.

Once we know how to compute first variations, we come back to the optimality conditions for the minimization problem (18). Which are these optimality conditions? roughly speaking, we should have

$$\frac{\delta F}{\delta \varrho}[\varrho^\tau_{k+1}] + \varphi = const$$

(where the reasons for having a constant instead of 0 is the fact that, in the space of probability measures, only zero-mean densities are considered as admissible perturbations, and the first variations are always defined up to additive constants). Note that here φ is the Kantorovich potential associated with the transport from ϱ^τ_{k+1} to ϱ^τ_k (and not viceversa), and with the transport cost $c(x, y) = \tau h(\frac{x-y}{\tau})$. It is related to the optimal transport map T through

$$\nabla h\left(\frac{x - T(x)}{\tau}\right) = \nabla\varphi(x) \quad \text{i.e.} \quad T(x) = x - \tau\nabla h^*(\nabla\varphi(x)).$$

Let us look at the consequences we can get from this optimality condition. Actually, if we combine the fact that the above sum is constant with the formula for the optimal T, we get

$$-\mathbf{v}(x) := \frac{T(x) - x}{\tau} = -\nabla h^*(\nabla\varphi(x)) = \nabla h^*\left(\nabla\frac{\delta F}{\delta \varrho}[\varrho]\right)(x), \tag{19}$$

where used the fact that we assume h^* to be even to remove the minus signs in the last expression.

Why did we denote by $-\mathbf{v}$ the ratio $\frac{T(x)-x}{\tau}$? because, as a ratio between a displacement and a time step, it has the meaning of a velocity, but since it is the displacement associated to the transport from ϱ^τ_{k+1} to ϱ^τ_k, it is better to view it rather as a backward velocity (which justifies the minus sign).

Since here we have $\mathbf{v} = -\nabla h^*(\nabla\frac{\delta F}{\delta \varrho}[\varrho])$, this suggests that at the limit $\tau \to 0$ we will find a solution of

$$\partial_t \varrho - \nabla\cdot\left(\varrho\,\nabla h^*\left(\nabla\left(\frac{\delta F}{\delta \varrho}[\varrho]\right)\right)\right) = 0. \tag{20}$$

This is a PDE where the velocity field in the continuity equation depends on the density ϱ itself.

We can see many interesting examples.

Let us start from the case where h is quadratic, i.e. $h(z) = |z|^2/2$ and $\nabla h^* = id$. In this case the equation is simpler.

First, suppose $F(\varrho) = \int f(\varrho(x))\mathrm{d}x$, with $f(s) = s\log s$. In such a case we have $f'(s) = \log s + 1$ and $\nabla(f'(\varrho)) = \frac{\nabla\varrho}{\varrho}$: this means that the gradient flow equation associated with the functional F would be the *Heat Equation*

$$\partial_t \varrho - \Delta \varrho = 0.$$

Using $F(\varrho) = \int f(\varrho(x))\,dx + \int V(x)\,d\varrho(x)$, we would have the *Fokker-Planck Equation*

$$\partial_t \varrho - \Delta \varrho - \nabla \cdot (\varrho \nabla V) = 0.$$

Changing the function f we can obtain other forms of diffusion. For instance, if one uses

$$F(\varrho) = \frac{1}{m-1} \int \varrho^m(x)\,dx + \int V(x)\varrho(x)\,dx,$$

the equation we obtain is

$$\partial_t \varrho - \Delta(\varrho^m) - \nabla \cdot (\varrho \nabla V) = 0.$$

When $m > 1$ this equation is called *Porous Media Equation* (see [53] for a complete monography and [46] for its treatment with the JKO scheme) and models the diffusion of a fluid into a material whose porosity slows down the diffusion. Among the properties of this equation there is a finite-speed propagation effect, different from linear diffusion (if ϱ_0 is compactly supported the same stays true for $\varrho_t, t > 0$, while this is not the case for the heat equation). It is also interesting to consider the case $m < 1$: the function $\varrho^m - \varrho$ is concave, but the negative coefficient $1/(m-1)$ finally gives a convex function (which is, unfortunately, not superlinear at infinity, hence more difficult to handle). The PDE that we get as a gradient flow is called *Fast diffusion* equation, and has opposite properties in terms of diffusion rate than the porous medium one.

We can then consider other choices of transport cost. When $h(z) = \frac{1}{p}|z|^p$ we have $\nabla h^*(z) = z^{q-1}$. When taking $F(\varrho) = \int f(\varrho(x))dx$ we obtain the following equation

$$\partial_t \varrho = \nabla \cdot (\varrho f''(\varrho)^{q-1} (\nabla \varrho)^{q-1}),$$

which can also be rewritten as

$$\partial_t \varrho = \Delta_q(\phi(\varrho)),$$

where Δ_q is the $q-$Laplace operator $\Delta_q u := \nabla \cdot ((\nabla u)^{q-1})$ and ϕ is defined via $\phi'(s) = s^{p-1} f''(s)$ (in this way, when $f$is convex ϕ is non-decreasing).

The case where we choose $f(s) = s \log s$ provides an interesting family of equations : those which are not linear, but are as linear as possible in this class, as they are one-homogeneous. In particular, among these equations we find $\partial_t \varrho = \Delta_q(\varrho^{p-1})$. The same happens when adding a potential V: $\partial_t \varrho = \nabla \cdot (\varrho(\nabla(\log \varrho + V))^{q-1})$.

3.4 Geodesic Convexity in $\mathbb{W}_p$

For several reasons in the theory of gradient flows in Wasserstein spaces it would be useful to know when a functional is geodesically convex. Being geodesically convex in $\mathbb{W}_p$ means that $t \mapsto F(\mu_t)$ is a convex function whenever $t \mapsto \mu_t$ is a constant-speed geodesic. Knowing the form of geodesics in W_p, we can say that this corresponds to the convexity of $t \mapsto F((T_t)_\#\mu)$ whenever $T_t := (1-t)id + tT$ and T is the optimal transport map from a measure μ to another measure ν. This can also be extended to the case where the transport map T is optimal for a transport cost of the form $c(x, y) = h(x-y)$ when h is not necessarily a power. In this case the curve $\mu_t = (T_t)_\#\mu)$ cannot be interpreted as a geodesic, but as an optimal displacement moving μ into ν. The notion of geodesical convexity is indeed also called *displacement convexity* and was first introduced by McCann in [43]. In the framework of these notes, geodesic convexity could be used in two—strongly related—aspects: one is the abstract notion of EVI gradient flows in metric spaces (only available for linear gradient flows, i.e. $p = 2$), the other is a sharp estimate of the rate of decrease of a functional G along the gradient flow of F which can be performed along the JKO scheme when G is displacement convex.

We will consider separately two interesting examples of functionals: $\mathcal{V}(\varrho) := \int V \mathrm{d}\varrho$ and $\mathcal{F}(\varrho) := \int f(\varrho(x))\mathrm{d}x$.

It is not difficult to check that the convexity of V is enough to guarantee geodesic convexity of $\mathcal{V}$, since

$$\mathcal{V}(\mu_t) = \int V \,\mathrm{d}\big((1-t)id + tT\big)_\#\mu = \int V\big((1-t)x + tT(x)\big)\,\mathrm{d}\mu,$$

The most interesting displacement convexity result is the one for functionals depending on the density. To consider these functionals, we need some technical facts.

The starting point is the computation of the density of an image measure, via standard change-of-variable techniques: if $T : \Omega \to \Omega$ is a map smooth enough[2] and injective, and $\det(DT(x)) \neq 0$ for a.e. $x \in \Omega$, if we set $\varrho_T := T_\#\varrho$, then ϱ_T is absolutely continuous with density given by

$$\varrho_T = \frac{\varrho}{\det(DT)} \circ T^{-1}.$$

Then, using a well-known fact in linear algebra, that the determinant raised to the power $1/d$ is a concave function on the set of positive-definite symmetric $d \times d$ matrices (which can be generalized to the case where the eigenvalues are real and non-negative), we can obtain the following result due to McCann

[2] We need at least T to be countably Lipschitz, i.e. Ω may be written as a countable union of measurable sets $(\Omega_i)_{i \geq 0}$ with Ω_0 negligible and $T \llcorner \Omega_i$ Lipschitz continuous for every $i \geq 1$.

Theorem 5 *Suppose that $f(0) = 0$ and that $s \mapsto s^d f(s^{-d})$ is convex and decreasing. Suppose that Ω is convex and take $1 < p < \infty$. Then $\mathcal{F}$ is displacement convex (hence geodesically convex in $\mathbb{W}_p$ for every $p > 1$).*

Remark 5 Note that the assumption that $s \mapsto s^d f(s^{-d})$ is convex and decreasing implies that f itself is convex (the reader can check it as an exercise), a property which can be useful to establish, for instance, lower semicontinuity of $\mathcal{F}$.

Here are some examples of convex functions satisfying the assumptions of Theorem 5:

- all the power functions $f(t) = t^q$ for $q > 1$ satisfy these assumptions, since $s^d f(s^{-d}) = s^{-d(q-1)}$ is convex and decreasing;
- the entropy function $f(t) = t \log t$ also satisfies it: $s^d f(s^{-d}) = -d \log s$ is convex and decreasing;
- the function $f(t) = -t^m$ is convex for $m < 1$, and if we compute $s^d f(s^{-d}) = -t^{m(1-d)}$ we get a convex and decreasing function as soon as $1 - \frac{1}{d} \leq m < 1$. Note that in this case f is not superlinear, which requires some attention for the semicontinuity of $\mathcal{F}$.

Finally, it is interesting to observe that the geodesic convexity of higher-order functionals such as $\varrho \mapsto \int |\nabla \varrho|^p$ generally fails, or is a very delicate matter, while these functionals are among the most standard examples of convex functionals in the usual sense. See [23] for some examples of first-order geodesically convex functionals (in dimension one).

4 Convergence of the JKO Scheme

Many possible proofs can be built for the convergence of the JKO scheme. In particular, one could follow the general theory developed in [4], i.e. checking all the assumptions to prove existence and uniqueness of an EVI gradient flow for the functional F in the space $\mathbb{W}_2(\Omega)$, and then characterizing the velocity field associated by Theorem 3 with the curve obtained as a gradient flow. In [4], it is proven, under suitable conditions, that such a vector field $\mathbf{v}_t$ must belong to what is defined as the *Wasserstein sub-differential* of the functional F, provided in particular that F is λ-geodescially convex. Then, the *Wasserstein sub-differential* is proven to be of the desired form (i.e. composed only of the gradient of the first variation of F, when F admits a first variation).

This approach has the advantage to use a general theory and to adapt it to the scopes of this particular setting. On the other hand, the important point when studying these PDEs is that the curves $(\varrho_t)_t$ obtained as a limit are weak solutions of the continuity equation; from this point of view, the metric notions of EDI and EVI solutions and the formalism developed in the first part of the book [4] are too general. Moreover, in the case of the space of measures, using optimal transport theory to select a suitable distance in the discrete scheme and choosing a suitable interpolation, the passage to the limit can be done by classical compactness techniques in functional analysis. Of course, there are often some difficulties in handling some non-linear terms, which are not always seen when using the theory of [4] (which is an advantage of such a general theory).

In this section we will first present the ideas to obtain convergence using compactness and weak convergence, and then an idea based on a specific adaptation of the EDI approach.

4.1 Passing to the Limit the PDE

Our goal is to find a curve $(\varrho_t)_t$ which is a solution (in the distributional sense on $\mathbb{R}^d$, which is the same as on Ω, with suitable no-flux boundary conditions on $\partial\Omega$) of the PDE

$$\partial_t \varrho_t + \nabla \cdot (\varrho_t \mathbf{v}_t) = 0, \quad \text{where we require } \mathbf{v}_t = -\nabla \left(\frac{\delta F}{\delta \varrho}[\varrho_t] \right).$$

We will set $E = \varrho \mathbf{v}$ (the variable E is called *momentum*, in physics) and require at least that E is a finite vector measure over $\Omega \times [0, T]$, acting on test functions ϕ via $\langle E, \phi \rangle := \int dt \int \phi(t, x) \cdot \mathbf{v}_t \, d\varrho_t$. Being a finite vector measure is equivalent to $\int_0^T ||\mathbf{v}_t||_{L^1(\varrho_t)} dt < +\infty$.

Starting from the discrete scheme (18), which defines a sequence $(\varrho_k^\tau)_k$, we also define a sequence of velocities $\mathbf{v}_k^\tau = (id - \mathrm{T})/\tau$, taking as T the optimal transport from ϱ_k^τ to ϱ_{k-1}^τ. The considerations in the previous sections guarantee that we have

$$\mathbf{v}_k^\tau = \nabla h^* \left(-\nabla \left(\frac{\delta F}{\delta \varrho}[\varrho_k^\tau] \right) \right).$$

This is proven thanks to the optimality conditions in (18) and the only delicate point is to guarantee the uniqueness of the Kantorovich potential in the transport from ϱ_k^τ to ϱ_{k-1}^τ. In some cases it is possible to prove a priori that the minimizers in (18) have strictly positive density a.e. whatever is ν (this is typically true when using the functional $\mathcal{F}$ with $f'(0) = -\infty$, as it is the case for the entropy functional), which provides uniqueness up to additive constants of the potential. In Section 8.3 of [49] full details are provided for the Fokker-Planck case, and it is indeed proved that the minimizers of the JKO scheme are strictly positive a.e., in this case. When positivity of the minimizer is not available, then one can obtain the same optimality conditions by first approximating ν with strictly positive densities, and then passing to the limit, or by adding an entropy term times a small parameter ε to the functional and then considering $\varepsilon \to 0$.

Then, we build at least two interesting curves in the space of measures:

- first we can define some piecewise constant curves, i.e. $\overline{\varrho}^\tau_t := \varrho^\tau_{k+1}$ for $t \in (k\tau, (k+1)\tau]$; associated with this curve we also define the velocities $\overline{\mathbf{v}}^\tau_t = \mathbf{v}^\tau_{k+1}$ for $t \in (k\tau, (k+1)\tau]$ and the momentum variable $\overline{E}^\tau = \overline{\varrho}^\tau \overline{\mathbf{v}}^\tau$;
- then, we can also consider the densities $\widehat{\varrho}^\tau_t$ that interpolate the discrete values $(\varrho^\tau_k)_k$ along geodesics:

$$\widehat{\varrho}^\tau_t = \left(id - (k\tau - t)\mathbf{v}^\tau_k\right)_\# \varrho^\tau_k, \quad \text{for } t \in ((k-1)\tau, k\tau[; \tag{21}$$

the velocities $\widehat{\mathbf{v}}^\tau_t$ are defined so that $(\widehat{\varrho}^\tau, \widehat{\mathbf{v}}^\tau)$ satisfy the continuity equation, taking

$$\widehat{\mathbf{v}}^\tau_t = \mathbf{v}^\tau_t \circ \left(id - (k\tau - t)\mathbf{v}^\tau_k\right)^{-1};$$

moreover, as before, we define: $\widehat{E}^\tau = \widehat{\varrho}^\tau \widehat{\mathbf{v}}^\tau$.

After these definitions we look for a priori bounds on the curves and the velocities that we defined. We already know that we have

$$\sum_k \tau \left(\frac{W_p(\varrho^\tau_k, \varrho^\tau_{k-1})}{\tau}\right)^p \leq C, \tag{22}$$

which is the discrete version of a $W^{1,p}$ estimate in time. As for $\widehat{\varrho}^\tau_t$, it is an absolutely continuous curve in the Wasserstein space and its velocity on the time interval $[(k-1)\tau, k\tau]$ is given by the ratio $W_p(\varrho^\tau_{k-1}, \varrho^\tau_k)/\tau$. Hence, the L^p norm of its velocity on $[0, T]$ is given by

$$\int_0^T |(\widehat{\varrho}^\tau)'|^p(t)\mathrm{d}t = \sum_k \tau \left(\frac{W_p(\varrho^\tau_k, \varrho^\tau_{k-1})}{\tau}\right)^p, \tag{23}$$

and, thanks to (22), it admits a uniform bound independent of τ. In our case, thanks to results on the continuity equation and the Wasserstein metric, this metric derivative is also equal to $||\widehat{\mathbf{v}}^\tau_t||_{L^p(\widehat{\varrho}^\tau_t)}$. This gives compactness of the curves $\widehat{\varrho}^\tau$, as well as Hölder estimates (since $W^{1,p} \subset C^{0,1/q}$). The characterization of the velocities $\overline{\mathbf{v}}^\tau$ and $\widehat{\mathbf{v}}^\tau$ allows to deduce bounds on these vector fields from the bounds on $W_p(\varrho^\tau_{k-1}, \varrho^\tau_k)/\tau$.

Considering all these facts, one obtains the following situation (see also [50] or Chapter 8 in [49]):

- The norm $\int ||\overline{\mathbf{v}}^\tau_t||^p_{L^p(\overline{\varrho}^\tau_t)}dt$ is τ-uniformly bounded. This quantity is equal to $\mathcal{B}_p(\overline{\varrho}^\tau, \overline{E}^\tau)$.
- In particular, the bound is valid in L^1 as well, which implies that $\overline{E}^\tau$ is bounded in the space of measures over $[0, T] \times \Omega$.
- The very same estimates are true for $\widehat{\mathbf{v}}^\tau$ and $\widehat{E}^\tau$.

- The curves $\widehat{\varrho}^\tau$ are bounded in $W^{1,p}([0,T],\mathbb{W}_p(\Omega))$ and hence compact in $C^0([0,T],\mathbb{W}_p(\Omega))$.
- Up to a subsequence, one has $\widehat{\varrho}^\tau \to \varrho$, as $\tau \to 0$, uniformly according to the W_p distance.
- From the estimate $W_p(\overline{\varrho}_t^\tau, \widehat{\varrho}_t^\tau) \le C\tau^{1/q}$ one gets that ϱ^τ converges to the same limit ϱ in the same sense.
- If we denote by E a weak limit of $\widehat{E}^\tau$, since $(\widehat{\varrho}^\tau, \widehat{E}^\tau)$ solves the continuity equation, by linearity, passing to the weak limit, also (ϱ, E) solves the same equation.
- It is possible to prove that the weak limits of $\widehat{E}^\tau$ and $\overline{E}^\tau$ are the same.
- From the semicontinuity of $\mathcal{B}_p$ and the bound $\mathcal{B}_p(\overline{\varrho}^\tau, \overline{E}^\tau) \le C$, one gets $\mathcal{B}_p(\varrho, E) < +\infty$, which means that E is absolutely continuous w.r.t. ϱ and has an L^p density, so that we have for a.e. time t a measure E_t of the form $\varrho_t \mathbf{v}_t$.
- We only need to prove that one has $\mathbf{v}_t = -\left(\nabla\left(\frac{\delta F}{\delta \varrho}[\varrho_t]\right)\right)^{q-1}$ ϱ_t-a.e. and for a.e. t. This means proving

$$\overline{E}^\tau = -\overline{\varrho}^\tau\left(\nabla\left(\frac{\delta F}{\delta \varrho}(\overline{\varrho}^\tau)\right)\right)^{q-1} \Rightarrow E = -\varrho\left(\nabla\left(\frac{\delta F}{\delta \varrho}[\varrho]\right)\right)^{q-1} \tag{24}$$

as a limit as $\tau \to 0$. It is crucial in this step to consider the limit of $(\overline{\varrho}^\tau, \overline{E}^\tau)$ instead of $(\widehat{\varrho}^\tau, \widehat{E}^\tau)$.

This last step is the most critical one in many cases. It requires passing to the limit (in the sense of distributions) the terms involving $\frac{\delta F}{\delta \varrho}$ on a sequence $\varrho_j \rightharpoonup \varrho$ (we do not care here where this sequence comes from). We can see what happens with the main class of functionals that we introduced so far.

The easiest case is that when $p = 1$ and the functional is $\mathcal{V}$: we have

$$\varrho\nabla\left(\frac{\delta \mathcal{V}}{\delta \varrho}[\varrho]\right) = \varrho\nabla V.$$

This term is linear in ϱ and $\varrho_j \rightharpoonup \varrho$ obviously implies $\varrho_j\nabla V \rightharpoonup \varrho\nabla V$ as soon as $V \in C^1$ (so that ∇V is a continuous function, which is exactly the functional space whose dual is the space of measures). In case $\nabla V \notin C^0$ it is possible to handle this term as soon as suitable bounds on ϱ_j provide a better weak convergence.

The case of the functional $\mathcal{F}$ (still with $p = 2$) is harder. In the case of the entropy $\mathcal{F}(\varrho) = \int f(\varrho)$ with $f(s) = s\log s$ then everything works fine because, again, the corresponding term is linear:

$$\varrho\nabla\left(\frac{\delta \mathcal{F}}{\delta \varrho}[\varrho]\right) = \varrho\frac{\nabla \varrho}{\varrho} = \nabla \varrho.$$

Then, the convergence of this term in the sense of distributions is straightforward when $\varrho_j \to \varrho$. By the way, the entropy term $\mathcal{F}$ is also enough to obtain suitable bounds to handle V or W which are only Lipschitz, as in this case we need to turn the weak convergence $\varrho_j \rightharpoonup \varrho$ from the sense of measures to the L^1 sense, which is exactly possible because the superlinear bound $\int \varrho_j \log(\varrho_j) \le C$ provides equi-integrability for ϱ_j.

Yet, for other functions f, there is no more linearity, and we need stronger bounds. For instance, for $f(s) = s^m/(m-1)$, we have

$$\varrho\nabla\left(\frac{\delta\mathcal{F}}{\delta\varrho}[\varrho]\right) = \varrho\frac{m}{m-1}\nabla(\varrho^{m-1}) = \nabla(\varrho^m).$$

This means that weak convergence of ϱ_j is no more enough, but one also needs weak convergence of ϱ_j^m, which means strong convergence of ϱ_j. This can be proven when we have stronger bounds (if possible, Sobolev bounds on ϱ_j).

The main ingredient is indeed an H^1 bound in space, which comes from the fact that we have

$$\begin{aligned}\int_0^T\int_\Omega |\nabla(\varrho^{m-1/2})|\,\mathrm{d}x\mathrm{d}t &\approx \int_0^T\int_\Omega |\nabla(\varrho^{m-1})|^2\,\mathrm{d}\varrho_t\mathrm{d}t\\ &\approx \int_0^T\int_\Omega \frac{|\nabla\varphi|^2}{\tau^2}\,\mathrm{d}\varrho_t\mathrm{d}t = \int_0^T |\varrho'|(t)^2\mathrm{d}t \le C. \qquad (25)\end{aligned}$$

This is only a bound on the H^1 norm in space, and does not involve time derivatives, so that we need to use the so-called *Aubin-Lions* lemma (see [5]), which interpolates between compactness in space and in time (roughly speaking: in our situation we have an L^2 bound in time with values in a strong space norm, H^1, but also a stronger bound in time, H^1, valued in the W_2 distance, which is weaker as it metrizes weak convergence; this information together can provide strong compactness).

However, so far we only considered the case $p = 2$. In this case the term E can be non-linear in ϱ but is in general linear in $\nabla\varrho$, which means that we only need to prove strong convergence of ϱ_j (and not of $\nabla\varrho_j$) in order to obtain (24). On the other hand, when $p \neq 2$ the relation between E and $\nabla\varrho$ is not any more linear, which means that we should obtain strong convergence not only of ϱ_j but also of its gradient, which is much harder (second-order Sobolev bounds would be needed, and they are not easy to obtain in the JKO scheme).

4.2 The EDI Formulation

In this section we present a very different strategy; more adapted to deal with cases where the strong convergence of the densities or of their gradients is not easy to prove. It has been used, for instance, in [1, 20, 22] to deal with non-linear gradient flows in W_p, and in [29] to deal with a system in $W_2 \times W_2$.

The starting point of this section is the following observation; finding a solution of $\partial \varrho_t = \nabla \cdot (\varrho \nabla h^*(\nabla \frac{\delta F}{\delta \varrho}))$ amounts to finding a pair $(\varrho, \mathbf{v})$ such that one has at the same time the continuity equation $\partial_t \varrho + \nabla \cdot (\varrho \mathbf{v})$ and $\mathbf{v} = -\nabla h^*(\nabla \frac{\delta F}{\delta \varrho})$. Assuming that $(\varrho, \mathbf{v})$ is regular enough so that we can differentiate in time $F(\varrho_t)$ and obtain the so-called *chain rule* $\frac{\mathrm{d}}{\mathrm{d}t} F(\varrho_t) = \int \nabla \frac{\delta F}{\delta \varrho} \cdot \mathbf{v}_t \mathrm{d}\varrho_t$ for a.e. t, this is equivalent to the continuity equation together with the following inequality

$$F(\varrho_T) + \int_0^T \mathrm{d}t \int_\Omega h^* \left(\nabla \frac{\delta F}{\delta \varrho} \right) \mathrm{d}\varrho_t + \int_0^T \mathrm{d}t \int_\Omega h(\mathbf{v}_t) \mathrm{d}\varrho_t \le F(\varrho_0). \tag{26}$$

Indeed, using the same rule we can replace $F(\varrho_T)$ with $F(\varrho_0) + \int_0^T \int_\Omega \nabla \frac{\delta F}{\delta \varrho} \cdot \mathbf{v}_t \mathrm{d}\varrho_t$ and hence see that this is equivalent to

$$\int_0^T \mathrm{d}t \int_\Omega \mathrm{d}\varrho_t \left(h^* \left(\nabla \frac{\delta F}{\delta \varrho} \right) + h(\mathbf{v}_t) - \nabla \frac{\delta F}{\delta \varrho} \cdot (-\mathbf{v}_t) \right) \le 0,$$

which is equivalent to the a.e. equality $-\mathbf{v}_t = \nabla h^*(\nabla \frac{\delta F}{\delta \varrho})$.

The goal becomes now to prove that the JKO scheme converges hence to a curve $(\varrho, \mathbf{v})$ satisfying (26). We set

$$S(\varrho) := \int_\Omega \mathrm{d}\varrho \left| \nabla \frac{\delta F}{\delta \varrho} \right|^q .$$

This functional is typically an integral functional involving convex functions of the gradient of ϱ and, as such, is lower semicontinuous for the weak convergence of probability measures (see, for instance, Chapter 3 in [51]). We will assume that it is indeed l.s.c. in the cases of our interest. The, in order to obtain (26), it is enough to find a suitable interpolation ϱ_t^τ (we dot choose here a specific one) of the solutions ϱ_k^τ such that we have

$$F(\varrho_T^\tau) + \frac{1}{q} \int_0^T S(\varrho_t^\tau) \mathrm{d}t + \sum_k \tau \frac{1}{p} \left(\frac{W_p(\varrho_k^\tau, \varrho_{k+1}^\tau)}{\tau} \right)^p \le F(\varrho_0). \tag{27}$$

Indeed, by semicontinuity of F the first term would provide at the limit the term $F(\varrho_T)$, by semicontinuity of S and Fatou's lemma the second would provide $\frac{1}{q} \int_0^T S(\varrho_t) \mathrm{d}t$, and the third one can be written in terms of the L^p norm of the velocity field of the piecewise geodesic interpolation and, by semicontinuity of the $\mathcal{B}_p$ functional, provides $\frac{1}{p} \int_0^T \mathrm{d}t \int_\Omega |\mathbf{v}_t|^p \mathrm{d}\varrho_t$.

A first natural question is whether the piecewise constant interpolation can be used to obtain (27). This would be the case if we had the following inequality involving one JKO step:

$$F(\varrho_{k+1}^\tau) + \tau \left(\frac{W_p(\varrho_k^\tau, \varrho_{k+1}^\tau)}{\tau} \right)^p \le F(\varrho_k^\tau). \tag{28}$$

Indeed, it is possible to see using the optimality conditions that we have

$$\left(\frac{x - T(x)}{\tau} \right)^{p-1} = \nabla \varphi = -\nabla \frac{\delta F}{\delta \varrho},$$

and we have, using $q = p/(p-1)$,

$$\left(\frac{W_p(\varrho_k^\tau, \varrho_{k+1}^\tau)}{\tau} \right)^p = \int \mathrm{d}\varrho_{k+1}^\tau \left| \frac{x - T(x)}{\tau} \right|^p = \int \mathrm{d}\varrho_{k+1}^\tau \left| \nabla \frac{\delta F}{\delta \varrho} \right|^q = S(\varrho_{k+1}^\tau), \tag{29}$$

which would allow to conclude writing

$$\tau \left(\frac{W_p(\varrho_k^\tau, \varrho_{k+1}^\tau)}{\tau} \right)^p = \frac{\tau}{q} S(\varrho_{k+1}^\tau) + \frac{\tau}{p} \left(\frac{W_p(\varrho_k^\tau, \varrho_{k+1}^\tau)}{\tau} \right)^p.$$

4.2.1 The Case of a Geodesically Convex Functional: The Flow Interchange Technique

The difficult point is that the condition (28) is in general very difficult to obtain, while it would be easy to obtain

$$F(\varrho_{k+1}^\tau) + \frac{\tau}{p} \left(\frac{W_p(\varrho_k^\tau, \varrho_{k+1}^\tau)}{\tau} \right)^p \le F(\varrho_k^\tau).$$

Yet, even if this is enough to obtain some bounds on the curve ϱ^τ, it is not enough to recover the desired EDI inequality.

Nonetheless, it is possible to improve this computation when F is geodesically convex. We start from an inequality.

Proposition 2 *Given two probability densities ϱ, g and a functional G which is geodesically convex, then we have*

$$G(g) \ge G(\varrho) - \tau \int_\Omega \nabla \frac{\delta G}{\delta \varrho}[\varrho] \cdot \nabla h^*(\nabla \varphi) \mathrm{d}\varrho,$$

where φ is the Kantorovich potential in the transport from ϱ to g for the cost $c(x, y) = \tau h(\frac{x-y}{\tau})$.

In particular, if $g = \varrho_k^\tau$, $\varrho = \Pi[g]$, $h(z) = |z|^p/p$ and $G = F$, we obtain (28)

Proof Let ϱ_s be a constant speed geodesic from $\varrho_0 = \varrho$ to $\varrho_1 = g$. It admits a velocity field $\mathbf{v}_s$ such that $\partial_s \varrho_s + \nabla \cdot (\varrho_s \mathbf{v}_s) = 0$. The geodesic convexity of G implies

$$G(g) \geq G(\varrho) + \frac{\mathrm{d}}{\mathrm{d}s} G(\varrho_s)_{|s=0}.$$

Let us compute

$$\frac{\mathrm{d}}{\mathrm{d}s} G(\varrho_s) = \int \frac{\delta G}{\delta \varrho}[\varrho_s] \partial_s \varrho = \int \nabla \frac{\delta G}{\delta \varrho}[\varrho_s] \cdot \mathbf{v}_s \varrho_s,$$

where the last equality comes from the continuity equation and integration by parts. Computing at time $s = 0$ we have $\mathbf{v}_0 = T - id = -\tau \nabla h^*(\nabla \varphi)$, which gives the claim.

The last part of the statement comes from the optimality conditions in the JKO scheme. This gives, using the explicit expression for ∇h^*,

$$F(\varrho_k^\tau) \geq F(\varrho_{k+1}^\tau) + \tau \int_\Omega \nabla \frac{\delta F}{\delta \varrho}[\varrho_{k+1}^\tau] \cdot \left(\nabla \frac{\delta F}{\delta \varrho}[\varrho_{k+1}^\tau] \right)^{q-1} \mathrm{d}\varrho_{k+1}^\tau,$$

which is the desired inequality. □

This kind of computation is known under the name of *flow interchange* technique and was introduced in [41]. Even if here we used it to prove a better estimate on the rate of decrease of F along the iterations of the JKO scheme for the same functional F, it could be used to obtain estimate on the rate of decrease of another functional G. In the case $p = 2$ the same estimate could also be obtained in a slightly different way: from the optimizer ϱ_{k+1}^τ start a curve ϱ_s such that at $s = 0$ it takes the value ϱ_{k+1}^τ and for $s > 0$ it follows the gradient flow in W_2 of G. This is a curve of possible competitors for the optimization problem defining ϱ_{k+1}^τ, so we should have

$$\frac{\mathrm{d}}{\mathrm{d}s} \left(F(\varrho_s) + \frac{W_2^2(\varrho_s, \varrho_k^\tau)}{2\tau} \right)_{|s=0} \geq 0,$$

by optimality. Then, we can use the EVI formulation of the gradient flow of G (which is geodesically convex), and obtain the upper bound

$$\frac{\mathrm{d}}{\mathrm{d}s} \left(\frac{W_2^2(\varrho_s, \varrho_k^\tau)}{2} \right)_{|s=0} \leq G(\varrho_k^\tau) - G(\varrho_{k+1}^\tau).$$

The conclusion follows by computing the derivative of F along the gradient flow of G (using as above the continuity equation satisfied by ϱ_s).

In practice, the flow interchange technique is a way to obtain in the discrete setting (the JKO scheme) the same computation that we would obtain in the continuous setting: if ϱ_t is a gradient flow of F, i.e. a solution of $\partial_t \varrho = \nabla \cdot (\varrho \nabla \frac{\delta F}{\delta \varrho})$ and we want to differentiate G we obtain

$$\frac{\mathrm{d}}{\mathrm{d}t}G(\varrho_t) = -\int \frac{\delta G}{\delta \varrho}[\varrho_t]\partial_t \varrho = \int \nabla \frac{\delta G}{\delta \varrho}[\varrho_t] \cdot \nabla \frac{\delta F}{\delta \varrho}[\varrho_t]\mathrm{d}\varrho_t.$$

A discrete analogue of this computation would be

$$G(\varrho_{k+1}^\tau) = G(\varrho_k^\tau) - \tau \int \nabla \frac{\delta G}{\delta \varrho}[\varrho_{k+1}^\tau] \cdot \nabla \frac{\delta F}{\delta \varrho}[\varrho_{k+1}^\tau]\mathrm{d}\varrho_{k+1}^\tau.$$

This is in general false, but the flow interchange technique allow to obtain it *as an inequality* as soon as G is geodesically convex.

As an other example of application of the flow interchange technique we cite the following.

Proposition 3 *Consider $F(\varrho) = \int f(\varrho(x))\mathrm{d}x$ and $G(\varrho) = \int g(\varrho(x))\mathrm{d}x$ and assume that G is geodesically convex in W_p. Consider the sequence ϱ_k^τ defined by the W_p-JKO scheme for the functional F. Then we have*

$$G(\varrho_k^\tau) \geq G(\varrho_{k+1}^\tau),$$

i.e. G decreases along the JKO scheme for F.

Proof We apply Proposition 2 and obtain

$$G(\varrho_k^\tau) \geq G(\varrho_{k+1}^\tau) + \tau \int_\Omega \nabla g'(\varrho_{k+1}^\tau) \cdot \left(\nabla f'(\varrho_{k+1}^\tau])\right)^{q-1} \mathrm{d}\varrho_{k+1}^\tau$$

and then use $\nabla g'(\varrho) \cdot (\nabla f'(\varrho))^{q-1} = g''(\varrho) f''(\varrho)^{q-1} |\nabla \varrho|^q \geq 0$. □

4.2.2 The General Case: The De Giorgi Variational Interpolation

In this section we want to prove that the limit of the JKO scheme solves the EDI condition also in the case when F is not geodesically convex, by choosing a better interpolation than the piecewise constant one.

The work has essentially been already done in the abstract metric setting in Sect. 2.3.

We define the De Giorgi variational interpolation as a curve $\hat{\varrho}_t^\tau$: for $t \in (k\tau, (k+1)\tau]$, we define $\theta = \frac{t-k\tau}{\tau} \in (0, 1]$ and

$$\hat{\varrho}_t^\tau = \operatorname{argmin} \frac{W_p^p(\varrho, \varrho_k^\tau)}{p(\theta\tau)^{p-1}} + F(\varrho). \tag{30}$$

In particular for $t = (k+1)\tau$ we have $\theta = 1$ and we do retrieve $\hat{\varrho}_{k+1}^\tau = \varrho_{k+1}^\tau$, and if $t = k\tau$, $\theta = 0$ and the minimizer has to be $\varrho = \varrho_k^\tau$.

Using the computations of Sect. 2.3 this interpolation allows us to derive the following precursor to the EDI interpretation of our gradient flow :

$$F(\varrho_k^\tau) \geq F(\varrho_{k+1}^\tau) + \frac{\tau}{q}\int_{k\tau}^{(k+1)\tau} \frac{W_p^p(\hat{\varrho}_t^\tau, \varrho_k^\tau)}{\theta(\theta\tau)^{p-1}}\mathrm{d}t + \frac{W_p^p(\varrho_{k+1}^\tau, \varrho_k^\tau)}{p\tau^{p-1}}. \quad (31)$$

The desired inequality is obtained once we observe that the equality (29) also holds when replacing τ with $\theta\tau$. This provides, summing over k, the desired inequality which then passes to the limit using lower semicontinuity.

We then finish this section by underlining that the techniques described here allow to obtain the inequality (26) which formally characterizes the solution of the PDE $\partial\varrho_t = \nabla\cdot(\varrho\nabla h^*(\nabla\frac{\delta F}{\delta\varrho}))$. Yet, the equivalence between the PDE and the inequality can only be established if we are able to prove the *chain rule*, i.e. that on the curve $(\varrho_t, \mathbf{v}_t)$ obtained at the limit $\tau \to 0$ the following relation holds

$$\frac{\mathrm{d}}{\mathrm{d}t}F(\varrho_t) = \int \nabla\frac{\delta F}{\delta\varrho}[\varrho_t]\cdot\mathbf{v}_t\mathrm{d}\varrho_t \quad \text{or} \quad F(\varrho_T) - F(\varrho_0) = \int_0^T \mathrm{d}t\int \nabla\frac{\delta F}{\delta\varrho}[\varrho_t]\cdot\mathbf{v}_t\mathrm{d}\varrho_t$$

This equality can easily be established at a formal level, but proving it rigorously when $(\varrho_t, \mathbf{v}_t)$ lack regularity is more delicate. In some cases it is possible to prove it by regularization (replacing ϱ and $\mathbf{v}$ by convolution with ϱ_ε and $\mathbf{v}_\varepsilon$, for which the formula holds, and passing to the limit). Another strategy discretizes the curve considering a finite family of time intervals $[t_k, t_{k+1}]$ and estimates the increments of F in terms of the velocity field $\mathbf{v}$ provided F is displacement convex. This is done in [1, 22]. An improvement is presented in [20], where it is proven that a large family of functionals of the form $\varrho \mapsto \int f(\varrho(x))\mathrm{d}x$ is approximated by functionals which are the difference of two displacement convex functionals, which allow to treat them separately. Anyway, we do not deal with the chain rule here in these notes and the results of this section require to be able to prove it by other means.

5 Higher-order Estimates

Proposition 3 is an example of result providing iterable estimates on the JKO scheme: it is proven that some quantities decrease from one step to the next one. In the statement of Proposition 3 the relevant quantities are of zero order, in the sense that they do not involve derivatives of the densities ϱ_k^τ. We will develop here techniques to deal with BV and Sobolev-like estimates.

5.1 The Five-gradients Inequality and BV Estimates

We start by presenting a general inequality, first studied in [26] which involves the gradients of two densities, of the corresponding Kantorovich potentials, and of an extra arbitrary convex function. Originally, the inequality was only stated in the case of a quadratic transport cost, and has later been extended to more general costs of the form $c(x, y) = h(x - y)$ in [18]. Extensions to the case of Riemannian manifolds are presented in [28].

Lemma 1 *Let $\Omega \subset \mathbb{R}^d$ be bounded and convex, let c be a transport cost given by $c(x, y) = h(x - y)$ for strictly convex h, let $\varrho, g \in W^{1,1}(\Omega)$ be two probability densities and $H \in C^2(\mathbb{R}^d)$ be a radially symmetric convex function. Then we have the following inequality*

$$\int_\Omega \Big(\nabla\varrho \cdot \nabla H(\nabla\varphi) + \nabla g \cdot \nabla H(\nabla\psi)\Big)\, \mathrm{d}x \geq 0, \tag{32}$$

where φ and ψ are the Kantorovich potentials in the optimal transport from ϱ to g for the cost c.

Proof We will assume that all functions are smooth, an assumption which can be removed by approximation.

We perform an integration by parts so that the left hand side of (32) becomes (we denote by $\mathbf{n}$ the exterior unit normal vector of the boundary $\partial\Omega$.

$$\int_{\partial\Omega} \Big(\varrho\, \nabla H(\nabla\varphi) \cdot \mathbf{n} + g\, \nabla H(\nabla\psi) \cdot \mathbf{n}\Big)\, \mathrm{d}\mathcal{H}^{d-1}$$
$$- \int_\Omega \Big(\varrho\, \nabla \cdot [\nabla H(\nabla\varphi)] + g\, \nabla \cdot [\nabla H(\nabla\psi)]\Big)\, \mathrm{d}x.$$

We first look at the boundary term. By the radial symmetry of H the vector $\nabla H(z)$ is always a scalar multiple of z. Since the gradients of the Kantorovich potentials $\nabla\varphi$ and $\nabla\psi$ calculated in boundary points are pointing outward Ω (since $T(x) = x - \nabla\varphi(x) \in \Omega$, and $S(x) = x - \nabla\psi(x) \in \Omega$) we have

$$\nabla H(\nabla\varphi(x)) \cdot \mathbf{n}(x) \geq 0 \qquad \text{and} \qquad \nabla H(\nabla\psi(x)) \cdot \mathbf{n}(x) \geq 0, \qquad \forall x \in \partial\Omega,$$

which proves that the boundary term is nonnegative.

We now write $g = T_\#\varrho$ for the optimal map T and expand the divergence, and obtain

$$\int_\Omega \Big(\varrho\, \nabla \cdot [\nabla H(\nabla\varphi)] + g\, \nabla \cdot [\nabla H(\nabla\psi)]\Big)\, \mathrm{d}x$$
$$= \int_\Omega \Big(\nabla \cdot [\nabla H(\nabla\varphi)] + (\nabla \cdot [\nabla H(\nabla\psi)]) \circ T\Big)\, \mathrm{d}\varrho$$
$$= \int_\Omega \left(\sum_{i,j} H_{ij}(\nabla\varphi)\varphi_{ij} + H_{ij}(\nabla\psi \circ T)\psi_{ij} \circ T\right) \mathrm{d}\varrho.$$

We then use the fact that $\varphi \oplus \psi - c$ is maximal on pairs (x, y) of the form $y = T(x)$ in order to obtain, from the first and the second order conditions when considering perturbations of the form $x' = x + v$, $y' = y + v$, the following information (we also use the equality $c(x', y') = c(x, y)$):

$$\nabla\phi(x) + \nabla\psi(T(x)) = 0, \quad D^2\phi(x) + D^2\psi(T(x)) \leq 0,$$

where he last inequality is to be intended in the sense of symmetric matrices. We then have

$$\sum_{i,j} H_{ij}(\nabla\varphi)\varphi_{ij} + H_{ij}(\nabla\psi \circ T)\psi_{ij} \circ T = \sum_{i,j} H_{ij}(\nabla\varphi)\left(\varphi_{ij} + \psi_{ij} \circ T\right).$$

This is the trace of the product of the matrices $D^2H(\varphi)$, which is positive definite because of the convexity of H, and of $D^2\phi(x) + D^2\psi(T(x))$, which is negative definite. The trace is then nonpositive, and this proves the claim. □

In the quadratic case $h(z) = |z|^2/2$ it is also possible to improve this inequality and obtain the following.

Lemma 2 *Let $\Omega \subset \mathbb{R}^d$ be bounded and convex, let $\varrho, g \in W^{1,1}(\Omega)$ be two probability densities and $H \subset C^2(\mathbb{R}^d)$ be a radially symmetric convex function. Then we have the following inequality*

$$\int_\Omega \Big(\nabla\varrho \cdot \nabla H(\nabla\varphi) + \nabla g \cdot \nabla H(\nabla\psi)\Big)\,\mathrm{d}x$$
$$\geq \int_\Omega \operatorname{tr}[D^2H(\nabla\varphi)(D^2\varphi)^2(I - D^2\varphi)^{-1}]\mathrm{d}\varrho, \tag{33}$$

where φ and ψ are the Kantorovich potentials in the optimal transport from ϱ to g for the quadratic cost.

Proof We just need to prove $D^2\phi(x)+D^2\psi(T(x)) = -(D^2\varphi(x))^2(I-D^2\varphi(x))^{-1}$ and use the previous computations. Consider the inverse map T^{-1}. We have $T^{-1}(y) = y - \nabla\psi(y)$ and, passing to the Jacobian, for $x = T^{-1}(y)$, we have $(DT(x))^{-1} = I - D^2\psi(y)$. This means $D^2\psi(y) = I - (I - D^2\varphi(x))^{-1}$. Setting $A = D^2\varphi(x)$ we now compute

$$D^2\phi(x) + D^2\psi(T(x)) = A + I - (I - A)^{-1} = -A^2(I - A)^{-1}.$$

□

By approximating $H(z) = |z|$ with $H(z) = \sqrt{\varepsilon^2 + |z|^2}$, Lemma 2 has the following important corollary, where we use the convention $\frac{z}{|z|} = 0$ for $z = 0$.

Corollary 1 *Let $\Omega \subset \mathbb{R}^d$ be a given bounded convex set and $\varrho, g \in W^{1,1}(\Omega)$ be two probability densities. Then the following inequality holds*

$$\int_\Omega \left(\nabla \varrho \cdot \frac{\nabla \varphi}{|\nabla \varphi|} + \nabla g \cdot \frac{\nabla \psi}{|\nabla \psi|} \right) \mathrm{d}x \geq 0, \tag{34}$$

where φ and ψ are the corresponding Kantorovich potentials.

This allows to obtain a first estimate on the evolution of the BV norm across the JKO scheme.

Proposition 4 *Consider $F(\varrho) = \int f(\varrho(x))\mathrm{d}x$ and the sequence ϱ_k^τ defined by the W_p-JKO scheme for the functional F. Then we have*

$$\int |\nabla \varrho_k^\tau| \mathrm{d}x \geq \int |\nabla \varrho_{k+1}^\tau| \mathrm{d}x,$$

i.e. the BV norm decreases along the JKO scheme for F.

Proof We apply Lemma 1 and obtain

$$\int_\Omega \left(\nabla \varrho_{k+1}^\tau \cdot \frac{\nabla \varphi}{|\nabla \varphi|} + \nabla \varrho_k^\tau \cdot \frac{\nabla \psi}{|\nabla \psi|} \right) \mathrm{d}x \geq 0.$$

Yet, from $\varphi = -f'(\varrho_{k+1}^\tau)$ and $f'' > 0$, we get that $\nabla\varphi$ and $\nabla \varrho_{k+1}^\tau$ are vectors with opposite directions. Hence we have

$$\int_\Omega |\nabla \varrho_{k+1}^\tau| \leq \int_\Omega \nabla \varrho_k^\tau \cdot \frac{\nabla \psi}{|\nabla \psi|} \mathrm{d}x \leq \int_\Omega |\nabla \varrho_k^\tau|,$$

which is the desired estimate. □

A corollary of this result about the JKO scheme is the following result on the corresponding PDE.

Corollary 2 *Consider the parabolic PDE $\partial_t \varrho = \Delta_q(\phi(\varrho))$ for an arbitrary exponent $q \in (1, \infty)$ and an arbitrary increasing nonlinearity $\phi : \mathbb{R}_+ \to \mathbb{R}$, with no-flux boundary condition on a convex domain $\Omega \subset \mathbb{R}^d$. Then, if $\varrho_0 \in BV(\Omega)$ there exists a solution ϱ_t such that $||\varrho_t||_{BV} \leq ||\varrho_0||_{BV}$. Moreover, if such an equation admits uniqueness of the solution for the Cauchy problem, then $t \mapsto ||\varrho_t||_{BV}$ is nonincreasing in time.*

Proof It is enough to choose a convex function f such that $sf''(s)^{q-1} = \phi'(s)^{q-1}$ and consider this equation as a W_p-gradient flow of $\mathcal{F}$. Then, Proposition 4 provides this uniform bound on the JKO scheme. In the limit $\tau \to 0$ this bound is preserved by lower semicontinuity. In case of uniqueness, it is possible to prove for any $t > t_0$ the inequality $||\varrho_t||_{BV} \leq ||\varrho_{t_0}||_{BV}$ by starting the JKO scheme from ϱ_{t_0}. □

5.2 Fisher Information Estimates

We focus here on the case of the functional $\mathscr{E}(\varrho) = \int f(\varrho(x))\mathrm{d}x + \int V\mathrm{d}\varrho$ for $f(s)+ s\log s$ (which is also a relative entropy w.r.t. the density e^{-V}), and consider transport costs of the form $c(x, y) = \tau h(\frac{x-y}{\tau})$ for a strictly convex and radial function h. We now look at the operator $\mathbf{\Pi}$ defined via

$$\mathbf{\Pi}[g] := \operatorname{argmin} \mathcal{T}_c(\varrho, g) + \mathscr{E}(\varrho). \tag{35}$$

The optimality conditions characterizing the optimal ϱ are $\log\varrho + V + \varphi = c$ and it is possible to see that ϱ depends on g in a 1-homogeneous and monotone way. Moreover, $\mathbf{\Pi}$ is also an L^1 contraction (see Theorem 1.3 from [36]).

We define the following Fisher information-type functional:

$$\mathscr{I}_{V,H}(\varrho) := \int_\Omega H(\nabla(\log\varrho + V))\mathrm{d}\varrho.$$

Using the five gradients inequality, one can prove the following :

Lemma 3 *Let H be a radially symmetric convex function, $g \in W^{1,1}(\Omega)$ be non negative and take $\varrho = \mathbf{\Pi}[g]$. Then we have*

$$\mathscr{I}_{V,H}(g) \geq \mathscr{I}_{V,H}(\varrho).$$

Proof We follow the proof of Proposition 5.1 from [27]. We have, from the fact that H is convex,

$$\begin{aligned}\int_\Omega H(\nabla(\log g + V))\mathrm{d}g \geq& \int_\Omega H(\nabla\psi)\mathrm{d}g\\ &+ \int_\Omega \nabla H(\nabla\psi)\cdot(\nabla\log g + \nabla V - \nabla\psi))\mathrm{d}g.\end{aligned}$$

Using $g = T_\#\varrho$ and $\nabla\psi\circ T = -\nabla\varphi$, together with the optimality condition of the optimization problem in (35) $\nabla\varphi = -\nabla(\log\varrho + V)$, we have

$$\int_\Omega H(\nabla\psi)\mathrm{d}g = \int_\Omega H(\nabla\psi\circ T)\mathrm{d}\varrho = \int_\Omega H(-\nabla\varphi)\mathrm{d}\varrho = \mathscr{I}_{V,H}(\varrho),$$

and

$$\begin{aligned}\int_\Omega \nabla H(\nabla\psi)\cdot\nabla\psi\mathrm{d}g &= -\int_\Omega \nabla H(-\nabla\varphi)\cdot\nabla\varphi\mathrm{d}\varrho\\ &= \int_\Omega \nabla H(-\nabla\varphi)\cdot(\nabla\varrho + \varrho\nabla V)\mathrm{d}x.\end{aligned}$$

Using the five-gradients inequality, we have

$$\int_\Omega \nabla H(\nabla\psi)\cdot(\nabla\log g)\,\mathrm{d}g + \int_\Omega \nabla H(\nabla\varphi)\cdot\nabla\varrho\,\mathrm{d}x$$
$$= \int_\Omega \nabla H(\nabla\psi)\cdot\nabla g\,\mathrm{d}x + \int_\Omega \nabla H(\nabla\varphi)\cdot\nabla\varrho\,\mathrm{d}x \geq 0,$$

so that we obtain

$$\mathscr{F}_{V,H}(g) \geq \mathscr{F}_{V,H}(\varrho) + \int \nabla H(\nabla\psi)\cdot\nabla V\mathrm{d}g - \int \nabla H(\nabla\varphi)\cdot\nabla V\mathrm{d}\varrho.$$

We can then use $\int \nabla H(\nabla\psi)\cdot\nabla V\mathrm{d}g = \int \nabla H(-\nabla\varphi)\cdot\nabla V\circ T\mathrm{d}\varrho$ to obtain

$$\int \nabla H(\nabla\psi)\cdot\nabla V\mathrm{d}g - \int \nabla H(\nabla\varphi)\cdot\nabla V\mathrm{d}\varrho = \int \nabla H(-\nabla\varphi)\cdot(\nabla V\circ T - \nabla V)\mathrm{d}\varrho.$$

Using the fact that H is radial and that $T(x)-x$ is a positive scalar multiple of $-\nabla\varphi$ we obtain, thanks to the convexity of V (which implies $(y-x)\cdot(\nabla V(y)-\nabla V(x)) \geq 0$), the positivity of this last integral, which gives the claim. □

When $V = 0$ and $H(z) = |z|^2$, the quantity $\mathscr{F}_{V,H}$ is the classical Fisher information. We are saying that this generalized Fisher information decreases along steps of the JKO scheme. Exactly as in Corollary 2 this can be transformed into a statement about the corresponding continous-in-time PDE, but only for 1-homogeneous equations of the form

$$\partial_t\varrho = \nabla\cdot(\varrho\nabla h^*(\nabla(\log\varrho + V))). \tag{36}$$

Such a result would be a consequence of the following statement about the JKO scheme (adapted from [27] and generalized in [21]):

Theorem 6 *Let $\varrho_0 \in L^1(\Omega)$. We assume V to be convex and Lipschitz continous. We have*

$$\int_\Omega H(\nabla(\log\varrho_0 + V))\,\mathrm{d}\varrho_0 \geq \int_\Omega H(\nabla(\log\varrho_k^\tau + V))\,\mathrm{d}\varrho_k^\tau$$

for all τ, k, where H is a radially symmetric and convex function. In particular, the following estimates are uniform in τ and k:

1. *if $\log(\varrho_0) + V$ is L-Lipschitz, then $\log(\varrho_k^\tau) + V$ is L-Lipschitz;*
2. *for $p > 1$, if $\varrho_0^{1/p} \in W^{1,p}(\Omega)$, then $(\varrho_k^\tau)^{1/p}$ is bounded in $W^{1,p}(\Omega)$;*
3. *if $\varrho_0 \in BV(\Omega)$, then ϱ_k^τ is bounded in $BV(\Omega)$;*
4. *if $\varrho_0 \in W^{1,1}(\Omega)$, all ϱ_k^τ belong to a weakly compact subset of $W^{1,1}(\Omega)$.*

Proof We only give the main ideas. The first statement is a direct consequence of Lemma 3, iterated along the steps of the scheme. For 1., we pick H to be the indicator of the centered ball of radius L (by indicator we mean here the function which is 0 on the ball and $+\infty$ outside it). For 2. we take $H(z) = |z|^p$ with $p > 1$, and 3. is covered by the case $p = 1$. For 4., one can use the Dunford-Pettis theorem so find a superlinear convex function H such that $H(\nabla \log(\varrho_0) + \nabla V) \in L^1(\varrho_0)$, which decreases at each step, thus providing the weak compactness. In all cases, the boundedness of ∇V allows to remove it from the integral. □

The above theorem includes the fact that Lipschitz bounds on the logarithm are preserved along the iterations of this JKO scheme. We want now to extend this to other moduli of continuity. We will use the following abstract fact. This statement (from [21], which we did not find stated in this very version elsewhere in the literature, seems interesting in itself.

Theorem 7 *Let $\pi : L^\infty(\Omega) \to L^\infty(\Omega)$ be an operator such that*

1. *For all $u, v \in L^\infty(\Omega)$, if $u \geq v$, then $\pi(u) \geq \pi(v)$.*
2. *For all $\lambda \in \mathbb{R}$ and $u \in L^\infty(\Omega)$, $\pi(\lambda + u) = \lambda + \pi(u)$.*
3. *For all $u \in L^\infty(\Omega)$, if u is k-Lipschitz, then $\pi(u)$ is also k-Lipschitz.*

Then, if u admits a concave modulus of continuity ω, then $\pi(u)$ admits the same modulus of continuity.

Proof If u admits ω as a modulus of continuity, we start by approximating it with L-Lipschitz functions : for $x \in \Omega$, set

$$u_L(x) = \inf_{y \in \Omega} L|x - y| + u(y),$$

so that u_L is L-Lipschitz. Of course u_L satisfies $u_L \leq u$ (by taking $y = x$). Furthermore, from the inequality

$$u(x) - u(y) \leq \omega(|x - y|),$$

we deduce that we have

$$L|x - y| + u(y) \geq L|x - y| + u(x) - \omega(|x - y|),$$

so that, passing to the inf in y, we obtain

$$u_L(x) \geq u(x) + \alpha(L),$$

where $\alpha(L) = \inf_{\mathrm{diam}(\Omega) > r > 0} Lr - \omega(r) \leq 0$. We can therefore conclude that we have

$$u \geq u_L \geq u + \alpha(L).$$

Applying π and using its properties 1 and 2 we obtain

$$\pi(u) \geq \pi(u_L) \geq \pi(u) + \alpha(L).$$

For $x, y \in \Omega$, since $\pi(u_L)$ is also L-Lipschitz, we can write

$$\pi(u)(x) - \pi(u)(y) \leq \pi(u_L)(x) - \pi(u_L)(y) - \alpha(L) \leq L|x-y| - \alpha(L).$$

Since ω is concave, we choose $L = \omega'(|x-y|)$ (note that ω is also non-decreasing, so we can take $L \geq 0$; in case ω is not differentiable one should use the super-differential of ω, which is non-empty because ω is finite on $\mathbb{R}_+$). This implies that $r \mapsto Lr - \omega(r)$ (which is a convex function) is minimized at $r = |x-y|$. Then we obtain for this L the equality $\alpha(L) = L|x-y| - \omega(|x-y|)$, and thus

$$\pi(u)(x) - \pi(u)(y) \leq \omega(|x-y|),$$

so that $\pi(u)$ admits ω as a modulus of continuity. □

Using the above theorem, we can then prove the following:

Theorem 8 *Let $\varrho_0 \in L^1(\Omega)$ be such that $\log(\varrho_0) + C$ admits ω as a modulus of continuity, then for all k and $\tau > 0$, $\log(\varrho_k^\tau) + V$ also admits ω as a modulus of continuity. As a consequence, this also holds for the solution of* (36).

Proof One only has to apply Theorem 7 where π is defined as follows: given u, solve one step of the JKO scheme with $g = \exp(u - V)$, call ϱ the solution, and then take $\pi(u) = \log \varrho + V$. □

Among particular examples of (36) we write not only equations of q–Laplacian type $\partial_t \varrho = \Delta_q(\varrho^{p-1})$ but also the so-called *relativistic heat equations* studied for instance in [44]. in such a paper the authors generalize the construction from [1] to "relativistic" cost functions, meaning cost functions where h is convex and only finite on a ball. The main equation of interest entering in this framework is the so called Relativistic Heat Equation

$$\partial_t \varrho = \nabla \cdot \left(\varrho \frac{\nabla \varrho}{\sqrt{\varrho^2 + |\nabla \varrho|^2}} \right),$$

where $h(z) = 1 - \sqrt{1 - |z|^2}$.

We finish this section with some improvements of Lemma 3 in the case $H(z) = |z|^2$

Lemma 4 *Let $g \in W^{1,1}(\Omega)$ be non negative and denote by ϱ the solution of* (35). *Suppose $D^2 V \geq \lambda I$. Then we have*

$$\mathcal{J}_{V,H}(g) \geq \mathcal{J}_{V,H}(\varrho) + 2\lambda\tau \int_\Omega |\nabla(\log \varrho + V)|^2 \, d\varrho$$

$$+2\tau\int_\Omega \mathrm{tr}\left[(D^2(\log\varrho+V))^2(I-A)^{-1}\right]\mathrm{d}\varrho,$$

where the matrix A is $D^2\varphi$, φ being the Kantorovich potential for the quadratic cost from ϱ to g, so that we also have $A=-\tau D^2(\log\varrho+V)$.

Proof The improvements w.r.t. Lemma 3 are obtained in two different ways. First, we estimate the term $\int\nabla H(-\nabla\varphi)\cdot(\nabla V\circ T-\nabla V)\,\mathrm{d}\varrho$ by using $(y-x)\cdot(\nabla V(y)-\nabla V(x))\geq\lambda|x-y|^2$ instead of just using that this scalar product is non-negative. Second, we use (33) in order to obtain the last line. □

6 Application to the Log-Sobolev Inequality

The goal of this section is to provide a proof of the classical Sobolev Inequality based on the use of the JKO scheme. Such an inequality reads

$$\int_\Omega f^2\log(f^2)\mathrm{d}\mu\leq\frac{1}{\lambda}\int_\Omega|\nabla f|^2\mathrm{d}\mu, \tag{37}$$

where we take $\mu=e^{-V}$ for some convex function $V$with $D^2V\geq\lambda I$, and assume that

$$\int_\Omega f^2e^{-V}\mathrm{d}x=1.$$

Setting $g=f^2$, this is equivalent to

$$\int_\Omega g\log g\,\mathrm{d}\mu\leq\frac{1}{2\lambda}\int_\Omega|\nabla\log g|^2\,g\,\mathrm{d}\mu.$$

Now we set $g=\varrho e^V$, so that ϱ is a probability density on Ω, and the inequality turns into

$$\int_\Omega\varrho e^V(\log\varrho+V)\,e^{-V}\mathrm{d}x=\mathscr{E}(\varrho)\leq\frac{1}{2\lambda}\mathscr{I}_{V,H}(\varrho).$$

This inequality can indeed be proved by using the so-called Bakry-Emery technique (see [6]), which consists in comparing the two quantities $\mathscr{E}$ and $\mathscr{I}_{V,H}$ (that we will write $\mathscr{I}$ for short) along the Fokker-Planck equation $\partial_t\varrho_t=\Delta\varrho_t+\nabla\cdot(\varrho\nabla V)$. Since the steady state, to which ϱ_t converges as $t\to\infty$ is exactly $\mu=e^{-V}$, and is such that $\mathscr{E}(\mu)=\mathscr{I}(\mu)=0$, it is enough to compare the time derivative of $\mathscr{E}(\varrho_t)$ and $\mathscr{I}(\varrho_t)$. We have

$$\frac{d}{dt}\mathscr{E}(\varrho_t) = -\int_\Omega |\nabla(\log\varrho_t + V)|^2 \, d\varrho_t = -\mathscr{I}(\varrho_t).$$

Then, it is possible to obtain

$$\frac{d}{dt}\mathscr{I}(\varrho_t) \leq -2\lambda\mathscr{I}(\varrho_t).$$

A precise computation leading to this inequality will be done in Sect. 7, but one can also see it as a consequence of Corollary 4. Note that here the second quantity that we differentiate happens to be the derivative of the first, so that we finally differentiate twice $\mathscr{E}(\varrho_t)$, but this is not crucial. The important point is being able to compare the derivatives of the desired functionals.

Thus, we have proven

$$\frac{d}{dt}(2\lambda\mathscr{E} - \mathscr{I})(\varrho_t) \geq 0,$$

which implies $(2\lambda\mathscr{E} - \mathscr{I})(\varrho_0) \leq \lim_{t\to\infty}(2\lambda\mathscr{E} - \mathscr{I})(\varrho_t) = 0$, which is the desired inequality (of course this computation requires to prove that the convergence as $t \to \infty$ implies the convergence of $\mathscr{E}$ and $\mathscr{I}$).

We now want to see if the same result can be obtained by looking at the evolution of $\mathscr{E}$ and $\mathscr{I}$ along the steps of the JKO scheme. This is so far just an interesting exercise, as it only re-proves a well-known result, but the technique can then be applied to other inequalities for which a continuous flow proof is not available. In particular, a work is in progress [19] about the case where V is convex but not uniformly convex (for instance $V(x) = |x|^4$).

Again, we need to prove that $2\lambda\mathscr{E} - \mathscr{I}$ increases from one step to the other. This means that fix a measure g and call ϱ the optimizer of one JKO step starting from g. In order to mimick what is done in the continuous proof we would need to have

$$\mathscr{I}(\varrho) \leq \mathscr{I}(g) - 2\lambda\tau\mathscr{I}(\varrho), \quad \text{and} \quad \mathscr{E}(\varrho) \geq \mathscr{E}(g) - \tau\mathscr{I}(\varrho).$$

The first of these inequalities is exactly contained in Lemma 4. As fir the second, since it amounts to differentiate a zero-order quantity which is indeed geodesically convex (since we assume V to be convex), the good tool is the flow interchange technique from Proposition 2. Yet, unfortunately, what could be proven in this way is the opposite inequality

$$\mathscr{E}(\varrho) \leq \mathscr{E}(g) - \tau\mathscr{I}(\varrho).$$

It is then necessary to work differently, and what can be proven is the following.

Lemma 5 *If* $\varrho = \Pi[g]$ *(using the functional* $\mathscr{E}$ *and the quadratic transport cost), then we have*

$$\mathcal{E}(\varrho) \geq \mathcal{E}(g) - \frac{\tau}{2}\mathcal{I}(\varrho) - \frac{\tau}{2}\mathcal{I}(g).$$

Proof We use Proposition 2 with $G = \mathcal{E}$, but this time inverting the roles of g and ϱ. We obtain, exploiting the precise choice of the quadratic cost, the following inequality

$$\mathcal{E}(\varrho) \geq \mathcal{E}(g) - \tau \int_\Omega \nabla(\log g + V) \cdot \nabla\psi dg,$$

where ψ is the Kantorovich potential from g to ϱ (and φ is the one from ϱ to g, which appears in the optimality conditions of the JKO step). We then apply Young's inequality and obtain

$$\mathcal{E}(\varrho) \geq \mathcal{E}(g) - \frac{\tau}{2}\int_\Omega |\nabla(\log g + V)|^2 dg - \frac{\tau}{2}\int_\Omega |\nabla\psi|^2 dg.$$

We now use $\int_\Omega |\nabla(\log g + V)|^2 dg = \mathcal{I}(g)$ and

$$\int_\Omega |\nabla\psi|^2 dg = \int_\Omega |\nabla\psi \circ T|^2 \, d\varrho = \int_\Omega |\nabla\varphi|^2 \, d\varrho = \int_\Omega |\nabla(\log \varrho + V)|^2 \, d\varrho = \mathcal{I}(\varrho),$$

where we used the optimality conditions on ϱ. This proves the claim. □

We then obtain the following.

Proposition 5 *Given an arbitrary ϱ_0 and starting the JKO scheme from it we have, for any $N \geq 0$*

$$2\lambda\mathcal{E}(\varrho_0) - \mathcal{I}(\varrho_0) \leq 2\lambda\mathcal{E}(\varrho_N^\tau) - \mathcal{I}(\varrho_N^\tau) + \lambda\tau(\mathcal{I}(\varrho_0) - \mathcal{I}(\varrho_N^\tau)).$$

This implies

$$2\lambda\mathcal{E}(\varrho_0) - \mathcal{I}(\varrho_0) \leq 0.$$

Proof We apply the previous lemma to $g = \varrho_k^\tau$ and $\varrho = \varrho_{k+1}^\tau$ for $k = 0, 1, \ldots, N-1$ and sum the inequalities, thus obtaining

$$\mathcal{E}(\varrho_0) \leq \mathcal{E}(\varrho_N^\tau) + \tau \sum_{k=1}^{N-1} \mathcal{I}(\varrho_k^\tau) + \frac{\tau}{2}(\mathcal{I}(\varrho_0) + \mathcal{I}(\varrho_N^\tau)).$$

Multiplying by 2λ and subtracting the inequality

$$\mathcal{I}(\varrho_0) \geq \mathcal{I}(\varrho_N^\tau) + 2\lambda\tau \sum_{k=1}^{N} \mathcal{I}(\varrho_k^\tau)$$

we obtain the desired inequality.

It is then enough to take the limit $N \to \infty$ in order to obtain

$$2\lambda\mathscr{E}(\varrho_0) - \mathscr{I}(\varrho_0) \le \lambda\tau\mathscr{I}(\varrho_0).$$

Note that this is even simpler to justify than in the continuous setting, as the convergence $\varrho_N^\tau \rightharpoonup \mu$ can be obtained by studying the fixed points of the JKO operator, and when $\tau > 0$ is fixed the convergence is automatically strong enough to pass to the limit both $\mathscr{E}$ and $\mathscr{I}$ as one can obtain second-order bounds of the order of τ^{-1}. Then, we take the limit as $\tau \to 0$, assuming $\mathscr{I}(\varrho_0) < +\infty$ (otherwise there is nothing to prove). □

7 Strong L^2H^2 Convergence of the JKO Scheme

The metric interpretation of the JKO scheme, according to the theory of [4], provides convergence as $\tau \to 0$ in the same distance which is used in the JKO scheme. This means that, in the case of the linear Fokker-Planck equation, one obtains uniform (in time) convergence in W_2, which means weak-* convergence (in space) of ϱ^τ to ϱ_t. In the original paper about the JKO scheme for the Fokker-Planck equation, [37], a slightly different convergence is proven: it is proven that ϱ^τ strongly converges in $L^1([0, T] \times \Omega)$.

In this section we use the notions developed so far in order to prove that, under some conditions on the initial datum, the convergence in the case of the linear Fokker-Planck equation is actually much stronger, and holds in $L^2([0, T]; H^2(\Omega))$. We will consider the case of a convex bounded domain $\Omega \subset \mathbb{R}^d$ whose boundary $\partial\Omega$ is smooth enough. We denote by $\mathbf{n}$ the exterior unit normal vector of the boundary $\partial\Omega$. We consider the Cauchy problem for the Fokker-Planck equation with no-flux boundary condition, i.e.,

$$\begin{cases} \partial_t \varrho(t,x) = \Delta\varrho(t,x) + \operatorname{div}(\varrho(t,x)\nabla V(x)), & (t,x) \in (0,T] \times \Omega, \\ \nabla\varrho(t,x)\cdot\mathbf{n}(x) + \varrho(t,x)\nabla V(x)\cdot\mathbf{n}(x) = 0, & (t,x) \in [0,T] \times \partial\Omega, \\ \varrho(0,x) = \varrho_0(x), & x \in \Omega, \end{cases} \tag{38}$$

where $\varrho_0 \in \mathcal{P}(\Omega) \cap L^1_+(\Omega)$.

We refer to classical texts on parabolic differential equations (see [39], [40]) for the existence, the uniqueness, and the regularity of the solution.

We now compute the derivative of the classical Fisher information along the flow. For shortness, we write u_ϱ for u_ϱ:

Lemma 6 *If Ω and V are smooth enough and ϱ is the solution of (38), then for $t > 0$ we have*

$$\frac{\mathrm{d}}{\mathrm{d}t}\mathcal{F}(\varrho_t) = -2\int_\Omega |D^2 u_\varrho|^2 \varrho\, dx - 2\int_\Omega (\nabla u_\varrho)^T \cdot D^2 V \cdot \nabla u_\varrho \varrho\, dx$$
$$+ \int_{\partial\Omega} (\nabla u_\varrho)^T \cdot D^2 u_\varrho \cdot \mathbf{n}\varrho\, d\mathcal{H}^{d-1}.$$

The last term in the previous formula can be re-written using the following lemma.

Lemma 7 *Suppose $\Omega = \{h < 0\}$ for a smooth function $h : \mathbb{R}^d \to \mathbb{R}$ with $\nabla h \neq 0$ on $\{h = 0\}$, so that the exterior normal vector at $x \in \partial\Omega$ is given by $\mathbf{n}(x) = \nabla h(x)/|\nabla h(x)|$. Let $\mathbf{v} : \Omega \to \mathbb{R}^d$ be a smooth vector field such that $\mathbf{v} \cdot \mathbf{n} = 0$ on $\partial\Omega$. Then we have the following equality for every $x \in \partial\Omega$*

$$\mathbf{v}(x)^T \cdot Dv(x) \cdot \mathbf{n}(x) = -\frac{\mathbf{v}(x)^T \cdot D^2 h(x) \cdot \mathbf{v}(x)}{|\nabla h(x)|}.$$

Proof Given $x \in \partial\Omega$, we consider a smooth curve $\gamma : (-t_0, t_0) \to \partial\Omega$ with $\gamma(0) = x$ and write the equality $\mathbf{v}(\gamma(t)) \cdot \nabla h(\gamma(t)) = 0$ for every t. Differentiating w.r.t. t we obtain

$$\gamma'(t)^T \cdot D\mathbf{v}(\gamma(t)) \cdot \nabla h(\gamma(t)) + \mathbf{v}(\gamma(t))^T \cdot D^2 h(\gamma(t)) \cdot \gamma'(t) = 0.$$

We can take $t = 0$ and choose a curve with $\gamma'(0) = \mathbf{v}(x)$ since $\mathbf{v}$ is tangent to the surface $\partial\Omega$, thus obtaining

$$\mathbf{v}(x)^T \cdot D\mathbf{v}(x) \cdot \nabla h(x) = -\mathbf{v}(x)^T \cdot D^2 h(x) \cdot \mathbf{v}(x).$$

It is then enough to divide by $|\nabla h(x)|$ in order to get the claim. □

We then obtain the following formula.

Corollary 3 *Let $\Omega = \{h < 0\}$ be a bounded domain defined as the negativity set of a convex function $h \in C^2$ with $\nabla h \neq 0$ on $\{h = 0\}$, and $V \in C^2(\bar{\Omega})$. Given a strictly positive $\varrho_0 \in H^1(\Omega)$ initial datum, let ϱ be the solution of (38). We then have*

$$\mathcal{F}(\varrho_T) - \mathcal{F}(\varrho_0) = -\int_0^T \mathrm{d}t \int_\Omega |D^2 u_\varrho|^2 \varrho\, \mathrm{d}x$$
$$- \int_0^T \mathrm{d}t \int_\Omega (\nabla u_\varrho)^T \cdot D^2 V \cdot \nabla u_\varrho \varrho\, \mathrm{d}x$$
$$- \int_0^T \mathrm{d}t \int_{\partial\Omega} (\nabla u_\varrho)^T \cdot D^2 h \cdot (\nabla u_\varrho)\varrho\, \mathrm{d}\mathcal{H}^{d-1}.$$

We now proceed to some uniform estimates on the minimizers of the JKO scheme and on the corresponding potentials. Note that here, in order to better see the dependance on τ of these terms, we write φ for the Kantorovich potentials w.r.t. the quadratic cost *without the coefficient* τ.

Proposition 6 *Suppose Ω is a bounded, smooth, and uniformly convex domain, and $V \in C^2(\bar{\Omega})$. Let (ϱ_k^τ) be the sequence obtained in the JKO scheme for the Fokker-Planck equation, (φ_k, ψ_k) denote the pair of Kantorovich potentials in the transport from ϱ_{k+1}^τ to ϱ_k^τ and T_k the corresponding optimal map, i.e. $T_k(x) = x - \nabla\varphi_k(x)$. Suppose that ϱ_0 is bounded from below and above by positive constants and denote by a, b two constants such that $a \le \log\varrho_0 + V \le b$. Suppose moreover that we have $\varrho_0 \in C^{0,\alpha} \cap H^1$. Then we have:*

1. *For each k we have $a \le \log(\varrho_k^\tau) + V \le b$. In particular, all ϱ_k^τ are bounded from below and above by some uniform positive constants.*
2. *All the potentials φ_k satisfy $||id - T_k||_{L^\infty} = ||\nabla\varphi_k||_{L^\infty} \le C\tau^{1/(d+2)}$.*
3. *If τ is small enough (depending on V and p), then the $C^{0,\alpha}$ seminorm of $\log(\varrho_k^\tau) + V$ is bounded in terms f that of $\log(\varrho_0) + V$ and the densities ϱ_k^τ are bounded in $C^{0,\alpha}$.*
4. *All the potentials φ_k belong to $C^{2+\alpha}(\bar{\Omega})$ and $||\varphi_k||_{C^{2+\alpha}(\bar{\Omega})}$ is bounded by a uniform constant.*
5. *The potentials φ_k also satisfy $||D^2\varphi_k||_{L^\infty} \le C\tau^\beta$ for a certain exponent $\beta > 0$.*

Proof

1. The uniform estimates on ϱ^τ can be proven as in Lemma 2.4 of [35].
2. Whenever μ, ν are two measures in a convex domain $\Omega \subset \mathbb{R}^d$ and T is the corresponding optimal map sending μ onto ν, if the density of μ is bounded from below by a constant $c_0 > 0$, then the following remarkable estimate is proven in [7] :

$$||T - id||_{L^\infty} \le C(d, c_0)W_2(\mu, \nu)^{2/(d+2)}.$$

If we combine this with

$$W_2^2(\varrho_{k+1}^\tau, \varrho_k^\tau) \le 2\tau\,(F(\varrho_k^\tau) - F(\varrho_{k+1}^\tau)) \le C\tau.$$

and the fact that the density ϱ_{k+1}^τ is bounded from below by a universal constant, we then have

$$||\nabla\varphi_k||_{L^\infty(\Omega)}^{2+d} = ||id - T_k||_{L^\infty(\Omega)}^{2+d} \le CW_2^2(\varrho_{k+1}^\tau, \varrho_k^\tau) \le C\tau. \tag{39}$$

3. If V is convex we already saw in Theorems 6 and 8 that the modulus of continuity of $\log\varrho + V$ are preserved along iterations. If $D^V \ge \lambda I$ but λ is negative this requires a slightly different argument. A proof the Lipschitz bound is contained in [31] and then we need to adapt the proof of Theorem 7 to the case where the Lipschitz constant is not preserved by π but increases by a quantified factor of order $1 + \tau$.

4. The bound on $||\varphi_k||_{C^{2+\alpha}(\bar{\Omega})}$ is a consequence of Caffarelli's regularity theory for the Monge-Ampère equation (see [12–17]), once we have proven that the densities are uniformly bounded from above, from below, and in $C^{0,\alpha}$, when the domain is uniformly convex and C^2.
5. If we apply standard interpolation arguments to $u = \nabla\varphi$ we get

$$||\nabla\varphi_k||_{C^1(\bar{\Omega})} \leq c||\nabla\varphi_k||^{\frac{1}{1+\alpha}}_{C^{1+\alpha}(\bar{\Omega})} ||\nabla\varphi_k||^{\frac{\alpha}{1+\alpha}}_{L^\infty(\bar{\Omega})}.$$

Using (39) and the uniform bounds on $||\varphi_k||_{C^{2+\alpha}(\bar{\Omega})}$ we obtain the claim with $\beta = \frac{\alpha}{(1+\alpha)(2+d)}$.

□

We now use the previous results to prove the following estimates, where $\varepsilon(\tau)$ denotes any quantity which depends only on $\bar{\Omega}$, ϱ_0, V and τ such that $\varepsilon(\tau) \to 0$ as $\tau \to 0$.

Lemma 8 *Under the same assumptions as Proposition 6, if Ω is of the form $\{h < 0\}$ for a convex and smooth function h with* $\min h < 0$, *then we have*

$$\begin{aligned}\mathcal{F}(\varrho_0) \geq \mathcal{F}(\varrho_T^\tau) + \int_0^T \int_\Omega |D^2 u_{\varrho_t^\tau}|^2 \varrho_t^\tau(x)\,\mathrm{d}x\mathrm{d}t \\ + \int_0^T \int_\Omega (\nabla u_{\varrho_t^\tau})^T \cdot D^2 V(x) \cdot \nabla u_{\varrho_t^\tau} \varrho_t^\tau(x)\,\mathrm{d}x\mathrm{d}t \\ + \int_0^T \int_{\partial\Omega} \frac{\varrho_t^\tau(x)}{|\nabla h(x)|} (\nabla u_{\varrho_t^\tau})^T \cdot D^2 h(x) \cdot (\nabla u_{\varrho_t^\tau}) \mathrm{d}\mathcal{H}^{d-1}\mathrm{d}t + \varepsilon(\tau).\end{aligned} \tag{40}$$

Proof The estimate can be obtained from Lemma 4 after improving some computations. First, we need to keep the Hessian of V instead of just using $D^V \geq \lambda$. The C^2 regularity of V is needed to use an inequality of the form

$$(x - y) \cdot (\nabla V(x) - \nabla V(y)) \geq (x - y)^T D^2 V(x)(x - y) + o(|x - y|^2).$$

Then, we apply this to $y = T(x)$ and use point 2. in Proposition 6 in order to estimate the error term with $\varepsilon(\tau)$.

We now look at the boundary integrals. This requires to use the boundary terms that we just bounded from below by 0 in the proof of the five-gradients-inequality.

Details for both arguments are presented in [52]. □

In order to conclude we need the following lemma.

Lemma 9 *Let Ω be a bounded Lipschitz domain of $\mathbb{R}^d$, $\{u_k\}_{k=1}^\infty$ be a sequence in $L^2([0, T]; H^2(\Omega))$ and bounded in $L^\infty([0, T] \times \Omega)$. If $u_k \to u$ strongly in $L^2([0, T]; H^2(\Omega))$, then, for any $f \in C^2(\mathbb{R})$, $\{f(u_k)\}_{k=1}^\infty$ converges to $f(u)$ in $L^2([0, T]; H^2(\Omega))$.*

Proof Using our assumptions and the Gagliardo-Nirenberg inequality (see [11, section 9])

$$||\nabla v||^4_{L^4(\Omega)} \leq C||v||^2_{H^2(\Omega)}||v||^2_{L^\infty(\Omega)} \tag{41}$$

applied to $v = u_k - u$ we obtain $u_k \to u$ in $L^4([0, T]; W^{1,4}(\Omega))$. A simple computation shows

$$D^2(f(u_k)) = f'(u_k)D^2u_k + f''(u_k)\nabla u_k \otimes \nabla u_k$$

and the L^2 convergence of this matrix-valued function to $D^2(f(u)) = f'(u)D^2u + f''(u)\nabla u \otimes \nabla u$ is due to the following facts:

- $D^2u_k \to D^2u$ in $L^2([0, T] \times \Omega)$;
- $\nabla u_k \otimes \nabla u_k \to \nabla u \otimes \nabla u$ in $L^2([0, T] \times \Omega)$;
- both $f''(u_k)$ and $f'(u_k)$ converge a.e. (to $f''(u)$ and $f'(u)$, respectively) as a consequence of the convergence of u_k to u; moreover, these terms are bounded in L^∞ as a consequence of the regularity of f and of the L^∞ bound on u_k.

□

We are now ready to prove the main theorem of this section.

Theorem 9 *Suppose* $0 < T < +\infty$, Ω *is a bounded, smooth, and convex domain. Let* $V \in C^2(\bar{\Omega})$, $\varrho_0 \in H^1(\Omega) \cap C^{0,\alpha}(\Omega)$ *be an initial datum bounded away from zero and infinity and* ϱ *be the solution of the Fokker-Planck equation (38). Then,* $\varrho_t^\tau \to \varrho$ *strongly in* $L^2([0, T]; H^2(\Omega))$ *as* $\tau \to 0$.

Proof Lemma 8 proves in particular that $\log \varrho_t^\tau + V$ is uniformly bounded in $L^2([0, T]; H^2(\Omega))$ with respect to τ. Since the space $L^2([0, T]; H^2(\Omega))$ is reflexive and ϱ_t^τ converges strongly in $L^2([0, T]; H^1(\Omega))$ to ϱ (use the H^2 bound together with the compactness in space and Aubin-Lions lemma), we get that $\log \varrho_t^\tau + V$ converges weakly to $\log \varrho + V$ in $L^2([0, T]; H^2(\Omega))$. This also implies that $\nabla \log \varrho_t^\tau + \nabla V$ converges weakly to $\nabla \log \varrho + \nabla V$ in $L^2([0, T]; L^2(\partial\Omega))$. If τ tends to zero then we use the lower semicontinuity of each term on the right hand side of (40) in order to obtain

$$\int_0^T \int_\Omega |D^2 u_{\varrho_t}|^2 \varrho_t(x)\,\mathrm{d}x\mathrm{d}t \geq \limsup_{\tau \to 0} \int_0^T \int_\Omega |D^2 u_{\varrho_t^\tau}|^2 \varrho_t^\tau(x)\,\mathrm{d}x\mathrm{d}t. \tag{42}$$

This condition and the lower bound on ϱ^τ show that $D^2 \log \varrho^\tau$ is bounded in $L^2([0, T]; L^2(\Omega))$. Since $L^2([0, T]; L^2(\Omega))$ is reflexive and ϱ^τ converges strongly in $L^2([0, T]; H^1(\Omega))$ to ϱ, then $D^2 \log \varrho^\tau$ converges weakly to $D^2 \log \varrho$ in $L^2([0, T]; L^2(\Omega))$. By adding D^2V and multiplying times $\sqrt{\varrho^\tau}$, which converges a.e. to $\sqrt{\varrho}$ and is bounded by a constant, we also have weak convergence in $L^2([0, T]; L^2(\Omega))$ of $\sqrt{\varrho^\tau} D^2(\log \varrho^\tau + V)$ to $\sqrt{\varrho} D^2(\log \varrho + V)$. Yet, this convergence becomes strong because of the convergence of the norm in (42). We can then multiply times $(\varrho^\tau)^{-1/2}$ and subtract D^2V and obtain strong convergence in $L^2([0,T]; H^2(\Omega))$ for $\log \varrho^\tau$ to $\log \varrho$.

We then apply Lemma 9 to obtain $\varrho^\tau \to \varrho$ in $L^2([0,T]; H^2(\Omega))$. □

Acknowledgments The author acknowledges the support of European union via the ERC AdG 101054420 EYAWKAJKOS. Moreover, discussions with many collaborators, and in particular recent discussions with Thibault Caillet and Fanch Coudreuse, were crucial for the preparation of this manuscript.

References

1. M. Agueh, Existence of solutions to degenerate parabolic equations via the Monge- Kantorovich theory. Adv. Differ. Equ. **10**(3):309–360 (2005)
2. L. Ambrosio, Movimenti minimizzanti. Rend. Accad. Naz. Sci. XL Mem. Mat. Sci. Fis. Natur. **113**, 191–246 (1995)
3. L. Ambrosio, P. Tilli, *Topics on Analysis in Metric Spaces*. Oxford Lecture Series in Mathematics and its Applications, vol. 25 (Oxford University Press, Oxford, 2004)
4. L. Ambrosio, N. Gigli, G. Savaré, *Gradient Flows in Metric Spaces and in the Spaces of Probability Measures*. Lectures in Mathematics, ETH Zurich (Birkhäuser, Basel, 2005)
5. J.-P. Aubin, Un théorème de compacité (French). C. R. Acad. Sci. Paris **256**, 5042–5044 (1963)
6. D. Bakry, I. Gentil, M. Ledoux, *Analysis and Geometry of Markov Diffusion Operators*. Grundlehren der Mathematischen Wissenschaften, GL, vol. 348 (Springer, Berlin, 2014)
7. G. Bouchitté, C. Jimenez, M. Rajesh, A new L^∞ estimate in optimal mass transport. Proc. Am. Math. Soc. **135**, 3525–3535 (2007)
8. J.-D. Benamou, Y. Brenier, A computational fluid mechanics solution to the Monge-Kantorovich mass transfer problem. Numer. Math. **84**, 375–393 (2000)
9. Y. Brenier, Décomposition polaire et réarrangement monotone des champs de vecteurs (French) C. R. Acad. Sci. Paris Sér. I Math. **305**(19), 805–808 (1987)
10. Y. Brenier, Polar factorization and monotone rearrangement of vector-valued functions. Commun. Pure Appl. Math. **44**, 375–417 (1991)
11. H. Brezis, *Functional Analysis, Sobolev Spaces and Partial Differential Equations* (Springer-Verlag, New York, 2011)
12. L.A. Caffarelli, Interior $W^{2,p}$ estimates for solutions of the Monge-Ampère equation. Ann. Math. (2) **131**(1), 135–150 (1990)
13. L.A. Caffarelli, A localization property of viscosity solutions to the Monge-Ampère equation and their strict convexity. Ann. Math. (2) **131**(1), 129–134 (1990)
14. L.A. Caffarelli, Some regularity properties of solutions of Monge-Ampère equation. Commun. Pure Appl. Math. **44**(8–9), 965–969 (1991)
15. L.A. Caffarelli, Boundary regularity of maps with convex potentials. Commun. Pure Appl. Math. **45**(9), 1141–1151 (1992)
16. L.A. Caffarelli, The regularity of mappings with a convex potential. J. Am. Math. Soc. **5**(1), 99–104 (1992)
17. L.A. Caffarelli, Boundary regularity of maps with convex potentials II. Ann. Math. (2) **144**(3), 453–496 (1996)
18. T. Caillet, The five gradients inequality for non quadratic costs. Comptes Rendus Math. **361**, 715–721 (2023)
19. T. Caillet, F. Coudreuse, Generalized Logarithmic Sobolev Inequality by the JKO Scheme Thibault Caillet, Fanch Coudreuse. Preprint (2026). arXiv:2601.16620

20. T. Caillet, F. Santambrogio, Doubly nonlinear diffusive PDEs: new existence results via generalized Wasserstein gradient flows. SIAM J. Math. An. **56**(6), 7043–7073 (2024)
21. T. Caillet, F. Santambrogio, Fisher information and continuity estimates for nonlinear but 1-homogeneous diffusive PDEs (via the JKO scheme). Bull. Hell. Math. Soc. **69**, 16–27 (2025)
22. C. Cancès, D. Matthes, F. Nabet, E.-M. Rott, Finite elements for Wasserstein W_p gradient flows. ESAIM: M2AN **59**, 1565–1600 (2025)
23. J.-A. Carrillo, D. Slepčev, Example of a displacement convex functional of first order Calc. Var. Partial Differ. Equ. **36**(4), 547–564 (2009)
24. S. Daneri, G. Savaré, Eulerian calculus for the displacement convexity in the wasserstein distance. SIAM J. Math. Anal. **40**, 1104–1122 (2008)
25. E. De Giorgi, New problems on minimizing movements. *Boundary Value Problems for PDE and Applications,* ed. by C. Baiocchi, J.L. Lions (Masson, Issy-les-Moulineaux, 1993), pp. 81–98
26. G. De Philippis, A.R. Mészáros, F. Santambrogio, B. Velichkov, BV estimates in optimal transportation and applications. Arch. Rati. Mech. Anal. **219**(2), 829–860 (2016)
27. S. Di Marino, F. Santambrogio, JKO estimates in linear and non-linear Fokker-Planck equations, and Keller-Segel: L^p and Sobolev bounds Ann. Inst. H. Poincaré Anal. Non Linéaire (39), 1485–1517 (2022)
28. S. Di Marino, S. Murro, E. Radici, The five gradients inequality on differentiable manifolds. J. Math. Pures Appl. 187(9), 294–328 (2024)
29. R. Ducasse, F. Santambrogio, H. Yoldas, A cross-diffusion system obtained via (convex) relaxation in the JKO scheme. Calc. Var. Partial Differ. Equ. **62**, Article number 29 (2023)
30. I. Ekeland, R. Temam, *Convex Analysis and Variational Problems.* North-Holland Pub. Co. Amsterdam (1976)
31. V. Ferrari, F. Santambrogio, Lipschitz estimates on the JKO scheme for the Fokker–Planck equation on bounded convex domains. Appl. Math. Lett. **112**, 106806 (2021)
32. W. Gangbo, An elementary proof of the polar factorization of vector-valued functions. Arch. Ration. Mech. Anal. **128**, 381–399 (1994)
33. W. Gangbo, The Monge mass transfer problem and its applications. Contemp. Math. **226**, 79–104 (1999)
34. W. Gangbo, R. McCann, The geometry of optimal transportation. Acta Math. **177**, 113–161 (1996)
35. M. Iacobelli, F.S. Patacchini, F. Santambrogio, Weighted ultrafast diffusion equations: from well-posedness to long-time behaviour. Arch. Ration. Mech. Anal. **232**(3), 1165–1206 (2019)
36. M. Jacobs, I.C. Kim, J. Tong, The L^1-contraction principle in optimal transport. Ann. Sc. Norm. Super. Cl. Sci. **23**(4), 1871–1919 (2022)
37. R. Jordan, D. Kinderlehrer, F. Otto, The variational formulation of the Fokker-Planck equation. SIAM J. Math. Anal. **29**(1), 1–17 (1998)
38. L. Kantorovich, On the transfer of masses. Dokl. Acad. Nauk. USSR **37**, 7–8 (1942)
39. O.A. Ladyzhenskaja, V.A. Solonnikov, N.N. Ural'ceva, *Linear and Quasilinear Equations of Parabolic Type* (Nauka, Moskow 1967) (Russian). English transl.: Translations of Mathematical Monographs (AMS, Providence, 1968)
40. A. Lunardi, *Analytic Semigroups and Optimal Regularity in Parabolic Problems* (Birkhäuser Verlag, Basel, 1995), xviii+424 pp. 2013 reprint of the 1995 original, *Modern Birkhäuser Classics* (Birkhäuser/Springer Basel AG, Basel, 1995), xviii+424 pp.
41. D. Matthes, R.J. McCann, G. Savaré, A family of nonlinear fourth order equations of gradient flow type. Commun. Partial Differ. Equ. **34**(10–12), 1352–1397 (2009)
42. R.J. McCann, Existence and uniqueness of monotone measure preserving maps. Duke Math. J. **80**, 309–323 (1995)
43. R.J. McCann, A convexity principle for interacting gases. Adv. Math. **128**(1), 153–159 (1997)
44. R.J. McCann, M. Puel, Constructing a relativistic heat flow by transport time steps. Ann. Inst. Henri Poincaré C Anal. Non linéaire **26**(6), 2539–2580 (2009)
45. G. Monge, Mémoire sur la théorie des déblais et des remblais. Hist. l'Acad. R. Sci. Paris 666–704 (1781)

46. F. Otto, The geometry of dissipative evolution equations: the porous medium equation. Commun. Partial Differ. Equ. **26**, 101–174 (2011)
47. R.T. Rockafellar, *Convex Analysis* (Princeton University Press, Princeton, 1970)
48. F. Santambrogio, Flots de gradient dans les espaces métriques et leurs applications (d'après Ambrosio-Gigli-Savaré). *Proceedings of the Bourbaki Seminar* (2013) (in French)
49. F. Santambrogio, *Optimal Transport for Applied Mathematicians*. Progress in Nonlinear Differential Equations and Their Applications, vol. 87 (Birkhäuser Basel, 2015)
50. F. Santambrogio, Euclidean, metric, and wasserstein. Gradient flows: an overview. Bull. Math. Sci. **7**(1), 87–154 (2017)
51. F. Santambrogio, *A Course in the Calculus of Variations* (Springer, Berlin, 2023)
52. F. Santambrogio, G. Toshpulatov, Strong $L^2 H^2$ convergence of the JKO scheme for the Fokker-Planck equation. Arch. Ration. Mech. Anal. **248**, Article number 99 (2024)
53. J.L. Vázquez, *The Porous Medium Equation. Mathematical Theory* (The Clarendon Press/Oxford University Press, Oxford, 2007)
54. C. Villani, *Topics in Optimal Transportation*. Graduate Studies in Mathematics (AMS, Providence, 2003)
55. C. Villani, *Optimal Transport: Old and New* (Springer Verlag, Berlin, 2008)

Resonances and Computations

Yvain Bruned, Frédéric Rousset, and Katharina Schratz

1 Introduction

The numerical computation of nonlinear partial differential equations (PDEs) for smooth solutions is nowadays in large parts well understood. A large toolbox of techniques to approximate effectively their time dynamics was established, involving splitting methods (with the intention of splitting the full equation into a series of simpler subproblems), exponential integrators (based on Duhamel's formula) and many more approaches. For an extensive overview of classical methods in context of dispersive PDEs we refer to [29, 31, 37, 37, 41, 68] and the references therein.

While classical techniques in many situations allow an efficient and well-suited numerical approximation, they in general fail to reproduce the behaviour of solutions once roughness comes into play. This is due to the fact that in general classical integration methods require a lot of regularity on the solutions in order to converge. If the solution is not regular enough, well-known theoretical error estimates break down and almost nothing is known on how to overcome this.

Y. Bruned
University of Lorraine, Boulevard des Aiguillettes, Vandœuvre-lès-Nancy, France
e-mail: yvain.bruned@univ-lorraine.fr

F. Rousset
Université Paris- Saclay, Orsay Cedex, France
e-mail: frederic.rousset@universite-paris-saclay.fr

K. Schratz (✉)
Sorbonne University, Paris, France
e-mail: katharina.schratz@sorbonne-universite.fr

© The Author(s), under exclusive license to Springer Nature Switzerland AG 2026
N. D. Alikakos, C. Villani (eds.), *Analysis Techniques for Mathematical Physics and Geometry – and Music*, Lecture Notes in Mathematics 2390,
https://doi.org/10.1007/978-3-032-11396-2_5

Break down of convergence for rough data is not only a theoretical artefact, but also severely observed in numerical simulations. If the solution does not meet the required smoothness assumptions, the numerical solution becomes unstable and eventually blows up. More details on the struggle of classical methods for rough data are given in Sect. 2.

The main question we want to address in this review is the following: How can we actually reproduce rough time dynamics in nonlinear PDEs an overcome classical smoothness assumptions?

To answer this question we will present a novel integration technique, so-called *resonances as a computational tool*, with the aim to reduce as far as possible the high regularity assumptions needed in classical schemes. The novel idea of this approach lies in building numerical schemes on the underlying structure of resonances of the PDE, instead of discretising the PDE itself. This is achieved by looking at iterations of Duhamel's formula and filtering out the dominant parts in the oscillations. To control the higher order oscillations, we use a decorated tree formalism with the frequencies as decorations. This will allow us to construct resonance based schemes up to arbitrary order. Details are given in Sect. 3 where we present the main ideas behind the derivation of low regularity schemes for dispersive equations. We also present some extensions that allow us to go deeper, e.g., preserving essential symmetries in the equation.

2 Classical Methods and Their Problem for Rough Data

To present the main challenges of classical methods for rough data, let us consider as a model problem the periodic Korteweg–de Vries equation

$$\partial_t u(t,x) + \partial_x^3 u(t,x) = \frac{1}{2}\partial_x u^2(t,x) \qquad (t,x) \in \mathbb{R} \times \mathbb{T} \tag{1}$$

with rough initial data

$$u(0,x) = u_0(x) \in H^\sigma(\mathbb{T}), \tag{2}$$

where we want to choose $\sigma > 0$ as small as possible (we will see later how far we can actually push down the Sobolev index σ). Here, $H^\sigma(\mathbb{T})$ denotes the classical Sobolev space on torus $\mathbb{T}$ with regularity $\sigma \geq 0$.

Before we present the main idea of *resonances as a computational tool*—a new method which ideally allows us to choose σ in (2) arbitrarily close to zero—let us first give a brief overview of classical numerical methods to solve the nonlinear PDE (1). This will allow us to better understand the regularity assumptions of classical approaches and get essential insight on how to overcome this.

2.1 Splitting Methods

One of the most famous numerical methods to approximate the time dynamics of KdV equation (1) and in general nonlinear dispersive models are splitting methods (see, e.g., [31, 37, 38, 44, 46, 48, 55, 57]). The main idea thereby lies in splitting the full equation into a series of simpler subproblems. One then solves these subproblems (either exactly or with a numerical scheme) and composes the sub-flows to obtain an approximation to the flow of the original—full—equation.

Looking at the KdV equation (1) we face two main challenges numerically: The differential operator ∂_x^3 and the non-linear Burger's term $\frac{1}{2}\partial_x u^2$. The idea of splitting is now the following: Instead of solving the full problem (1) we consider its linear and nonlinear part separately, i.e.,

$$\text{(L)} \quad \partial_t u(t,x) = -\partial_x^3 u(t,x) \quad \text{and} \quad \text{(N)} \quad \partial_t u(t,x) = \frac{1}{2}\partial_x u^2(t,x). \tag{3}$$

The main advantage thereby lies in the fact that the time dynamics in both subproblems can be solved exactly: (L) in Fourier space (in case of a spatial discretisation we can couple this with for instance a Fourier pseudo spectral method) and (N) with the method of characteristics.

Remark 1 In the end we are of course interested in a full discretisation as we have a PDE both in time and in space. This will introduce an additional spatial error. In case of splitting for KdV we will have a spectral error for the linear problem (L) as well as an interpolation error for the nonlinear problem (N) as we do not necessarily end up on the grid points any more when using the method of characteristics. The latter requires an interpolation, e.g., Lagrange interpolation, or the use of WENO schemes, e.g., upwind scheme, see for instance [1, 53]. In this review we only focus on the time dynamics and hence neglect the additional error introduced by the spatial discretisation of PDEs. For fully discrete resonance based schemes in context of Navier–Stokes equation and Schrödinger equations we refer for instance to [47, 51].

A simple composition of the kinetic flow ϕ_L^t and nonlinear flow ϕ_N^t leads at first-order to the so-called first order Lie splitting method (with time step size τ)

$$\phi_L^\tau \circ \phi_N^\tau \tag{4}$$

and its symetrised version the classical second-order Strang splitting

$$\phi_L^{\tau/2} \circ \phi_N^\tau \circ \phi_L^{\tau/2}. \tag{5}$$

We refer to an overview of splitting methods to [37, 44, 48, 57].

Here we are in interested in the local error structure the splitting approach introduces for KdV, as this will allow us to answer the question what kind of regularity assumptions this imposes on the solution and more importantly, how small can we actually choose the regularity σ in the initial value (2) in case of solving KdV with Strang splitting?

First rigorous error estimates where achieved in [39]. Estimates requiring less regularity on the solution have been established in [40] showing that the local error of the Strang splitting (5) for the KdV equation (1) is driven by the triple commutator

$$[\partial_x^3, [\mathcal{N}, \partial_x^3]] \tag{6}$$

where $\mathcal{N}$ denotes the Burgers nonlinearity $\mathcal{N}(u) = \frac{1}{2}\partial_x u^2$. In this commutator structure the "worst-order" derivatives ∂_x^7 and ∂_x^6 drop such that formally $[\partial_x^3, [\mathcal{N}, \partial_x^3]]$ is a fifth-order differential operator (cf. [40, Lemma 6.1]) which implies a local error behaviour at order

$$\mathcal{O}\left(\tau^3 \partial_x^5 u\right).$$

Hence, in order to reach second-order convergence Strang splitting requires at least the boundedness of five additional derivatives of the solution.

2.2 Exponential Integrators

Another well-known method to solve KdV-type equations are exponential integrators (see, e.g., [17, 41, 50, 61] and the references therein). The main idea thereby lies in discretising Duhamel's formula which for KdV (1) takes the form

$$u(t) = e^{-t\partial_x^3} u(0) + \frac{1}{2}\partial_x e^{-t\partial_x^3} \int_0^t e^{s\partial_x^3} u^2(s) ds. \tag{7}$$

At time $t_{n+1} = t_n + \tau$ we find (considering Duhamels formula on the time interval $[0, \tau]$ with initial value $u(t_n)$) that

$$u(t_{n+1}) = e^{-t\partial_x^3} u(t_n) + \frac{1}{2}\partial_x e^{-\tau\partial_x^3} \int_0^\tau e^{s\partial_x^3} u^2(t_n + s) ds. \tag{8}$$

Exponential integrator schemes are based on Taylor series expansion of the solution within the integral. At first order they build on the first-order Taylor series expansion

$$u(t_n + s) = u(t_n) + \mathcal{O}(su'). \tag{9}$$

Plugging the approximation (9) into Duhamel's formula (8) leads to the first-order exponential integrator scheme

$$u^{n+1} = e^{-\tau\partial_x^3} u^n + \frac{1}{2}\tau\partial_x \varphi_1(-\tau\partial_x^3)(u^n)^2 \quad \text{with} \quad \varphi_1(z) = \frac{e^z - 1}{z}. \tag{10}$$

From the Taylor series expansion (9) we easily see that the local error of the first-order exponential integrator method (10) is driven by the time derivative u', where by the KdV equation (1) in the sense of regularity we have

$$\mathcal{O}(u') = \mathcal{O}(\partial_x^3 u). \tag{11}$$

Hence, first-order convergence requires the boundedness of at least three additional derivatives. Higher-order exponential integrators are based on higher-order Taylor series expansion of the solution within the integral, i.e., at second-order one takes

$$u(t_n + s) = u(t_n) + su'(t_n) + \mathcal{O}(s^2 u'') \tag{12}$$

and replaces the time derivative $u'(t_n)$ by the equation itself $u'(t_n) = -\partial_x^3 u(t_n) + \frac{1}{2}\partial_x u^2(t_n)$. Due to the local error scaling

$$\mathcal{O}(u'') = \mathcal{O}(\partial_x^6 u)$$

we see that second-order exponential integrator method requires the boundedness of at least six additional derivatives.

2.3 *Structure of the Solution and Classical Methods*

From the local error structure (6) and (11) we see that both splitting and exponential integrator methods require sufficiently smooth solutions to guarantee the boundedness of their local error. A natural question therefore arises:

What happens for rough data? And can we actually construct numerical schemes which allow convergence for rougher data than classical methods do?

To answer this question we first have to understand the underlying structure of the solution to KdV (1) in a better way. For this purpose we turn back to Duhamel's formula (7). However, instead of approximating its solution $u(s)$ by a classical Taylor series expansion (as done in exponential integrator schemes) we look at iterations of Duhamel's formula: Using that

$$u(s) = e^{-s\partial_x^3} u(0) + \int_0^s \ldots \, \mathrm{d}s_1$$

we find

$$u(t) = e^{-t\partial_x^3} u(0) + \frac{1}{2}\partial_x e^{-t\partial_x^3} \int_0^t e^{s\partial_x^3} \left(e^{-s\partial_x^3} u(0)\right)^2 \mathrm{d}s + \int_0^t \int_0^s \ldots \, \mathrm{d}s_1 \mathrm{d}s. \tag{13}$$

If we want to get a rough idea of the dynamics of the solution, we can at first forget about the higher order iterations, and neglect the double integral $\int_0^t \int_0^s \mathrm{d}s_1 \mathrm{d}s$ in

(13). We then see that the underlying structure of the solution $u(t)$ is driven by the nonlinear frequency interaction of ∂_x^3 and $-\partial_x^3$ with leading oscillations

$$\mathrm{Osc}(s, \partial_x^3, u(0)) = \partial_x e^{s\partial_x^3} \left(e^{-s\partial_x^3} u(0)\right)^2 . \tag{14}$$

Numerical schemes stay close to the structure of the solution if they resolve the leading oscillations (14).

A closer look, however, shows that splitting methods and exponential integrators in general neglect the nonlinear frequency interactions in (14): Lie splitting (4) is based on the frequency approximation

$$\mathrm{Osc}(s, \partial_x^3, u(0)) \approx \partial_x u^2(0) \tag{15}$$

while exponential integrator methods swallow all frequencies within the nonlinearity based on

$$\mathrm{Osc}(s, \partial_x^3, u(0)) \approx \partial_x e^{s\partial_x^3} u^2(0). \tag{16}$$

In case of smooth solutions, for which $\partial_x^3 u$ is well defined in the space of interest linearisation of frequencies such as (15) and (16) in general lead to good approximations of the exact solution (13). This can be seen by a simple Taylor series expansion of the oscillations

$$e^{s\partial_x^3} u = u + \mathcal{O}(s\partial_x^3 u). \tag{17}$$

Expansion (17) introduces a small remainder of order s as long as u is sufficiently smooth, i.e., $\partial_x^3 u$ is bounded.

For rough solutions, on the other hand, for which

$$\partial_x^3 u$$

is unbounded, approximations such as (15) and (16) break down as the linearisation of the frequency interactions (17) is no longer valid. The latter is not only a theoretical artefact stemming from error analysis, but also drastically observed in numerical experiments, where we observe instability and loss of convergence for rough data.

The main aim of resonances as a computational tool is to overcome this by stepping away from linearising the frequency interactions as done by classical methods (e.g., (15) and (16)) towards new schemes which deeply embed the nonlinear frequency interactions in PDEs into the numerical discretisation. In general this allows us to achieve reliable approximations for much rougher data than classical schemes can handle.

In Sect. 3 we will explain this idea of tackling nonlinear frequency interactions in a nonlinear way in detail on the concrete example of periodic KdV (1).

We refer to [43, 60, 69] for first results on resonance based schemes for KdV, NLS and Dirac equations and the recent review article on first- and second-order methods [67].

3 Resonances as a Computational Tool: The Main Idea

We rewrite in Fourier space the Duhamel's formula given in (7) and we obtain the following equation for the k-th Fourier coefficient of the solution:

$$u_k(t) = e^{itk^3} v_k + ik \sum_{\substack{k_1,k_2\in\mathbb{Z}\\ k_1+k_2=k}} e^{itk^3} \int_0^t e^{-isk^3} u_{k_1}(s) u_{k_2}(s) ds, \tag{18}$$

where the linear operators ∂_x, $e^{-t\partial_x^3}$, $e^{t\partial_x^3}$ are sent respectively in Fourier space to ik, e^{itk^3}, e^{-itk^3}. The splitting of k into k_1, k_2 comes from the fact that moving to Fourier space, pointwise products become convolution products on the frequency. We fix $r = 0$ and iterate (18) by replacing $u_{k_j}(t)$ by

$$u_{k_j}(t) = e^{itk_j^3} v_{k_j} + \mathcal{O}(t),$$

with $j \in \{1, 2\}$. We obtain the following first order approximation of the k-th Fourier coefficient $u_k(t)$:

$$u_k(t) = e^{itk^3} v_k + ik \sum_{\substack{k_1,k_2\in\mathbb{Z}\\ k_1+k_2=k}} e^{itk^3} \int_0^t e^{-isk^3} (e^{isk_1^3} v_{k_1})(e^{isk_2^3} v_{k_2}) ds + \mathcal{O}(t^2).$$

One can encode the previous Duhamel iterates using a decorated tree series. We denote by $U_k^r(v, t)$ the first iterated integrals of size r of the Duhamel expansion. These are integrals with r integrations in time. One has

$$|u_k(t) - U_k^r(v, t)| = \mathcal{O}(t^{r+1})$$

where the regularity asked on the initial data hidden in the notation $\mathcal{O}$ corresponds to the regularity needed to define the first iterated integrals up to order r. Decorated trees are used to provide a precise description of $U_k^r(v, t)$ and to implement in a systematic way the resonance analysis. One uses the following B-series type formula

$$U_k^r(v, t) = \sum_{T \in \mathcal{T}_k^r} \frac{\Upsilon(T)(v)}{S(T)} (\Pi T)(t) \tag{19}$$

where $\mathcal{T}_k^r$ is a suitable set of decorated trees of size r, $S(T)$ is a symmetry factor and $\Upsilon(T)$ is an elementary differential associated with T depending on the initial data v. The map Π sends T to an oscillatory integral. Such a formalism is reminiscence of the B-series (named after Butcher) that describes numerical schemes for ODEs and PDEs. We refer to [16, 19, 37] for references on B-series and to [18, 32, 35, 42, 54] for references on tree series used in the context of dispersive equations . The formalism (19) is largely inspired from the treatment of singular stochastic partial differential equations (SPDEs) via regularity structures in [9, 11, 36] where a local ansatz for the solutions takes a similar form. It is also connected to rough paths [33, 34, 56], one of the foundational ideas of regularity structure. Note that the main difference with the classical B-series comes from the decorations and the application from set of trees to differential objects. Indeed, the decorated trees encode iterated integrals $(\Pi T)(t)$ but also elementary differentials $\Upsilon(T)(v)$. Below, we describe the series for $r = 2$. One has

$$\mathcal{T}_k^2 = \{T_0, T_1, T_2, \ k_1, k_2, k_3 \in \mathbb{Z}\},$$

where the decorated trees T_i are given by

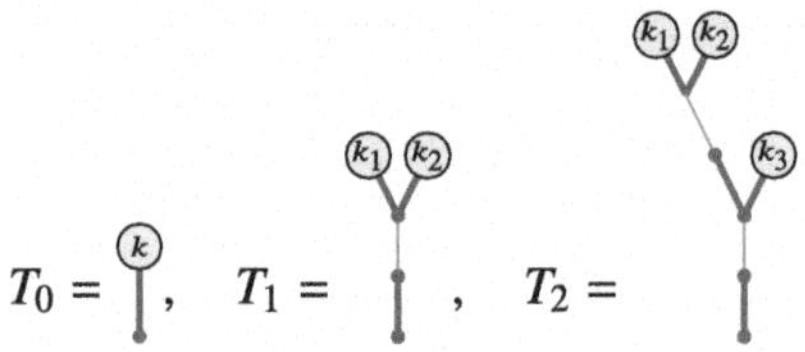

where for T_1, one has $k = k_1 + k_2$ and for T_2 one has $k = k_1 + k_2 + k_3$. For the decorated trees written above, we have used the following coding: An edge | corresponds to an integral $ik \int_0^t e^{-isk^3} \cdots ds$. The decorations on the leaves correspond to the frequencies that add up in the intern node up to k which is the decoration of the node connected to the root of the tree. We have omitted these node decorations. One has

$$(\Pi T_0)(t) = e^{itk^3}, \quad (\Pi T_1)(t) = ike^{itk^3} \int_0^t e^{-is(k^3-k_1^3-k_2^3)} ds$$

$$(\Pi T_2)(t) = -kk_{12}e^{itk^3} \int_0^t e^{-is(k^3-k_{12}^3-k_3^3)} v \int_0^s e^{-i\bar{s}(k_{12}^3-k_1^3-k_2^3)} d\bar{s} ds$$

where $k_{12} = k_1 + k_2$. The symmetry factor $S(T)$ corresponds to the number of internal symmetries of the tree T taking the edge decorations into account but not the node decorations. One obtains

$$S(T_0) = 1, \quad S(T_1) = 2, \quad S(T_2) = 2.$$

The 2 in $S(T_1)$ and $S(T_2)$ comes from the fact that one can permute the two leaves decorated by k_1 and k_2. The elementary differential $\Upsilon(T)(v)$ corresponds to a product of initial data associated with the leaves of a decorated tree. One has to take into account also a factor connected to the structure of the tree: In the case of the KdV equation, this factor is 2^m where m is the number of inner nodes in the tree. One gets

$$\Upsilon(T_0)(v) = v_k, \quad \Upsilon(T_1)(v) = 2v_{k_1}v_{k_2}, \quad \Upsilon(T_2)(v) = 4v_{k_1}v_{k_2}v_{k_3}.$$

Let us remark that the sum over elements $\mathcal{T}_k^r$ is infinite. One has a finite number of tree shapes and the infinity comes from the frequencies k_1, k_2, k_3 on which one performs an infinite summation. The term $U_k^r(t)$ gives a first approximation of order $r+1$ but one has to discretise each $(\Pi T_i)(t)$ such that its discretisation is easy to rewrite in physical space via usual differential operators. To produce a better scheme, one has to embed the resonance into the discretisation. Let us consider the first iterated integral

$$(\Pi T_1)(t) = ike^{itk^3}\int_0^t e^{-is(k^3-k_1^3-k_2^3)}ds, \quad k = k_1 + k_2.$$

One observes the following

$$k^3 - k_1^3 - k_2^3 = 3k_1k_2(k_1+k_2) - \mathcal{L}_{\text{low}}. \tag{20}$$

We use the notation $\mathcal{L}_{\text{low}}$ with the subscript $_{\text{low}}$ to stress that the term obtained is of order lower than the operator meaning that one has at most a factor k_i^2 which is lower than k_i^3. Then, by exact integration when $k_i \neq 0$ and $k_1 + k_2 \neq 0$, one gets

$$\int_0^t e^{-is\mathcal{L}_{\text{low}}}ds = -\frac{e^{-it\mathcal{L}_{\text{low}}}-1}{i3k_1k_2(k_1+k_2)}.$$

One can notice that the factor at the denominator in the previous term can be easily mapped back to physical space. The same is true for $e^{-it\mathcal{L}_{\text{low}}}$. Indeed, one has

$$\frac{1}{k_1k_2(k_1+k_2)}v_{k_1}v_{k_2} \longrightarrow \partial_x^{-1}\left(\partial_x^{-1}v\right)^2$$

and by (20) (using it from the right to the left)

$$e^{-it\mathcal{L}_{\text{low}}}v_{k_1}v_{k_2} = e^{-itk^3}e^{itk_1^3}e^{itk_2^3}v_{k_1}v_{k_2} \longrightarrow e^{t\partial_x^3}\left(e^{-t\partial_x^3}v\right)^2.$$

This is a crucial observation as the first iterated integral can be computed exactly without any further regularity on the initial data. Therefore, if we denote by $(\Pi^r T_1)(t)$ the approximation of order r of $(\Pi T_1)(t)$ then one has

$$(\Pi^r T_1)(t) = (\Pi T_1)(t) = -ik_{12}e^{itk^3}\frac{e^{-it\mathcal{L}_{\text{low}}}-1}{i3k_1k_2(k_1+k_2)},$$

Now, we have to face the discretisation of the second order integral $(\Pi T_2)(t)$. One can maybe hope for the same type of trick but it is not the case. Indeed, if we substitute the previous approximation, one has

$$(\Pi T_2)(t) = k k_{12} e^{itk^3} \int_0^t e^{-is(k^3-k_{12}^3-k_3^3)} \frac{e^{-is\mathcal{L}_{\text{low}}}-1}{i3k_1k_2(k_1+k_2)} ds.$$

Then, one has two contributions to the resonance analysis which are given by

$$\begin{aligned} k^3 - k_{12}^3 - k_3^3 &= 3k_{12}k_3k, \\ k^3 - k_{12}^3 - k_3^3 + \mathcal{L}_{\text{low}} &= k^3 - k_1^3 - k_2^3 - k_3^3. \end{aligned} \tag{21}$$

The first term in the computation above can lead to an exact integration. Unfortunately, the second term does not have a nice factorisation. In this case, we have to Taylor expand everything which is still better than classical schemes. As we want a very simple scheme, we do not perform any exact integration and Taylor expand everything as the local error is the same. This gives the following approximation:

$$\begin{aligned} (\Pi T_2)(t) &= -k k_{12} e^{itk^3} \int_0^t e^{-is(k^3-k_{12}^3-k_3^3)} \int_0^s \left(1 + O(\bar{s}(k_1^2k_2 + k_2^2k_1))\right) d\bar{s} ds \\ &= -k k_{12} e^{itk^3} \int_0^t s\left(1 + O(s(k_{12}^2k_3 + k_3^2k_{12}))\right) ds \\ &\quad + O(t^3 k k_{12}(k_1^2k_2 + k_2^2k_1)) \\ &= -k k_{12} e^{itk^3} \frac{t^2}{2} + O(t^3 k k_{12}(k_{12}^2k_3 + k_3^2k_{12})) \\ &\quad + O(t^3 k k_{12}(k_1^2k_2 + k_2^2k_1)) \end{aligned}$$

where the local error term with a polynomial in the frequency inside the $O(\cdot)$ means that we have to have enough regularity on the initial data v in some suitable Sobolev space. One can observe that the two error terms ask the same regularity on the initial data which is 5 derivatives. We therefore set

$$(\Pi^2 T_2)(t) = -k k_{12} e^{itk^3} \frac{t^2}{2}$$

and one has

$$\left| \frac{\Upsilon(T)(v)}{S(T)} (\Pi T_2)(t) - \frac{\Upsilon(T)(v)}{S(T)} (\Pi^2 T_2)(t) \right| = O(t^3 P(k_1, k_2, k_3) \Upsilon(T)(v))$$

where P is a polynomial with degree at most equal to 5 in each variable k_i. Computing by hand without any structure for higher-order schemes is difficult.

This is where the decorated trees' formalism is useful providing a framework that organises these computations.

One important remark is that in general in the computation of the resonance, one does not expect to see a decrease in the order of the polynomial. Indeed, in (21), we can observe that one moves degree three to degree two. This is not the case for the Schrödinger equation and other dispersive equations. In general, one has the following structure:

$$\mathcal{L} = P(k) + \sum_{i=1}^{n} a_i P(k_i)$$

where the a_i belongs to $\{-1, 1\}$ and the polynomial P is the differential operator of the equation translated into Fourier space. One expects a decomposition of the form

$$P(k) + \sum_{i=1}^{n} a_i P(k_i) = \mathcal{L}_{\text{dom}} + \mathcal{L}_{\text{low}} \tag{22}$$

where $\mathcal{L}_{\text{dom}}$ contains the higher monomials in some frequency k_i. If $\mathcal{L}_{\text{dom}} = 0$ then we are in a case similar to the KdV equation. Otherwise, we want $\mathcal{L}_{\text{dom}}$ to be in a specific form for being able to map $\int_0^t e^{is\mathcal{L}_{\text{dom}}} ds$ back into physical space with usual differential operators. This is the case for $\mathcal{L}_{\text{dom}} = k_i^n$ or any linear combination of these monomials with coefficients in $\{-1, 1\}$. Then, one proceeds as follows

$$\begin{aligned} \int_0^t e^{is\mathcal{L}} ds &= \int_0^t e^{is\mathcal{L}_{\text{dom}}} e^{is\mathcal{L}_{\text{low}}} ds \\ &= \int_0^t e^{is\mathcal{L}_{\text{dom}}} (1 + O(s\mathcal{L}_{\text{low}})) ds = \int_0^t e^{is\mathcal{L}_{\text{dom}}} ds + O(s^2 \mathcal{L}_{\text{low}}) \end{aligned} \tag{23}$$

which gives a discretisation requiring a regularity given by $\mathcal{L}_{\text{low}}$ instead of $\mathcal{L}$. The approximation above could be seen as Taylor expanding the lower part of the operator $\mathcal{L}$ denoted by $\mathcal{L}_{\text{low}}$ and integrating exactly its dominant part $\mathcal{L}_{\text{dom}}$. Let us stress that in the deterministic setting, one has an exact expression for the integral $\int_0^t e^{is\mathcal{L}_{\text{dom}}} ds$ In a stochastic context, one has to deal with stochastic iterated integrals as in [6]. For those, one does not have pathwise calculus for these iterated integrals and therefore, one is limited to a few iterations for a low regularity scheme.

The splitting observed in (22) happens in fact in the context of the nonlinear cubic Schrödinger equation (NLS) given for $(t, x) \in \mathbb{R}_+ \times \mathbb{T}^d$ by

$$i\partial_t u + \Delta u = |u|^2 u, \quad u(0, x) = v(x). \tag{24}$$

The Duhamel formula for the k-th Fourier coefficient of the solution is given by

$$u_k(t) = e^{-itk^2} v_k - \sum_{\substack{k_1, k_2, k_3 \in \mathbb{Z}^d \\ -k_1 + k_2 + k_3 = k}} i e^{-itk^2} \int_0^t e^{isk^2} \overline{u_{k_1}}(s) u_{k_2}(s) u_{k_3}(s) ds.$$

Now as we work for $d \geq 1$, the notations k^2 and $k\ell$ mean for $k, \ell \in \mathbb{Z}^d$

$$k^2 = \sum_{i=1}^{d} (k^{(i)})^2, \quad k\ell = \sum_{i=1}^{d} k^{(i)} \ell^{(i)}$$

where $k = (k^{(1)}, ..., k^{(d)})$ and $\ell = (\ell^{(1)}, ..., \ell^{(d)})$. The first iterated integral is

$$-ie^{-itk^2} \int_0^t e^{isk^2} e^{isk_1^2} e^{-isk_2^2} e^{-isk_3^2} ds\, \overline{v}_{k_1} v_{k_2} v_{k_3} \tag{25}$$

where $k = -k_1 + k_2 + k_3$. Then, the resonance inside the integral is given by

$$k^2 + k_1^2 - k_2^2 - k_3^2 = 2k_1^2 - 2k_1(k_2 + k_3) + 2k_2 k_3.$$

In this case, one has $\mathcal{L}_{\text{dom}} \neq 0$ and

$$\mathcal{L}_{\text{dom}} = 2k_1^2, \quad \mathcal{L}_{\text{low}} = -2k_1(k_2 + k_3) + 2k_2 k_3.$$

We are therefore in the case where exact integration and full Taylor expansion are needed at the same time. The discretisation of (25) is given by

$$-ie^{-itk^2} \int_0^t e^{isk_1^2} ds + \mathcal{O}(t^2(-2k_1(k_2 + k_3) + 2k_2 k_3)).$$

The decorated trees for NLS are of different nature and are given for $r = 2$ by

$$\mathcal{T}_k^2 = \left\{ T_0, T_1, T_2, T_3,\ k_i \in \mathbb{Z}^d \right\},$$

$T_0 =$ 

, $T_1 =$ 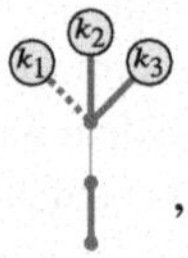

, $T_2 =$ 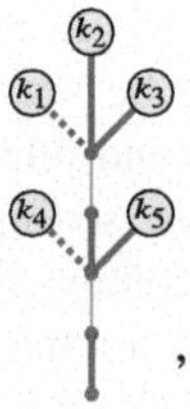

, $T_3 =$ 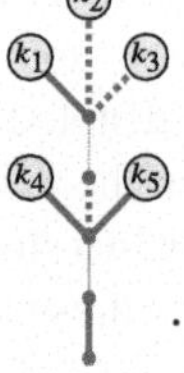

.

where for T_1, $k = -k_1 + k_2 + k_3$, for T_2, $k = -k_4 - k_1 + k_2 + k_3 + k_5$ and for T_3, $k = k_4 + k_1 - k_2 - k_3 + k_5$. An edge (resp.) corresponds to a factor e^{-itk^2} (resp. e^{itk^2}), while an edge (resp.) corresponds to an integral $-i \int_0^t e^{isk^2} \cdots ds$ (resp. $i \int_0^t e^{-isk^2} \cdots ds$). The dotted edges can be seen as taking the complex conjugate of the operator. Also, the frequencies add up to the root with a minus sign for dotted edges. Indeed, for the decorated tree T_3, one has

$$k = k_4 - \ell + k_5, \quad -\ell = k_1 - k_2 - k_3$$

where ℓ corresponds to the node decoration of the inner nodes not connected to the root. This shows that our decorated tree formalism is very robust for encoding many iterated integrals originating from dispersive PDEs.

One can further simplify the scheme by using a nice trick. The idea is that one may know in advance the regularity a priori asked on the initial data in some Sobolev space. If one has enough regularity on the initial data, then the computation (23) can be replaced by a classical discretisation without a resonance analysis

$$\int_0^t e^{is\mathcal{L}} ds = \int_0^t (1 + O(s\mathcal{L}))ds = t + O(t^2\mathcal{L}). \tag{26}$$

Then, given a decorated tree T, we denote by $(\Pi^{n,r}T)(t)$ the discretisation of $(\Pi T)(t)$ following these rules. Here, the index n corresponds to the a priori regularity of the initial data and r is the order of the discretisation. If we let n vary, one gets an entire family of resonance-based schemes. The resonance-based scheme takes the form

$$U_k^{n,r}(v,t) = \sum_{T \in \mathcal{T}_k^r} \frac{\Upsilon(T)(v)}{S(T)} (\Pi^{n,r}T)(t). \tag{27}$$

One has the following general result for the computation of the local error

Theorem 1 *For every decorated tree T one has,*

$$\left(\Pi T - \Pi^{n,r}T\right)(t) = \mathcal{O}\left(t^{r+1}\mathcal{L}_{\text{low}}^r(T,n)\right).$$

Then, one has

$$U_k^{n,r}(v,t) - U_k(t) = \sum_{T \in \mathcal{T}_k^r} \mathcal{O}\left(t^{r+1}\mathcal{L}_{\text{low}}^r(T,n)\Upsilon(T)(v)\right)$$

where there exists an operator $\mathcal{L}_{\text{low}}^r(T,n)$ which can be explicitly computed embeds the necessary regularity of the solution.

The operator $\mathcal{L}_{\text{low}}^r(T,n)$ is uniquely determined by the choice performed in the discretisation meaning the splitting of the various operators $\mathcal{L}$ connected to the decorated tree T into $\mathcal{L}_{\text{low}}$ and $\mathcal{L}_{\text{dom}}$. In [14, Def. 3.11], one has a recursive formula for computing these operators that relies crucially on the Hopf algebraic construction via a well-chosen coproduct on decorated trees.

As an application of the previous theorem, one can get a low regularity scheme for the KdV equation:

Corollary 1 *For the KdV equation* (1) *the general scheme* (27) *takes at first order the form*

$$u^{\ell+1} = e^{-\tau\partial_x^3}u^\ell + \frac{1}{6}\left(e^{-\tau\partial_x^3}\partial_x^{-1}u^\ell\right)^2 - \frac{1}{6}e^{-\tau\partial_x^3}\left(\partial_x^{-1}u^\ell\right)^2 \tag{28}$$

with a local error of order $\mathcal{O}\left(\tau^2\partial_x^2 u\right)$ *and at second-order*

$$\begin{aligned} u^{\ell+1} &= e^{-\tau\partial_x^3}u^\ell + \frac{1}{6}\left(e^{-\tau\partial_x^3}\partial_x^{-1}u^\ell\right)^2 - \frac{1}{6}e^{-\tau\partial_x^3}\left(\partial_x^{-1}u^\ell\right)^2 \\ &+ \frac{\tau^2}{4}e^{-\tau\partial_x^3}\Psi(i\tau\partial_x^2)\Big(\partial_x\Big(u^\ell\partial_x(u^\ell u^\ell)\Big)\Big) \end{aligned} \tag{29}$$

with a local error of order $\mathcal{O}\left(\tau^3\partial_x^4 u\right)$ *and a suitable filter function* Ψ *satisfying*

$$\Psi = \Psi\left(i\tau\partial_x^2\right), \quad \Psi(0) = 1, \quad \|\tau\Psi\left(i\tau\partial_x^2\right)\partial_x^2\|_r \leq 1.$$

With the aid of stability one can obtain a global error estimate from the local error of Theorem 1 with the aid of Lady Windamere's fan argument [37]. The necessary stability estimates in general rely on the algebraic structure of the underlying space. In the stability analysis of dispersive PDEs set in Sobolev spaces H^r one classically exploits bilinear estimates of type

$$\|vw\|_r \leq c_{r,d}\|v\|_r\|w\|_r.$$

The latter only holds for $r > d/2$ and thus restricts the analysis to sufficiently smooth Sobolev spaces H^r with $r > d/2$. To obtain (sharp) L^2 global error estimates one needs to exploit discrete Strichartz estimates and discrete Bourgain spaces in the periodic setting, see, e.g., [45, 62, 64] and Sect. 4.

One can study the composition and substitution of numerical methods of the form (27) repeating the construction performed for B-series (see [19] for a survey). The substitution is connected to the renormalisation in quantum field theory [22]. A first work in this direction is [12] which computes such operations in the context of regularity structures from which the schemes borrow their form.

Theorem 1 has been obtained in [14] for a large class of dispersive PDEs. It computes the local error for a resonance-based scheme. Let us briefly explain how $\mathcal{L}_{\text{low}}^r(T,n)$ can be computed via Hopf algebra techniques. One can define a coproduct on $\mathcal{F}$ the forest formed of decorated trees in $\mathcal{T}$. This is the free commutative monoid generated by $\mathcal{T}$. The coproduct $\Delta_{\text{BCK}} : \mathcal{F} \to \mathcal{F} \otimes \mathcal{F}$ is defined via admissible cuts and it is very similar to the Butcher-Connes-Kreimer coproduct [16, 21] used for numerical analysis and renromalisation in quantum field theory. We will not give a precise definition of this map but just a computation on one example coming from the KdV equation. One has

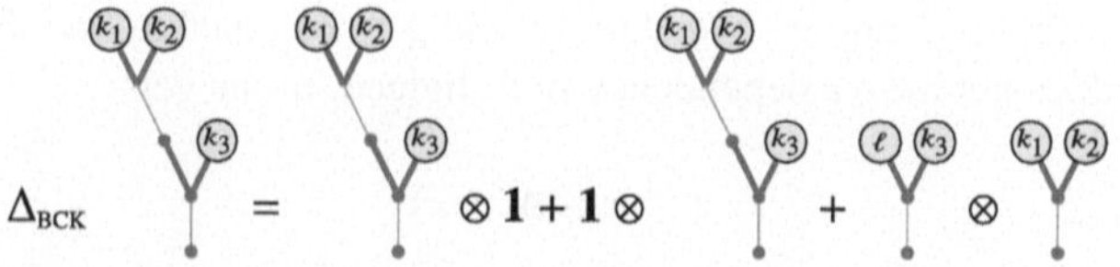

where $\ell = k_1 + k_2$. On the right part of the tensor product, we have collected the branches of the decorated tree which have been cut. On the left part, we have put the trunk. Here, $\mathbf{1}$ denotes the empty tree. The edges that can be cut are in blue and there are only two choices: Cutting the edge connected to the root or the one above. It is not possible to cut them simultaneously as they are on the path to the root (not admissible cut). Why such a cutting algorithm can be used in the context of dispersive PDEs? This comes from the computation of the dominant parts and their exact integration in the resonance analysis. Indeed, one has

$$\int_0^t e^{is\mathcal{L}_{\text{dom}}} ds = \frac{e^{it\mathcal{L}_{\text{dom}}} - 1}{i\mathcal{L}_{\text{dom}}}$$

that gives two different contributions $e^{is\mathcal{L}_{\text{dom}}}$ and 1 to plug into a new resonance problem for computing the next steps of the discretisation. Then, the contribution 1 corresponds to a branch we detach and $e^{is\mathcal{L}_{\text{dom}}}$ to the trunk. One can define a new map on decorated trees $\hat{\Pi}^{n,r}$ which is of the form

$$(\hat{\Pi}^{n,r} T)(t) = b(n, r, T, t) e^{it\mathscr{F}_{\text{dom}}(T)}.$$

where $\mathscr{F}_{\text{dom}}(T)$ is computed recursively and corresponds to the dominant resonance coming from $\Pi^{n,r} T$. Here, $b(n, r, T, t)$ is a coefficient depending on many paremeters. In fact, one is able to write a Birkhoff type factorisation of the form

$$\hat{\Pi}^n = \left(\Pi^n \otimes \left(Q \circ \Pi^n \mathcal{A} \cdot\right)(0)\right) \hat{\Delta}_{\text{BCK}}.$$

Here, both Π^n and $\hat{\Pi}^n$ are defined on a larger set of decorated trees where the order of the discretisation is now part of the root decoration. The coprodcut $\hat{\Delta}_{\text{BCK}}$ is an extension of Δ_{BCK} closer in spirit to the recentering coproduct introduced in [11, 36]. The map $\mathcal{A}$ is a recursive map called antipode that performs the various non-admissible cuts. The map Q is a projector that keeps only the mode 1 removing all the terms of the form $e^{iP(k_1,\ldots,k_n)}$. This Hopf algebra approach has been pushed forward in [4] by proposing a forest formula for the resonance-based schemes to find symmetric schemes. It is an open problem to get very explicit formulae for these schemes although we know that they depend on various lower and dominant parts that can be computed explicitly.

Before moving to structure-preserving schemes, one can generalise the resonance approach to non-polynomial nonlinearities of type

$$f(u) g(\overline{u})$$

for smooth functions f and g. We allow f and g to be equal to one. For example, in KdV, one does not have a dependency in $\bar{u}$. Indeed, if one gets

$$e^{is\mathcal{L}}v + A(v,s)$$

where $A(v,s)$ a linear combination of iterated integrals. Then, we perform a Taylor expansion around the point $e^{is\mathcal{L}}v$:

$$f(e^{is\mathcal{L}}v + A(v,s)) = \sum_{m\leq r} \frac{A(v,s)^m}{m!} f^{(m)}(e^{is\mathcal{L}}v) + O(A(v,s)^{r+1}).$$

Performing the same type of computations on g, one obtains

$$\frac{A(v,s)^m}{m!} f^{(m)}(e^{is\mathcal{L}}v) \frac{\overline{A(v,s)}^n}{n!} g^{(n)}(e^{-is\mathcal{L}}\overline{v}).$$

We need to pull the oscillatory phases $e^{\pm is\mathcal{L}}$ out of f and g. This is achieved via expansions of the form

$$f(e^{is\mathcal{L}}v) = \sum_{\ell\leq r} \frac{s^\ell}{\ell!} e^{is\mathcal{L}} C^\ell[f,\mathcal{L}](v) + O(s^{r+1}C^{r+1}[f,\mathcal{L}](v))$$

where $C^\ell[f,\mathcal{L}]$ denote nested commutators which in general require less regularity than powers of the full operator $\mathcal{L}^\ell$. Then, one can perform the resonance analysis in Fourier space or continue directly in physical space by using again the same commutators. This approach is fairly general and allows us to cover the time dynamics of a large class of equations, including parabolic and hyperbolic problems, as well as dispersive equations, up to arbitrarily high order on general domains (see [3, 65]).

With the naive choices performed for deriving our schemes, it is quite likely that the scheme does not preserve any structure. There are various places where one can improve the discretisation:

- One can perform a different Duhamel iteration. Indeed, one has the general iteration for $s \in [0,t]$

$$I(k,u,s,t) = e^{i(t-s)k^3} u_k(s) + ik \sum_{\substack{k_1,k_2\in\mathbb{Z}\\ k_1+k_2=k}} ie^{itk^3} \int_s^t e^{-i\tilde{s}k^3} u_{k_1}(\tilde{s}) u_{k_2}(\tilde{s}) d\tilde{s},$$

We can take a weighted sum of these Duhamel iterations

$$u_k(t) = \frac{1}{2}\left(I(k,u,0,t) + I(k,u,s,t)\right).$$

which produced the midpoint iteration. One has to work with an implicit scheme as the value of both $u_k(s)$ and $u_k(0)$ are needed. Let us mention that this trick could be used outside numerical analysis as it gives a certain degree of freedom in the way one iterates Duhamel's formula.

- Another degree of freedom is when we perform Taylor expansions. Instead of doing them around 0, we can take an interpolation. We consider $r+1$ distinct interpolation points $0 \leq a_0 < a_1 < \cdots < a_r \leq 1$ which are symmetrically distributed such that $a_j = 1 - a_{r-j}$, $j = 0, \ldots, r$. Let us denote the corresponding nodal polynomials by $p_{j,r}$ such that

$$p_{j,r}(a_m t) = \delta_{j,m}.$$

Then, we define the following approximation

$$\tilde{p}_r(f,s) = \sum_{j=0}^{r} f(a_j) p_{j,r}(s), \quad f(a_j) = e^{i a_j t \mathcal{L}_{\text{low}}}.$$

With the previous ideas coming from [4, 58], one can derive a low regularity symmetric scheme for the KdV equation. We recall briefly what is the definition of a symmetric scheme. It is defined by considering its adjoint method: For a given method $u^\ell \mapsto u^{\ell+1} = \Phi_t(u^\ell)$ its adjoint method is defined as $\hat{\Phi}_t := \Phi_{-t}^{-1}$.

Definition 1 The method Φ_t is called *symmetric* if $\Phi_t = \widehat{\Phi}_t$.

Corollary 2 *For the KdV equation* (1) *the general midpoint scheme* (27) *takes at first order the form*

$$\begin{aligned} u^{\ell+1} &= e^{-\tau\partial_x^3} u^\ell + \frac{1}{24}\left(e^{-\tau\partial_x^3}\partial_x^{-1} u^\ell + \partial_x^{-1} u^{\ell+1}\right)^2 \\ &\quad - \frac{1}{24} e^{-\tau\partial_x^3}\left(\partial_x^{-1} u^\ell + e^{\tau\partial_x^3}\partial_x^{-1} u^{\ell+1}\right)^2 \end{aligned} \tag{30}$$

with a local error of order $\mathcal{O}\left(\tau^2\partial_x^2 u\right)$ *at first-order and with a local error of order* $\mathcal{O}\left(\tau^3\partial_x^4 u\right)$ *at second order.*

Symmetric schemes have been understood in a great degree of generality in [4]. One may also want to get low regularity *symplectic* schemes for Hamiltonian systems. The Poisson structure of the KdV equation in Fourier variables for instance takes the form (see, e.g., [2])

$$\{P, Q\}(u) = \sum_{k\in\mathbb{Z}^*} (\partial_{u_{-k}} P)(2i\pi k)(\partial_{u_k} Q)$$

and it is an interesting question in how far we can reproduce this structure at the discrete (numerical) level.

For doing so, one may want to exploit another degree of freedom which is the splitting of the differential operator $\mathcal{L}$ into

$$\mathcal{L} = \mathcal{S}_{\text{dom}} + \mathcal{S}_{\text{low}}$$

where the splitting has to preserve the symplectic structure of the equation. In the case of the KdV equation one does not have the choice for the first iterated integral as one performs directly an exact integration that preserves the symplectic structure. For other equations such as for instance cubic NLS, one changes the splitting:

$$\mathcal{S}_{\text{dom}}(k, k_1) = -2k_1k, \quad \mathcal{S}_{\text{low}}(k_2, k_3) = 2k_2k_3.$$

What is crucial is that now, one has

$$e^{is\mathcal{S}_{\text{dom}}(k,k_1)}e^{is\mathcal{S}_{\text{low}}(k_2,k_3)} = \overline{e^{is\mathcal{S}_{\text{dom}}(k_2,k_3)}e^{is\mathcal{S}_{\text{low}}(k_1,k_2)}} \tag{31}$$

which corresponds to the symmetry coming from the symplectic nature of the NLS equation. Then in the descritisation, one has to perform almost identical steps on $e^{is\mathcal{S}_{\text{dom}}(k,k_1)}$ and $e^{is\mathcal{S}_{\text{low}}(k_2,k_3)}$ to respect (31). The discretisation is limited to the dimension one ($d = 1$) (see [58] and [7] in the SPDE context). It is an open problem to get low regularity symplectic schemes for any dimension and a large class of dispersive equations. The main restriction is on the dimension as one cannot go back to physical space for $\frac{1}{kk_1}$ with usual operators when d is larger than one.

Let us finish this section by providing some perspectives. As we have already mentioned, structure-preserving low regularity schemes are quite challenging to get. However, many recent techniques have allowed some significant progress in this direction. On the combinatorial side, the structures proposed for describing the schemes have been used in a more analytical context.

Indeed, Poincaré-Dulac normal forms for dispersive PDEs have been derived in [13] with the same decorated trees and an arborification map. This gives a new derivation of this normal form [32] often used in the context of dispersive PDEs. This derivation is also very natural as it is reminiscence of structures that have been used in a more general context such as mould calculus introduced by Jean Ecalle [26, 27] for dynamical systems. The same arborification could be found there [28, 30] but also in the context of rough analysis in [71]. This formalism can even explain some of the most recent progress for dispersive PDEs with random initial data. In [15], one can compute hidden cancellations with a similar arborification for dispersive PDEs. It covers cancellations coming from Wave turbulence [23, 24] and the proof of the invariance of the Gibbs measure under the dynamics of the three-dimensional cubic wave equation [10]. Let us mention low regularity schemes in the context of Wave turbulence have been derived in [5]. The decorated formalism could also cover the recent theory of random tensors [25] that needs to be compared with some word formalism developed in [59].

The combinatorial formalism proposed for low regularity schemes for dispersive PDEs is very robust and it connects to various other fields such as dynamical systems, noncommutative geometry, numerical analysis with B-series, renormalisation in quantum field theory, rough analysis, singular SPDEs. Then, some ideas can be borrowed from these fields for organising computations connected to dispersive PDEs but also beyond.

4 Error Analysis at Low Regularity with Discrete Bourgain Spaces

In this section, we shall survey the main ideas which allows to perform error estimates at low regularity by taking the KdV equation (1) as the main example. We shall focus on error estimates in L^2 and thus restrict ourselves to solutions with positive regularity.

A flexible tool for the study of the local well-posedness of dispersive partial differential equations at low regularity is the use of Bourgain spaces introduced in [8]. We refer for more details to the books [52, 70].

To make use of these spaces at the discrete—numerical—level we make the following observation.

Remark 2 In order to exploit discrete Bourgain spaces in numerical analysis, we have to introduce frequency cut-off in our numerical schemes. This is crucial in our estimates to be able to push down the regularity assumptions on our solution. Indeed in the discrete case, the dispersion relation is periodic of period $2\pi/\tau$ with respect to σ, this create additional frequency interactions, compared to the continuous case, which could not be controlled without loosing derivatives. The presence of the cut-off allows us to have at the discrete level exactly the same type of frequency interactions as in the continuous case and in particular to be able to use the factorization property (20) which is crucial to recover the derivative in KdV.

The introduction of a frequency cut-off Π_τ in the scheme (see (47) for its precise structure) does not change the resonance analysis given in the previous section. Indeed, the filtered scheme can be seen as a resonance based discretisation of the corresponding filtered PDE, i.e.,

$$\partial_t u^\tau + \partial_x^3 u^\tau + \Pi_\tau(\Pi_\tau u^\tau \Pi_\tau \partial_x u) = 0$$

in case of KdV.

It is also interesting to mention that frequency cut-offs are natural in the numerical discretisation of PDEs (thinking for instance about spectral methods) as we can not reproduce all frequencies on the computer. Our approach is to choose this frequency cut-off in such a way that we can exploit discrete Bourgain spaces and thus optimise the error estimates in sense of regularity.

A tempered distribution $u(t,x)$ on $\mathbb{R} \times \mathbb{T}$ belongs to the Bourgain space $X^{s,b}$ adapted to the Airy equation (the linear part of (1)) if its following norm is finite

$$\|u\|_{X^{s,b}} = \left(\int_{\mathbb{R}} \sum_{k\in\mathbb{Z}} (1+|k|)^{2s} \left(1+|\sigma-k^3|\right)^{2b} |\widetilde{u}(\sigma,k)|^2 \, d\sigma \right)^{\frac{1}{2}},$$

where $\widetilde{u}$ is the space-time Fourier transform of u:

$$\widetilde{u}(\sigma,k) = \int_{\mathbb{R}\times\mathbb{T}} e^{-i\sigma t - ikx} u(t,x)\, dt dx.$$

This norm can be also written as

$$\|u\|_{X^{s,b}} = \|e^{t\partial_x^3} u\|_{H_t^b H_x^s}$$

where $H_t^b H_x^s$ is the space time Sobolev space, whose norm is defined by $\|f\|_{H_t^b H_x^s} := \|(1+|\sigma|)^b (1+|k|)^s \tilde{f}\|_{L^2(\mathbb{R}\times\mathbb{T})}$.

Elementary properties of the Bourgain spaces are the following.

For $\eta \in C_c^\infty(\mathbb{R})$, we have that

$$\|\eta(t) e^{-t\partial_x^3} f\|_{X^{s,b}} \lesssim_\eta \|f\|_{H^s}, \quad s\in\mathbb{R},\ f\in H^s(\mathbb{T}), \tag{32}$$

$$\|\eta(\tfrac{t}{T}) u\|_{X^{s,b'}} \lesssim_{\eta,b,b'} T^{b-b'} \|u\|_{X^{s,b}}, \quad s\in\mathbb{R}, -\tfrac{1}{2} < b' \le b < \tfrac{1}{2},\ 0 < T \le 1, \tag{33}$$

$$\left\| \eta(t) \int_0^t e^{-(t-s)\partial_x^3} F(s)\, ds \right\|_{X^{s,b}} \lesssim_{\eta,b} \|F\|_{X^{s,b-1}}, \quad s\in\mathbb{R}. \tag{34}$$

In order to solve by fixed point an equation under the form

$$\partial_t u + \partial_x^3 u = B[u,u]$$

where $B[\cdot,\cdot]$ is a bilinear term, one can notice that a solution in short times solves the modified Duhamel Formula

$$u(t) = \eta(t) e^{-t\partial_x^3} u_0 + \eta(t) \int_0^t e^{-(t-s)\partial_x^3} \eta(\frac{s}{\delta}) B[u,u](s)\, ds.$$

By using the above elementary properties, local well-posedness for data in H^s would follow from the Banach fixed point if the bilinear estimate

$$\|B[u,v]\|_{X^{s,b'}} \lesssim \|u\|_{X^{s,b}} \|v\|_{X^{s,b}} \tag{35}$$

for any $0 > b' > b-1$, $b > 1/2$ holds.

Unfortunately, for the periodic KdV equation, the bilinear term being $B[u,u] = -\partial_x(u^2/2)$, in order to recover one full derivative, one needs to take $b = 1/2$, $b' = -1/2$ [20, 49] which is a borderline case for which (34) for example does not hold. To deal with a space which keeps the good property of being a subspace of

$C(\mathbb{R}, H^s)$, we can use the framework of [20] and the smaller space X^s, which has the same scaling properties in time as $X^{s,\frac{1}{2}}$, defined by the following norm:

$$\|u\|_{X^s} = \|u\|_{X^{s,\frac{1}{2}}} + \|\langle k\rangle^s \tilde{u}\|_{l^2(k)L^1(\sigma)}. \tag{36}$$

We define more precisely X^s as the space of space-time tempered distributions such that $\tilde{u}(\sigma, 0) = 0$ and the above norm is finite. The norm $\|\langle k\rangle^s \tilde{u}\|_{l^2(k)L^1(\sigma)}$ is defined by

$$\|\langle k\rangle^s \tilde{u}\|_{l^2(k)L^1(\sigma)} = \left(\sum_k \left|\int_{\mathbb{R}} |\tilde{u}(k, \sigma)|\, d\sigma\right|^2\right)^{\frac{1}{2}}.$$

A localized version $X^s(I)$ for I an interval of time is obtained by setting

$$\|u\|_{X^s(I)} = \inf\{\|\overline{u}\|_{X^s},\ \overline{u}|_I = u\}.$$

A well-posedness result for (1) at the L^2 level reads:

Theorem 2 *For every $T > 0$ and $u_0 \in L^2$, $\int_{\mathbb{T}} u_0 = 0$, there exists a unique solution u of* (1) *such that $u \in X^0(0, T)$. Moreover, if $u_0 \in H^{s_0}$, $s_0 > 0$, then $u \in X^{s_0}(0, T)$.*

Note that we have $X^0(0, T) \subset C([0, T], L^2)$.

We can now introduce the discrete (in time) counter part of the above spaces.

For sequences of functions $(u^n(x))_{n\in\mathbb{Z}}$, we define the Fourier transform $\widetilde{u^n}(\sigma, k)$ by

$$\mathcal{F}_{n,x}(u^n)(\sigma, k) = \widetilde{u^n}(\sigma, k) = \tau \sum_{m\in\mathbb{Z}} \widehat{u^m}(k)\, e^{im\tau\sigma},$$

$$\widehat{u^m}(k) = \frac{1}{2\pi}\int_{-\pi}^{\pi} u^m(x)\, e^{-ikx}\, dx.$$

Parseval's identity then reads

$$\|\widetilde{u^n}\|_{L^2 l^2} = \|u^n\|_{l^2_\tau L^2}, \tag{37}$$

where

$$\|\widetilde{u^n}\|^2_{L^2 l^2} = \int_{-\frac{\pi}{\tau}}^{\frac{\pi}{\tau}} \sum_{k\in\mathbb{Z}} |\widetilde{u^n}(\sigma, k)|^2\, d\sigma, \quad \|u^n\|^2_{l^2_\tau L^2} = \tau \sum_{m\in\mathbb{Z}} \int_{-\pi}^{\pi} |u^m(x)|^2\, dx.$$

We define the discrete Bourgain spaces $X^{s,b}_\tau$ for $s \geq 0$, $b \in \mathbb{R}$, $\tau > 0$ by

$$\|u^n\|_{X^{s,b}_\tau} = \left\| \langle k\rangle^s \langle d_\tau(\sigma + k^3)\rangle^b \widetilde{u^n}(\sigma, k) \right\|_{L^2 l^2}, \tag{38}$$

where $d_\tau(\sigma) = \frac{e^{i\tau\sigma}-1}{\tau}$. Note that d_τ is $2\pi/\tau$ periodic and that uniformly in τ, we have $|d_\tau(\sigma)| \sim |\sigma|$ for $|\tau\sigma| \leq \pi$. Since $|d_\tau(\sigma)| \lesssim \tau^{-1}$, we also have that the discrete spaces satisfy the embeddings

$$\|u^n\|_{X_\tau^{0,b}} \lesssim \frac{1}{\tau^{b-b'}} \|u^n\|_{X_\tau^{0,b'}}, \quad b \geq b' \tag{39}$$

note that this estimate is not uniform in the time discretization parameter.

With this definition, we have the discrete counter part of the properties (32), (33), (33):

Lemma 1 *For $\eta \in C_c^\infty(\mathbb{R})$ and $\tau \in (0, 1]$, we have that*

$$\|\eta(n\tau)e^{-n\tau\partial_x^3} f\|_{X_\tau^{s,b}} \lesssim_{\eta,b} \|f\|_{H^s}, \quad s \in \mathbb{R},\ b \in \mathbb{R},\ f \in H^s, \tag{40}$$

$$\|\eta(n\tau)u^n\|_{X_\tau^{s,b}} \lesssim_{\eta,b} \|u^n\|_{X_\tau^{s,b}}, \quad s \in \mathbb{R},\ b \in \mathbb{R},\ u^n \in X_\tau^{s,b}, \tag{41}$$

$$\left\|\eta\left(\frac{n\tau}{T}\right)u^n\right\|_{X_\tau^{s,b'}} \lesssim_{\eta,b,b'} T^{b-b'} \|u^n\|_{X_\tau^{s,b}}, \quad s \in \mathbb{R},$$

$$-\frac{1}{2} < b' \leq b < \frac{1}{2},\ 0 < T = N\tau \leq 1,\ N \geq 1. \tag{42}$$

In addition, for

$$U^n(x) = \eta(n\tau)\tau \sum_{m=0}^{n} e^{-(n-m)\tau\partial_x^3} u^m(x),$$

we have

$$\|U^n\|_{X_\tau^{s,b}} \lesssim_{\eta,b} \|u^n\|_{X_\tau^{s,b-1}}, \quad s \in \mathbb{R},\ b > 1/2. \tag{43}$$

We stress that all the above estimates are uniform in τ.

The next step that we shall need in order to handle the KdV equation is to adapt (34) in the case $b = 1/2$. We thus define the discrete counterpart X_τ^s of the X^s space. We say that a sequence of function $(u^n(x))_n \in l_\tau^2 L^2$ such that $\int_{\mathbb{T}} u^n = 0,\ \forall n$ is in X_τ^s for $s \geq 0$ if the following norm is finite

$$\|u^n\|_{X_\tau^s} = \|u^n\|_{X_\tau^{s,\frac{1}{2}}} + \|\langle k\rangle^s \tilde{u}(\sigma, k)\|_{l^2(k)L^1(\sigma)}$$

and in the same way, we also define Y_τ^s by

$$\|F^n\|_{Y_\tau^s} = \|F^n\|_{X_\tau^{s,-\frac{1}{2}}} + \left\| \frac{\langle k\rangle^s}{\langle d_\tau(\sigma + k^3)\rangle} \widetilde{F^n}(\sigma, k) \right\|_{l^2(k)L^1(\sigma)}.$$

Lemma 2 *We have the following properties:*

1. *We have the embedding $X^s_\tau \subset l^\infty(\mathbb{Z}, H^s(\mathbb{T}))$:*

$$\sup_n \|u^n\|_{H^s(\mathbb{T})} \lesssim \|u^n\|_{X^s_\tau}, \quad s \in \mathbb{R},\ (u^n)_n \in X^s_\tau; \tag{44}$$

2. *Let us define for $(u^n)_n \in Y^s_\tau$, and $\eta \in C^\infty_c(\mathbb{R})$*

$$U^n(x) := \eta(n\tau)\tau \sum_{m=0}^{n} e^{-(n-m)\tau\partial_x^3} u^m(x), \tag{45}$$

then, we have

$$\|U^n\|_{X^s_\tau} \lesssim_\eta \|u^n\|_{Y^s_\tau}, \quad s \in \mathbb{R}. \tag{46}$$

The above estimates are uniform for $\tau \in (0, 1]$.

The last step in order to use these spaces for the analysis of numerical scheme at low regularity is to establish bilinear estimates like (35). A convenient one for the KdV equation is the following:

For every $s \geq 0$, there exists $C > 0$ such that for every $(u^n)_n, (v^n)_n \in X^s_\tau$, we have the estimate

$$\|\partial_x \Pi_\tau \left(\Pi_\tau u^n\, \Pi_\tau v^n\right)\|_{Y^s_\tau} \leq C \left(\|u^n\|_{X^{s,\frac{1}{2}}_\tau} \|v^n\|_{X^{s,\frac{1}{3}}_\tau} + \|v^n\|_{X^{s,\frac{1}{2}}_\tau} \|u^n\|_{X^{s,\frac{1}{3}}_\tau} \right) \tag{47}$$

where Π_τ is a projection on frequencies $|k|$ smaller than $\tau^{-\frac{1}{3}}$, see also Remark 2

Once these estimates are established, the convergence analysis at the L^2 level of a numerical scheme for KdV, for example a filtered version of (30) follows the following steps:

- Since the filtered scheme can be seen as a scheme for a filtered PDE,

$$\partial_t u^\tau + \partial_x^3 u^\tau + \Pi_\tau(\Pi_\tau u^\tau \Pi_\tau \partial_x u) = 0, \tag{48}$$

  the first step is to establish an error estimate for

$$\|u - u^\tau\|_{L^2}.$$

  This can be performed at the continuous level by using the continuous Bourgain spaces.
- The second step is to estimate the consistency error of the scheme $\mathcal{E}_n$ in the discrete Bourgain space Y^0_τ.

- The last step is to estimate $e^n = u^n - u^\tau(t_n)$ in the Bourgain space X_τ^0 which allows us to get a convergence result for $\sup_n \|e^n\|_{L^2}$. This is done by writing the discrete Duhamel formula and by using (46) and (47).

We refer for example to [47, 63, 64, 66, 66] for the implementation of this strategy in the analysis of various time discretizations of dispersive partial differential equations.

Acknowledgments Y. B. gratefully acknowledges funding support from the European Research Council (ERC) through the ERC Starting Grant Low Regularity Dynamics via Decorated Trees (LoRDeT), grant agreement No. 101075208. K.S. and F.R. gratefully acknowledges funding from the European Research Council (ERC) under the European Union's Horizon 2020 research and innovation programme (grant agreement No. 850941).

References

1. J-P. Berrut, L.N. Trefethen, Barycentric lagrange interpolation. SIAM Rev. **46** (2004)
2. J. Bernier, B. Grébert, Long time dynamics for generalized Korteweg–de Vries and Benjamin–ono equations. Arch. Ration. Mech. Anal. **241**, 1139–1241 (2021)
3. Y. Alama Bronsard, Y. Bruned, K. Schratz, Low regularity integrators via decorated trees. arXiv:2202.01171
4. Y. Alama Bronsard, Y. Bruned, G. Maierhofer, K. Schratz, Symmetric resonance based integrators and forest formulae. arXiv:2305.16737
5. Y. Alama Bronsard, Y. Bruned, K. Schratz, Approximations of dispersive PDEs in the presence of low-regularity randomness. Found. Comput. Math. **24**, 1819–1869 (2024)
6. J. Armstrong-Goodall, Y. Bruned, Resonance based schemes for SPDEs. arXiv:2312.16690
7. J. Armstrong-Goodall, Y. Bruned, Low regularity symplectic schemes for stochastic NLS. arXiv:2410.22359
8. J. Bourgain, Fourier transform restriction phenomena for certain lattice subsets and applications to nonlinear evolution equations. Part II: The KDV Equation. Geom. Funct. Anal. **3**, 209–262 (1993)
9. Y. Bruned, A. Chandra, I. Chevyrev, M. Hairer, Renormalising SPDEs in regularity structures. J. Eur. Math. Soc. (JEMS) **23**(3), 869–947 (2021)
10. B. Bringmann Y. Deng, A. Nahmod, H. Yue , Invariant Gibbs measures for the three dimensional cubic nonlinear wave equation. Invent. Math. **236**, 1133–1411 (2024)
11. Y. Bruned, M. Hairer, L. Zambotti, Algebraic renormalisation of regularity structures. Invent. Math. **215**(3), 1039–1156 (2019)
12. Y. Bruned, Composition and substitution of Regularity Structures B-series. arXiv:2310.14242
13. Y. Bruned, Derivation of normal forms for dispersive PDEs via arborification. arXiv:2409.03642
14. Y. Bruned, K. Schratz, Resonance based schemes for dispersive equations via decorated trees. Forum Math. Pi, 10, E2 (2022)
15. Y. Bruned, L. Tolomeo, Cancellations for dispersive PDEs with random initial data. arXiv:2412.17051
16. J.C. Butcher, An algebraic theory of integration methods. Math. Comput. **26**, 79–106 (1972)
17. E. Celledoni, D. Cohen, B. Owren, Symmetric exponential integrators with an application to the cubic Schrödinger equation. Found. Comput. Math. **8**, 303–317 (2008)
18. M. Christ, Power series solution of a nonlinear Schrödinger equation, in *Mathematical Aspects of Nonlinear Dispersive Equations*. Annals of Mathematics Studies, vol. 163 (Princeton University Press, Princeton, 2007), pp. 131–155

19. P. Chartier, E. Hairer, G. Vilmart, Algebraic structures of B-series. Found. Comput. Math. **10**(4), 407–427 (2010)
20. J. Colliander, M. Keel, G. Staffilani, H. Takaoka, T. Tao, Sharp global well-posedness for KdV and modified KdV on $\mathbb{R}$ and $\mathbb{T}$. J. Am. Math. Soc. **16**(3), 705–749 (2003)
21. A. Connes, D. Kreimer, Hopf algebras, renormalization and noncommutative geometry. Commun. Math. Phys. **199**(1), 203–242 (1998)
22. A. Connes, D. Kreimer, Renormalization in quantum field theory and the Riemann-Hilbert problem I: the Hopf algebra structure of graphs and the main theorem. Commun. Math. Phys. **210**, 249–273 (2000)
23. Y. Deng, Z. Hani, Full derivation of the wave kinetic equation. Invent. math. **233**, 543–724 (2023)
24. Y. Deng, Z. Hani, Derivation of the wave kinetic equation: Full range of scaling laws. arXiv:2301.07063
25. Y. Deng, A.R. Nahmod, H. Yue, Random tensors, propagation of randomness, and nonlinear dispersive equations. Invent. math. **228**, 539–686 (2022)
26. J. Ecalle, *Les fonctions résurgentes*, Tome I, II and III [Mathematical Publications of Orsay 81 and 85] (Université de Paris-Sud, Département de Mathématique, Orsay, 1981/1985)
27. J. Ecalle, Singularités non abordables par la géométrie (French) [Singularities that are inaccessible by geometry]. Ann. Inst. Fourier (Grenoble) **42**(1–2), 73–164 (1992)
28. J. Ecalle, B. Vallet, The arborification-coarborification transform: analytic, combinatorial, and algebraic aspects. Ann. Fac. Sci. Toulouse (6) **13**(3), 575–657 (2004)
29. B. Engquist, A. Fokas, E. Hairer, A. Iserles, *Highly Oscillatory Problems* (Cambridge University Press, Cambridge, 2009)
30. F. Fauvet, F. Menous, Ecalle's arborification coarborification transforms and Connes Kreimer Hopf algebra. Ann. Sc. l'Ecole Normale Sup. **50**(1), 39–83 (2017)
31. E. Faou, *Geometric Numerical Integration and Schrödinger Equations* (European Mathmetical Society, Publishing House, Zürich, 2012)
32. Z. Guo, S. Kwon, T. Oh, Poincaré-Dulac normal form reduction for unconditional well-posedness of the periodic cubic NLS. Commun. Math. Phys. **322**(1), 19–48 (2013)
33. M. Gubinelli, Controlling rough paths. J. Funct. Anal. **216**(1), 86–140 (2004)
34. M. Gubinelli, Ramification of rough paths. J. Differ. Equ. **248**(4), 693–721 (2010)
35. M. Gubinelli, Rough solutions for the periodic Korteweg-de Vries equation. Commun. Pure Appl. Anal. **11**(4), 709–733 (2012)
36. M. Hairer, A theory of regularity structures. Invent. Math. **198**(2), 269–504 (2014)
37. E. Hairer, C. Lubich, G. Wanner, *Geometric Numerical Integration. Structure-preserving Algorithms for Ordinary Differential Equations*. Springer Series in Computational Mathematics, vol. 31, 2nd edn. (Springer-Verlag, Berlin, 2006)
38. H. Holden, K.H. Karlsen, K.-A. Lie, N.H. Risebro, *Splitting for Partial Differential Equations with Rough Solutions* (European Mathmetical Society, Publishing House, Zürich, 2010)
39. H. Holden, K.H. Karlsen, N.H. Risebro, T. Tao, Operator splitting methods for the Korteweg-de Vries equation. Math. Comput. **80**, 821–846 (2011)
40. H. Holden, C. Lubich, N.H. Risebro, Operator splitting for partial differential equations with Burgers nonlinearity. Math. Comput. **82**, 173–185 (2012)
41. M. Hochbruck, A. Ostermann, Exponential integrators. Acta Numer. **19**, 209–286 (2010)
42. M. Hochbruck, J. Leibold, A. Ostermann, On the convergence of Lawson methods for semilinear stiff problems. Numer. Math. **145**, 553–580 (2020)
43. M. Hofmanová, K. Schratz, An oscillatory integrator for the KdV equation. Numer. Math. **136**, 1117–1137 (2017)
44. W. Hundsdorfer, J. Verwer, *Numerical Solution of Time-Dependent Advection- Diffusion-Reaction Equations* (Springer, Berlin, 2003)
45. L. Ignat, E. Zuazua, Numerical dispersive schemes for the nonlinear Schrödinger equation. SIAM J. Numer. Anal. **47**(2), 1366–1390 (2009)
46. T. Jahnke, C. Lubich, Error bounds for exponential operator splittings. BIT **40**, 735–744 (2000)

47. L. Ji, A. Ostermann, F. Rousset, K. Schratz, Low regularity full error estimates for the cubic nonlinear Schrödinger equation. SIAM J. Numer. Anal. **62**, 2071–2086 (2024)
48. S. Jin, *Schrödinger Equation: Computation*. Invited contribution to Springer "Encyclopedia of Applied and Computational Mathematics", ed. by B. Engquist (2015), pp. 1299–1301
49. C.E. Kenig, G. Ponce, L. Vega, A bilinear estimate with applications to the KdV equation. J. Am. Math. Soc. **9**(2), 573–603 (1996)
50. C. Klein, Fourth order time-stepping for low dispersion Korteweg-de Vries and nonlinear Schrödinger equation. ETNA **29**, 116–135 (2008). http://eudml.org/doc/117659
51. B. Li, S. Ma, K. Schratz, A semi-implicit low-regularity integrator for Navier-Stokes equations. SIAM J. Numer. Anal. **60**, 2273–2292 (2022)
52. F. Linares, G. Ponce, *Introduction to Nonlinear Dispersive Equations*, 2nd edn. (Springer, New York, 2015)
53. D. Levy, G. Puppo, G. Russo, Compact central WENO schemes for multidimensional conservation laws. SIAM J. Sci. Comput. **22**, 656–672 (2000)
54. V.T. Luan, A. Ostermann, Exponential B-series: the stiff case SIAM J. Numer. Anal. **51**(6), 3431–3445 (2013)
55. C. Lubich, On splitting methods for Schrödinger–Poisson and cubic nonlinear Schrödinger equations. Math. Comput. **77**, 2141–2153 (2008)
56. T.J. Lyons, Differential equations driven by rough signals. Rev. Mat. Iberoamericana **14**(2), 215–310 (1998)
57. R.I. McLachlan, G.R.W. Quispel, splitting methods. Acta Numer. **11**, 341–434 (2002)
58. G. Maierhofer, K. Schratz. Bridging the gap: symplecticity and low regularity in Runge-Kutta resonance-based schemes. arXiv:2205.05024.
59. A. Murua, J.M. Sanz-Serna, Word Series for dynamical systems and their numerical integrators. Found. Comput. Math. **17**, 675–712 (2017)
60. A. Ostermann, K. Schratz, Low regularity exponential-type integrators for semilinear Schrödinger equations. Found. Comput. Math. **18**, 731–755 (2018)
61. A. Ostermann, C. Su, Two exponential-type integrators for the "good" Boussinesq equation. Numer. Math. **143**, 683–712 (2019)
62. A. Ostermann, F. Rousset, K. Schratz, Error estimates of a Fourier integrator for the cubic Schrödinger equation at low regularity. Found. Comput. Math. **21**, 725–765 (2021)
63. A. Ostermann, F. Rousset, K. Schratz, Error estimates at low regularity of splitting schemes for NLS. Math. Comput. **91**(333), 169–182 (2021)
64. A. Ostermann, F. Rousset, K. Schratz, Fourier integrator for periodic NLS: low regularity estimates via discrete Bourgain spaces. J. Eur. Math. Soc. **25**(10), 3913–3952 (2023)
65. F. Rousset, K. Schratz, A general framework of low regularity integrators. SIAM J. Numer. Anal. **59**(3), 1735–1768 (2021)
66. F. Rousset, K. Schratz, Convergence error estimates at low regularity for time discretizations of KdV. Pure Appl. Anal. **4**(1), 127–152 (2022)
67. F. Rousset, K. Schratz, Resonances as a computational tool. Found. Comput. Math. (2024) (online first). https://doi.org/10.1007/s10208-024-09665-8
68. J.M. Sanz-Serna, M.P. Calvo, *Numerical Hamiltonian Problems* (Chapman and Hall, London, 1994)
69. K. Schratz, Y. Wang, X. Zhao, Low-regularity integrators for nonlinear Dirac equations. Math. Comput. **90**, 189–214 (2021)
70. T. Tao, *Nonlinear Dispersive Equations. Local and Global Analysis* (American Mathematical Society, Providence 2006)
71. J. Unterberger. Hölder-continuous rough paths by fourier normal ordering. Commun. Math. Phys. **298**, 1–36 (2010)

Part III
Advances in Specific Programs

Splitting Maps in Type I Ricci Flows

Panagiotis Gianniotis

1 Introduction

Ricci flow is a geometric evolution equation for Riemannian metrics on a smooth manifold M that deforms the metrics in a manner analogous to the heat equation, introduced by Hamilton in his seminal paper [63] in 1982. In particular, a smooth 1-parameter family of Riemannian metrics $g(t)$, $t \in I$, on M is a Ricci flow if

$$\frac{\partial g}{\partial t} = -2 \operatorname{Ric}_{g(t)} . \tag{1}$$

In this introduction, we will give an overview of the main ideas behind the Ricci flow, try to give some intuition on the kind of problems that it has addressed so far and describe some of its past and future challenges. Then, in Sects. 2–8 we will delve into a more detailed description of the theory, starting from a brief presentation of the main concepts required from Riemannian Geometry, in Sect. 2. Then, in Sects. 3–8, we will describe some of the fundamental concepts of the theory of Ricci flow, aiming mostly to build the necessary intuition and background that will allow us to discuss some of the results of the author in [59] regarding the behaviour of almost splitting maps under Ricci flow.

Before Hamilton's groundbreaking work on the Ricci flow in 1982, Eells–Sampson in [54] had introduced the harmonic map heat flow, a non-linear evolution equation for maps between Riemannian manifolds, which is the negative gradient flow of an energy functional. Eells–Sampson used this flow to prove that if the target manifold has non-positive sectional curvature then any homotopy class of maps has

P. Gianniotis (✉)
National and Kapodistrian University of Athens, Zografou, Athens, Greece
e-mail: pgianniotis@math.uoa.gr

© The Author(s), under exclusive license to Springer Nature Switzerland AG 2026
N. D. Alikakos, C. Villani (eds.), *Analysis Techniques for Mathematical Physics and Geometry – and Music*, Lecture Notes in Mathematics 2390,
https://doi.org/10.1007/978-3-032-11396-2_6

a unique optimal (harmonic) representative, that minimizes the energy functional in its homotopy class. This was the first application of heat-like evolution equations to answer questions in Geometry and Topology, and Ricci flow is its natural analogue in the realm of Riemannian Geometry.

Using the Ricci flow, one hopes to exploit the averaging nature of a diffusion equation to even-out the irregularities of the curvature of a given Riemannian g_0. Under a Ricci flow $g(t)$, $t \in [0, T)$, with $g(0) = g_0$, the curvature in fact satisfies a heat-like non-linear evolution equation of the form

$$\frac{\partial \operatorname{Rm}}{\partial t} = \Delta_{g(t)} \operatorname{Rm} + Q(\operatorname{Rm}), \tag{2}$$

where Q is a quadratic expression. It is thus expected that the Ricci flow will average out the geometric characteristics of g_0 that are not relevant to the underlying topology of the manifold M, and the flow will eventually converge to a special Riemannian metric that best captures topological information. In other words, we may think the Ricci flow as a way to construct a deformation retraction of the space $\operatorname{Met}(M)$ of the Riemannian metrics on M, or a large subset of it, into a much smaller collection of special geometric configurations which reveal topological information.

Hamilton's first application of the Ricci flow was to prove that three dimensional compact, simply connected, manifolds M that carry a Riemannian metric g_0 with positive Ricci curvature should necessarily be diffeomorphic to $\mathbb{S}^3$. To prove this result, the Ricci flow is used to deform any Riemannian metric with positive Ricci curvature to one with constant positive sectional curvature, and only $\mathbb{S}^3$ can carry such a Riemannian metric. Thus, here, Ricci flow deforms the large subset of Riemannian metrics with positive Ricci curvature to the subset of Riemannian metrics with constant positive curvature, showing that it is non-empty. In fact, since any two Riemannian metrics with constant positive curvature are related by a diffeomorphism and a scaling, from a geometric point of view they correspond to a unique geometry. Therefore, the space of metrics on M with positive Ricci curvature, modulo diffeomorphisms and scalings, is contractible.

1.1 3d Ricci Flow with Surgery

Ricci flow proved to be a powerful tool to tackle problems in Geometry and Topology after its success in the proof of Thurston's Geometrization Conjecture for oriented three dimensional compact manifolds and the Poincaré Conjecture, by Perelman [80–82].

In Hamilton's result in [63], any Ricci flow $g(t)$, $t \in [0, T)$, on a compact manifold of dimension three that starts from a Riemannian metric with positive Ricci curvature encounters a finite time *global* singularity, i.e. for every $x \in M$, $|\operatorname{Rm}_{g(t)}|(x) \to +\infty$ as $t \to T$, and it is shown that, after an appropriate scaling, the metrics converge to a metric with constant curvature. However, to prove

Thurston's Geometrization Conjecture one needs to start the Ricci flow from an *arbitrary* Riemannian metric g_0 on a three dimensional manifold M and deal with the manifestation of *local* finite time singularities at some time $t = T$, where the curvature of $g(t)$ blows up only at a subset of M. Once such singularities appear, the smooth Ricci flow stops and can not be continued further.

Perelman developed a surgery procedure that allows the Ricci flow to continue even after such singularities develop, but now on a different manifold obtained by M by cutting necks diffeomorphic to $\mathbb{S}^2 \times \mathbb{R}$ and capping the remaining ends to form a new manifold with compact connected components, and restarting the Ricci flow in each component. This procedure repeats in each connected component, whenever a new local singularity occurs. As a result, the initial manifold breaks up into various pieces, some of which may eventually become *extinct*, forming global singularities. These pieces can only be diffeomorphic to quotients of $\mathbb{S}^3$ or $\mathbb{S}^2\times\mathbb{R}$. In the remaining pieces the Ricci flow can be continued up to $t \to +\infty$, and leads to a collection of *thick* parts on which the Ricci flow converges to a hyperbolic metrics, and *thin* parts that collapse and are *graph manifolds*.[1]

A consequence of Perelman's work on the Ricci flow is that compact orientable three manifolds admitting a metric with positive scalar curvature are connected sums of manifolds diffeomorphic to $\mathbb{S}^3$ and $\mathbb{S}^2 \times \mathbb{S}^1$. The Ricci flow with surgery performs the necessary connected sum decomposition, deforming any metric with positive scalar curvature to the standard homogenous metrics in each piece diffeomorphic to $\mathbb{S}^3$ or $\mathbb{S}^2 \times \mathbb{S}^1$. Marques [75] in fact used this to prove that the space of Riemannian metrics with positive scalar curvature, on orientable compact three dimensional manifolds, if non-empty, is path-connected.

Perelman's Ricci flow with surgery is a piecewise smooth, in time, evolution, since the flow needs to stop and then continue after the surgery procedure. The surgery happens at very small, but never zero, scale, and depends on many parameters. As such it does not result in a canonical Ricci flow through singularities. However, Perelman conjectured that as the parameters that determine the quality of the surgery improve, it should be possible to obtain a canonical Ricci flow through singularities in the limit. This was achieved by Kleiner–Lott in [70], where given any Riemannian metric g_0 on a three dimensional manifold M they construct what they call a *singular Ricci flow*, starting from g_0. Singular Ricci flows were proven to be unique and to depend continuously on the initial data by Bamler–Kleiner in [15].

A very interesting topological application of the theory of singular Ricci flows in 3d, was the proof of the Generalized Smale Conjecture, by Bamler–Kleiner [16, 17], which asserts that the space of metrics on a three dimensional manifold with constant curvature $\pm$ 1 is contractible.[2] The basic principle is to use Ricci flow to deform any metric in a k-disc of Riemannian metrics, with the metrics

[1] See Appendix I [69] for a brief description of graph manifolds and their role in Thurston's Geometrization conjecture.

[2] This is equivalent to the statement that the inclusion of the isometry group of such a manifold into the group of its diffeomorphism is a homotopy equivalence.

corresponding to its boundary having constant curvature (this exists because the space of Riemannian metrics is contractible), to a metric with constant curvature. Of course singularities will develop, and the singular Ricci flow will not lead to a constant curvature metric on the whole initial manifold, however the continuous dependence on the initial data is crucial.

1.2 Dealing with Singularities

As should be clear by now, Ricci flow is expected to develop singularities in finite time. Moreover, singularities are not to been seen as a deficiency of the Ricci flow, but rather its strength as a tool to probe topological information; in 3d Ricci flow singularities are responsible for the decomposition of the manifold into simpler pieces, proving the Geometrization Conjecture. Thus, hoping to apply the Ricci flow to understand higher dimensional manifolds—in particular dimension four—we have to

- Understand the local geometry and topology of the manifold around regions that become singular.
- Understand, in as precise terms as possible, the mechanism that is responsible for the formation of singularities.
- Construct a weak Ricci flow, in an appropriate sense, that flows through singularities, analogous to Perelman's Ricci flow with surgery and the Kleiner–Lott singular Ricci flows in dimension three.

The main method to understand the local geometry and topology of singularities that a Ricci flow $(M, g(t))_{t\in[0,T)}$ develops, is to take a *blow-up sequence* $(p_j, t_j) \in M \times [0, T)$ such that $Q_j = |\operatorname{Rm}|(p_j, t_j) \to +\infty$ and consider the *pointed* sequence of rescaled Ricci flows $(M, g_j(t), p_j)_{t\in[-Q_j t_j, 0]}$, where $g_j(t) = Q_j g(t_j + Q_j^{-1} t)$. By the scaling behaviour of the Riemann tensor, $|\operatorname{Rm}_{g_j}|(p_j, 0) = 1$. A smooth limit of $(M, g_j(t), p_j)_{t\in[-Q_j t_j, 0]}$, if it exists, is called a *blow-up limit* of the flow, and is a pointed *ancient* Ricci flow $(N, g_\infty(t), p_\infty)_{t\in(-\infty,0]}$.

The underlying hope is that, due to the parabolic nature of Ricci flow, ancient Ricci flows have evolved for enough time to diffuse away any irregularities and can be classified. In dimension three this is indeed the case. Smooth blow up limits exist, have non-negative curvature operator and are non-collapsed.[3] We also have a detailed description of their geometry. See Sect. 1.3 for a brief discussion.

In higher dimensions, such program has only been successfully carried out under additional curvature assumptions in Brendle [21]. Apart from the difficulty of the classification of the possible blow-up limits, in higher dimensions—even dimension four—the blow up limits may not even be smooth Ricci flows. In fact, concrete examples in dimension four by Appleton [7] show that blow-up limits can have

[3] See Sect. 7 for more on non-collapsing in Ricci flow.

isolated orbifold singularities. It is however conceivable that the singularities of a 'generic' Ricci flow are much simpler and easier to classify. This issue is related to the stability of the possible blow-up limits, in the sense that unstable singularity models can be 'perturbed' away [26]. Although the state of the art in the theory of Ricci flow is still far from understanding the properties of a generic Ricci flow, analogous issues for the mean curvature flow of surfaces in $\mathbb{R}^3$ have recently been more or less settled, see Colding–Minicozzi [46], Chodosh–Choi–Mantoulidis–Schulze [36–38], Chodosh–Choi–Schulze [39] and Chodosh–Mantoulidis–Schulze [40].

1.3 Perelman's Bounded Diameter Conjecture for 3d Ricci Flow

Let $(M, g(t))_{t\in[0,T)}$ be a smooth compact Ricci flow on an three dimensional manifold, $T < +\infty$ being the maximal time of existence of the flow. The picture of the singularity formation in this situation had already been conjectured by Hamilton in [66], and finally rigorously described by Perelman in [80–82].

The most prevalent behaviour of the flow in high curvature regions resembles that of the neck-pinch singularity (Fig. 1), which will be discussed in more detail in Sect. 5. In fact, Perelman in [80] states:

> Now suppose that $g_{ij}(t)$ is defined on $M \times [1, T)$, $T < \infty$, and goes singular as $t \to T$. Then using 12.1 we see that, as $t \to T$, either the curvature goes to infinity everywhere, and then M is a quotient of either $\mathbb{S}^3$ or $\mathbb{S}^2 \times \mathbb{R}$, or the region of high curvature in $g_{ij}(t)$ is the union of several necks and capped necks, which in the limit turn into horns (the horns most likely have finite diameter, but at the moment I don't have a proof of that).

Bounding the diameter of the horns, or of the high curvature necks that appear near the singular time, say at $t = 0$ for convenience, in a 3d Ricci flow, is not an easy task and it is a question that remains unresolved to this day. Locally, each neck evolves within each time interval of the form $[-4^j, -4^{-j-1}]$ *approximately* like a piece of a thin cylinder $\mathbb{S}^2 \times \mathbb{R}$ of length $\sim 2^{-j}$, whose $\mathbb{S}^2$ factor shrinks homothetically under Ricci flow. However, the errors that are present could potentially introduce enough negative Ricci curvature in the directions along

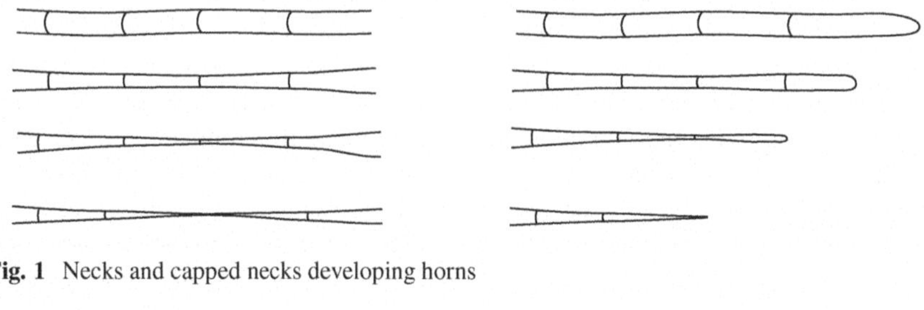

Fig. 1 Necks and capped necks developing horns

the neck, stretching it by a factor $1+\varepsilon$ in the duration of each of the time intervals $[4^{-j}, 4^{-j-1}]$, thus stretching the length of the neck to $+\infty$ as $j \to +\infty$.

The analogous question for the mean curvature flow of 2-convex hypersurfaces in $\mathbb{R}^n$, where the picture near the singularity is strikingly similar, has been resolved by the author and Haslhofer in [60]. The main ingredient in [60] was the Łojasiewicz inequality of Colding–Minicozzi [47], indicating that the issue is related to that of the uniqueness of tangent flows and the rectifiability of the singular set, as will be discussed in Sect. 8.2. The general theme, both for the mean curvature flow as well as the 3d Ricci flow, is that although at each scale there is a local approximation of the flow in a high curvature region by some model (in this case the shrinking cylinders), the total approximation errors may accumulate, preventing the passing of information from large scales down to arbitrarily small scales, near the singular time.

The diameter along Ricci flow has been shown by Topping in [90] to be controlled in terms of the $L^{\frac{n-1}{2}}$ norm of the scalar curvature. In fact, in a 3d Ricci flow $(M, g(t))_{t\in[-T,0)}$ it turns out that

$$\sup_{t\in[-T,0)} \operatorname{diam}(g(t)) < +\infty \quad \Longleftrightarrow \quad \sup_{t\in[-T,0)} \int_M R_{g(t)} d\mu_{g(t)} < +\infty.$$

Bounding the L^1 norm of the scalar curvature along a 3d Ricci flow at first seems on par with the general picture of the geometry of high curvature regions: a thin cylindrical piece $\mathbb{S}^2 \times [-1, 1]$, where the $\mathbb{S}^2$ factor has diameter $\sim r$, will have volume $\sim r^2$ and scalar curvature $\sim r^{-2}$.

Kleiner–Lott in [70] proved that along a 3d singular Ricci flow, the $L^{1-\eta}$ norm of the scalar curvature is uniformly bounded for any $\eta \in (0, 1)$. The estimates in [70] however degenerate as $\eta \to 1$, and the L^1 bound on the scalar curvature remains elusive.

Integral regularity estimates have recently attracted a lot of interest in the study of many other geometric PDEs, like non-collapsed Ricci limit spaces and harmonic maps. The theme there is quite similar to the one we have described so far. Examples usually indicate that a certain L^p, or weak L^p, regularity estimate should be valid, however techniques that rely only on approximating the solution at each scale with some model—typically invariant under dilations, because of some monotonicity formula of a scale invariant quantity—do not suffice to prove the sharp estimate. For instance, the quantitative differentiation arguments introduced by Cheeger–Naber [33] in the context of non-collapsed Ricci limit spaces have been remarkably successful in extracting as much information as possible to prove such regularity estimates, by merely using the Bishop–Gromov monotonicity formula. The main principles of [33] have then found applications in many different settings [30, 31], including Ricci flows under a Type I bound on the curvature (see Sect. 8 for the definition of Type I bound) by the author in [58] and Bamler in [11] in general.

Obtaining the sharp regularity estimates in each of the settings above turns out to require more sophisticated arguments, able to estimate in a more precise way the accumulation of errors passing from larger scales to smaller scales. This has been successfully achieved for stationary and energy minimizing harmonic maps by Naber–Valtorta in [79] and for non-collapsed Ricci limit spaces by Cheeger–Jiang–Naber in [32], in which they also prove that the respective singular sets are rectifiable.

Being able to carry over the ideas in [32] to the setting of the Ricci flow has the potential to not only affirmatively answer Perelman's conjecture, but also shed light to the following question.

Question 1 Suppose that $(M, g(t))_{t\in[0,T)}$ is a Ricci flow on a compact manifold that becomes singular as $t \to +\infty$, and let $d_{g(t)}$ denote the associated Riemannian distance. Is there a limit, in some topology—for instance the Gromov–Haussdorff topology—of $(M, d_{g(t)})$ as $t \to T$? If such a limit exists, how large is its singular set?

In Sects. 9 and 10 of these notes, we will describe recent work of the author in [59]. The main results in [59] are parabolic counterparts of the main estimates in [32], for Ricci flows that satisfy Type I bounds on the curvature. The work in [59] is part of an ongoing program aiming to prove Perelman's conjecture and explore Question 1, for a Ricci flow with Type I singularities.

1.4 Splitting Maps

A central concept in the analysis in [32] is that of an almost splitting map. In order to describe what it is, why it is useful, and motivate analogous considerations for the Ricci flow, we first need to recall the fundamental splitting theorem of Cheeger–Gromoll, in Riemannian Geometry.

Theorem 1 *Let (M, g) be a complete Riemannian manifold with $\mathrm{Ric}_g \geq 0$ and suppose that it contains a line, namely a unit speed geodesic $\gamma : \mathbb{R} \to M$ such that for every $s, t \in \mathbb{R}$, $d_g(\gamma(s), \gamma(t)) = |t - s|$. Then (M, g) splits isometrically as a product $(M', g') \times (\mathbb{R}, dr^2)$.*

For a proof of the splitting theorem see [27]. Using the line γ we can construct some special functions called Buseman functions $b_\pm$ on M, such that $b_+ + b_- = 0$. These functions turn out to be harmonic, i.e. $\Delta b_\pm = 0$, and satisfy $|\nabla b_\pm| = 1$.[4] Now, the Böchner formula (18) states that

$$\frac{1}{2}\Delta|\nabla b_\pm|^2 = \langle \nabla \Delta b_\pm, \nabla b_\pm \rangle + |\operatorname{Hess} b_\pm|^2 + \mathrm{Ric}(\nabla b_\pm, \nabla b_\pm). \tag{3}$$

[4] Such functions are called distance functions.

Since $\Delta b_{\pm} = \nabla|\nabla b_{\pm}|^2 = 0$, it thus follows that $\operatorname{Hess} b_{\pm} = 0$, i.e. $\nabla b_{\pm}$ is parallel. This suffices to prove that M is diffeomorphic to $N \times \mathbb{R}$, where N is any level set of $b_{\pm}$, and that the metric splits as $g_N \oplus dr^2$.

Cheeger–Colding in [29] proved the following quantitative version of this result, the almost splitting theorem.

Theorem 2 *Let (M, g) be a complete Riemannian manifold of dimension n such that $\mathrm{Ric} \geq -L^{-1}$, and suppose that there are points $p, q_{\pm} \in M$ such that*

$$d_g(p, q_{\pm}) \geq L, \tag{4}$$

$$d_g(p, q_+) + d_g(p, q_-) \leq d_g(q_+, q_-) + L^{-1}, \tag{5}$$

i.e. the equality case of the triangle inequality between the points p and $q_{\pm}$ is almost satisfied. Given $r > 0$ and $\varepsilon > 0$, if $L > 0$ is large, depending only on n, r, ε, then there is a metric space (N, d_N) and $o \in N \times \mathbb{R}$ such that

$$d_{GH}(B_{r/2}(o), B_{r/2}(p)) \leq \varepsilon, \tag{6}$$

where d_{GH} denotes the Gromov–Hausdorff distance.

In the proof, if $L > 0$ is large, Cheeger–Colding construct a harmonic function v in $B_r(p)$ such that

$$\frac{1}{\mathrm{Vol}(B_r(p))} \int_{B_r(p)} ||\nabla v|^2 - 1|^2 d\mu_g < \delta, \tag{7}$$

$$\frac{r^2}{\mathrm{Vol}(B_r(p))} \int_{B_r(p)} |\operatorname{Hess} v|^2 d\mu_g < \delta. \tag{8}$$

Note that the quantities on the left hand side in (7) and (8) are scale invariant. Such a function v is called a δ-splitting map.

In the setup of Theorem 2, we would like to imitate the proof of Theorem 1. A similar construction as in the proof of Theorem 1 results to functions $b_{\pm}$ with $|\nabla b_{\pm}| = 1$. However, these functions are no longer harmonic to apply the Böchner formula to estimate their Hessian. The δ-splitting map v is constructed by approximating b_+ by a harmonic function, which now only satisfies the L^2 gradient estimate (7). This is enough, though, to use the Böchner formula to obtain the L^2 Hessian estimate (8). The δ-splitting map is then used to establish a quantitative version of the Pythagorean Theorem for three points $x, y, z \in B_{r/2}(p)$ such that $v(x) = v(z) = a$, and z is the closest point of the level set $\{v(x) = a\}$ to y, which is the crucial step that leads to (6). We refer the reader to [27] for more details on the proof.

1.4.1 Almost Splitting Maps in the Study of Ricci Limit Spaces

Let (M_j, g_j, p_j) be a sequence of complete Riemannian manifolds of dimension n satisfying

$$\mathrm{Ric}_{g_j} \geq -(n-1), \tag{9}$$

$$\mathrm{Vol}(B_1(p_j)) \geq v_0 > 0. \tag{10}$$

By Gromov's compactness theorem we know that a subsequence converges in the Gromov–Hausdorff sense to a pointed metric space (X, d_X, q).

One of the consequences of Cheeger–Colding [28] is that for any $x \in X$ and sequence $r_i \to 0$, the sequence $(X, r_i^{-1} d_X, x)$ converges, in the Gromov–Hausdorff sense to a metric cone, called a tangent cone at x. Regular points are the points where every tangent cone is the Euclidean n-space, and the singular set $\mathcal{S}$ is its complement. By Cheeger–Colding [29], the singular set $\mathcal{S}$ of (X, d_X) has Hausdorff dimension at most $n-2$.[5]

A natural question that arises is whether the singular set $\mathcal{S}$ is an $n-2$-rectifiable subset of X. Namely, whether there is a countable collection $S_l \subset \mathcal{S}$ such that $\mathcal{H}^k(\mathcal{S} \setminus \cup_l S_l) = 0$ and there exist bi-Lipschitz maps $f_l : S_l \to \mathbb{R}^{n-2}$.

Cheeger–Jiang–Naber [32] prove that this is indeed the case,[6] and (k, δ)-splitting maps (k-dimensional analogues of a δ-splitting map, which are harmonic maps $v : B_r(p) \to \mathbb{R}^k$ with small Hessian and with components with almost orthonormal gradients, in the L^2 sense) play a central role in constructing the bi-Lipschitz maps required to show the rectifiability of the singular set.

The main issues regarding (k, δ)-splitting maps considered in [32] are the following. Let (M_j, g_j, p_j) be a sequence of complete Riemannian manifolds satisfying (9) and (10).

1. Given a ball $B_2(p_j) \subset M_j$, under what conditions is there an optimal (k, ε)-splitting map that best captures the local approximate splitting of the geometry, uniformly as $j \to +\infty$?
2. Given a (k, ε)-splitting map v in the ball $B_2(p_j) \subset M_j$, is it true that $v|_{B_r(p_j)}$ is also an almost splitting map, for $r < 2$? The issue here is that the gradients of the components of v may no longer be approximately orthonormal at smaller scales as $j \to +\infty$, in the scale invariant L^2 sense. For instance, the gradient of a $(1, \varepsilon)$-splitting map in $B_2(p_j)$ may degenerate in smaller scales, i.e. $\frac{1}{\mathrm{Vol}(B_r(p_j))} \int_{B_r(p_j)} |\nabla v|^2 d\mu_{g_j} \to 0$, for r small j large enough. When can we exclude such degeneration? Excluding degeneration is essential if we are to use almost splitting maps to prove the rectifiability of the singular set.

[5] In fact there is a filtration $\mathcal{S}_0 \subset \cdots \subset \mathcal{S}_{n-2} = \mathcal{S}$, with $\dim_{\mathrm{H}} \mathcal{S}_k \leq k$.

[6] In fact each $\mathcal{S}_k$ is k-rectifiable.

Almost Splitting Maps in Ricci Flow

Harmonic almost splitting maps were used by Bamler [8, 9] to understand the structure of the singularities of a Ricci flow $(M, g(t))_{t\in[0,T)}$ with uniformly bounded scalar curvature. Later in [11] he introduced a parabolic analogue of an almost splitting map, which solves the heat equation along the Ricci flow, in the development of a structure theory for limits of sequences of non-collapsed sequences of Ricci flows.

In Sects. 9 and 10, we consider a similar notion of almost splitting maps for the Ricci flow, see Definition 4, and describe a few results from [59] which address the two issues above in the context of Ricci flow under Type I bounds on the curvature, Theorems 9, 10 and 11.

2 Overview of Riemannian Geometry

A Riemannian metric g on a smooth manifold M is a smooth assignment of a positive definite inner product $g_p(\cdot,\cdot)$ on each tangent space T_pM, $p \in M$. Here smooth means that for any two smooth vector fields V, W on M, the function $p \mapsto g_p(V(p), W(p))$ is smooth.

Any Riemannian metric g extends to a bundle metric—an inner product in each of the fibres—of the dual bundle T^*M, as well as of every bundle that is a multiple tensor product of the bundles TM and T^*M.

In local coordinates $\{x^i\}_{i=1}^n$, the partial derivatives $\left\{\frac{\partial}{\partial x^i}\right\}_{i=1}^n$ form a basis of T_pM, while the differentials $\{dx^i\}_{i=1}^n$ are the dual basis of $(T_pM)^* = T_p^*M$. We can then express the Riemannian metric as

$$g = g_{ij}dx^i \otimes dx^j,$$

using the Einstein summation convention, where we sum over repeated upper and lower indices.

Any Riemannian metric g induces a unique connection ∇ which is compatible with g and torsion free, called the Levi–Civita connection, on the tangent bundle $TM = \sqcup_{p\in M} T_pM$. By definition a connection is a way to differentiate vector fields on M. In local coordinates, if $V = V^i\frac{\partial}{\partial x^i}$, $W = W^i\frac{\partial}{\partial x^i}$, the Levi–Civita connection is given by

$$\nabla_V W = \left(V^i\frac{\partial W^k}{\partial x^i} + V^iW^j\Gamma_{ij}^k\right)\frac{\partial}{\partial x^k},$$

where the Christoffel symbols Γ_{ij}^k are given by

$$\Gamma_{ij}^k = \frac{1}{2} g^{kl} \left(\frac{\partial g_{jl}}{\partial x^i} + \frac{\partial g_{il}}{\partial x^j} - \frac{\partial g_{ij}}{\partial x^l} \right).$$

Here $(g^{ij})_{i,j=1}^n$ denotes the inverse of the matrix $(g_{ij})_{i,j}^n$.

The Levi–Civita connection also extends as a connection on any tensor bundle, by the requirement that it commutes with contractions and satisfies the Leibniz rule with respect to tensor products. The metric compatibility of the Levi–Civita connection then is equivalent to $\nabla g = 0$, i.e. the Riemannian metric is parallel with respect to ∇.

In particular this allows for the definition of the second covariant derivative $\nabla^2_{X,Y} Z$ of any vector field Z on M, in the directions of vector fields X, Y. In fact,

$$\nabla^2_{X,Y} Z = \nabla_X(\nabla_Y Z) - \nabla_{\nabla_X Y} Z.$$

The curvature tensor $R(\cdot, \cdot) \cdot$ then of (M, g) is defined as a commutator, in the sense

$$R(X, Y)Z = \nabla^2_{X,Y} Z - \nabla^2_{Y,X} Z,$$

for any smooth vector fields X, Y, Z on M. We also define the $(0, 4)$-Riemann curvature tensor of (M, g) as

$$\mathrm{Rm}(X, Y, Z, W) = \langle R(X, Y)Z, W \rangle.$$

In local coordinates we write $R_{ijkl} = \mathrm{Rm}\left(\frac{\partial}{\partial x^i}, \frac{\partial}{\partial x^j}, \frac{\partial}{\partial x^k}, \frac{\partial}{\partial x^l} \right)$.

The Ricci tensor Ric is a trace of the Riemann tensor and defines a symmetric bilinear form on each tangent space of M, so that in local coordinates

$$\mathrm{Ric}(V, W) = V^i W^j R_{ij}$$

where $R_{ij} = g^{ab} R_{iabj}$. The scalar curvature R is then defined as the trace of Ric, so that $R = g^{ij} R_{ij}$.

Let's denote by $\mathrm{Met}(M)$ the space of smooth Riemannian metrics on M. The diffeomorphism group

$$\mathrm{Diff}(M) = \{\phi : M \to M, \text{ diffeomorphism}\},$$

then acts on $\mathrm{Met}(M)$ by pullback, namely $(\phi, g) \mapsto \phi^* g$.

The Riemann, Ricci and Scalar curvatures are all second order quasilinear differential operators on g, and are geometric invariants of the Riemannian manifold, in the sense that if $\phi : M \to M$ is a diffeomorphism and g is a Riemannian metric on M, then the pullback metric $\phi^* g$ satisfies

$$
\begin{aligned}
\mathrm{Rm}_{\phi^* g} &= \phi^* \mathrm{Rm}_g, \\
\mathrm{Ric}_{\phi^* g} &= \phi^* \mathrm{Ric}_g, \\
R_{\phi^* g} &= \phi^* R_g = R_g \circ \phi.
\end{aligned} \tag{11}
$$

Recall that in Riemannian geometry, there are no first order geometric invariants, since choosing appropriate (normal) coordinates around any point, we can make the first derivatives of g_{ij} vanish, at that point.

3 The Ricci Flow

In this section we describe some foundational ideas behind Ricci flow and study the first simple examples, those induced by Einstein metrics. We will present three reasons why the Ricci flow is a very natural way to evolve Riemannian metrics. The first comes from the PDE aspects of the Ricci flow, requiring to define an analogue for Riemannian metrics of the heat equation. The other two, come from analyzing the behaviour of solutions of the heat equation and the backwards conjugate heat equation along the flow. It turns out that Ricci flow evolves the geometry of the manifold in such a way that these solutions enjoy some very nice properties, revealing also an interesting link with Optimal Transport Theory.

3.1 The Ricci Flow as a Diffusion of Riemannian Metrics

Let M be a compact manifold and $g(t)$, $t \in [0, T)$, be a smooth 1-parameter family of Riemannian metrics on M. We say that $g(t)$ evolves by the Ricci flow if

$$
\frac{\partial g}{\partial t} = -2\,\mathrm{Ric}_g\,. \tag{12}
$$

Lets try to understand why the Ricci flow is the most natural way to evolve Riemannian metrics with a diffusion-like process. First of all both sides of equation (12) are of the same type, namely at each point of the manifold they define a symmetric bilinear form. On the other hand, it may not be clear from the beginning that the differential operator $g \mapsto -2\,\mathrm{Ric}_g$ is the correct choice for the right-hand side of (12), in the sense that it might look more natural to consider instead the term $\Delta_g g$, namely the Laplacian of the Riemannian metric g with respect to the Levi-Civita connection of g. However, the compatibility of ∇ with the Riemannian metric g implies that $\nabla g = 0$, hence $\Delta_g g = \mathrm{tr}_g \nabla^2 g = 0$. Thus, the flow we would obtain by deforming Riemannian metrics in the direction $\Delta_g g$ would keep any Riemannian metric fixed.

The key insight that explains why $-2\,\mathrm{Ric}_g$ is the correct analogue of a Laplacian of a Riemannian metric comes from expressing the Ricci curvature of g in coordinates as $\mathrm{Ric}_g = R_{jk}dx^j \otimes dx^k$, and in particular when the coordinate functions are chosen to be harmonic, namely $\Delta_g x^i = 0$, $i = 1, \ldots, n$. These coordinates are called g-harmonic coordinates, and they have proven to be very useful when doing Analysis on a Riemannian manifold, since the metric in this coordinates has the heightest degree of regularity, see [53]. It is then a classic computation that

$$R_{jk} = -\frac{1}{2}\Delta_g(g_{jk}) + Q(g, \partial g)_{jk}, \tag{13}$$

where $Q(g, \partial g)$ is a quadratic expression, both on g and ∂g.

The (almost) elliptic nature of the differential operator $P(g) = -2\,\mathrm{Ric}_g$ becomes more evident once we compute its linearization $D_g P(h) = \frac{d}{dt}\big|_{t=0} -2\,\mathrm{Ric}_{g+th}$. Then

$$D_g P(h) = \Delta_g h + \mathcal{L}_{\beta_g(h)}g + 2\,\mathrm{Rm}(h) - \mathrm{Ric} \circ h - h \circ \mathrm{Ric}, \tag{14}$$

where $\beta_g(h) = \mathrm{div}_g\, h - \frac{1}{2}\nabla\, \mathrm{tr}\, h$ is the so called Bianchi operator and $\mathrm{Rm}(h)_{jk} = R_{ijkl}h_{ab}g^{ai}g^{bl}$, see [45] for more details.

By the diffeomorphism invariance of the Ricci tensor (11) we know that for any vector field X

$$D_g P(\mathcal{L}_X g) = -2\mathcal{L}_X \,\mathrm{Ric}\,. \tag{15}$$

This follows from applying (11) to the flow of any vector field X and differentiating.

An implication of equation (15) is that $D_g P$ is not an elliptic operator. To see this, observe that the left-hand side of (15) is a-priori a third order differential operator on X, while the right-hand side is a first order operator. The ellipticity of $D_g P$ would imply, via the standard elliptic regularity estimates that if X is a $C^{1,\alpha}$ vector field then $\mathcal{L}_X g$ is $C^{2,\alpha}$, and thus X would gain an extra regularity of 2 derivatives!

On the other hand, for any $h \in \ker \beta_g$ expression (14) becomes

$$D_g P(h) = \Delta_g h + \text{lower order terms}$$

Intuitively, we might say that $D_g P$ is elliptic when restricted to variations $g+th$ that are transverse to the action of the diffeomorphism group on the space of Riemannian metrics. Considering such variations is similar to choosing the right coordinates to express each Riemannian metric on M, such as the harmonic coordinates used to obtain (13).

This lack of (full) ellipticity, due to the diffeomorphism invariance, prevents us to apply standard PDE theory to prove the short-time existence of the Ricci flow starting from a given smooth Riemannian metric g_0, even on a compact manifold

M. However, Hamilton in [63] obtains the following short-time existence result for the Ricci flow on compact manifolds.

Theorem 3 *Let M be a compact manifold and g_0 a smooth Riemannian metric on M. Then, there is $0 < T \leq +\infty$ and a unique Ricci flow $g(t)$, $t \in [0, T)$ such that $g(0) = g_0$. Moreover, if $T < +\infty$ and there is no continuation of the Ricci flow for $t \geq T$,*

$$\lim_{t \to +\infty} \max_M |\operatorname{Rm}|(\cdot, t) = +\infty.$$

In the case $T < +\infty$ we say that $g(t)$ exhibits a finite time singularity at $t = T$.

Hamilton in [63] had to use the Nash–Moser implicit function Theorem in order to get around the lack of the full ellipticity of the Ricci tensor. Later, however, DeTurck in [52] showed that the Ricci flow is equivalent—after modification by diffeomorphisms—to a quasilinear parabolic PDE for which standard theory applies, thus providing a more elegant proof of Theorem 3. Once we know that solutions exist, at least for short time, then the blow-up of the curvature if $T < +\infty$ follows from the fact that bounds on the curvature along Ricci flow imply bounds on any derivative of the curvature. Thus, if the curvature where to remain bounded up to $t = T$, $g(t)$ would converge smoothly to a Riemannian metric $g(T)$, and we could restart the flow to obtain a continuation past time $t = T$, see [45].

3.2 Diffeomorphism and Scale Invariance of the Ricci Flow

Let (M, g) be a Riemannian manifold, $\phi : M \to M$ a diffeomorphism and $\lambda > 0$. Then, by (11) we know that $\operatorname{Ric}_{\phi^* g} = \phi^* \operatorname{Ric}_g$. Moreover, a direct computation shows that, for any $\lambda > 0$, $\operatorname{Ric}_{\lambda^2 g} = \operatorname{Ric}_g$.

From this, it is clear that if $(M, g(t))_{t \in [0,T)}$ is a Ricci flow then $\tilde{g}(t) = \lambda^2 \phi^* g(\lambda^{-2} t)$, $t \in [0, \lambda^2 T]$, is also a Ricci flow, since

$$\frac{\partial \tilde{g}}{\partial t} = -2\phi^* \operatorname{Ric}_{g(t)} = -2 \operatorname{Ric}_{\tilde{g}(t)}.$$

The conceptual importance of this observation is that the Ricci flow doesn't just define a path in the space $\operatorname{Met}(M)$ of Riemannian metrics on M, but also a path in the space of Riemannian metrics modulo diffeomorphisms and scalings. We can thus view the Ricci flow as a dynamical system in the quotient $\operatorname{Met}(M)/\sim$ where $g_1 \sim g_2$ if and only if $g_2 = \lambda^2 \phi^* g_1$ for some diffeomorphism ϕ and $\lambda > 0$. This property, in particular the diffeomorphism invariance, is what classifies the Ricci flow as a *geometric flow* of Riemannian metrics.

3.3 The First Examples

To get a first intuition on the behaviour of the Ricci flow it is very informative to consider solutions generated by Einstein manifolds, namely Riemannian manifolds $(M, \bar{g})$, $\dim M = n$, that satisfy

$$\mathrm{Ric}_{\bar{g}} = (n-1)\lambda \bar{g}, \tag{16}$$

for some constant $\lambda \in \mathbb{R}$.

Every Riemannian manifold with constant curvature K is an Einstein manifold with $\lambda = K$. In particular, the Euclidean space $(\mathbb{R}^n, g_{\mathbb{R}^n})$ is Einstein with $\lambda = 0$, the sphere $(\mathbb{S}^n, g_{\mathbb{S}^n})$ of radius 1 is Einstein with $\lambda = 1$, and any hyperbolic manifold with curvature $K < 0$. Although every Einstein metric in dimensions 2 and 3 has constant curvature, in higher dimensions there are many Einstein metrics with non-constant curvature [19]. In any case, Einstein manifolds can be viewed as generalizations of constant curvature metrics in higher dimensions, and can be thought of as canonical—or optimal—Riemannian metrics.

A direct computation shows that if $(M, \bar{g})$ satisfies (16) then

$$g(t) = (1 - 2(n-1)\lambda t)\bar{g},$$

defined for $t \in \left(-\infty, \frac{1}{2(n-1)\lambda}\right)$, $t \in (-\infty, +\infty)$ and $t \in \left(\frac{1}{2(n-1)\lambda}, +\infty\right)$, depending on whether λ is positive, zero or negative, satisfies

$$\frac{\partial g}{\partial t} = -2(n-1)\lambda \bar{g} = -2\,\mathrm{Ric}_{\bar{g}} = -2\,\mathrm{Ric}_g\,.$$

In particular, if $\bar{g} = g_{\mathbb{R}^n}$ we will refer to the corresponding Ricci flow as the *static Ricci flow* on $\mathbb{R}^n$, whereas if $\bar{g} = g_{\mathbb{S}^n}$ we will call the corresponding Ricci flow the *round shrinking sphere*.

Since the Riemannian product of two Ricci flows is also a Ricci flow, we can construct the *generalized shrinking cylinder* Ricci flows on $\mathbb{S}^{n-k} \times \mathbb{R}^k$ as

$$g(t) = (1 - 2(n-k-1)t) g_{\mathbb{S}^{n-k}} \oplus g_{\mathbb{R}^k}. \tag{17}$$

These examples, although quite simple, allow us to make the following observations.

1. We can not expect a solution to the Ricci flow to exist for all time, in general. Even starting with a round sphere of curvature 1 the corresponding Ricci flow will not continue after $t = \frac{1}{2(n-1)}$. It will evolve by spheres of constant sectional curvature $\frac{1}{1-2(n-1)t}$, which blows up as $t \to \frac{1}{2(n-1)}$.

2. Viewing the Ricci flow as a dynamical system in the space of Riemannian metrics modulo diffeomorphism and scalings, then *all* Einstein metrics, not just those with $\lambda = 0$, are fixed points of the Ricci flow.

3.4 The Heat Equation Under Ricci Flow

Let (M, g) be a compact Riemannian manifold and denote $\Delta_g = g^{ij}\nabla^2_{ij}$ the Laplace-Beltrami operator of (M, g). We say that a smooth function $v : M \times [0, +\infty) \to \mathbb{R}$ evolves by the heat equation if

$$\left(\frac{\partial}{\partial t} - \Delta_g\right) v = 0.$$

A standard computation, involving commuting derivatives and using the definition of the curvature tensor, leads to the Böchner formula

$$\frac{1}{2}\Delta_g|\nabla u|^2 = \langle \nabla \Delta_g u, \nabla u\rangle + |\operatorname{Hess}_g u|^2 + \operatorname{Ric}_g(\nabla u, \nabla u), \tag{18}$$

which is crucial in the study of the effect that Ricci curvature bounds have on the analysis of (M, g).

Applying (18) we easily obtain the following evolution equation for $|\nabla v|^2$

$$\left(\frac{\partial}{\partial t} - \Delta_g\right) |\nabla v|^2 = -2|\operatorname{Hess} v|^2 - 2\operatorname{Ric}(\nabla v, \nabla v). \tag{19}$$

Notice that if (M, g) satisfies $\operatorname{Ric} \geq 0$, (19) becomes

$$\left(\frac{\partial}{\partial t} - \Delta_g\right) |\nabla v|^2 \leq -2|\operatorname{Hess} v|^2 \leq 0. \tag{20}$$

By the maximum principle then we obtain the gradient bound

$$|\nabla v(x, t)|^2 \leq \max_M |\nabla v(\cdot, 0)|^2. \tag{21}$$

In fact, if the gradient bound (21) holds for every solution to the heat equation then (M, g) mush have non-negative Ricci curvature, thus characterizing non-negative Ricci curvature in terms of the behaviour of the diffusion on (M, g).

Now, suppose that $g(t)$, $t \in I$, is a smooth 1-parameter family of Riemannian metrics on M. A smooth function $v : M \times I \to \mathbb{R}$ evolves by the heat equation if in $M \times I$

$$\left(\frac{\partial}{\partial t} - \Delta_{g(t)}\right) v = 0.$$

Since in local coordinates $|\nabla v|^2 = g^{ij}\frac{\partial v}{\partial x^i}\frac{\partial v}{\partial x^j}$ and $\frac{\partial g^{ij}}{\partial t} = -\frac{\partial g_{pq}}{\partial t} g^{pi} g^{qj}$ we obtain

$$\begin{aligned}\frac{\partial|\nabla v|^2}{\partial t} &= -\frac{\partial g_{pq}}{\partial t} g^{pi} g^{qj} \frac{\partial v}{\partial x^i}\frac{\partial v}{\partial x^j} + g^{ij}\frac{\partial v}{\partial x^i}\frac{\partial^2 v}{\partial x^j \partial t} \\ &= -\frac{\partial g}{\partial t}(\nabla v, \nabla v) + \langle \nabla v, \nabla \frac{\partial v}{\partial t}\rangle_{g(t)}.\end{aligned}$$

Combining this with the Böchner formula then gives

$$\left(\frac{\partial}{\partial t} - \Delta_{g(t)}\right)|\nabla v|^2 = -2|\operatorname{Hess} v|^2 - \left(\frac{\partial g}{\partial t} + 2\operatorname{Ric}\right)(\nabla v, \nabla v). \tag{22}$$

Arguing like in the case of a static Riemannian manifold, the maximum principle gives that if

$$\frac{\partial g}{\partial t} \geq -2\operatorname{Ric} \tag{23}$$

then estimate (21) holds. A 1-parameter family of Riemannian metrics that satisfies (23) is called a *super Ricci flow*.

Moreover, by McCann–Topping [77], the validity of the gradient estimate (21) for any solution to the heat equation fully characterizes super Ricci flows, namely $g(t)$ is a super Ricci flow if and only if (21) always holds.

3.5 The Conjugate Heat Equation

Let $(M, g(t))_{t\in I}$ be a smooth 1-parameter family of Riemannian metrics on a compact manifold with $h = \frac{\partial g}{\partial t}$, and let $u, v \in C^\infty(M \times I)$. Using that

$$d\mu_{g(t)} = \frac{\operatorname{tr}_g h}{2} d\mu_{g(t)}, \tag{24}$$

where $\operatorname{tr}_g h = g^{ij}h_{ij}$, in local coordinates, a direct computation using integration by parts gives

$$\frac{d}{dt}\int_M vu\, d\mu_{g(t)} = \int_M u\Box v\, d\mu_{g(t)} + \int_M v\Box^* u\, d\mu_{g(t)}, \tag{25}$$

where $\Box = \frac{\partial}{\partial t} - \Delta_{g(t)}$ is the heat operator along $g(t)$, and $\Box^* = \frac{\partial}{\partial t} + \Delta_{g(t)} + \frac{\operatorname{tr}_g h}{2}$ the (backwards) conjugate heat operator along $g(t)$.

Observe that if $\Box v = \Box^* u = 0$ on $M \times I$ then the quantity $\int_M vu\, d\mu_{g(t)}$ is constant along Ricci flow. Moreover,

1. Choosing $v = 1$ on $M \times I$, gives that $\int_M u d\mu_{g(t)}$ is constant, if $\Box^* u = 0$. If we further assume that

$$\int_M u d\mu_{g(t)} = 1,$$

then u defines a family of probability measures $d\nu_t = u(\cdot, t) d\mu_{g(t)}$ on M, for every $t \in I$, which we call conjugate heat flow.

2. If $u_{(y,s)}$ is the conjugate heat kernel starting at (y, s), namely $\Box^* u_{(y,s)} = 0$ and

$$\lim_{t \to s^-} \int_M f(x) u_{(y,s)}(x, t) d\mu_{g(t)} = f(y),$$

then formula (25) gives the following representation formula for any function $v \in C^\infty(M \times I)$, and any $t, s \in I$ with $t < s$,

$$v(y, s) = \int_M v(x, t) u_{(y,s)}(x, t) d\mu_{g(t)} + \int_t^s \int_M \Box v(x, \tau) u_{(y,s)}(x, \tau) d\mu_{g(\tau)} d\tau.$$

If $u_{(y,s)}$ is the conjugate heat kernel starting at (y, s) we will denote the corresponding family of probability measures by $d\nu_{(y,s),t} = u_{(y,s)}(\cdot, t) d\mu_{g(t)}$, for $t < s$.

3.6 Super Ricci Flow and Optimal Transport

Let $(M, g(t))_{t \in I}$ be a smooth 1-parameter family of Riemannian metrics on a compact manifold M, and let $W_2(\mu, \nu, t)$ denote the 2-Wasserstein distance of the probability measures μ, ν with respect to the metric $d_{g(t)}$ induced by $g(t)$, for any $t \in I$.

McCann–Topping in [77] investigate the behaviour of the 2-Wasserstein distance $W_2(\mu_t, \nu_t, t)$ between two families of probability measures μ_t, ν_t which correspond to solutions of the backwards conjugate heat equation along $g(t)$. They showed that $W_2(\mu_t, \nu_t, t)$ is non-decreasing if and only if $g(t)$ evolves by super Ricci flow (23). In particular, applying this result to the probability measures defined by conjugate heat kernels, then under super Ricci flow we obtain that

$$W_2(\nu_{(y_1,s),t}, \nu_{(y_2,s),t}, t) \leq d_{g(s)}(y_1, y_2),$$

for any $t < s$ in I.

Bamler in [10] discovered that a similar monotonicity property also holds for the 1-Wasserstein distance along a super Ricci flow, which is easier to describe compared to the monotonicity of the 2-Wasserstein distance. Let $\mu_t, \nu_t, t \in I$, be two conjugate heat flows on M. We can express the 1-Wasserstein distance with respect to $d_{g(t)}$ as

$$W_1(\mu_t, \nu_t, t) = \sup\left\{\int_M f\mu_t - \int_M f\nu_t,\ f \in C^\infty(M),\ |\nabla f|_{g(t)} \leq 1\right\}.$$

By the discussion in Sect. 3.4, given any $f \in C^\infty(M)$ with $|\nabla f|_{g(t_1)} \leq 1$, if $g(t)$ evolves by the super Ricci flow then the solution v of the heat equation with $v(\cdot, t_1) = f$ satisfies $|\nabla v(\cdot, t_2)|_{g(t_2)} \leq 1$, for any $t_1 < t_2$ in I.

It follows that for any $f \in C^\infty(M)$ with $|\nabla f|_{g(t_1)} \leq 1$, the corresponding solution v of the heat equation with $v(\cdot, t_1) = f$ has $\int_M v d\mu_t$ and $\int_M v d\nu_t$ constant in t, therefore

$$\int_M f d\mu_{t_1} - \int_M f d\nu_{t_1} = \int_M v(\cdot, t_2) d\mu_{t_2} - \int_M v(\cdot, t_2) d\nu_{t_2} \leq W_1(\mu_{t_2}, \nu_{t_2}, t_2).$$

Therefore, taking the supremum over f gives that

$$W_1(\mu_{t_1}, \nu_{t_1}, t_1) \leq W_1(\mu_{t_2}, \nu_{t_2}, t_2),$$

for any $t_1 < t_2$ in I. Applying this to conjugate heat kernels gives

$$W_1(\nu_{(y_1,s),t}, \nu_{(y_2,s),t}, t) \leq d_{g(s)}(y_1, y_2),$$

for any $t < s$ in I.

4 The First Applications

In his seminal paper [63] Hamilton, after introducing the Ricci flow, he used it to prove the following theorem, regarding the topology of compact 3-manifolds admitting metrics with positive Ricci curvature.

Theorem 4 *Let (M, g_0) be a simply connected compact three dimensional Riemannian manifold with $\mathrm{Ric}_{g_0} > 0$. Then M is diffeomorphic to $\mathbb{S}^3$.*

The idea is to use the Ricci flow $g(t), t \in [0, T)$, starting at $g(0) = g_0$ to deform g_0 to a Riemannian metric on M that has constant positive curvature. The only simply connected compact manifold in dimension three that can carry such a Riemannian metric, up to diffeomorphism, is $\mathbb{S}^3$.

It can be shown that $g(t)$ will have a finite time singularity, i.e. $T < +\infty$. In fact, any Ricci flow starting at a Riemannian metric with positive scalar curvature, has to encounter a finite time singularity [45]. However, normalizing the volume along the flow to be 1 by considering $\tilde{g}(t) = \mathrm{Vol}(g(t))^{-2/3} g(t)$, Hamilton shows that $\tilde{g}(t)$ converges to a metric with constant positive curvature and unit volume.[7]

The key aspects of the behaviour of $g(t)$ that lead to the results are the following.

1. The positivity of Ricci curvature is preserved along the Ricci flow, namely $\mathrm{Ric}_{g(t)} > 0$ for every $t \in [0, T)$.
2. The trace-free part of the Ricci curvature $\mathrm{Ric}_0 = \mathrm{Ric} - \frac{R}{3} g$ satisfies a scale invariant estimate of the from

$$\frac{|\,\mathrm{Ric}_0\,|^2}{R^2} \leq C R^{-\delta}.$$

This estimate tells us that when the scalar curvature of $g(t)$ is large, then after rescaling $\tilde{g}(t)$ tends to be an Einstein metric, with positive scalar curvature.

Remark 1 If we assume that $\mathrm{Ric}_{g_0} \geq 0$ then there are only two possibilities. Either $\mathrm{Ric}_{g(t)} > 0$, for every $t > 0$, so by Theorem 4, M is diffeomorphic to $\mathbb{S}^3$, or by the strong maximum principle $(M, g(t))$ splits isometrically as $(\mathbb{S}^2 \times \mathbb{S}^1, g'(t) \oplus d\theta^2)$, where $g'(t)$ is a Ricci flow on $\mathbb{S}^2$, which evolves according to Theorem 5 below.

Understanding the behaviour of the Ricci flow on compact two dimensional manifold, although at first it may seem simpler, it turned out to be harder to prove than Theorem 4, and it was the combined work of Hamilton [65] and Chow [43] that finally gave a complete understanding of the Ricci flow on compact surfaces.

On any two dimensional Riemannian manifold $\mathrm{Ric} = 2Rg$, so the Ricci flow becomes

$$\frac{\partial g}{\partial t} = -Rg. \tag{26}$$

It follows that the Ricci flow on surfaces preserves the conformal class, namely for every $t \in [0, T)$, $g(t) = e^{2u(\cdot,t)} g(0)$, for some smooth function $u \in M \times [0, T)$.

Using (24) and the Gauss–Bonnet Theorem we can easily establish the following simple evolution equation for the volume (area) of a compact surface evolving under Ricci flow. In fact,

$$\frac{d}{dt} \mathrm{Vol}(g(t)) = -\int_M R\, d\mu = -4\pi \chi(M),$$

where $\chi(M)$ is the Euler characteristic of M. Therefore the behaviour of the volume of $(M, g(t))$ under Ricci flow on compact surfaces is completely determined by topology, namely

[7] In fact Hamilton considers the volume normalized Ricci flow. $\tilde{g}(t)$ here is just volume normalized Ricci flow, after a time reparameterization.

$$\mathrm{Vol}(g(t)) = \mathrm{Vol}(g(0)) - 4\pi\chi(M)t.$$

This is a special fact that holds only in two dimensions. In particular, if $\chi(M) > 0$, then $\mathrm{Vol}(g(t)) \to 0$ as $t \to \frac{\mathrm{Vol}(g(0))}{8\pi\chi(M)}$, and the flow has to form a finite time singularity at, or before, time $t = \frac{\mathrm{Vol}(g(0))}{4\pi\chi(M)}$.

The final picture established by Hamilton in [65] and Chow [43] is described in the following theorem.

Theorem 5 (Hamilton [65] and Chow [43]) *Let (M, g_0) be compact Riemannian manifold with* $\dim M = 2$, *and $g(t)$, $t \in [0, T)$, the maximal Ricci flow with $g(0) = g_0$. Then, if $\chi(M) > 0$, $T = \frac{\mathrm{Vol}(g(0))}{4\pi\chi(M)}$, otherwise $T = +\infty$. Moreover,*

1. *If $\chi(M) > 0$, $\tilde{g}(t) = \frac{1}{T-t}g(t)$ converges to a metric on M with constant positive curvature, as $t \to T$.*
2. *If $\chi(M) = 0$, $g(t)$ converges to a metric of zero curvature as $t \to +\infty$.*
3. *If $\chi(M) < 0$, $\tilde{g}(t) = \frac{1}{t}g(t)$ converges to a metric with constant negative curvature, as $t \to +\infty$.*

The hardest case in Theorem 5 was the case of the 2-sphere. Hamilton proved the result in [65] when the scalar curvature is initially positive, while later Chow showed in [43] that given any initial metric on the 2-sphere, the scalar curvature will eventually become positive and Hamilton's analysis from [65] applies. This step required a Harnack inequality for the scalar curvature.

The behaviour of the Ricci flow in higher dimensions is much more complicated. The non-negativity of the Ricci tensor is no longer preserved in dimensions greater than three, see for instance [76]. However, Hamilton in [64], used the (normalized) Ricci flow to show that simply connected 4-manifolds with positive curvature operator are diffeomorphic to 4-spheres. Moreover, he discovered a parabolic strong maximum principle satisfied by the curvature operator, provided that the curvature operator is non-negative. Because of this strong maximum principle, any possible vanishing of the curvature operator should occur globally on the manifold along parallel directions, which eventually reduces the holonomy of the evolving Riemannian metrics. In dimension 4 this allowed Hamilton to understand the full topological classification of manifolds with non-negative curvature operator. In dimensions greater than four, Böhm–Wilking [20] also proved that a metric with positive curvature operator will evolve under normalized Ricci flow to a metric with constant positive curvature on the sphere. Finally, Brendle–Schoen in [22] use the Ricci flow to prove the long-standing 1/4-pinched differentiable sphere theorem, that states that any simply connected n-manifold that admits a Riemannian metric whose sectional curvature takes values in the interval $(1/4, 1]$ must be diffeomorphic to $\mathbb{S}^n$. That it must be homeomorphic to $\mathbb{S}^n$ had been established earlier by Berger [18] and Klingenberg [71].

5 The Neck-Pinch Singularity

The examples we have encountered so far should convince us that it is common for a Ricci flow to encounter a finite time singularity. We saw in Sect. 3.3 that any Ricci flow starting from an Einstein metric with positive scalar curvature must become singular in finite time. Similarly, starting from any metric on $\mathbb{S}^2$, or any metric with positive Ricci curvature on $\mathbb{S}^3$, the Ricci flow will encounter a finite time singularity. These are all "global" singularities, in the sense that $\min_M |\operatorname{Rm}|(\cdot, t) \to +\infty$, as t approaches the maximal time of existence, i.e. the curvature blows up everywhere.

What if we start with a metric on $\mathbb{S}^3$ that does not have positive Ricci curvature? Assuming for simplicity that the metric is rotationally symmetric with diameter d, we can express it in the form

$$g_0 = ds^2 + (\psi(s))^2 g_{\mathbb{S}^2}. \tag{27}$$

where $\psi : (-d/2, d/2) \to \mathbb{R}_+$ is a smooth even function. Note that since g_0 represents a metric on $\mathbb{S}^3$, we know that $\lim_{s\to\pm d/2} \psi(s) = 0$ and $\lim_{s\to\pm d/2} \psi'(s) = \mp 1$.

The Ricci curvature of g_0 can be explicitly computed in terms of ψ, see for instance [83].

$$\operatorname{Ric}\left(\frac{\partial}{\partial s}, \frac{\partial}{\partial s}\right) = -2\frac{\psi''}{\psi}, \tag{28}$$

$$\operatorname{Ric}\left(\frac{\partial}{\partial s}, X\right) = 0, \tag{29}$$

$$\operatorname{Ric}(X, Y) = \left(\frac{1-(\psi')^2}{\psi^2} - \frac{\psi''}{\psi}\right) g_0(X, Y), \tag{30}$$

where X, Y here are tangent to the $\mathbb{S}^2$ factor, in the product $\mathbb{R} \times \mathbb{S}^2$.

Since Ricci curvature is allowed to be negative, we can imagine ψ to have a local minimum at $s = 0$, and two global maxima at $s = \pm d/4$, with $\psi(-d/4) = \psi(d/4) \gg \psi(0)$. Geometrically this means that $(\mathbb{S}^3, g_0)$ has the shape of a dumbbell, with the two lobes connected by a thin neck.

Moreover, by (30), since $\psi(0)$ is very small, the Ricci curvature of g_0 in directions perpendicular to $\frac{\partial}{\partial s}$, around $s = 0$, is going to be strictly positive, so the waist of the neck will tend to contract under Ricci flow. Presumably, this configuration will lead to the waist contracting to a point, leading to a singularity called a "neck-pinch". Such behaviour had been conjectured by Hamilton in [66], and later Angenent–Knopf proved in [3] that there is an open set of metrics of the form (27) that exhibit exactly this behaviour. The neckpinch is a "local" singularity, in the sense that the curvature blows up only along the $\mathbb{S}^2$ for $s = 0$. Moreover, the local picture of the Ricci flow around $s = 0$, at the scale of curvature and near the singular time, resembles the shrinking cylinder Ricci flow on $\mathbb{S}^2 \times \mathbb{R}$ (Fig. 2).

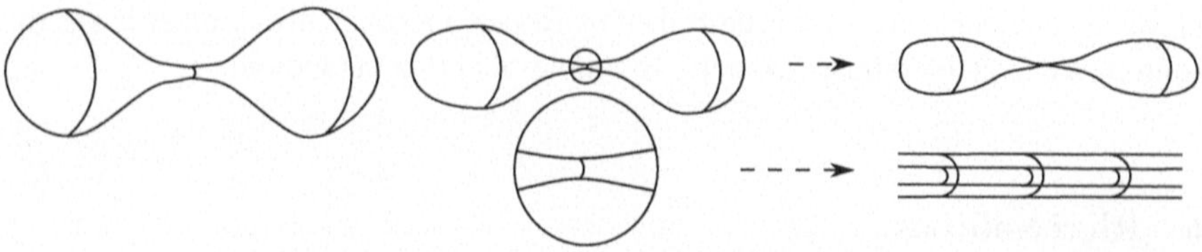

Fig. 2 A two dimensional representation of a rotationally symmetric neck-pinch singularity on $\mathbb{S}^3$, with reflection symmetry. The waist of the neck pinches to a point, while the Ricci flow at small scales near the singular time looks like the shrinking cylinder Ricci flow on $\mathbb{S}^2 \times \mathbb{R}$

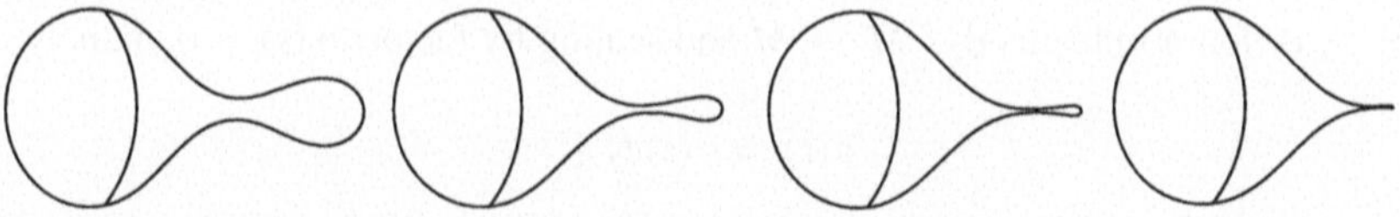

Fig. 3 Two dimensional representation of a Ricci flow on $\mathbb{S}^3$ developing a rotationally symmetric degenerate neckpinch singularity on $\mathbb{S}^3$

The existence of the neck-pinch singularity should lead us to the following conclusions.

1. The positive Ricci curvature assumption in Hamilton's 3d Theorem is necessary, in the sense that we can not otherwise exclude singularities to occur before the Ricci flow has enough time to become round.
2. Using Ricci flow to improve general Riemannian metrics will require a strategy to deal with local singularities, like the neck-pinch singularity. A neck-pinch singularity seems to indicate that the Ricci flow wants to cut the manifold into two pieces, and this is what Perelman's Ricci flow with surgery achieves.

To conclude this section we briefly mention another important singularity that is possible to develop, even in the rotationally symmetric Ricci flow in $\mathbb{S}^3$, the *degenerate neck-pinch* singularity. One can think of this kind of behaviour occurring as a boundary case, namely as the limit behaviour of a sequence of rotationally symmetric Ricci flows on $\mathbb{S}^3$ that all develop a neck-pinch singularity, but the size of one of the two lobes of the dumbbell becomes smaller and smaller. If one of the two lobes of the initial metric was too small, with size comparable to the width of the neck, then the metric would have positive Ricci curvature, and the Ricci flow would develop a global finite time singularity that would look like the round shrinking sphere, by Hamilton's Theorem 4. Therefore, there should be some initial rotationally symmetric dumbbell metric on $\mathbb{S}^3$ with the size of the two lobes just right in order to develop the type of singularity described in Fig. 3.

The existence of degenerate neck-pinch singularities has been confirmed by the work of Gu–Zhu [62] and Angenent–Isenberg–Knopf [4]. Near the singularity, its local geometry around the neck is still described by the shrinking cylinder Ricci flow. However, at the tip of the degenerate neck-pinch where the curvature also blows up, the local geometry is described by Bryant's rotationally symmetric steady soliton (see Sect. 6), whose geometry resembles that of a paraboloid.

6 Ricci Solitons

As we saw in Sect. 3 it is natural to view a Ricci flow on M as a path in the space of Riemannian metrics on M modulo diffeomorphisms and scalings. From this viewpoint, a fixed point of the flow will be a Ricci flow $g(t), t \in I$, that evolves only by diffeomorphisms $\phi_t : M \to M$ and scaling by factors $\sigma(t) > 0$, namely,

$$g(t) = \sigma(t)\phi_t^* \bar{g},$$

for a fixed Riemannian metric $\bar{g}$ on M. Any Ricci flow of this form will be called a *self-similar* solution of the Ricci flow.

Without loss of generality, we may assume that $0 \in I$, $\sigma(0) = 1$ and $\phi_0 = \mathrm{id}_M$, in other words that $\bar{g} = g(0)$. Differentiating at $t = 0$ we obtain that

$$-2\,\mathrm{Ric}_{g(0)} = -2\,\mathrm{Ric}_{\bar{g}} = \sigma'(0)\bar{g} + \mathcal{L}_X \bar{g},$$

where $X = \frac{d}{dt}\big|_{t=0} \phi_t$. Therefore, the triplet $(M, \bar{g}, X)$ satisfies the equation

$$\mathrm{Ric}_{\bar{g}} + \frac{1}{2}\mathcal{L}_X \bar{g} = \lambda \frac{\bar{g}}{2}, \tag{31}$$

where $\lambda = -\sigma'(0)$.

Definition 1 A triplet $(M, \bar{g}, X)$, where $\bar{g}$ is a complete Riemannian metric on M and X a complete vector field on M, that satisfies (31) is called a Ricci soliton. Moreover, we will say that $(M, \bar{g}, X)$ is a

- *shrinking Ricci soliton at scale* $\tau > 0$ if $\lambda = \tau^{-1}$.
- *steady Ricci soliton* if $\lambda = 0$.
- *expanding Ricci soliton* at scale $\tau > 0$ if $\lambda = -\tau^{-1}$.

If $(M, \bar{g}, X)$ is a Ricci soliton and there is a smooth function f on M such that $X = \nabla f$ then we say that $(M, \bar{g}, f)$ is a *gradient* Ricci soliton and f is a soliton function.

On the other hand, let $(M, \bar{g}, X)$ be a Ricci soliton. Let $\phi_t : M \to M$ be the 1-parameter family of diffeomorphisms defined by

$$\frac{d}{dt}\phi_t = \frac{1}{1 - \lambda t} X \circ \phi_t,$$
$$\phi_0 = \mathrm{id}_M.$$

A direct computation using (31) shows that $g(t) = (1 - \lambda t)\phi_t^* \bar{g}$ is a self-similar Ricci flow on M, defined in $(-\infty, \lambda^{-1})$, $(-\infty, +\infty)$ or $(\lambda^{-1}, +\infty)$, if $(M, \bar{g}, X)$ is a shrinking, steady or expanding Ricci soliton respectively.

6.1 Examples of Ricci Solitons

In this subsection we will present the most fundamental examples of Ricci solitons, aiming mostly to build the necessary intuition, and to describe the current state of the art.

Every Einstein manifold is a shrinking, steady or expanding gradient Ricci soliton, depending on whether its scalar curvature is positive, zero or negative, respectively, choosing the corresponding soliton function $f = 1$. Moreover, by the classical Bonnet–Myers theorem in Riemannian Geometry, any shrinking Ricci soliton which is Einstein should be compact.

On the other hand, compact steady or expanding Ricci solitons should necessarily be Einstein manifolds, see for instance [45]. There are, however, examples of compact non-Einstein (Kähler) gradient shrinking solitons, constructed by Koiso [72], Cao [24] and Wang–Zhu [91].

The simplest, but extremely illuminating, examples of non compact, complete, Ricci solitons are defined in the Euclidean space $(\mathbb{R}^n, g_{\mathbb{R}^n})$, which we can view as any type of *gradient* Ricci soliton, steady, shrinking or expanding. In all cases, the corresponding self-similar Ricci flow is just the static Ricci flow on $\mathbb{R}^n$, $g(t) = g_{\mathbb{R}^n}$.

Since $\mathrm{Ric}_{\mathbb{R}^n} = 0$, $(\mathbb{R}^n, g_{\mathbb{R}^n})$ is an Einstein manifold with zero scalar curvature, hence it defines a steady gradient Ricci soliton on $\mathbb{R}^n$. However, we can also view $(\mathbb{R}^n, g_{\mathbb{R}^n})$ both as a shrinking as well as an expanding gradient Ricci soliton with an appropriate choice of a soliton function. The resulting shrinking/expanding gradient Ricci solitons are called Gaussian shrinking/expanding solitons in $\mathbb{R}^n$, with the soliton function $f : \mathbb{R}^n \to \mathbb{R}$ defined as

$$f(x) = \begin{cases} \frac{|x|^2}{4}, & \text{shrinking case} \\ -\frac{|x|^2}{4}, & \text{expanding case} \end{cases}$$

Thus, the Gaussian shrinking/expanding Ricci solitons are the triplets $(\mathbb{R}^n, g_{\mathbb{R}^n}, \frac{|x|^2}{4})$ and $(\mathbb{R}^n, g_{\mathbb{R}^n}, -\frac{|x|^2}{4})$, respectively.

A direct computation gives that $\nabla f = \pm \frac{x^i}{2} \frac{\partial}{\partial x^i} = \pm \frac{r}{2} \frac{\partial}{\partial r}$ and $\mathrm{Hess}_{g_{\mathbb{R}^n}} f = \pm \frac{g_{\mathbb{R}^n}}{2}$. Therefore,

$$\mathrm{Ric}_{g_{\mathbb{R}^n}} + \mathrm{Hess}_{g_{\mathbb{R}^n}} f = \pm \frac{g_{\mathbb{R}^n}}{2}.$$

Taking the Riemannian product of any two gradient Ricci solitons (M_i, g_i, f_i) of the same type -steady, shrinking or expanding- yields a new gradient Ricci soliton

$(M_1 \times M_2, g_1 \oplus g_2, f_1 + f_2)$, of the same type. For instance, if (M', g', f') is a gradient shrinking Ricci soliton at scale 1 and $(\mathbb{R}^k, g_{\mathbb{R}^k}, \frac{|x|^2}{4})$ is the Gaussian shrinking soliton described above, the product $(M' \times \mathbb{R}^k, g' \oplus g_{\mathbb{R}^k}, f)$ where

$$f(q, x) = \frac{|x|^2}{4} + f'(q),$$

for every $(q, x) \in M' \times \mathbb{R}^k$, is a gradient shrinking Ricci soliton at scale 1. In particular, this yields a standard shrinking Ricci soliton structure at scale 1 on the generalized cylinders $\mathbb{S}^{n-k} \times \mathbb{R}^k$, with associated self-similar Ricci flow the shrinking cylinder Ricci flow described by (17).

The Euclidean space, as a non-compact, complete, steady soliton is Einstein. There are however important non-Einstein steady Ricci solitons.

1. The cigar soliton. This is a steady Ricci soliton in $\mathbb{R}^2$, described by the metric $g_{\text{cigar}} = \frac{dx^2+dy^2}{1+x^2+y^2}$, and soliton function $f(x, y) = -\log(1 + x^2 + y^2)$. The metric g_{cigar} is invariant under rotations centered at the origin, and its geometry at infinity resembles that of the cylinder $(\mathbb{R} \times \mathbb{S}^1, dr^2 + d\theta^2)$. The reader may consult [45] for more details.
2. Bryant's steady soliton. This is a rotationally symmetric steady Ricci soliton in $\mathbb{R}^n$, with Riemannian metric $g_{Br} = ds^2 + (\psi(s))^2 g_{\mathbb{S}^{n-1}}$, where $\psi(s) \sim \sqrt{s}$ as $s \to +\infty$. As such, its geometry resembles that of a paraboloid.

More, non-compact, collapsed, steady (Kähler) Ricci solitons were constructed by Cao [24], on $\mathbb{C}^n$ and on certain line bundles over $\mathbb{C}P^{n-1}$. Non-Kähler, non-collapsed, examples have recently been constructed by Appleton [6].

Cao in [25] constructed a 1-parameter family of $U(n)$-invariant expanding Kähler gradient expanding Ricci solitons, each asymptotic to a conical metric on $\mathbb{C}^n \setminus \{0\}$. Bryant constructed rotationally symmetric expanding Ricci solitons on $\mathbb{R}^n$, $n \geq 3$. These have positive curvature and are asymptotic to positively curved cones over round $\mathbb{S}^{n-1}$ at infinity. Deruelle in [51] shows that there exist positively curved expanders asymptotic at infinity to any cone over $\mathbb{S}^{n-1}$ with positive Riemann curvature operator, while Conlon–Deruelle in [49] prove the existence of Kähler expanding Ricci solitons asymptotic to some Kähler cones. Later Bamler–Chen in [13] prove that given any cone over $\mathbb{S}^{n-1}$ with positive scalar curvature, there exist expanding Ricci solitons with positive scalar curvature, asymptotic to that cone. Finally, Feldman–Ilmanen–Knopf [57] construct many Kähler, asymptotically conical expanding Ricci solitons on line bundles over $\mathbb{C}P^{n-1}$, asymptotic to metric cones on $\mathbb{C}^n/\mathbb{Z}^k$, while very recently Angenent–Knopf in [5] constructed expanding Ricci solitons on $\mathbb{S}^{p_1} \times \mathbb{R}^{p_2+1}$ and $\mathbb{R}^{p_1+1} \times \mathbb{S}^{p_2}$, for any $p_1, p_2 \geq 2$ with $p_1 + p_2 \leq 8$, which are all asymptotic to cones over $\mathbb{S}^{p_1} \times \mathbb{S}^{p_2}$.

Although expanding Ricci solitons seem to be very flexible, the world of shrinking Ricci solitons is much more rigid. The first non-compact examples were constructed by Feldman–Ilmanen–Knopf in [57]. These are Kähler, $U(n)$-invariant, gradient shrinking Ricci solitons on line bundles over $\mathbb{C}P^{n-1}$, $n \geq 2$, which are asymptotic at infinity to a conical metric on $\mathbb{C}^n \setminus \{0\}/\mathbb{Z}_k$, for any $0 < k < n$. In fact, in complex dimension $n = 2$, $k = 1$ and their construction gives a particular Kähler gradient Ricci soliton asymptotic to a conical metric on $\mathbb{C}^2 \setminus \{0\}$. Very recently, Bamler–Cifarelli–Conlon–Deruelle in [14] discovered another shrinking Ricci soliton in complex dimension 2 on the blow-up at one point of $\mathbb{C} \times \mathbb{C}P^1$. In fact, in [14], the authors complete the classification of all Kähler shrinking Ricci solitons with bounded scalar curvature in complex dimension 2. Finally, in [5] Angenent–Knopf also construct many shrinking Ricci solitons in $\mathbb{S}^{p_1} \times \mathbb{R}^{p_2+1}$, $p_1, p_2 \geq 2$, $p_1 + p_2 \leq 8$, asymptotic to cones over $\mathbb{S}^{p_1} \times \mathbb{S}^{p_2}$.

6.2 *Gradient Shrinking Ricci Solitons*

Let $(M, \bar{g}, f)$ be a shrinking gradient Ricci soliton at scale 1. Performing a time translation, we can express the corresponding self-similarly shrinking Ricci flow as $g(t) = |t|\phi_t^* \bar{g}$, for $t < 0$, where ϕ_t is the 1-parameter family of diffeomorphisms generated by

$$\frac{d}{dt}\phi_t = \frac{1}{|t|}\nabla^{\bar{g}} f \circ \phi_t, \tag{32}$$

$$\phi_{-1} = \mathrm{id}_M. \tag{33}$$

It is then easy to see that for any $t < 0$, $(M, g(t), f_t = f \circ \phi_t)$ is also a gradient shrinking Ricci soliton satisfying

$$\mathrm{Ric}_{g(t)} + \mathrm{Hess}_{g(t)}\, f_t = \frac{g(t)}{2|t|}.$$

Namely, it is a gradient shrinking Ricci soliton at scale $|t|$.

The flow $(M, g(t))_{t<0}$ satisfies the following scale invariance property. Given a $\lambda > 0$, let $g_\lambda(t) = \lambda^{-1} g(\lambda t)$ be the parabolic rescaling of $g(t)$ by the factor λ^{-1}, which is still a Ricci flow by Sect. 3.2. Then we can compute that

$$g_\lambda(t) = |t|\phi_{\lambda t}^* \bar{g} = \phi_\lambda^*(|t|\phi_t^* \bar{g}) = \phi_\lambda^* g(t). \tag{34}$$

Therefore, there is a fixed isometry ϕ_λ such that $g_\lambda(t)$ is isometric to $g(t)$ for every $t < 0$.[8] In this sense, the self-similar Ricci flow induced by a (gradient) shrinking Ricci soliton is the parabolic (Ricci flow) analogue of a cone in Euclidean space, since cones are characterized by their invariance under dilations with respect to a fixed point, the *vertex* of the cone. In fact, the whole cone is generated by dilating the *link* of the cone, namely the intersection of the cone with the sphere of unit

[8] This in fact also holds for non-gradient shrinking Ricci solitons.

radius centred at a vertex. In the context of Ricci flow, it is the shrinking soliton $(M, \bar{g}, f)$ that plays the role of a link, since it generates the self-similar Ricci flow $g(t)$ by dilations. Finally, note that a cone can have multiple vertices, which form an affine subspace of Euclidean space. Let's call this set of vertices the *spine* of the cone. We will see below that it is possible to give a reasonable, and useful, definition of an analogous concept for gradient shrinking Ricci solitons.

The Gaussian shrinking Ricci soliton $(\mathbb{R}^n, g_{\mathbb{R}^n}, \frac{|x|^2}{4})$ is a good example to explicitly demonstrate the behaviour we described above. Since the solution of the ODE

$$\frac{dr}{dt} = \frac{r}{2|t|},$$
$$r(-1) = r_0 \geq 0,$$

for $t < 0$, is just $r(t) = r_0/\sqrt{|t|}$, the corresponding self-similar Ricci flow is defined by the diffeomorphisms $\phi_t(x) = x/\sqrt{|t|}$, $t < 0$. These are just dilations pushing to infinity as $t \to 0$, and one can check that $g(t) = |t|\phi_t^* g_{\mathbb{R}^n} = g_{\mathbb{R}^n}$, for every $t < 0$. Moreover, setting $f(x) = \frac{|x|^2}{4}$, we see that $f_t(x) = f(\phi_t(x)) = \frac{|x|^2}{4|t|}$ and we can check that $(\mathbb{R}^n, g_{\mathbb{R}^n}, \frac{|x|^2}{4|t|})$ is a gradient shrinking soliton at scale $|t|$.

6.2.1 Properties of Soliton Functions

Let $(M, \bar{g}, f)$ be a gradient shrinking Ricci soliton at scale τ, namely suppose that

$$\mathrm{Ric}_{\bar{g}} + \mathrm{Hess}_{\bar{g}} f = \frac{\bar{g}}{2\tau}. \tag{35}$$

Taking a trace in (35) we obtain

$$R_{\bar{g}} + \Delta_{\bar{g}} f = \frac{n}{2\tau}, \tag{36}$$

while taking the divergence and using the contracted second Bianchi identity we obtain that

$$\nabla(\tau(|\nabla^{\bar{g}} f|_{\bar{g}}^2 + R_{\bar{g}}) - f) = 0.$$

Thus, $\tau(|\nabla^{\bar{g}} f|_{\bar{g}}^2 + R_{\bar{g}}) - f = C(\bar{g}, f)$ is always constant on a gradient shrinking Ricci soliton.

The soliton function $\frac{|x|^2}{4}$ of the Gaussian shrinking Ricci soliton has quadratic growth. It turns out that this is a general property of soliton functions of shrinking solitons. By Haslhofer–Buzano [67], if $(M, \bar{g}, f)$, $n = \dim M$, is a complete gradient shrinking Ricci soliton at scale $\tau > 0$ then f attains at least a minimum point $p \in M$ and

$$\frac{1}{4}\left(\frac{d(p,x)}{\sqrt{\tau}}-5n\right)_+^2 \leq f(x)+C(\bar{g},f) \leq \frac{1}{4}\left(\frac{d_{\bar{g}}(p,x)}{\sqrt{\tau}}+\sqrt{2n}\right)^2. \tag{37}$$

Moreover, if $p, p' \in M$ are two points where f attains its minimum value,

$$d_{\bar{g}}(p,p') \leq A(n)\sqrt{\tau},$$

where $A(n) = 5n+\sqrt{2n}$. In particular, if $(M, g(t))_{t<0}$ is the self-similarly shrinking Ricci flow induced by a gradient shrinking Ricci soliton $(M, \bar{g}, f)$ then the subset $\mathcal{K} \subset M$ where f, and thus each f_t, attains its minimum value satisfies

$$\operatorname{diam}_{g(t)}(\mathcal{K}) \leq A(n)\sqrt{|t|}. \tag{38}$$

Finally, by [67] any complete gradient shrinking Ricci soliton $(M, \bar{g}, f)$ at scale τ has polynomial volume growth. In fact there is a positive constant $c(n)$ such that for every $p \in M$ and $r > 0$,

$$\operatorname{Vol}(B(p,r)) \leq c(n) r^n.$$

Thus, the quadratic growth estimate (37) implies that

$$\int_M e^{-f} \, d\mu_{\bar{g}} < +\infty.$$

Since equation (35) is invariant under the transformation $f \mapsto f + c$, we can always assume that

$$\int_M \frac{e^{-f}}{(4\pi\tau)^{n/2}} d\mu_{\bar{g}} = 1. \tag{39}$$

We will call $(M, \bar{g}, f)$ a *normalized gradient shrinking Ricci soliton*, if (39) holds.

Finally, it can be shown [67] that there is a constant $C(\bar{g})$ such that if $(M, \bar{g}, f)$ is a normalized gradient shrinking Ricci soliton then

$$\tau(|\nabla^{\bar{g}} f|_{\bar{g}}^2 + R_{\bar{g}}) - f = C(\bar{g}). \tag{40}$$

6.2.2 The Spine of a Gradient Shrinking Ricci Soliton

Now let $(M, \bar{g}, f_1)$, $(M, \bar{g}, f_2)$ be both normalized gradient shrinking solitons at the same scale $\tau > 0$. By (35), we obtain that $\operatorname{Hess}_{\bar{g}}(f_2 - f_1) = 0$. The existence of a non-constant function $L = f_2 - f_1$ with vanishing Hessian on a Riemannian manifold $(M, \bar{g})$ is easily seen to imply that $(M, \bar{g})$ splits isometrically as a product $M' \times \mathbb{R}$, namely M is diffeomorphic to $M' \times \mathbb{R}$ and $\bar{g}$ is isometric to a product metric $g' \oplus dr^2$, where g' is a Riemannian metric on M.

Exploiting this phenomenon, we can conclude that given any normalized gradient shrinking soliton $(M, \bar{g}, f)$ at scale $\tau > 0$, there is a maximal non-negative integer k, a *unique* normalized gradient shrinking Ricci soliton (M', g', f') at scale τ and $a \in \mathbb{R}^k$, so that $(M, \bar{g})$ is isometric to the product $M' \times \mathbb{R}^k$, $\bar{g} = g' \oplus g_{\mathbb{R}^k}$ and for every $(q, x) \in M' \times \mathbb{R}^k$

$$f(q, x) = \frac{|x - a|^2}{4\tau} + f'(q). \tag{41}$$

Moreover, for every $a \in \mathbb{R}^k$, (41) determines a soliton function for $(M, \bar{g})$.

Any soliton function f determined by (41) will attain its minimum value at points of the form $(a, q) \in M' \times \mathbb{R}^k$, where q is a point where f' attains its minimum. By (38), the collection $\mathcal{K} \subset M'$ of the points were f' attains its minimum value has $\operatorname{diam}_{\bar{g}}(\mathcal{K}) \leq A(n)\sqrt{\tau}$.

Therefore, given a gradient shrinking soliton $(M, \bar{g}, f)$, $n = \dim M$, at scale $\tau > 0$, there is a maximal $0 \leq k \leq n$ such that $(M, \bar{g})$ isometrically splits as the product $(M', g') \times (\mathbb{R}^k, g_{\mathbb{R}^k})$ and

1. The collection of all possible soliton functions forms a k-dimensional subspace of the smooth functions, parameterized by $\mathbb{R}^k$, as in (41). We will call this collection the *spine* of the gradient shrinking Ricci soliton, and denote it by $\mathcal{S}$.
2. The subset of M that consists of the points where some normalized soliton function at scale τ attains its minimum is a set of the form $\mathcal{K} \times \mathbb{R}^k$, where $\mathcal{K} \subset M'$ satisfies $\operatorname{diam}_{g'}(\mathcal{K}) \leq A(n)\sqrt{\tau}$. We will call this set the *point-spine* of the gradient shrinking Ricci soliton, and denote it by $\mathcal{S}_{\text{point}}$.

Now, let $(M, g(t))_{t<0}$ be the self-similarly shrinking Ricci flow induced by a gradient shrinking Ricci soliton $(M, \bar{g}, f)$ at scale 1, $n = \dim M$, and let ϕ_t be the family of diffeomorphisms generated by (32) and (33). Since any $p \in \mathcal{S}_{\text{point}}$ is a critical point of some soliton function, which we may assume it is f itself, we know that $\phi_t(p) = p$, for any $t < 0$. Therefore, the dilation invariance property (34) gives that for any $\lambda > 0$, ϕ_λ is an isometry between $g_\lambda(t)$ and $g(t)$, and maps p to itself. In this sense, any $p \in \mathcal{S}_{\text{point}}$ is the analogue of a vertex of a cone in Euclidean space. Although $\mathcal{S}_{\text{point}} = \mathcal{K} \times \mathbb{R}^k$ is not a k-dimensional set—for instance in shrinking cylinder Ricci solitons in $\mathbb{S}^{n-k} \times \mathbb{R}^k$ we have that $\mathcal{S}_{\text{point}} = \mathbb{S}^{n-k} \times \mathbb{R}^k$—we do have the estimate $\operatorname{diam}_{g'(t)}(\mathcal{K}) \leq A(n)\sqrt{|t|}$, which says that the *large scale geometry* of $\mathcal{S}_{\text{point}}$ is indeed that of $\mathbb{R}^k$, and in fact $\mathcal{S}_{\text{point}}$ converges in the pointed Gromov–Hausdorff sense to $\mathbb{R}^k$, as $t \to 0$. A more detailed discussion on the spine and point-spine of a gradient shrinking Ricci soliton can be found in [59].

Curvature Blow Up at the Point-spine

Let $(M, g(t))_{t<0}$ be a self-similar Ricci flow induced by a gradient shrinking Ricci soliton $(M, \bar{g}, f)$, let $p \in \mathcal{S}_{\text{point}}$ be a point where f attains its minimum value, and let ϕ_t be the family of diffeomorphisms generated by (32) and (33). Then the scalar curvature at p satisfies

$$R_{g(t)}(p) = R_{|t|\phi_t^*\bar{g}}(p) = |t|^{-1} R_{\bar{g}}(\phi_t(p)) = |t|^{-1} R_{\bar{g}}(p).$$

It is thus clear that $R_{g(t)}(p)$ either blows up at a rate $\sim|t|^{-1}$ or $R_{g(t)}(p) = 0$ for every $t < 0$. We will see below that the latter case can only happen if $(M, \bar{g}, f)$ is isometric to the Gaussian shrinking Ricci soliton.

Under Ricci flow, the scalar curvature satisfies the evolution equation

$$\frac{\partial R_{g(t)}}{\partial t} = \Delta_{g(t)} R_{g(t)} + 2|\operatorname{Ric}_{g(t)}|^2_{g(t)} \geq \Delta_{g(t)} R_{g(t)}, \tag{42}$$

see [45]. Moreover, by Chen [34] we know that $R_{g(t)} \geq 0$, for every $t < 0$. Then, the strong maximum principle implies that if $R_{g(t)}(p) = 0$ for some $t < 0$, then $R_{g(t)} \equiv 0$ on M for every $t < 0$. Therefore, (42) gives that $\operatorname{Ric}_{g(t)} \equiv 0$ on M for every $t < 0$, and thus the gradient shrinking soliton equation gives that

$$\operatorname{Hess}_{\bar{g}} f = \frac{\bar{g}}{2}.$$

From this we can conclude that $(M, \bar{g}, f)$ is isometric to the Gaussian shrinking soliton, see [84] for more details.

6.2.3 Evolution of Soliton Functions

Now suppose that $(M, \bar{g}, f)$ is a normalized gradient shrinking soliton at scale 1. The family of diffeomorphisms ϕ_t defined by (32) and (33) then defines a self-similarly shrinking Ricci flow $g(t) = |t|\phi_t^*\bar{g}$, with $g(-1) = \bar{g}$ such that $(M, g(t), f_t = f \circ \phi_t)$ is a normalized gradient shrinking Ricci soliton at scale $|t|$. In particular, for every $t < 0$

$$\int_M (4\pi|t|)^{-n/2} e^{-f_t} d\mu_{g(t)} = 1.$$

Moreover, by the definition of f_t and equation (36) we obtain that

$$\frac{\partial f_t}{\partial t} = |\nabla^{g(t)} f_t|^2_{g(t)} = -\Delta_{g(t)} f_t + |\nabla^{g(t)} f_t|^2_{g(t)} - R_{g(t)} + \frac{n}{2|t|}, \tag{43}$$

which is equivalent to $u = (4\pi|t|)^{-n/2} e^{-f_t}$ evolving by the backwards conjugate heat flow, namely $\square^* u = 0$.

Note that equation (43) is a backwards diffusion equation. It turns out that the evolution of a soliton function can be described by a forward diffusion equation as well. This follows from identities (36) and (40). Namely,

$$\frac{\partial f_t}{\partial t} = |\nabla^{g(t)} f_t|^2_{g(t)} = -R_{g(t)} + \frac{f_t + C(\bar{g})}{|t|} = \Delta_{g(t)} f_t - \frac{n}{2|t|} + \frac{f_t + C(\bar{g})}{|t|},$$

hence $w = |t|(f_t + C(\bar{g}))$ satisfies

$$\frac{\partial w}{\partial t} = \Delta_{g(t)} w - \frac{n}{2}. \tag{44}$$

7 Monotone Scale Invariant Quantities Under Ricci Flow

Given a compact Riemannian manifold (M, g), a smooth function f and $\tau > 0$, define Perelman's $\mathcal{W}$ functional [80]

$$\mathcal{W}(g, f, \tau) = \int_M [\tau(R + |\nabla f|^2) + f - n] \frac{e^{-f}}{(4\pi\tau)^{n/2}} d\mu.$$

This is a functional on $\mathrm{Met}(M) \times C^\infty(M) \times \mathbb{R}_+$ which is diffeomorphism and scale invariant in the sense that for every $\lambda > 0$ and any diffeomorphism $\phi : M \to M$

$$\mathcal{W}(\lambda^2 g, f, \lambda^2 \tau) = \mathcal{W}(g, f, \tau) = \mathcal{W}(\phi^* g, \phi^* f, \tau).$$

The remarkable observation of Perelman in [80] was that if $g(t)$, $t \in I$ evolves by the Ricci flow, $\tau(t)$ satisfies $\frac{d\tau}{dt} = -1$ and $f(t) \in C^\infty(M)$ is such that $u(t) = (4\pi\tau)^{-n/2} e^{-f(t)}$ satisfies $\Box^* u = 0$ then

$$\frac{d}{dt}\mathcal{W}(g(t), f(t), \tau(t)) = \int_M 2\tau \left| \mathrm{Ric}_{g(t)} + \mathrm{Hess}_{g(t)} f(t) - \frac{g(t)}{2\tau(t)} \right|^2 u(t) d\mu_{g(t)}.$$

If $u_{(p,s)}(t) = (4\pi(s-t))^{-n/2} e^{-f_{(p,s)}(t)}$ is the conjugate heat kernel starting at $(p, s) \in M \times I$ then for any $\tau > 0$ such that $s - \tau \in I$ define the pointed entropy at (p, s) as

$$\mathcal{W}_{(p,s)}(\tau) = \mathcal{W}(g(s-\tau), f_{(p,s)}(s-\tau), \tau).$$

It follows that

$$\frac{d}{d\tau}\mathcal{W}_{(p,s)}(\tau) =$$

$$= -\int_M 2\tau \left| \mathrm{Ric}_{g(s-\tau)} + \mathrm{Hess}_{g(s-\tau)} f(s-\tau) - \frac{g(s-\tau)}{2\tau} \right|^2 u_{(p,s)}(s-\tau) d\mu_{g(s-\tau)}. \tag{45}$$

It is a consequence of the log-Sobolev inequality in $\mathbb{R}^n$ that $\mathcal{W}_{(p,s)}(\tau) \le 0$. Moreover, it follows from (45) that if $\mathcal{W}_{(p,s)}(\tau) = 0$ for every $\tau > 0$ with $s - \tau \in I$ then

$$\mathrm{Ric}_{g(t)} + \mathrm{Hess}_{g(t)} f(t) - \frac{g(t)}{2(s-t)} = 0,$$

for every $t \in I, t < s$. Therefore $g(t)$ is the Ricci flow induced by a shrinking Ricci soliton. Since $g(t)$ remains smooth up to $t = 0$ it follows that $(M, g(t), f(t))$ is in fact the Gaussian shrinking Ricci soliton.

On the other hand, integrating (45), we obtain for any $\tau_1 < \tau_2$ with $s - \tau_2 \in I$ that

$$\begin{aligned} &\mathcal{W}_{(p,s)}(\tau_1) - \mathcal{W}_{(p,s)}(\tau_2) = \\ &= \int_{\tau_1}^{\tau_2} \int_M 2\tau \left| \mathrm{Ric}_{g(s-\tau)} + \mathrm{Hess}_{g(s-\tau)} f(s-\tau) - \frac{g(s-\tau)}{2\tau} \right|^2 \\ &\quad \times u_{(p,s)}(s-\tau) d\mu_{g(s-\tau)} d\tau. \end{aligned} \tag{46}$$

Intuitively this means that if the difference $\mathcal{W}_{(p,s)}(\tau_1) - \mathcal{W}_{(p,s)}(\tau_2)$ is small, then $g(t), t \in [s - \tau_2, s - \tau_1]$ is close to be a gradient shrinking Ricci soliton, with soliton function given by $f(t)$.

Given a Riemannian metric g on M and $\tau > 0$ Perelman also defined the following functionals on $\mathrm{Met}(M) \times \mathbb{R}_+$

$$\begin{aligned} \mu(g, \tau) &= \inf \left\{ \mathcal{W}(g, f, \tau), \int_M (4\pi\tau)^{-n/2} e^{-f} d\mu_g = 1 \right\}, \\ \nu(g, \tau) &= \inf\{\mu(g, \tau'), 0 < \tau' < \tau\}. \end{aligned}$$

These functionals are also diffeomorphism and scale invariant

$$\begin{aligned} \mu(\lambda^2 g, \lambda^2 \tau) &= \mu(\phi^* g, \tau) = \mu(g, \tau), \\ \nu(\lambda^2 g, \lambda^2 \tau) &= \nu(\phi^* g, \tau) = \nu(g, \tau) \end{aligned}$$

and are monotone increasing under Ricci flow, in the sense that if $t_1 < t_2$ and $\tau > 0$

$$\begin{aligned} \mu(g(t_1), \tau + t_2 - t_1) &\le \mu(g(t_2), \tau), \\ \nu(g(t_1), \tau + t_2 - t_1) &\le \nu(g(t_2), \tau). \end{aligned}$$

One of the most important implications that a lower bound on ν has on the geometry of a Riemannian manifold is the following non-collapsing result. It says that at the scale of the scalar curvature, the volume of a metric ball is controlled from below by the volume of a Euclidean ball of the same radius. This is due to Perelman [80], however the precise statement below can be found in [2].

Theorem 6 *Let (M, g) be a Riemannian manifold, $r > 0$, and $p \in M$ such that the scalar curvature of (M, g) satisfies $R_g \leq r^{-2}$ on $B_g(p, r)$. Then, for any $s \in (0, r]$*

$$\frac{\mathrm{Vol}_g(B(p, s))}{s^n} \geq C(n)e^{\nu(g, r^2)}.$$

Now, coupling this estimate with the monotonicity of ν under a Ricci flow $(M, g(t))_{t\in[0,T)}$ and using that $\nu(g(0), \tau)$ is bounded below due to the Log-Sobolev inequality, Perelman obtained in [80] his remarkable no-local collapsing Theorem.

Theorem 7 *If $(M, g(t))_{t\in[0,T)}$, $\dim M = n$, is a smooth compact Ricci flow then there is $\kappa = \kappa(n, g(0), T)$ such that*

$$\mathrm{Vol}_{g(t)}(B_{g(t)}(p, s)) \geq \kappa s^n,$$

whenever the scalar curvature satisfies $R_{g(t)} \leq s^{-2}$ in $B_{g(t)}(p, s)$.

Perelman's no local collapsing Theorem played a key role in developing his Ricci flow with surgery, removing (at least) two major stumbling blocks to understand the singularities of the Ricci flow:

1. It allowed to use Hamilton's sequential compactness theorem to obtain smooth limits of blow-up sequences of Ricci flows, essential to understand singularity formation, see Sect. 1.2. The no local collapsing prevents the geometry from collapsing into a lower dimensional space during that process.
2. It excludes Ricci flows that are collapsed at large scales to appear as limits of blow-up sequences, and thus as singularity models. Here, collapsed roughly means that there are metric balls of large enough radius r where the curvature satisfies $|\,\mathrm{Rm}\,| \leq r^{-2}$, but their volume is arbitrarily small (for large r) compared to the volume of the Euclidean ball of the same radius. The cigar soliton (see Sect. 6) is collapsed, as well as its product with $\mathbb{R}$. The possible appearance of this as a singularity model in 3d Ricci flow would make it impossible to perform a surgery procedure and continue the flow.

8 Type I Singularities

Self-similar shrinking Ricci flows, namely Ricci flows induced by gradient shrinking Ricci solitons, are examples of Ricci flows that form a singularity in finite time. In particular, recall from Sect. 6.2.2, that if $(M, g(t))_{t<0}$ is induced by a complete gradient shrinking Ricci soliton which is not the Gaussian shrinking soliton, then along the point-spine $\mathcal{S}_{\mathrm{point}}$ its curvature blows up as $t \to 0$.

Thus gradient shrinking Ricci solitons provide a model for a mechanism that can drive the formation of the singularities of a Ricci flow by homothetically shrinking the metric in several regions of the manifold.

Moreover, if this is the only mechanism that drives the singularity formation of a Ricci flow $(M, g(t))_{t\in[-T,0)}$ that becomes singular at $t = 0$, then it is expected that its curvature will satisfy

$$\max_M |\operatorname{Rm}(\cdot, t)| \leq \frac{C_I}{|t|}, \tag{47}$$

for some large positive constant C_I, merely due to scaling considerations. Therefore, we naturally arrive at the following definition.

Definition 2 We say that compact Ricci flow $(M, g(t))_{t\in(-T,0)}$ has Type I singularities at $t = 0$ if it can not be smoothly extended past time $t = 0$ and there is a large positive constant $C_I < +\infty$ such that (47) holds.

If a Ricci flow becomes singular but

$$\sup_{M\times[-T,0]} |t||\operatorname{Rm}| = +\infty,$$

we say that it is a Type II singular Ricci flow. Of course, a Type II singular Ricci flow may only exhibit "badly" behaved (non Type I singularities) only at certain parts of the manifold. A more refined analysis of the type of singularities that may develop locally is done by Buzano–Di Matteo in [23].

Given a Ricci flow $(M, g(t))_{t\in(-T,0)}$ that becomes singular at $t = 0$ a reasonable definition of its singular set Σ, as a subset of M, is that $x \in \Sigma$ if there is no neighbourhood U of x on which $g(t)$ converges smoothly to a Riemannian metric $g(T)$ on U, as $t \to T$.

8.1 Conjugate Heat Kernel Bounds

Let $(M, g(t))_{t\in[-T,0]}$ be a compact Ricci flow and suppose that there exists a large positive constant $C_I < +\infty$ such that

$$|\operatorname{Rm}| \leq \frac{C_I}{|t|} \tag{48}$$

on $M \times [-T, 0]$. Therefore, although we assume that the Ricci flow remains smooth up to time $t = 0$, we don't assume that the curvature satisfies an a-priori bound as $t \to 0$, but rather the Type I bound (48). Such Ricci flows will appear, after a time translation, if we restrict a Ricci flow with Type I singularities at $t = 0$ into intervals of the form $[-T, t_j]$, for $t_j \to 0$, so we can think of them as smooth approximations of Ricci flows that exhibit Type I singularities.

For any $p \in M$, denote by $u_{(p,0)} = (4\pi|t|)^{-n/2}e^{-f}$ the backwards conjugate heat kernel starting at $(p, 0)$. If $M = \mathbb{R}^n$ and $g(t) = g_{\mathbb{R}^n}$, then $u_{(p,0)}$ is just the (backwards) heat kernel in Euclidean space, namely for $t < 0$

$$u_{(p,0)}(x,t) = (4\pi|t|)^{-n/2}e^{-\frac{|x-p|^2}{4|t|}}.$$

Obtaining bounds of the heat kernel on Riemannian manifold is a problem with long history, and it has been understood that such bounds are intimately tied to the geometry of the underlying manifold, see for instance the works of Davies [50], Grigor'yan [61], for an non-exhaustive list. For evolving Riemannian manifolds, like the Ricci flow, it had been possible to adapt some of these techniques to obtain Gaussian upper and lower bounds for the heat, and conjugate heat, kernels, under the assumption that the curvature is uniformly bounded under the flow, see for instance [44].

To study the singularities of the Ricci flow we would like however to understand what bounds do the heat and conjugate heat kernels satisfy in situations that the curvature is not a-priori bounded. The crucial result in this direction has been the work of Hein–Naber [68]. In [68] they first make the remarkable observation that given a smooth compact Ricci flow $(M, g(t))_{t\in[-T,0]}$, the probability measures $d\nu_{(p,0),t} = u_{(p,0)}(\cdot, t)d\mu_{g(t)}$ automatically satisfy a Poincaré and a log-Sobolev inequality. Namely, for any $s < 0$ and $f \in C^\infty(M)$ with $\int_M f d\nu_{(p,0),s} = 0$,

$$\int_M f^2 d\nu_{(p,0),s} \leq 2|s| \int_M |\nabla^{g(s)} f|^2_{g(s)} d\nu_{(p,0),s}$$

and for every $f \in C^\infty(M)$ with $\int_M f^2 d\nu_{(p,0),s} = 1$

$$\int_M f^2 \log f^2 d\nu_{(p,0),s} \leq 4|s| \int_M |\nabla^{g(s)} f|^2_{g(s)} d\nu_{(p,0),s}.$$

It is then well understood, see Ledoux [73], that the validity of a log-Sobolev inequality implies Gaussian concentration of the measure, via the so-called Herbst argument. Exploiting this, Hein–Naber in [68] discover that conjugate heat kernel measures satisfy the Gaussian concentration inequality for any $A, B \subset M$

$$\nu_{(p,0),s}(A)\nu_{(p,0),s}(B) \leq e^{-\frac{1}{8|s|}d_{g(s)}(A,B)^2}, \tag{49}$$

where $d_{g(s}(A, B) = \inf\{d_{g(s)}(x, y), x \in A, y \in B\}$. Hein-Naber then apply (49) to obtain an integral gaussian decay bound on conjugate heat kernels.

In the setting of compact Ricci flows satisfying the Type I bound on the curvature (48), Buzano–Mantegazza in [74] are able to exploit the Gaussian concentration bound (49) to prove that the conjugate heat kernel always satisfies upper Gaussian bounds, namely, there is a large positive constant $C_1 < +\infty$, depending on $g(-T)$ such that

$$u_{(p,0)}(x,s) \leq \frac{C_1}{(4\pi|s|)^{n/2}} e^{-\frac{d_{g(s)}(p,x)^2}{C_1|s|}}. \tag{50}$$

Buzano–Mantegazza also show that the Type I bound on the curvature implies lower Gaussian bounds on conjugate heat kernels, namely

$$u_{(p,0)}(x,s) \geq \frac{C_2}{(4\pi|s|)^{n/2}} e^{-\frac{d_{g(s)}(p,x)^2}{C_2|s|}}, \tag{51}$$

for some positive constant C_2 depending only on the dimension and C_I.

Expressing $u_{(p,0)}(x,s) = (4\pi|s|)^{-n/2}e^{-f}$, we can combine (50) and (51) into the estimate

$$\frac{d_{g(s)}(p,x)^2}{H|s|} - H \leq f(x,s) \leq H\left(\frac{d_{g(s)}(p,x)^2}{|s|} + 1\right). \tag{52}$$

8.2 Tangent Flows

Let $(M, g(t))_{t\in[-T,0)}$ be a smooth compact Ricci flow that exhibits Type I singularities at $t = 0$. Fix $x \in M$ and a sequence $\lambda_j > 0$ with $\lambda_j \to 0$. Then $g_j(t) = \lambda_j^{-2} g(\lambda_j^2 t)$, $t \in [-\lambda_j^{-2}T, 0)$ is a sequence of Ricci flows on M, generated by parabolically dilating the initial Ricci flow. We wish to understand the local picture around x of the sequence $(M, g_j(t))_{t\in[-\lambda_j^{-2}T,0)}$ as $j \to +\infty$.

The rate that the curvature of a Ricci flow with Type I singularities blows up respects the natural parabolic scaling of the Ricci flow equation. Indeed, observe that the bound (47) implies that $g_j(t) = \lambda_j^{-2} g(\lambda_j^2 t)$, for any $\lambda_j \to 0$, satisfies

$$\max_M |\mathrm{Rm}_{g_j}(\cdot, t)| \leq \frac{C_I}{|t|}. \tag{53}$$

for every $t \in [-\lambda_j^{-2}T, 0)$.

Combining this bound with Perelman's no local collapsing theorem allows us to apply a standard result in Ricci flow theory—Hamilton's compactness theorem—to assert that passing to a subsequence, we may assume that the *pointed* sequence of Ricci flows $(M, g_j(t), x)_{t\in[-\lambda_j^{-2}T,0)}$ has a smooth limit $(M_\infty, g_\infty(t), x_\infty)_{t\in(-\infty,0)}$. We call any such limit *a tangent flow* of $(M, g(t))_{t\in[-T,0)}$ at x.

In more precise terms, this means that for every $r > 0$ and j there is a smooth map $F_j : B_{g_\infty(-1)}(x_\infty, r) \to M_j$, diffeomorphism onto its image, such that $F_j(x_\infty) = x_j$ and the sequence of Ricci flows $F_j^* g_j(t)$ converges to $g_\infty(t)$, smoothly and uniformly on compact subsets of $B_{g_\infty(-1)}(x_\infty, r) \times (-\infty, 0)$.

For instance, in the case of a Ricci flow on the 2-sphere, or a three dimensional Riemannian manifold with positive Ricci curvature, any such limit will be the shrinking sphere solution. However, a tangent flow need not be compact and fixing different points on the manifold will typically give rise to different tangent flows. A typical example of this situation is the rotationally symmetric neckpinch of Angenent–Knopf. Figure 2 there, most points are regular points and the corresponding tangent flows are just the static Ricci flow in Euclidean space. On the other hand, if x is a singular point, then any tangent flow around that point is going to be a shrinking cylinder Ricci flow on $\mathbb{S}^{n-1} \times \mathbb{R}$.

Notice that in the examples above, the shrinking spheres and cylinders that arise as tangent flows are both self-similarly shrinking Ricci flows induced by gradient shrinking Ricci solitons. Intuitively that is to be expected, given that the pointed entropy described in Sect. 7 is scale invariant, monotone under Ricci flow, and if it is constant then the Ricci flow is induced by a gradient shrinking Ricci soliton, by (45). This expectation turns out to be true.

Theorem 8 *In a Type I singular Ricci flow, tangent flows are self-similar Ricci flows induced by gradient shrinking Ricci solitons. Moreover, x is a singular point if and only if there is a tangent flow at x which is not trivial, namely not induced by the Gaussian shrinking Ricci soliton.*

Theorem 8 has been proven by Sesum [86] for compact tangent flows. The non-compact case was first proven by Naber [78] and Enders [55], with the non-triviality statement at singular points being settled by Buzano–Enders–Topping in [56]. Buzano–Mantegazza also prove Theorem 8 in [74] with different methods, exploiting a version of the pointed entropy of Sect. 7. In their proof, the conjugate heat kernel bounds are crucial, as they ensure that conjugate heat kernels along the rescaled sequence $g_j(t)$ also converge smoothly to a conjugate heat flow $u_\infty = (4\pi|t|)^{-n/2} e^{-f_\infty}$ on the tangent flow. They then prove that the function f_∞ is a soliton function.

In the above examples, the tangent flows at any given point do not depend on the choice of the rescaling sequence λ_j. However, in general, different rescaling sequences may a priori give rise to different tangent flows. This is a recurring theme in the singularity analysis of geometric PDE, and the question whether blowup limits are unique has a long history. Most influential has been the remarkable work of L. Simon [87] (see also [88]) in which the uniqueness of blow ups at isolated singularities of energy minimizing harmonic maps $u : M \to N$ was established if N is an analytic Riemannian manifold. Uniqueness however fails, once the analyticity assumption is dropped, by an example constructed by White in [92]. Simon's approach to uniqueness exploited a Łojasiewicz inequality for an appropriate scale-invariant energy functional, and has found application to establish uniqueness of blow ups for many other geometric PDE. For instance, for the mean curvature flow of hypersurfaces in Euclidean space, Schulze in [85] establishes the uniqueness of tangent flows if at least one of them is compact, using a Łojasiewicz inequality argument. Later, Colding–Minicozzi [47] proved a Łojasiewicz inequality to establish the uniqueness of tangent flows if at least one of them is a generalized cylinder $\mathbb{S}^{n-k} \times \mathbb{R}^k \subset \mathbb{R}^{n+1}$. Finally, Chodosh–Schulze in [41], again proving an appropriate Łojasiewicz inequality, establish the uniqueness of asymptotically conical tangent flows. The uniqueness of blow-up limits has been understood to have significant impact on the regularity of the singular set, since one can then prove that it is rectifiable, see for instance [48, 88].

Unfortunately, to this date, no similar result exists for the Ricci flow, except for the case of compact tangent flows [1, 42, 86].

We conclude this section with a brief description of several examples of compact Ricci flows exhibiting Type I singularities that have been constructed in the literature.

1. The rotationally symmetric Ricci flows on $\mathbb{S}^{n+1}$ of Angenent–Knopf [3], exhibiting a neck-pinch singularity. The tangent flow at a singular point is the shrinking cylinder Ricci flow on $\mathbb{S}^n \times \mathbb{R}$.
2. Maximo in [76] constructs a (Kähler) Ricci flow in real dimension 4, whose tangent flow at a singular point is given by the gradient shrinking Ricci soliton constructed by Feldman–Ilmanen–Knopf in [57].
3. Stolarski in [89], given any asymptotically conical gradient Ricci soliton, constructs a Ricci flow exhibiting Type I singularities with the corresponding tangent flow at the singular points given by that shrinking Ricci soliton.
4. The gradient shrinking Ricci soliton constructed by Bamler–Ciffarelli–Conlon–Deruelle in [14] appears as the tangent flow of a Ricci flow on the blow-up at one point of $\mathbb{C}P^1 \times \mathbb{C}P^1$ that exhibits a Type I singularity.

8.3 Beyond Type I Singularities

The first concrete example of a Ricci flow exhibiting Type II singular behaviour is the degenerate neck-pinch singularity, in which the steady Ricci soliton of Bryant can appear as a singularity model. In dimension three we know, by Perelman, that any Type II singular behaviour will resemble the degenerate neck-pinch. Examples of 4d Ricci flow exhibiting Type II singularities were recently constructed by Appleton [6]. In these examples the Eguchi–Hanson space, a complete Kähler Ricci flat metric on the tangent bundle $T\mathbb{S}^2$ which is asymptotic to a flat cone over $\mathbb{R}P^3$, can appear as a blow-up limit.

Although the infinitesimal picture of Type I singularity formation is described by smooth gradient shrinking Ricci solitons, this is no longer, necessarily, the case for Type II singularities. One no longer has the curvature bound (53) when performing a sequence of parabolic rescalings around a singular point, so the existence of a smooth limit is no longer guaranteed. Another serious obstacle is that the conjugate heat kernel bounds (52) are no longer available in general.

Bamler, in the series of groundbreaking papers [10–12] managed to resolve some of the issues above

- In [10], he shows that upper Gaussian bounds for the conjugate heat kernels $u_{(p,s)}$ are always available, however the Gaussian bounds do not depend on the distance from p, but the distance from the so-called H_n-centers of the conjugate heat kernel.
- In [12], he defines a weak notion of (super)-Ricci flows, called *metric flows*, and also a weak enough topology that allows one to extract convergent subsequences of any non-collapsed sequence of Ricci flows. Their limit will not be smooth in general.
- In [11], he proves that the parabolic rescalings $g_j(t)$, always converge—in the appropriate sense—to a self-similar shrinking metric flow, which of course is not going to be induced by a smooth shrinking Ricci soliton.

9 Almost Self-similarity and Almost Splitting

Let $(M, g(t))_{t<0}$ be a self-similarly shrinking Ricci flow induced by a gradient shrinking Ricci soliton $(M, \bar{g}, \bar{f})$. If $(M, g(t))_{t<0}$ isometrically splits a maximal number k of Euclidean factors, where $0 \leq k \leq n = \dim M$, then:

1. The Euclidean coordinates $\{x^i\}_{i=1}^k$ of the isometric splitting $M = M' \times \mathbb{R}^k$ satisfy

$$\begin{aligned} \langle \nabla^{g(t)} x^i, \nabla^{g(t)} x^j \rangle_{g(t)} &= \delta^{ij}, \\ \operatorname{Hess}_{g(t)} x^i &= 0, \end{aligned} \tag{54}$$

for every $t < 0$ and $i, j = 1, \ldots, k$. In particular, the functions x^i evolve by the heat equation:

$$\left(\frac{\partial}{\partial t} - \Delta_{g(t)}\right) x^i = 0.$$

2. There is a k-parameter collection of conjugate heat flows induced by the k-parameter collection of soliton functions of $(M, \bar{g})$.

Recall that if a Ricci flow exhibits Type I singularities, tangent flows at any singular point will be self-similarly shrinking Ricci flows induced by gradient shrinking Ricci solitons. We thus expect that around a singular point p, after dilating by a large factor, the flow around p will resemble a self-similar Ricci flow, in a large region of space-time. In more precise terms, it will be (k, δ)-selfsimilar around p, in the sense of the following definition.

Definition 3 Let $(M, g(t))_{t\in[-\delta^{-1},0)}$ be a smooth compact Ricci flow and $p \in M$. We will say that $(M, g(t), p)_{t\in[-\delta^{-1},0)}$ is (k, δ)-selfsimilar if there exists a Ricci flow $(\bar{M}, \bar{g}(t))_{t<0}$ induced by a gradient shrinking Ricci soliton that isometrically splits at least k Euclidean factors, a point $q \in \mathcal{S}_{\text{point}}$—i.e. q is a minimum point of some soliton function—and a smooth map

$$F : B_{\bar{g}(-1)}(q, \delta^{-1}) \to M,$$

diffeomorphism onto its image, with $F(q) = p$ and

$$|\nabla^l (F^* g(t) - \bar{g}(t))| < \delta,$$

in $B_{\bar{g}(-1)}(q, \delta^{-1}) \times [-\delta^{-1}, -\delta]$, for every integer $0 \le l \le \delta^{-1}$.

Moreover, we will say that $(M, g(t), p)_{t\in[-\delta^{-1}r^2,0)}$ is (k, δ)-selfsimilar at scale $r > 0$ if $(M, r^{-2}g(r^2 t), p)_{t\in[-\delta^{-1},0)}$ is (k, δ)-selfsimilar.

The notion of (k, δ)-selfsimilarity quantifies, *at the level of Riemannian metrics*, how close is a Ricci flow, locally, to a the flow induced by a gradient shrinking Ricci soliton that splits k Euclidean factors. Therefore, locally, the geometry of the flow has an *approximate* translational $\mathbb{R}^k$-symmetry. Moreover, there exist coordinate functions that satisfy (54) *up to some error*, as well as a k-parameter family of functions that satisfy the gradient shrinking soliton equation *approximately*, in analogy to (1) and (2) in the discussion above.

One way to quantify directly how close a given Ricci flow is, locally, to isometrically split k Euclidean factors is the notion of a (k, δ)-splitting map. Roughly speaking, we ask that along the flow there are solutions to the heat equation that *approximately* and *locally* behave like the Euclidean coordinates in a product $M' \times \mathbb{R}^k$, satisfying (54) up to some error.

Definition 4 Let $r > 0$ and $(M, g(t))_{t\in[-r^2,0]}$ be a smooth compact Ricci flow and let $p \in M$. We say that $v : M \times [-r^2, 0] \to \mathbb{R}^k$, $v = (v^1, \ldots, v^k)$ is a (k, δ)-splitting map if v^i, $i = 1, \ldots, k$, evolve by the heat equation

$$\left(\frac{\partial}{\partial t} - \Delta_{g(t)}\right) v^i = 0,$$

and if $d\nu_{(p,0),t} = u_{(p,0)}(\cdot, t) d\mu_{g(t)}$ is the conjugate heat kernel measure starting at $(p, 0)$ then

$$\int_{-r^2}^{-\delta r^2} \int_M |\langle \nabla v^i, \nabla v^j \rangle_{g(t)} - \delta^{ij}|^2 d\nu_{(p,0),t} dt < \delta r^2, \tag{55}$$

$$\int_{-r^2}^{-\delta r^2} \int_M |\operatorname{Hess}_{g(t)} v^i|^2 d\nu_{(p,0),t} dt < \delta. \tag{56}$$

Notice that in the definition above, although the estimates (55) and (56) hold globally on M, the use of the conjugate heat kernel measures as a weight effectively localizes the estimates.

On the other hand, to quantify how close a given Ricci flow is to have a k-parameter family of approximate soliton functions, recall that by the monotonicity formula of the pointed entropy (46), if $(M, g(t))_{t\in[-T,0]}$ is a smooth compact Ricci flow, $x \in M$, $u_{(x,0)} = (4\pi|t|)^{-n/2}e^{-f_{(x,0)}}$ is the conjugate heat kernel starting at $(x, 0)$, and $d\nu_{(x,0),t}$ is the corresponding probability measure, then

$$\int_{-T}^{-1}\int_M 2|t|\left|\mathrm{Ric}_{g(t)}+\mathrm{Hess}_{g(t)}\, f_{(x,0)}-\frac{g(t)}{2|t|}\right|^2 d\nu_{(x,0),t}dt=\mathcal{W}_{(x,0)}(1)-\mathcal{W}_{(x,0)}(T). \tag{57}$$

Hence, the entropy drop $\mathcal{W}_{(x,0)}(1) - \mathcal{W}_{(x,0)}(T)$ between the two scales, 1 and T, controls how close $f_{(x,0)}$ is to be a soliton function in the interval $[-1, -T]$, around x.

If we could find points $\{x_i\}_{i=0}^k$ that spread enough into "k dimensions", in a sense that we will describe below, such that the quantity

$$\sum_{i=0}^k \mathcal{W}_{(x_i,0)}(1) - \mathcal{W}_{(x_i,0)}(T)$$

is small, then the Ricci flow would be approximately self-similar around the "centers" x_i, indicating that it has a local approximate translation symmetry. This idea leads to the following definition.

Definition 5 Let $(M, g(t))_{t\in[-\delta^{-1}r^2,0]}$ be smooth compact Ricci flow, $p \in M$, $R < +\infty$ and $\mu > 0$. We define the (k, δ, μ, R)-entropy-pinching around p at scale $r > 0$ to be the quantity

$$\mathcal{E}_r^{(k,\delta,\mu,R)}(p) = \inf_{\{x_i\}_{i=0}^k}\left\{\sum_{i=0}^k \mathcal{W}_{(x_i,0)}(r^2) - \mathcal{W}_{(x_i,0)}(Tr^2)\right\},$$

where the infimum is taken among all subsets $\{x_i\}_{i=0}^k \subset B_{g(-r^2)}(p, Rr)$ that satisfy

1. $(M, g(t), x_i)_{t\in[-\delta^{-1}r^2,0)}$ is $(0, \delta)$-selfsimilar.
2. There is no embedding of $\{x_i\}_{i=0}^k$ inside a metric space of the form $\mathcal{K} \times \mathbb{R}^l$, with $l < k$, $\mathrm{diam}(\mathcal{K}) \le A(n)r$, and distortion less than $(A(n) + R\mu)r$, where $A(n)$ is the maximum distance between two points in which a soliton function attains its minimum.

Remark 2 Observe that in the case $k = 1$, Condition 2 above just means that x_0, x_1 must have distance at least $(A(n) + R\mu)r$. Recall that, by Sect. 6.2.2, if a gradient shrinking Ricci soliton at scale r^2 does not split any Euclidean factor, i.e $k = 0$, then the distance between two minimum points of its soliton function is at most $A(n)r$. This is the reason that in the definition of the 1-entropy pinching we must require that x_0, x_1 to have distance uniformly greater than $A(n)r$.

The following theorem in [59] asserts that, under a Type I bound on the curvature and a non-collapsing assumption in the form of a lower bound on the ν entropy, if a Ricci flow is sufficiently (k, δ^2)-selfsimilar around p then there is a (k, ε)-splitting map around p. Moreover, the square of the L^2 norm of the Hessian, weighted by the corresponding conjugate heat kernel, is *linearly* controlled by entropy pinching. In a sense, if a Ricci flow is sufficiently self-similar, the following theorem detects the best almost splitting map allowed by the geometry of the flow, as described by the entropy pinching.

Theorem 9 *Fix $\varepsilon > 0$, $\mu \in (0, 1/6)$ and $R \geq A(n)/\mu$. Let $(M, g(t), p)_{t\in[-10\delta^{-2},0]}$ be a smooth compact Ricci flow satisfying*

$$\max_M |\operatorname{Rm}|(\cdot, t) \leq \frac{C_I}{|t|},$$

$$\nu(g(-10\delta^{-2}, 20\delta^{-2}) \geq -\Lambda,$$

for every $t \in [-10\delta^{-2}, 0)$. Suppose that $(M, g(t), p)_{t\in[-2\delta^{-2},0)}$ is (k, δ^2)-selfsimilar, for some $1 \leq k \leq n - \dim M$. Then we can choose $T = T(n, C_I)$ in the definition of (k, μ, δ, R)-entropy pinching such that the following holds.

If $0 < \delta \leq \delta(n, C_I, \Lambda|R, \mu, \varepsilon)$, then there is a (k, ε)-splitting map $v = (v^1, \ldots, v^k) : M \times [-1, 0] \to \mathbb{R}^k$ at scale 1 around p such that for every $i = 1, \ldots, k$

$$\int_M |\operatorname{Hess}_{g(t)} v^i|^2 d\nu_{p,t} \leq C(n, C_I, \Lambda|R, \mu, \varepsilon)\mathcal{E}_1^{(k,\mu,\delta,R)}(p),$$

for every $t \in [-1, -\varepsilon]$.

The idea of the proof is to use the (k, δ^2)-selfsimilarity to find points $\{x_i\}_{i=0}^k \subset B_{g(-1)}(p, R)$ that satisfy Conditions 1 and 2 in Definition 5 and consider the corresponding conjugate heat kernels $u_i = (4\pi|t|)^{-n/2}e^{-f_i}$ starting at $(x_i, 0)$. Then, motivated by (44), we define w_i by solving

$$\frac{\partial w_i}{\partial t} = \Delta_{g(t)} w_i - \frac{n}{2}, \tag{58}$$

with appropriate initial conditions given by the functions Tf_i at time $t = -T$. We can view the functions w_i as regularizations of the functions $|t|f_i$, and it can be shown that as $\delta \to 0$ they converge to a soliton function; here it is important that w_i satisfies the same equation as the soliton functions $|t|f_t$, by (44).

We then obtain a weighted-L^2 estimate on $\operatorname{Hess} w_i$ of the form

$$\int_M \left| |t| \operatorname{Ric}_{g(t)} + \operatorname{Hess}_{g(t)} w_i - \frac{g(t)}{2} \right|^2 dv_{(x_i,0),t} \le C(\mathcal{W}_{(x_i,0)}(1) - \mathcal{W}_{(x_i,0)}(T)), \tag{59}$$

for any $t \in [-T, -\varepsilon]$. Observe that a similar but weaker estimate holds for the functions f_i by (57), in the sense that it gives some control only in the time interval $[-T, -1]$. The improved estimate (59) is a consequence of the regularizing properties of the *forward* heat equation (58), compared to the *backwards* heat equation $\square^* u_i = 0$ satisfied by u_i. The required almost splitting map is then constructed by suitably normalizing, in the spirit of the Gramm-Schmidt process, the difference $(w_1 - w_0, \dots, w_k - w_0)$, which evolves by the heat equation.

10 The Small Scale Behaviour of Almost Splitting Maps in Type I Ricci Flows

Suppose that $(M, g(t))_{t\in[-\delta^{-2},0]}$ is a smooth compact Ricci flow which stops right before it encounters a finite time singularity at $p \in M$, and that there is some $0 < r < 1$ such that $(M, g(t), p)$ is (k, δ^2)-selfsimilar at every scale $s \in [r, 1]$. This is a reasonable assumption to make, since we saw in Sect. 8.2 that tangent flows—at least for Type I singularities—are induced by gradient shrinking Ricci solitons.

We may then invoke Theorem 9 to construct good almost splitting maps around p at each scale $s \in [r, 1]$. However, it is unclear whether an almost splitting map v around p at scale 1 restricts to an almost splitting map at all smaller scales down to scale r. It may be that v behaves in an uncontrollable way in small scales very close to the singularity. This could happen, for instance, in the case of a 3d Ricci flow if the length of a neck around p stretches rapidly to infinity as the flow approaches the singular time, causing $|\nabla v|$ to become very small along the neck (Fig. 4). Even in this scenario, however, it could be that by stretching v, at each scale $s \in [r, 1]$, by a large enough factor $a_s \ge 1$, then $v_s = a_s v$ is going to be an almost splitting map at that scale. Performing this stretching though may result in a loss on the smallness of the Hessian of v_s. The first theorem below, from [59], essentially says that this does not happen, and that in fact the Hessian of v_s is controlled in terms of the sum of the entropy pinching among all scales, which is purely geometric information of the flow.

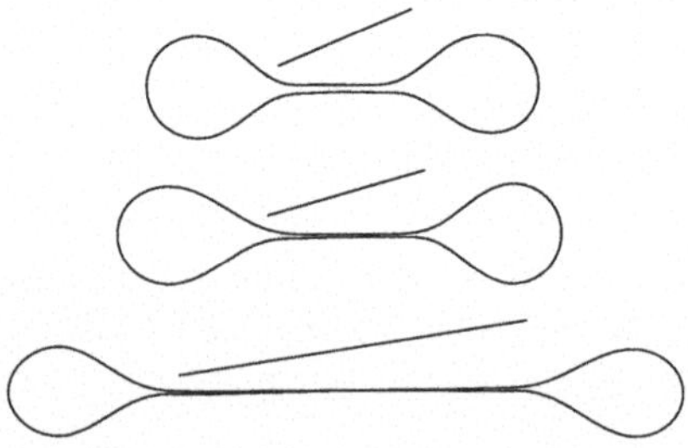

Fig. 4 The slope of an almost splitting map degenerating as the length of the neck stretches rapidly

To state the result, let $(M^n, g(t), p)_{t\in[-10\delta^{-2},0]}$ be a smooth compact Ricci flow satisfying

$$\max_M |\operatorname{Rm}|(\cdot, t) \leq \frac{C_I}{|t|},$$

$$\nu(g(-10\delta^{-2}), 20\delta^{-2}) \geq -\Lambda,$$

for every $t \in [-10\delta^{-2}, 0)$, and let v be a (k, δ)-splitting map v around p at scale 1.

Theorem 10 *Fix $\varepsilon > 0$, $\eta > 0$, $\mu \in (0, 1/6)$ and $R \geq \frac{A(n)}{\mu}$. Suppose that $(M, g(t), p)_{t\in(-2\delta^{-2},0)}$ is (k, δ^2)-selfsimilar but not $(k+1, \eta)$-selfsimilar at every scale $s \in [r, 1]$, for some $r \in (0, 1)$ and $1 \leq k \leq n$.*

If $0 < \delta \leq \delta(n, C_I, \Lambda, \eta|R, \mu, \varepsilon)$, then for every $s \in [r, 1]$ there is a lower triangular $k \times k$ matrix T_s such that

1. $v_s := T_s v$ is a (k, ε)-splitting map around p at scale s normalized so that for any $a, b = 1, \ldots, k$

$$\frac{4}{3s^2}\int_{-s^2}^{-s^2/4}\int_M \langle \nabla v_s^a, \nabla v_s^b\rangle d\nu_{(p,0),t} dt = \delta^{ab}. \tag{60}$$

2. Let $r_j = \frac{1}{2^j}$. There is $\theta = \theta(n, C_I, \Lambda) \in (0, 1)$ and $T = T(n, C_I)$ in the definition of (k, μ, δ, R)-entropy pinching, such that for every $s \in [r, 1]$ and $a = 1, \ldots, k$

$$\int_{-s^2}^{-\frac{s^2}{4}}\int_M |\operatorname{Hess} v_s^a|^2 d\nu_{(p,0),t} dt \leq$$
$$\leq C(n, C_I, \Lambda, \eta|R, \mu)\left(\sum_{s\leq r_j\leq 1}\left(\frac{s}{r_j}\right)^{\theta}\mathcal{E}_{r_j}^{(k,\mu,\delta,R)}(p) + \delta s^{\theta}\right). \tag{61}$$

The matrices T_s are essentially constructed by performing the Gramm-Schmidt process to achieve the normalization (60) at each scale s. It is not hard to prove that if these matrices exist—so that v_s are (k, ε)-splitting maps—then

$$||T_{s_1} \circ T_{s_2}^{-1}|| \leq \left(\frac{s_2}{s_1}\right)^{C(n)\sqrt{\varepsilon}},$$

whenever $r \leq s_1 < s_2$.

If $s_0 \in [r, 1]$ is the smallest scale for which Part 1 of the theorem holds, then for any $s > s_0$ we know that $v_{s_0} = T_{s_0} \circ T_s^{-1} v_s$. This leads to an estimate of the form

$$\int_M |\nabla v_{s_0}|^2 d\nu_{p,-s^2} \leq \left(\frac{s}{s_0}\right)^{C(n)\sqrt{\varepsilon}} \int_M |\nabla v_s|^2 d\nu_{p,-s^2} \sim \left(\frac{s}{s_0}\right)^{C(n)\sqrt{\varepsilon}}, \tag{62}$$

due to the normalization (60). Such an estimate indicates that $|\nabla v_{s_0}|^2$ has a relatively slow growth towards larger scales.

Assuming now, towards a contradiction, that $s_0 > r$, and sending $\delta \to 0$, we necessarily have that $s_0 \to 0$. After rescaling, we can pass the whole configuration to a subsequential limit, to obtain a solution to the heat equation $v_\infty : \bar{M} \times (-\infty, 0) \to \mathbb{R}^k$ on a self-similarly shrinking Ricci flow induced by a gradient Ricci soliton $(\bar{M}, \bar{g}, \bar{f})$, that splits k-Euclidean factors and is η-far from splitting $k+1$ Euclidean factors.

The key point now is to show that the slow growth estimate (62) forces v_∞ to be a non-constant function with vanishing Hessian. Once we succeed in this, it is then not hard to see that s_0 can not be the smallest scale for which the theorem holds, since we can use v_∞ to extend our estimates in smaller scales.

The behaviour of v_∞ is governed by the spectrum of the drift Laplacian

$$\Delta_{\bar{f}} = \Delta_{\bar{g}} - \nabla_{\nabla \bar{f}}$$

on $(\bar{M}, \bar{g})$, which is self-adjoint with respect to the measure $d\nu_{\bar{f}}{=}(4\pi)^{-n/2}e^{-\bar{f}}\, d\mu_{\bar{g}}$. By [35] the drift Laplacian has discrete infinite spectrum $\lambda_0 = 0 > -\frac{1}{2} \geq \lambda_1 \geq \cdots \geq, \lambda_j \to -\infty$, and $-\frac{1}{2}$ is an eigenvalue of multiplicity k if and only if $(M, \bar{g})$ splits isometrically k Euclidean factors. In this case, the $-\frac{1}{2}$-eigenspace has dimension k and is spanned by the k-linearly independent non-constant linear $\mathbb{R}^k$-coordinate functions.

In [59] we prove a spectral gap result which asserts that if a gradient shrinking soliton splits k Euclidean factors, but is η-far from splitting any additional Euclidean factor, then $\lambda_{k+1} < -\frac{1}{2} - \tau(n, C_I, \Lambda, \eta)$. This gap forces solutions of the heat equation along the corresponding Ricci flow with non-vanishing hessian to grow, backwards in time, fast enough to violate the estimate (62). Thus the Hessian of v_∞ should vanish.

The Hessian estimate in Part 2 of the theorem is also a consequence of the spectral gap. We split each $v_s = v_{s,\text{linear}} + v_{s,0}$ where $v_{s,\text{linear}}$ is a multiple of the optimal almost splitting map provided by Theorem 9, at scale s. The spectral gap shows that the Hessian of $v_{s,0}$ decays in Hölder rate, towards smaller scales, while $v_{s,\text{linear}}$ is controlled in terms of the entropy pinching. This eventually leads to estimate (61).

Finally, recall from equation (22) that the evolution of each $|\nabla v_s|^2$ is controlled by the Hessian of v_s. We can thus expect to get much finer control of the matrices T_s using (61), once the entropy pinching is summable over all scales. This leads to the following theorem in [59].

Theorem 11 *Fix $\varepsilon > 0$, $\eta > 0$, $\mu \in (0, 1/6)$, $R \geq \frac{A(n)}{\mu}$. Suppose that $(M, g(t), p)_{t\in(-2\delta^{-2},0)}$ is (k, δ^2)-selfsimilar but not $(k+1, \eta)$-selfsimilar at every scale $s \in [r, 1]$, for some $r \in (0, 1)$ and $1 \leq k \leq n$. Then, there is a $T = T(n, C_I)$ in the definition of (k, μ, δ, R)-entropy pinching such that the following holds. If for $r_j = 2^{-j}$, $j \geq 0$ integer, we have*

$$\sum_{r\leq r_j\leq 1} \mathcal{E}_{r_j}^{(k,\mu,\delta,R)}(p) < \delta, \tag{63}$$

and $0 < \delta \leq \delta(n, C_I, \Lambda, \eta|R, \mu, \varepsilon)$, then for every $s \in [r, 1]$, $v : M \times [-s^2, 0] \to \mathbb{R}^k$ is a (k, ε)-splitting map around p at scale s.

Acknowledgments The author was supported by the Hellenic Foundation for Research and Innovation (H.F.R.I.) under the "2nd Call for H.F.R.I. Research Projects to support Faculty Members & Researchers" (Project Number: HFRI-FM20-2958).

Competing Interests The author has no conflicts of interest to declare that are relevant to the content of this chapter.

References

1. A.G. Ache, *On the Uniqueness of Asymptotic Limits of the Ricci Flow* (2013). https://arxiv.org/abs/1211.3387
2. B. Andrews, C. Hopper, *The Ricci Flow in Riemannian Geometry*. Lecture Notes in Mathematics, vol. 2011 (Springer, Heidelberg, 2011). A complete proof of the differentiable 1/4-pinching sphere theorem
3. S. Angenent, D. Knopf, An example of neckpinching for Ricci flow on S^{n+1}. Math. Res. Lett. **11**(4), 493–518 (2004)
4. S.B. Angenent, J. Isenberg, D. Knopf, Degenerate neckpinches in Ricci flow. J. Reine Angew. Math. **709**, 81–117 (2015)
5. S.B. Angenent, D. Knopf, Ricci solitons, conical singularities, and nonuniqueness. Geom. Funct. Anal. **32**(3), 411–489 (2022)
6. A. Appleton, *A Family of Non-collapsed Steady Ricci Solitons in Even Dimensions Greater or Equal to Four* (2022). https://arxiv.org/abs/1708.00161
7. A. Appleton, Eguchi-Hanson singularities in U(2)-invariant Ricci flow. Peking Math. J. **6**, 1–141 (2023)
8. R. Bamler, Convergence of Ricci flows with bounded scalar curvature. Ann. Math. (2) **188**(3), 753–831 (2018)
9. R.H. Bamler, Structure theory of singular spaces. J. Funct. Anal. **272**(6), 2504–2627 (2017)
10. R.H. Bamler, *Entropy and Heat Kernel Bounds on a Ricci Flow Background* (2021). https://arxiv.org/abs/2008.07093
11. R.H. Bamler, *Structure Theory of Non-collapsed Limits of Ricci Flows* (2021). https://arxiv.org/abs/2009.03243
12. R.H. Bamler, *Compactness theory of the space of super Ricci flows*. Invent. Math. **233**(3), 1121–1277 (2023)
13. R.H. Bamler, E. Chen, *Degree theory for 4-dimensional asymptotically conical gradient expanding solitons*. Commun. Pure Appl. Math. (2025). https://doi.org/10.1002/cpa.70024. https://onlinelibrary.wiley.com/doi/abs/10.1002/cpa.70024

14. R.H. Bamler, C. Cifarelli, R.J. Conlon, A. Deruelle, *A new complete two-dimensional shrinking gradient Kähler-Ricci soliton*. Geom. Funct. Anal. **34**(2), 377–392 (2024). https://doi.org/10.1007/s00039-024-00668-9
15. R.H. Bamler, B. Kleiner, Uniqueness and stability of Ricci flow through singularities. Acta Math. **228**(1), 1–215 (2022)
16. R.H. Bamler, B. Kleiner, Ricci flow and diffeomorphism groups of 3-manifolds. J. Am. Math. Soc. **36**(2), 563–589 (2023)
17. R.H. Bamler, B. Kleiner, Diffeomorphism groups of prime 3-manifolds. J. Reine Angew. Math. **806**, 23–35 (2024)
18. M. Berger, Les variétés Riemanniennes $(1/4)$*-pincées*. Ann. Scuola Norm. Sup. Pisa Cl. Sci. (3) **14**, 161–170 (1960)
19. A.L. Besse, *Einstein Manifolds*. Classics in Mathematics (Springer, Berlin, 2008). Reprint of the 1987 edition
20. C. Böhm, B. Wilking, Manifolds with positive curvature operators are space forms. Ann. Math. (2) **167**(3), 1079–1097 (2008)
21. S. Brendle, Ricci flow with surgery on manifolds with positive isotropic curvature. Ann. Math. (2) **190**(2), 465–559 (2019)
22. S. Brendle, R. Schoen, Manifolds with $1/4$-pinched curvature are space forms. J. Am. Math. Soc. **22**(1), 287–307 (2009)
23. R. Buzano, G. Di Matteo, A local singularity analysis for the Ricci flow and its applications to Ricci flows with bounded scalar curvature. Calc. Var. Partial Differ. Equ. **61**(2), Paper No. 65, 36 (2022)
24. H.-D. Cao, *Existence of Gradient Kähler-Ricci Solitons*. Elliptic and Parabolic Methods in Geometry (Minneapolis, MN, 1994) (A K Peters, Wellesley, 1996), pp. 1–16
25. H.-D. Cao, Limits of solutions to the Kähler-Ricci flow. J. Differ. Geom. **45**(2), 257–272 (1997)
26. H.-D. Cao, R.S. Hamilton, T. Ilmanen, *Gaussian Densities and Stability for Some Ricci Solitons* (2004). https://arxiv.org/abs/math/0404165
27. J. Cheeger, *Degeneration of Riemannian metrics under Ricci curvature bounds*, Lezioni Fermiane. [Fermi Lectures] (Scuola Normale Superiore, Pisa, 2001)
28. J. Cheeger, T.H. Colding, Lower bounds on Ricci curvature and the almost rigidity of warped products. Ann. Math. (2) **144**(1), 189–237 (1996)
29. J. Cheeger, T.H. Colding, On the structure of spaces with Ricci curvature bounded below. I. J. Differ. Geom. **46**(3), 406–480 (1997)
30. J. Cheeger, R. Haslhofer, A. Naber, Quantitative stratification and the regularity of mean curvature flow. Geom. Funct. Anal. **23**(3), 828–847 (2013)
31. J. Cheeger, R. Haslhofer, A. Naber, Quantitative stratification and the regularity of harmonic map flow. Calc. Var. Partial Differ. Equ. **53**(1–2), 365–381 (2015)
32. J. Cheeger, W. Jiang, A. Naber, Rectifiability of singular sets of noncollapsed limit spaces with Ricci curvature bounded below. Ann. Math. (2) **193**(2), 407–538 (2021)
33. J. Cheeger, A. Naber, Lower bounds on Ricci curvature and quantitative behavior of singular sets. Invent. Math. **191**(2), 321–339 (2013)
34. B.-L. Chen, Strong uniqueness of the Ricci flow. J. Differ. Geom. **82**(2), 363–382 (2009)
35. X. Cheng, D. Zhou, Eigenvalues of the drifted Laplacian on complete metric measure spaces. Commun. Contemp. Math. **19**(1), 1650001, 17 (2017)
36. O. Chodosh, K. Choi, C. Mantoulidis, F. Schulze, Mean curvature flow with generic initial data. Invent. Math. **237**(1), 121–220 (2024)
37. O. Chodosh, K. Choi, C. Mantoulidis, F. Schulze, Mean curvature flow with generic low-entropy initial data. Duke Math. J. **173**(7), 1269–1290 (2024)
38. O. Chodosh, K. Choi, C. Mantoulidis, F. Schulze, Revisiting generic mean curvature flow in $\mathbb{R}^3$. J. reine angew. Math. (2024). https://doi.org/10.1515/crelle-2025-0064
39. O. Chodosh, K. Choi, F. Schulze, *Mean Curvature Flow with Generic Initial Data II* (2023). https://arxiv.org/abs/2302.08409
40. O. Chodosh, C. Mantoulidis, F. Schulze, *Mean curvature flow with generic low-entropy initial data II*. Duke Math. J. **174**(13), 2765–2819 (2024). https://doi.org/10.1215/00127094-2024-0080

41. O. Chodosh, F. Schulze, Uniqueness of asymptotically conical tangent flows. Duke Math. J. **170**(16), 3601–3657 (2021)
42. K. Choi, Y. Lai, *Convergence of Ricci Flow and Long-Time Existence of Harmonic Map Heat Flow* (2025). https://arxiv.org/abs/2504.02804
43. B. Chow, The Ricci flow on the 2-sphere. J. Differ. Geom. **33**(2), 325–334 (1991)
44. B. Chow, S.-C. Chu, D. Glickenstein, C. Guenther, J. Isenberg, T. Ivey, D. Knopf, P. Lu, F. Luo, L. Ni, *The Ricci Flow: Techniques and Applications. Part III. Geometric-Analytic Aspects*. Mathematical Surveys and Monographs, vol. 163 (American Mathematical Society, Providence, 2010)
45. B. Chow, D. Knopf, *The Ricci Flow: An Introduction*. Mathematical Surveys and Monographs, vol. 110 (American Mathematical Society, Providence, 2004)
46. T.H. Colding, W.P. Minicozzi, II, Generic mean curvature flow I: generic singularities. Ann. Math. (2) **175**(2), 755–833 (2012)
47. T.H. Colding, W.P. Minicozzi, II, Uniqueness of blowups and łojasiewicz inequalities. Ann. Math. (2) **182**(1), 221–285 (2015)
48. T.H. Colding, W.P. Minicozzi, II, The singular set of mean curvature flow with generic singularities. Invent. Math. **204**(2), 443–471 (2016)
49. R.J. Conlon, A. Deruelle, Expanding Kähler-Ricci solitons coming out of Kähler cones. J. Differ. Geom. **115**(2), 303–365 (2020)
50. E.B. Davies, *Heat Kernels and Spectral Theory*. Cambridge Tracts in Mathematics, vol. 92 (Cambridge University Press, Cambridge, 1989)
51. A. Deruelle, Smoothing out positively curved metric cones by Ricci expanders. Geom. Funct. Anal. **26**(1), 188–249 (2016)
52. D.M. DeTurck, Deforming metrics in the direction of their Ricci tensors. J. Differ. Geom. **18** (1983), no. 1, 157–162.
53. D.M. DeTurck, J.L. Kazdan, Some regularity theorems in Riemannian geometry. Ann. Sci. École Norm. Sup. (4) **14**(3), 249–260 (1981)
54. J. Eells, Jr., J.H. Sampson, Harmonic mappings of Riemannian manifolds. Am. J. Math. **86**, 109–160 (1964)
55. J. Enders, *Generalizations of the Reduced Distance in the Ricci Flow—Monotonicity and Applications*. (ProQuest LLC, Ann Arbor, 2008). Thesis (Ph.D.)–Michigan State University
56. J. Enders, R. Müller, P.M. Topping, On type-I singularities in Ricci flow. Commun. Anal. Geom. **19**(5), 905–922 (2011)
57. M. Feldman, T. Ilmanen, D. Knopf, Rotationally symmetric shrinking and expanding gradient Kähler-Ricci solitons. J. Differ. Geom. **65**(2), 169–209 (2003)
58. P. Gianniotis, Regularity theory for type I Ricci flows. Calc. Var. Partial Differ. Equ. **58**(6), Paper No. 200, 24 (2019)
59. P. Gianniotis, *Splitting maps in type I Ricci flows*. Calc. Var. Partial Differential Equations, **64**(7), 212–298 (2024). https://doi.org/10.1007/s00526-025-03044-z
60. P. Gianniotis, R. Haslhofer, Diameter and curvature control under mean curvature flow. Am. J. Math. **142**(6), 1877–1896 (2020)
61. A. Grigor'yan, *Heat Kernel and Analysis on Manifolds*. AMS/IP Studies in Advanced Mathematics, vol. 47 (American Mathematical Society/International Press, Providence/Boston, 2009)
62. H.-L. Gu, X.-P. Zhu, The existence of type II singularities for the Ricci flow on S^{n+1}. Commun. Anal. Geom. **16**(3), 467–494 (2008)
63. R.S. Hamilton, Three-manifolds with positive Ricci curvature. J. Differ. Geom. **17**(2), 255–306 (1982)
64. R.S. Hamilton, Four-manifolds with positive curvature operator. J. Differ. Geom. **24**(2), 153–179 (1986)
65. R.S. Hamilton, *The Ricci Flow on Surfaces*. Mathematics and General Relativity (Santa Cruz, CA, 1986). Contemporary Mathematics, vol. 71 (American Mathematical Society, Providence, 1988), pp. 237–262
66. R.S. Hamilton, *The Formation of Singularities in the Ricci Flow*. Surveys in Differential Geometry, Vol. II (Cambridge, MA, 1993) (Int. Press, Cambridge, MA, 1995), pp. 7–136

67. R. Haslhofer, R. Müller, A compactness theorem for complete Ricci shrinkers. Geom. Funct. Anal. **21**(5), 1091–1116 (2011)
68. H.-J. Hein, A. Naber, *New Logarithmic Sobolev Inequalities and an ϵ-Regularity Theorem for the Ricci Flow*. Commun. Pure Appl. Math. **67**(9), 1543–1561 (2014)
69. B. Kleiner, J. Lott, Notes on Perelman's papers. Geom. Topol. **12**(5), 2587–2855 (2008)
70. B. Kleiner, J. Lott, Singular Ricci flows I. Acta Math. **219**(1), 65–134 (2017)
71. W. Klingenberg, über Riemannsche Mannigfaltigkeiten mit positiver Krümmung. Comment. Math. Helv. **35**, 47–54 (1961)
72. N. Koiso, *On Rotationally Symmetric Hamilton's Equation for Kähler-Einstein Metrics*. Recent Topics in Differential and Analytic Geometry. Advanced Studies in Pure Mathematics, vol. 18-I (Academic, Boston, 1990), pp. 327–337
73. M. Ledoux, *The Concentration of Measure Phenomenon*. Mathematical Surveys and Monographs, vol. 89 (American Mathematical Society, Providence, 2001)
74. C. Mantegazza, R. Müller, Perelman's entropy functional at Type I singularities of the Ricci flow. J. Reine Angew. Math. **703**, 173–199 (2015)
75. F.C. Marques, Deforming three-manifolds with positive scalar curvature. Ann. Math. (2) **176**(2), 815–863 (2012)
76. D. Máximo, On the blow-up of four-dimensional Ricci flow singularities. J. Reine Angew. Math. **692**, 153–171 (2014)
77. R.J. McCann, P.M. Topping, Ricci flow, entropy and optimal transportation. Am. J. Math. **132**(3), 711–730 (2010)
78. A. Naber, Noncompact shrinking four solitons with nonnegative curvature. J. Reine Angew. Math. **645**, 125–153 (2010)
79. A. Naber, D. Valtorta, Rectifiable-Reifenberg and the regularity of stationary and minimizing harmonic maps. Ann. Math. (2) **185**(1), 131–227 (2017)
80. G. Perelman, *The Entropy Formula for the Ricci Flow and Its Geometric Applications* (2002). https://arxiv.org/abs/math/0211159
81. G. Perelman, *Finite Extinction Time for the Solutions to the Ricci Flow on Certain Three-Manifolds* (2003). https://arxiv.org/abs/math/0307245
82. G. Perelman, *Ricci Flow with Surgery on Three-Manifolds* (2003). https://arxiv.org/abs/math/0303109
83. P. Petersen, *Riemannian Geometry*. Graduate Texts in Mathematics, vol. 171, 3rd edn. (Springer, Cham, 2016)
84. S. Pigola, M. Rimoldi, A.G. Setti, Remarks on non-compact gradient Ricci solitons. Math. Z. **268**(3–4), 777–790 (2011)
85. F. Schulze, Uniqueness of compact tangent flows in mean curvature flow. J. Reine Angew. Math. **690**, 163–172 (2014)
86. N. Sesum, Convergence of the Ricci flow toward a soliton. Commun. Anal. Geom. **14**(2), 283–343 (2006)
87. L. Simon, Asymptotics for a class of nonlinear evolution equations, with applications to geometric problems. Ann. Math. (2) **118**(3), 525–571 (1983)
88. L. Simon, *Theorems on Regularity and Singularity of Energy Minimizing Maps*. Lectures in Mathematics ETH Zürich (Birkhäuser, Basel, 1996). Based on lecture notes by Norbert Hungerbühler
89. M. Stolarski, *Closed Ricci Flows with Singularities Modeled on Asymptotically Conical Shrinkers* (2024). https://arxiv.org/abs/2202.03386
90. P. Topping, Diameter control under Ricci flow. Commun. Anal. Geom. **13**(5), 1039–1055 (2005)
91. X.-J. Wang, X. Zhu, Kähler-Ricci solitons on toric manifolds with positive first Chern class. Adv. Math. **188**(1), 87–103 (2004)
92. B. White, Nonunique tangent maps at isolated singularities of harmonic maps. Bull. Am. Math. Soc. (N.S.) **26**(1), 125–129 (1992)

Fisher Information in Kinetic Theory

Cédric Villani

These are the notes of the course which I gave at the summer school *Mathemata* in Chania, Crete in July 2024 as part of the Festum Pi 2024 festival. My course was triggered by a spectacular theorem of Guillen & Silvestre, which I wanted to present, expand and generalize. The core of these notes is based on my current research to extend this theorem, in collaboration with Cyril Imbert and Luis Silvestre. This course is also inspired by my long-term past collaborations in kinetic theory with Laurent Desvillettes, Giuseppe Toscani, Eric Carlen and Clément Mouhot; and by my long-term past interactions about functional inequalities with Michel Ledoux. It is a pleasure to thank them all warmly, as well as the main organisers of the summer school, Dionysios Dervis-Bournias and Bertrand Maury. These thanks extend to the talented students attending and participating in the course, and to the Mediterranean Agronomic Institute of Chania (MAICH), which provided the inspiring location and outstanding support. My gratitude goes to Stéphane Mischler for a very careful reading of these notes. Finally, part of this work was done as I was holding a joint chair between Université Claude Bernard Lyon 1 and Institut des hautes études scientifiques (IHES) in Bures-sur-Yvette, France, whose support is also gratefully acknowledged.

In several respects these lectures are the sequel of my course at Institut Henri Poincaré in 2001, *Entropy production and convergence to equilibrium* [186], but this is also my first research memoir after ten years of leave. So I consider it a key step bridging my past and future research. Awkward as this may seem, on this occasion I consider it fair to thank the French politicians (EM, PM, JLM) who ensured, willingly or unwillingly, my defeat at the 2022 French Parliamentary elections, even

C. Villani (✉)
Institut Camille Jordan, Université Claude Bernard Lyon 1, Villeurbanne, France
e-mail: cv@cedricvillani.org

© The Author(s), under exclusive license to Springer Nature Switzerland AG 2026

N. D. Alikakos, C. Villani (eds.), *Analysis Techniques for Mathematical Physics and Geometry – and Music*, Lecture Notes in Mathematics 2390,
https://doi.org/10.1007/978-3-032-11396-2_7

though by a statistically nonsignificant margin, making way for my comeback in research.

This work is dedicated to the memory of the great mathematician Henry P. McKean, visionary pioneer, among many other subjects, of Fisher information in kinetic theory. Henry passed away in April 2024, as this work was in preparation. Our encounter, organised by the late Paul Malliavin, remains a dear memory.

1 Fisher Information

Definition 1.1 Let (M, g, ν) be a complete Riemannian manifold equipped with a Borel reference measure ν. Then for any Borel probability measure μ on M, define

$$I_\nu(\mu) = \begin{cases} \displaystyle\int_M \frac{|\nabla f|^2}{f}\, d\nu & \text{if } \mu = f\nu, \\ +\infty & \text{if } \mu \text{ is singular to } \nu. \end{cases} \tag{1.1}$$

This formally 1-homogeneous, convex, lower semi-continuous functional is always well-defined (by convexity), even if f is not differentiable. It has a long and rich history, appearing in particular in the following five fields, roughly ranked by chronological order, all related to the combination of statistics and measurement (repeated measurement, or typical measurement).

1.1 Statistics

Fisher introduced the functional I hundred years ago in relation to his notion of *efficient statistics*. I shall develop the idea a little bit, both for historical reasons and for the intuition.

Suppose $X_1, \ldots, X_N, \ldots$ are independent identically distributed (i.i.d.) random variables with law $f\nu$ on some Polish space $\mathcal{X}$, where $f = f(x, \theta)$ is a parametric density with parameter $\theta \in \Theta$. An estimator $\widehat{\theta}$ is a map $\mathcal{X}^N \to \Theta$. The most famous such estimator is the maximum likelihood estimate (MLE): θ such that the density of the observed sample is as large as possible; in other words

$$\widehat{\theta}_{\mathrm{MLE}}(x_1, \ldots, x_N) \text{ maximizes } \quad f(x_1, \ldots, x_N; \theta) = \prod_{i=1}^N f(x_i, \theta).$$

As $N \to \infty$ we expect $\widehat{\theta}$ to converge almost surely (a.s.) to θ if $X_1, \ldots, X_N$ are drawn according to parameter θ; meaning that the estimator is asymptotically unbiased. But what about the fluctuation of $\widehat{\theta}$ around θ?

Proposition 1.2 (Fisher's Fluctuation Estimate for MLE) *Let $\widehat{\theta}$ be an unbiased estimator valued in Θ, which near θ is a differentiable manifold. Then under $f(x, \theta)$,*

$$\text{As } N \to \infty, \quad \widehat{\theta}(X_1, \ldots, X_N) \sim N\left(\theta, \frac{\mathbb{I}_\theta^{-1}(f)}{N}\right),$$

meaning that $\sqrt{N}(\widehat{\theta} - \theta) \sim N(0, \mathbb{I}_\theta^{-1}(f))$, the centered normal (Gaussian) law whose covariance matrix is the inverse of $\mathbb{I}$, where

$$\mathbb{I}_\theta(f)_{ij} = \int \frac{1}{f} \frac{\partial f}{\partial \theta_i} \frac{\partial f}{\partial \theta_j}(x, \theta)\, \nu(dx) = \int \frac{\partial}{\partial \theta_i}(\log f)\, \frac{\partial}{\partial \theta_j}(\log f)\, f(x, \theta)\, \nu(dx). \tag{1.2}$$

In particular,

$$\mathbb{I}_\theta(f) = \mathbb{E}\Big[\nabla_\theta \log f \otimes \nabla_\theta \log f\Big]; \tag{1.3}$$

$$I_\theta(f) := \operatorname{tr} \mathbb{I}_\theta(f) = \int \frac{|\nabla_\theta f|^2}{f}\, \nu(dx) \tag{1.4}$$

Before going further, note that if $f(x, \theta) = f(x - \theta)$ on $(\mathbb{R}^d, dx)$ (fixed profile, but centred at an unknown parameter), then I_θ is just the standard Fisher information with respect to Lebesgue measure ($\nu = \mathcal{L}^d$ in (1.1)), independently of θ.

Sketch of Proof of Proposition 1.2 First note that

$$\mathbb{I}_\theta(f) = -\int f \nabla^2 \log f(x, \theta)\, \nu(dx).$$

(If $f(x, \theta) = f(x - \theta)$ this results from integration by parts, but the formula holds true regardless of this assumption.) Indeed,

$$0 = \frac{\partial}{\partial \theta_i} \int f\, d\nu = \int \frac{\partial f}{\partial \theta_i}\, d\nu = \int f\, \partial_{\theta_i}(\log f)\, d\nu.$$

So

$$0 = \frac{\partial}{\partial \theta_j} \int f\, \partial_{\theta_i}(\log f) = \int \frac{\partial f}{\partial \theta_j} \frac{\partial}{\partial \theta_i}(\log f) + \int f\, \frac{\partial^2}{\partial \theta_j\, \partial \theta_i}(\log f)$$

$$= \int f \frac{\partial}{\partial \theta_j}(\log f) \frac{\partial}{\partial \theta_i}(\log f) + \int f \frac{\partial^2}{\partial \theta_j \, \partial \theta_i}(\log f),$$

as announced. Now, $\widehat{\theta}_{\mathrm{MLE}}$ is obtained by solving the equation

$$\forall k \in \{1, \dots, d\}, \qquad \frac{\partial}{\partial \theta_k} \prod_i f(x_i, \theta) = 0,$$

where d is the dimension of Θ (locally near θ); by taking logarithm,

$$\frac{1}{N} \sum_{i=1}^{N} \frac{\partial}{\partial \theta_k} \big(\log f(x_i, \theta) \big) = 0. \tag{1.5}$$

Let $\ell(x, \theta') = \log f(x, \theta')$. Then

$$\frac{\partial}{\partial \theta_k} \ell(x, \theta') = \frac{\partial}{\partial \theta_k} \ell(x, \theta) + \sum_{k,m} \frac{\partial^2}{\partial \theta_k \, \partial \theta_m} \ell(x, \theta)(\theta' - \theta)_m + o(|\theta' - \theta|),$$

and for simplicity let us assume that the latter $o(|\theta' - \theta|)$ is uniform in x; so

$$\frac{1}{N} \sum_i \partial_{\theta_k} \ell(x_i, \widehat{\theta}) - \frac{1}{N} \sum_i \partial_{\theta_k} \ell(x_i, \theta)$$
$$= \frac{1}{N} \sum_{i,m} \frac{\partial^2}{\partial \theta_k \, \partial \theta_m} \ell(x_i, \theta) \, (\widehat{\theta} - \theta)_m + o(|\widehat{\theta} - \theta|), \tag{1.6}$$

where $\widehat{\theta} = \theta(x_1, \dots, x_N)$. The first term on the left hand side of (1.6) vanishes by (1.5), and by the law of large numbers and central limit theorem, the second term is distributed near

$$- \mathbb{E} \, \partial_{\theta_k} \ell(X, \theta) = - \int f(x, \theta) \, \partial_{\theta_k} \log f(x, \theta) \, \nu(dx) = 0,$$

with fluctuation $N^{-1/2} \mathcal{N}(0, C)$, where C is the covariance matrix

$$C = \mathbb{E}\Big[\nabla_\theta \log f(X, \theta) \otimes \nabla_\theta \log f(X, \theta) \Big] = \mathbb{I}_\theta(f). \tag{1.7}$$

As for the first term in the right-hand side of (1.6), by the law of large numbers again it is approximately

$$\sum_m \left(\mathbb{E} \frac{\partial^2}{\partial \theta_k \, \partial \theta_m} \ell(X, \theta) \right) (\widehat{\theta} - \theta)_m = -\mathbb{I}_\theta(f)(\widehat{\theta} - \theta)_k.$$

To summarise: Asymptotically, $-\sqrt{N}\mathbb{I}_\theta(f)(\widehat{\theta}-\theta) \sim \mathcal{N}(0,\mathbb{I}_\theta)$, so $\sqrt{N}(\widehat{\theta}-\theta) \sim \mathcal{N}(0,\mathbb{I}_\theta^{-1})$. □

It turns out that this estimate is best possible, so that MLE is essentially "the best" estimator, one which fluctuates the least around the mean. If the reader wonders why this estimator is not systematically used, the answer is easy: A major drawback of MLE is that in general it is so difficult to evaluate! In any case its asymptotic optimality is expressed by the following uncertainty principle.

Proposition 1.3 (Cramér–Rao Inequality) *If* $\theta \in \mathbb{R}^d$, $\widehat{\theta} : (\mathbb{R}^d)^N \to \mathbb{R}^d$, *and* $(X_i)_{i\in\mathbb{N}}$ *are i.i.d. with law* $f(x,\theta)\,\nu(dx)$, *then*

$$\mathrm{Var}\,(\widehat{\theta}) \geq \frac{\left(d+\nabla_\theta\cdot\left(\mathbb{E}\widehat{\theta}-\theta\right)\right)^2}{N\,I_\theta(f)} - \left|\mathbb{E}\widehat{\theta}-\theta\right|^2.$$

In particular, if $\mathbb{E}\widehat{\theta}-\theta = o(1/\sqrt{N})$ and $\nabla\cdot\mathbb{E}(\widehat{\theta}-\theta) \longrightarrow 0$ as $N\to\infty$, then $\liminf N\mathrm{Var}\,(\widehat{\theta}) \geq d^2/I_\theta(f)$.

Sketch of Proof of Proposition 1.3 Write $f^N(x_1,\dots,x_N;\theta) = f^N(x;\theta) = \prod_i f(x_i,\theta)$. The integral is with respect to $\nu^{\otimes N}$. Then for any $k\in\{1,\dots,d\}$,

$$\begin{aligned}\frac{\partial}{\partial\theta_k}\int f^N(x;\theta)(\widehat{\theta}-\theta)_k &= \int \partial_{\theta_k} f^N(x;\theta)\,(\widehat{\theta}-\theta)_k - \int f^N(x;\theta)\\ &= \int \partial_{\theta_k} f^N(x;\theta)\,(\widehat{\theta}-\theta)_k - 1.\end{aligned}$$

Summing up for $1\leq k\leq d$, we obtain

$$\begin{aligned}d+\nabla_\theta\cdot\mathbb{E}(\widehat{\theta}-\theta) &= \sum_k\int \partial_{\theta_k} f^N(x;\theta)(\widehat{\theta}-\theta)_k\\ &= \sum_k\int f^N(x;\theta)\,\partial_{\theta_k}\log f^N(x,\theta)\,(\widehat{\theta}-\theta)_k\\ &\leq \sum_k\left(\int f^N(x;\theta)\left|\partial_{\theta_k}\log f^N(x;\theta)\right|^2\right)^{1/2}\\ &\quad\times\left(\int f^N(x;\theta)(\widehat{\theta}-\theta)_k^2\right)^{1/2}\\ &\leq \left(\int f^N(x;\theta)\sum_k\left|\partial_{\theta_k}\log f^N(x;\theta)\right|^2\right)^{1/2}\end{aligned}$$

$$\times \left(\int f^N(x;\theta) \sum_k (\widehat{\theta} - \theta)_k^2 \right)^{1/2}$$
$$= I_\theta(f^N)^{1/2} \big(\mathbb{E}|\widehat{\theta} - \theta|^2 \big)^{1/2}.$$

Then on the one hand,

$$I_\theta(f^N) = N I_\theta(f);$$

this comes from the fact that $\log f^N = \sum \log f_j$, where $f_j = f(x_j;\theta)$ and in the Fisher information of f^N the cross-products $\int f^N \nabla_\theta \log f_j \cdot \nabla_\theta \log f_\ell$ simplify into $\int f_j f_\ell \nabla_\theta \log f_j \cdot \nabla_\theta \log f_\ell = (\nabla_\theta \int f_j) \cdot (\nabla_\theta \int f_\ell) = 0$. On the other hand

$$\mathbb{E}|\widehat{\theta} - \theta|^2 = \operatorname{Var} \widehat{\theta} + |\mathbb{E}\widehat{\theta} - \theta|^2.$$

So in the end

$$d + \nabla_\theta \cdot \mathbb{E}(\widehat{\theta} - \theta) \leq \sqrt{N} I_\theta(f) \Big(\operatorname{Var} \widehat{\theta} + |\mathbb{E}\widehat{\theta} - \theta|^2 \Big)^{1/2},$$

which leads to the desired conclusion. □

The meaning is the following, at least for an unbiased estimator: If we wish to estimate θ from N independent observations, then however clever we are, we can be sure that fluctuations around the true value will be at least $d\, I_\theta(f)^{-1/2}/\sqrt{N}$. The higher I, the most efficient the estimate may be.

While Fisher was a devout frequentist, this interpretation led Jeffreys to use the same functional in a Bayesian context. If Θ, the set of parameters, is a differentiable manifold, let us bet that θ is all the more likely when it is easier to estimate. (This has a flavor of the maximum likelihood estimate, in which one bets that θ is all the more likely when it makes it easier to produce the observation.) Then Jeffreys' procedure, similar to Riemannian geometry, is to use $\mathbb{I}_\theta(f)$ as a metric on Θ, and the prior $\mathcal{J}(d\theta)$ will be the associated (diffeomorphism-invariant) volume:

$$\mathcal{J}(d\theta_1 \dots d\theta_d) = \sqrt{\det(\mathbb{I}_\theta(f))}\, d\theta_1 \dots d\theta_d.$$

For example, if $f(x,\theta) = \mathcal{N}(\mu, \sigma^2)$ and

- σ fixed, μ variable: $\mathcal{J}(d\mu) = d\mu$ (Lebesgue on $\mathbb{R}$);
- μ is fixed, σ variable: $\mathcal{J}(d\sigma) = d\sigma/\sigma$.

Or, if $f(x,\theta) = (\theta, 1-\theta)$ is the Bernoulli trial, then $\mathcal{J}(d\theta) = d\theta/\sqrt{\theta(1-\theta)}$. (In this example the observation space is discrete, $\nu = \delta_0 + \delta_1$, but the Fisher information makes sense as the parameter space is continuous.)

1.2 Information Theory

Information theory is the science of complexity, signal processing and transmission. The most important pillar of that theory is the Boltzmann–Shannon information, introduced by Boltzmann in he nineteenth century to statistically found thermodynamics, and rediscovered by Shannon in his cult 1948 book on information theory.

Definition 1.4 Let $(\mathcal{X}, d, \nu)$ be a complete, separable metric space equipped with a Borel reference measure. Then

$$H_\nu(\mu) = \begin{cases} \displaystyle\int_{\mathcal{X}} f \log f \, d\nu & \text{if } \mu = f\nu, \\ +\infty & \text{if } \mu \text{ is singular to } \nu. \end{cases}$$

Here log is the natural logarithm with base e; but other choices, like base 10, or more frequently base 2, are also used without any real trouble.

In the fifties, the functional I became another hero of information theory, thanks primarily to **de Bruijn's identity:** If the reference measure ν is Lebesgue on $\mathbb{R}^d$ and Δ is the usual Laplace operator, then

$$\left.\frac{d}{dt}\right|_{t=0} H\big(e^{t\Delta}\mu\big) = -I(\mu). \tag{1.8}$$

More generally, if $\nu(dx) = e^{-V(x)}\, dx$ and $\mathcal{L}\mu = \Delta\mu + \nabla\cdot(\mu\nabla V)$ is the generator of the natural semigroup on probability measures having ν as an invariant measure, then

$$\left.\frac{d}{dt}\right|_{t=0} H_\nu\big(e^{t\mathcal{L}}\mu\big) = -I_\nu(\mu). \tag{1.9}$$

Note that $\mathcal{L}(f\nu) = (Lf)\nu$ where $Lf = \Delta f - \nabla V \cdot \nabla f$ is the generator defined on functions (densities) rather than measures; going from one formula to the other is a popular game.

The intuition is as follows. In either Boltzmann's or Shannon's theory, H measures how μ is exceptional and may carry significant information by itself: The higher H, the more the distribution is rare and precious. But any observation of μ will imply some dose of fluctuation. Adding a small Gaussian noise to the signal is a way to measure the size of these fluctuations. The volume of configurations is proportional to $e^{-NH_\nu(\mu)}$, and its logarithmic derivative along the heat flow is $NI_\nu(\mu)$: the higher I, the larger the increase. If Boltzmann's theory is about the volume in the space of microstates, then Fisher's information is about the surface. The finite-dimensional analogue of the perturbation by small Gaussian noise would actually be the classical formula relating volume to surface via enlargement by a small ball:

$$S(A) = \frac{d}{dt}\bigg|_{t=0} V(A + B_t),$$

where $B_t = B(0, t)$ is the ball of radius t and the sum is in the sense of Minkowski.

By abuse of notation, I shall often write $I(f)$ for $I(f\,\mathcal{L}^d)$ ($\mathcal{L}^d$= Lebesgue measure on $\mathbb{R}^d$) or $I(X) = I(\mu)$ where X is a random variable with law μ. The four most famous properties of I in information theory are as follows; they all have counterparts for H.

(a) Scaling: For all random variable X and $\lambda > 0$,

$$I(\lambda X) = \frac{1}{\lambda^2} I(X);$$

equivalently, if $f_\lambda(x) = \lambda^{-d} f(x/\lambda)$, then

$$I(f_\lambda) = \frac{1}{\lambda^2} I(f).$$

The entropic counterpart is

$$H(f_\lambda) = H(f) - d \log \lambda.$$

(b) Marginal superadditivity: If X_1, X_2 are valued in $\mathbb{R}^{d_1}, \mathbb{R}^{d_2}$ respectively, then

$$I(X_1, X_2) \geq I(X_1) + I(X_2),$$

with equality when X_1, X_2 are independent. (There may be correlations between X_1 and X_2, making the joint observation easier to detect.) Equivalently, if $f(x_1, x_2)$ is a joint distribution,

$$I(f) \geq I\left(\int f\,dx_2\right) + I\left(\int f dx_1\right).$$

The entropic counterpart is

$$H(f) \geq H\left(\int f\,dx_2\right) + H\left(\int f dx_1\right).$$

(c) Gaussian minimisers: Let $X \in \mathbb{R}^d$ be a random vector with finite variance, and let G be a Gaussian vector with the same variance as X; then

$$I(X) \geq I(G) = \frac{d^2}{\mathrm{Var}\,(G)}.$$

Alternatively, if $\gamma_\sigma(x) = (2\pi\sigma^2)^{-d/2}\exp(-|x|^2/2\sigma^2)$, and σ is chosen to be the variance of f, which means

$$\int f(x)|x-u|^2\,dx = \int \gamma_\sigma(x)|x|^2\,dx = d\,\sigma^2, \qquad u = \int f(x)\,x\,dx,$$

then

$$I(f) \geq I(\gamma_\sigma) = \frac{d}{\sigma^2}.$$

Note that Gaussians also saturate the Cramér–Rao inequality. The entropic counterpart is

$$H(f) \geq H(\gamma_\sigma) = -\frac{d}{2}\log(2\pi e\sigma^2).$$

(d) Stam inequalities: These are two functional inequalities expressing a subadditivity property; there is the Blachman–Stam inequality dealing with I, and the Shannon–Stam inequality dealing with H.

First, **the Blachman–Stam inequality** states that: If X and Y are two independent random vectors in $\mathbb{R}^d$, then for all $\alpha \in [0,1]$,

$$I(\sqrt{1-\alpha}\,X + \sqrt{\alpha}\,Y) \leq (1-\alpha)\,I(X) + \alpha\,I(Y).$$

Alternatively, for all f, g, α,

$$I\big(f_{\sqrt{1-\alpha}} * g_{\sqrt{\alpha}}\big) \leq (1-\alpha)\,I(f) + \alpha\,I(g).$$

By optimisation over α and the scaling property (b), there is an equivalent form for the Blachman–Stam inequality: For all f, g,

$$\frac{1}{I(f*g)} \geq \frac{1}{I(f)} + \frac{1}{I(g)}.$$

Then, the entropic counterpart consists in either of the two forms of the **Shannon–Stam inequality**: for all densities f, g and $\alpha \in [0,1]$,

$$H\big(f_{\sqrt{1-\alpha}} * g_{\sqrt{\alpha}}\big) \leq (1-\alpha)\,H(f) + \alpha\,H(g),$$

or equivalently for all probability densities f, g,

$$\mathcal{N}(f*g) \geq \mathcal{N}(f) + \mathcal{N}(g), \qquad \mathcal{N}(f) = \frac{e^{-\frac{2}{d}H(f)}}{2\pi e};$$

here $\mathcal{N}$ is Shannon's **entropy power** functional. All these inequalities are saturated by Gaussian functions and then reduce to the additivity of the variance of independent random variables. There is a parallel between Shannon–Stam and Brunn–Minkowski inequalities, which was spectacularly explored by Dan Voiculescu in his theory of free probability.

Without providing here complete proofs of the above claims, three arguments will be worth sketching for future reference.

Gaussian Minimisers: Sketch of Proof Let u and σ be the mean and variance of f, and g the Gaussian with same mean and variance. So

$$\int f(x)\,x\,dx = u = \int g(x)\,x\,dx, \quad \int f(x)|x-u|^2\,dx = d\,\sigma^2 = \int g(x)|x-u|^2\,dx,$$

$$g(x) = \frac{e^{-|x-u|^2/(2\sigma^2)}}{(2\pi\sigma^2)^{d/2}}.$$

Then

$$\begin{aligned}
0 &\le \int f(x)\big|\nabla \log f - \nabla \log g\big|^2\,dx \\
&= \int f(x)|\nabla \log f|^2 + 2\int f(x)\nabla \log f(x)\cdot\left(\frac{x-u}{\sigma^2}\right)\,dx \\
&\quad + \int f(x)\frac{|x-u|^2}{\sigma^4}\,dx \\
&= I(f) + \frac{2}{\sigma^2}\int \nabla f(x)\cdot(x-u)\,dx + \frac{1}{\sigma^4}\int f(x)|x-u|^2\,dx \\
&= I(f) - \frac{2d}{\sigma^2} + \frac{d}{\sigma^2} = I(f) - \frac{d}{\sigma^2} = I(f) - I(g).
\end{aligned}$$

□

Blachman–Stam Inequality: Sketch of Proof Let $Q_\alpha(f,g) = f_{\sqrt{1-\alpha}} * g_{\sqrt{\alpha}}$. Then, distributing ∇ on either function in the convolution product,

$$\begin{aligned}
\nabla Q_\alpha(f,g) &= (1-\alpha)\,\nabla Q_\alpha(f,g) + \alpha\nabla Q_\alpha(f,g) \\
&= (1-\alpha)\,\nabla f_{\sqrt{1-\alpha}} * g_{\sqrt{\alpha}} + \alpha\, f_{\sqrt{1-\alpha}} * \nabla g_{\sqrt{\alpha}} \\
&= \int \left[(1-\alpha)\,\frac{\nabla f_{\sqrt{1-\alpha}}(y)}{f_{\sqrt{1-\alpha}}(y)} + \alpha\,\frac{\nabla g_{\sqrt{\alpha}}(x-y)}{g_{\sqrt{\alpha}}(x-y)}\right] \\
&\qquad \times f_{\sqrt{1-\alpha}}(y)\,g_{\sqrt{\alpha}}(x-y)\,dy
\end{aligned}$$

(evaluated at x, of course). By Jensen's inequality, or Cauchy–Schwarz, for each $x \in \mathbb{R}^d$,

$$
\begin{aligned}
|\nabla Q_\alpha(f,g)|(x) \le \Bigg(\int \left| (1-\alpha)\frac{\nabla f_{\sqrt{1-\alpha}}(y)}{f_{\sqrt{1-\alpha}}(y)} + \alpha \frac{\nabla g_{\sqrt{\alpha}}(x-y)}{g_{\sqrt{\alpha}}(x-y)} \right|^2 \\
\times f_{\sqrt{1-\alpha}}(y)\, g_{\sqrt{\alpha}}(x-y)\, dy \Bigg)^{1/2} \\
\left(\int f_{\sqrt{1-\alpha}}(y)\, g_{\sqrt{\alpha}}(x-y)\, dy \right)^{1/2},
\end{aligned}
$$

which is the same as

$$
\begin{aligned}
\frac{|\nabla Q_\alpha(f,g)|^2}{Q_\alpha(f,g)}(x) \le \int \left| (1-\alpha)\frac{\nabla f_{\sqrt{1-\alpha}}(y)}{f_{\sqrt{1-\alpha}}(y)} + \alpha \frac{\nabla g_{\sqrt{\alpha}}(x-y)}{g_{\sqrt{\alpha}}(x-y)} \right|^2 \\
\times f_{\sqrt{1-\alpha}}(y)\, g_{\sqrt{\alpha}}(x-y)\, dy.
\end{aligned}
$$

Upon integration,

$$
\begin{aligned}
I(Q_\alpha(f,g)) &\le \iint \left| (1-\alpha)\nabla \log f_{\sqrt{1-\alpha}}(y) + \alpha \nabla \log g_{\sqrt{\alpha}}(x-y) \right|^2 \\
&\qquad \times f_{\sqrt{1-\alpha}}(y)\, g_{\sqrt{\alpha}}(x-y)\, dx\, dy \\
&= (1-\alpha)^2 \left(\int f_{\sqrt{1-\alpha}} |\nabla \log f_{\sqrt{1-\alpha}}|^2 \right) \left(\int g_{\sqrt{\alpha}} \right) \\
&\quad + \alpha^2 \left(\int f_{\sqrt{1-\alpha}} \right) \left(\int g_{\sqrt{\alpha}} |\nabla \log g_{\sqrt{\alpha}}|^2 \right) \\
&\quad + 2\alpha(1-\alpha) \left(\int f_{\sqrt{1-\alpha}} \nabla \log f_{\sqrt{1-\alpha}} \right) \left(\int g_{\sqrt{\alpha}} \nabla \log g_{\sqrt{\alpha}} \right).
\end{aligned}
$$

The last term vanishes since $\int f \nabla \log f = \int \nabla f = 0$, and of course $\int g \nabla \log g = 0$ as well (as in Cramér–Rao, the key is to apply Cauchy–Schwarz *before* expanding the Fisher information) and we are left with

$$
I(Q_\alpha(f,g)) \le (1-\alpha)^2 I(f_{\sqrt{1-\alpha}}) + \alpha^2 I(g_{\sqrt{\alpha}}) = (1-\alpha) I(f) + \alpha I(g),
$$

after a final application of the scaling property. □

Shannon–Stam Inequality: Sketch of Proof While the proof of Blachman–Stam involved symmetries and scalings, the proof of Shannon–Stam is completely different and invokes a semigroup argument. First note that the heat semigroup commutes with Q_α: Indeed, $e^{t\Delta} f = \gamma_{\sqrt{2t}} * f$,

$$Q_\alpha\big(e^{t\Delta} f, e^{t\Delta} g\big) = (\gamma_{\sqrt{2t}} * f)_{\sqrt{1-\alpha}} * (\gamma_{\sqrt{2t}} * g)_{\sqrt{\alpha}}$$
$$= (\gamma_{\sqrt{(1-\alpha)2t}} * \gamma_{\sqrt{\alpha 2t}}) * (f_{\sqrt{1-\alpha}} * g_{\sqrt{\alpha}})$$
$$= \gamma_{\sqrt{2t}} * Q_\alpha(f, g) = e^{t\Delta} Q_\alpha(f, g).$$

(This is even easier to see with random variables.) In particular, by (1.8),

$$-\frac{d}{dt} H\Big(Q_\alpha(e^{t\Delta} f, e^{t\Delta} g)\Big) = -\frac{d}{dt} H(e^{t\Delta} Q_\alpha(f, g)) = I\big(Q_\alpha(e^{t\Delta} f, e^{t\Delta} g)\big)$$
$$\leq (1-\alpha) I(e^{t\Delta} f) + \alpha I(e^{t\Delta} g) = -\frac{d}{dt}\Big[(1-\alpha) H(e^{t\Delta} f) + \alpha H(e^{t\Delta} g)\Big].$$

The idea then is to integrate this inequality from $t = 0$ to $t = \infty$. However integration in t is a bit tricky since $I(e^{t\Delta} f)$ is typically $O(1/t)$ as $t \to \infty$ along the heat flow, and likewise for the other terms. So it is better to replace the heat semigroup $e^{t\Delta}$ by its variant localised by rescaling, the semigroup $(S_t)_{t\geq 0}$ solving the linear drift-diffusion equation

$$\frac{\partial f}{\partial t} = \Delta f + \nabla \cdot (f x). \tag{1.10}$$

Equation (1.10) is known as the linear Fokker–Planck, or adjoint Ornstein–Uhlenbeck semigroup, and its behaviour as $t \to \infty$ is convergence to the Gaussian $\gamma = \gamma_1$. This semigroup preserves the class of Gaussian distributions and commutes with Q_α. Then the inequality we obtained can be rewritten

$$-\frac{d}{dt} H\big(Q_\alpha(S_t f, S_t g)\big) \leq -\frac{d}{dt}\Big[(1-\alpha)\, H(S_t f) + \alpha H(S_t g)\Big], \tag{1.11}$$

and all three quantities $H(Q_\alpha(S_t f, S_t g)) = H(S_t Q_\alpha(f, g))$, $H(S_t f)$ and $H(S_t g)$ converge to $H(\gamma)$ as $t \to \infty$ (actually the convergence is exponentially fast), so there is no problem in integrating (1.11) from $t = 0$ to ∞, and the Shannon–Stam inequality follows. □

Stam's inequalities are precious because they have the right scaling with respect to summation of independent variables (this is no accident, as the correct rescaling is the one which preserves the variance). For instance, if $X_1, X_2, \ldots$ are i.i.d. with finite variance, then $S_n = (X_1 + \ldots + X_n)/\sqrt{n}$ satisfies

$$I(S_{2n}) \leq I(S_n), \qquad H(S_{2n}) \leq H(S_n).$$

In other words, H and I are Lyapunov functionals along the rescaled summation operation, at least along powers of 2 ($n = 2^k$). Linnik and Barron and others used this and related estimates to get explicit rates of convergence for the central limit theorem. Even for the intuition it is precious, identifying the universal Gaussian fluctuation profile as a consequence of Stam's inequalities and the Gaussian minimisation property.

1.3 Large Deviations

To present this line of thought it will be useful to compare the meanings of H and I in the spirit of large deviations. If Boltzmann's H functional quantifies how rare a distribution function is and how difficult it is to realize, Fisher's I functional is about how visible this distribution is and how difficult it is to measure.

Boltzmann's historical justification of the H function was the following. Take N particles occupying K possible states; if the frequencies $f_k = N_k/N$ are given (N_k = number of particles in state k), then how many ways are there to realize them? The result is

$$W_N(f) = \frac{N!}{(Nf_1)! \dots (Nf_K)!}.$$

(Obviously each f_k has to be an integer multiple of $1/N$ for the problem to have a solution.) Then

$$\frac{1}{N} \log W_N(f) \xrightarrow{N \to \infty} -\sum_{k=1}^{K} f_k \log f_k.$$

This is Boltzmann's celebrated formula, $S = k \log W$, in its barest expression. Here is a neat quantitative bound, recast by Cover and Thomas in the language of large deviations: If all particles are independent and drawn according to $\nu = (f_1, \dots, f_K)$, and $\widehat{\mu} = (\widehat{f_1}, \dots, \widehat{f_K})$ is the collection of observed frequencies, then $\widehat{\mu} \longrightarrow \nu$ as $N \to \infty$ (law of large numbers) and for any $\varepsilon > 0$,

$$\mathbb{P}_{\nu^{\otimes N}}\big[H_\nu(\widehat{\mu}) \geq \varepsilon\big] \leq e^{-N\left[\varepsilon - (K-1)\frac{\log(N+1)}{N}\right]}. \tag{1.12}$$

Note that $H_\nu(\mu) \geq 0$ if μ, ν are both probability measures, since $H_\nu(\mu) = \int f \log f \, d\nu \geq (\int f \, d\nu) \log(\int f \, d\nu) = 0$. Inequality (1.12) is an instance of **Sanov's theorem**, which states roughly speaking that

$$\mathbb{P}_{\nu^{\otimes N}}\big[\widehat{\mu}^N \in O\big] \simeq \exp\left[-N \inf_{\mu \in O} H_\nu(\mu)\right] \qquad \text{as } N \to \infty$$

(large deviations of the empirical measure), under various assumptions and refinements.

A more precise formulation would be: If $(\varphi_k)_{k\in\mathbb{N}}$ is a dense sequence of observables (test functions in some well-chosen function space, for instance bounded and going to zero at infinity, or bounded Lipschitz), then

$$\lim_{K\to\infty}\lim_{\varepsilon\to 0}\lim_{N\to\infty} -\frac{1}{N}\log \mathbb{P}_{\nu^{\otimes N}} \left[\forall k\le K,\quad \left|\frac{1}{N}\sum_{n=1}^{N}\varphi_k(X_n) - \int \varphi\, d\mu\right| \le \varepsilon\right] = H_\nu(\mu). \tag{1.13}$$

(Actually one should write inequalities involving lim sup and lim inf, but let me remain sloppy in this broad presentation.)

Now, what about the large deviation meaning of Fisher? One intuition is that fluctuation of particles states will inevitably blur the observation of the empirical measure, and Fisher's information gives a lower bound for that. Another intuition pertains to reconstruction of trajectories. To make a statement, we need a dynamical model, so let us assume that particles are subject to Gaussian fluctuations and feel a potential field V: this gives the standard Langevin stochastic differential equation

$$\frac{dX}{dt} = \sqrt{2}\,\frac{dB}{dt} - \nabla V(X_t), \tag{1.14}$$

where (B_t) is the standard Brownian motion and the equilibrium measure is $\nu(dx) = e^{-V(x)}\,dx/Z$, with Z a normalising constant. The transition kernel would be something like

$$a_t(x,y) \simeq \frac{e^{-|y-(x-t\nabla V(x))|^2/4t}}{(4\pi t)^{d/2}}.$$

The likelihood of a time-discretised trajectory may be proportional to

$$f(x_0)\,a(x_0,x_1)\ldots a(x_{N-1},x_N) = f(x_0)\prod e^{\log a(x_i,x_{i+1})},$$

so the log likelihood of that trajectory, after dividing by time, should be something like

$$\frac{1}{N}\sum \frac{\left|x_{i+1} - \big(x_i - \delta t\nabla V(x_i)\big)\right|^2}{4\,\delta t}.$$

When we try to reconstruct V, we might assume $dX_t = \sqrt{2}\,dB_t + \xi(X_t)\,dt$,

$$\frac{1}{4N}\sum(\delta t)\left|\frac{x_{i+1}-x_i}{\delta t} + \nabla V(x_i)\right|^2 \simeq \frac{1}{4N}\sum \delta t\,|\xi(x_i) + \nabla V(x_i)|^2$$
$$+ \quad \text{fixed divergent Brownian part}$$

(the cross product involving $(B_{t_{i+1}} - B_{t_i}) \cdot \nabla V(X_i)$ will disappear by martingale property). All in all, the variational problem which emerges is

$$\inf\left\{\frac{1}{4T}\int_0^T f(t,x)\left|\xi(t,x)+\nabla V(x)\right|^2 dx; \qquad \frac{\partial f}{\partial t} = \Delta f - \nabla\cdot(\xi f)\right\}.$$

This tells us about the probability to observe a time-dependent profile $f(t, x)$ which, by some small probability event, would not be the solution of $\partial_t f = \Delta f + \nabla \cdot (f \nabla V)$. Let us particularise this to the static case. As $T \to \infty$ the average of particle properties along particle trajectories will certainly look like those of the equilibrium measure e^{-V}, but it may be that we are tricked by an exponentially small probability event and observe another profile f_∞, then the infimum should be over ξ such that $0 = \Delta f_\infty - \nabla \cdot (\xi f_\infty) = \nabla \cdot (f_\infty(\nabla \log f_\infty - \xi))$. The infimum of $\int f_\infty|\xi + \nabla V|^2$ is obtained for $\xi = \nabla \log f_\infty$, and the value is $\int f_\infty|\nabla \log f_\infty + \nabla V|^2 = I_{e^{-V}}(f_\infty\, dx)$. After this long chain of approximations and reductions, we finally arrived at a specific instance of the Donsker–Varadhan large deviation principle, which is parallel to Sanov's problem, but now the problem is to empirically estimate the equilibrium distribution not from independent samples but from trajectories: when particles satisfy the stochastic differential Eq. (1.14),

$$\lim_{K\to\infty}\lim_{\varepsilon\to 0}\lim_{T\to\infty} -\frac{1}{T}\log\mathbb{P}\left[\ \forall k \le K,\quad \left|\frac{1}{T}\int_0^T \varphi_k(X_t)\,dt - \int \varphi\, d\mu\right| \le \varepsilon\right]$$
$$= \frac{1}{4} I_\nu(\mu).$$

To summarise, Fisher's information appears in large deviation estimates when the problem is the reconstruction of the density by the trajectorial empirical mean. Compare with Sanov's formula (1.13).

1.4 Logarithmic Sobolev Inequalities

Introduced by Leonard Gross as a substitute for Sobolev inequalities in infinite dimension, these inequalities read as follows: Given (M, g, ν) a complete Riemannian manifold with a reference measure ν, for all $u : M \to \mathbb{R}$,

$$\int_M |u|^2\log|u|^2\,d\nu \le \frac{2}{K}\int_M |\nabla u|^2\,d\nu + \left(\int_M |u|^2\,d\nu\right)\log\left(\int_M |u|^2\,d\nu\right).$$

Or equivalently, if $P(M)$ stands for the space of Borel probability measures on M,

$$\forall \mu \in P(M), \qquad H_\nu(\mu) \leq \frac{1}{2K} I_\nu(\mu). \qquad \text{(LSI}(K)) \tag{1.15}$$

This property may be true or not (more rigorously, the optimal K may be positive or 0), depending on ν. If true, this imposes stringent localisation properties on ν: at least Gaussian decay at ∞. The archetype of such a property is the Stam–Gross logarithmic Sobolev inequality (first proven by Stam in an equivalent form, rediscovered by Gross) : If γ is the standard Gaussian with identity covariance matrix on $\mathbb{R}^d$, then

$$\forall \mu \in P(\mathbb{R}^d), \qquad H_\gamma(\mu) \leq \frac{I_\gamma(\mu)}{2},$$

independently of the dimension d.

Logarithmic Sobolev inequalities play tremendous service in the study of diffusion processes, large systems of particles, concentration of measure and isoperimetry, in particular. So it is a classical and important problem to know convenient criteria for them to hold. The most important of those is due to Bakry & Émery: If $\nu(dx) = e^{-V(x)} \operatorname{vol}(dx)$ on (M, g) and $\nabla^2 V + \mathrm{Ric} \geq Kg$ for some $K > 0$, with vol being the Riemannian volume, ∇^2 the Riemannian Hessian and Ric the Ricci curvature, then ν satisfies LSI(K). Their argument uses a semigroup again: Identifying the density f and the measure $f \operatorname{vol}$, writing as before S_t for the semigroup solving $\partial_t f = \Delta f + \nabla \cdot (f \nabla V)$, they prove

$$-\frac{d}{dt} I_\nu(S_t\mu) \geq 2K\, I_\nu(S_t\mu) = -2K \frac{d}{dt} H_\nu(S_t\mu),$$

and integration from $t = 0$ to ∞ yields (1.15).

I shall come back in more detail to log Sobolev inequalities and to the Bakry–Émery argument later in this course.

1.5 *Optimal Transport Theory*

The Monge–Kantorovitch theory of optimal transport is about the most economical way to rearrange a mass distribution from initial to final state. At the end of the nineties, a dynamical version has been explored at length by Brenier–Benamou and Otto, in the formalism of fluid mechanics and Riemannian geometry. If an infinitesimal variation of measure, $\partial_t \mu$, is given, then the associated quadratic cost is

$$\|\partial_t \mu\|^2 = \inf \left\{ \int |\xi|^2 \, d\mu; \quad \partial_t \mu + \nabla \cdot (\mu \xi) = 0 \right\}.$$

This provides a formal Riemannian structure on $P(M)$, for which geodesics are dynamical realisations of the optimal transport problem with quadratic cost function. (If one wishes to draw an analogy with Jeffreys' prior, parameterize $\mu \simeq \mu_0$ by vector fields ξ, each vector field ξ on the time interval $[0, \varepsilon]$ generates a flow T_ε which pushes μ_0 forward to μ. Then take quotient to define a geometry on the space of probability measures.)

A major insight by Felix Otto is that this procedure identifies the heat flow (on probability measures) as the gradient flow for the H functional. Moreover, it is not difficult to identify Fisher's information with the square norm of the gradient (in the sense of Riemannian geometry) of the H functional,

$$I_\nu(\mu) = \| \operatorname{grad} H_\nu(\mu) \|^2.$$

This was the start of a series of works, in which I was strongly involved around the turn of the previous century, providing new interpretations and discoveries in the interplay between Riemannian geometry, entropy, Fisher information and optimal transport. For instance the HWI inequality reinforces Bakry–Émery's theorem as follows: If ν is a probability measure on M and $K > 0$ then

$$\nabla^2 V + \mathrm{Ric} \geq K g \Longrightarrow \forall \mu \in P(M),$$
$$H_\nu(\mu) \leq W_2(\mu, \nu)\sqrt{I_\nu(\mu)} - \frac{K}{2} W_2(\mu, \nu)^2, \tag{1.16}$$

where W_2 is the 2-Wasserstein distance,

$$W_2(\mu, \nu) = \Big\{ \inf \mathbb{E} d(X_0, X_1)^2; \quad \mathrm{law}(X_0) = \mu, \quad \mathrm{law}(X_1) = \nu \Big\}^{1/2}.$$

Note that by $ab - Ka^2/2 \leq b^2/(2K)$, the conclusion of (1.16) implies (1.15).

1.6 And Then...

After this overview, it is time to turn to the use of Fisher information in mathematical physics. There are at least two fields in which it plays an important role. One is quantum chemistry: if ψ is a wave function, then $\rho = |\psi|^2$ is a probability density. Writing $\psi = \sqrt{\rho}\, e^{i\varphi}$, we have

$$\int |\nabla \psi|^2 = \int \left(|\nabla \sqrt{\rho}|^2 + \rho\, |\nabla \varphi|^2 \right) = \frac{I(\rho)}{4} + \int \rho\, |\nabla \varphi|^2,$$

so that Fisher's information is naturally involved in the computation of the "kinetic energy" of the particle.

Another field of powerful application for the Fisher information is the kinetic theory of gases and plasmas, and this will be the focus of all the remaining of these notes. As we shall see, in particular, Fisher's information is monotonous (nonincreasing) along solutions of the most fundamental model in collisional kinetic theory, and this has important consequences.

Bibliographical Notes

Some of the important founding papers are those by Fisher [79], Shannon and Weaver [158], Jeffreys [112], Stam [164], Linnik [123], Donsker–Varadhan [76], Gross [95], Bakry–Émery [15], Brenier–Benamou [21], Jordan–Kinderlehrer–Otto [114], Otto [145], Otto–Villani [146].

The time-honoured reference course on information theory is the book by Cover and Thomas [53], containing itself hundreds of references. The discussion on large deviations and Boltzmann's formula there is illuminating. My version of the classical Cramér–Rao bound is a very slight generalisation of the classical bound, for coherence of the presentation.

The parallel between Brunn–Minkowski and information theory is analysed by Dembo, Cover and Thomas [57]. Free probability is exposed in Voiculescu's lecture notes [189]. From there Voiculescu and Szarek [167] also devised a brilliant proof of Shannon–Stam resting on a large-dimensional analogue of Brunn–Minkowski (a proof which is considerably more complicated than the one by Stam, but very meaningful).

Explicit central limit theorems were developed by Barron [19] and further authors.

A classical course on large deviations is the book by Dembo and Zeitouni [56]. Some quantitative variants of Sanov's theorem are in my work with Bolley and Guillin [27]. The Donsker–Varadhan theorem can be found in various levels of generality [153] but it is not so easy to find a neat reference for the instance presented here, and it would be good to have some quantitative estimates.

Logarithmic Sobolev inequalities are the object of many works, among which the synthesis book by the Toulouse research group [6] and the one by Bakry–Gentil–Ledoux [16].

My own books can be consulted for optimal transport theory [183, 188]. Chapter 21 in the latter reference is also largely devoted to logarithmic Sobolev inequalities.

An introduction to Fisher's information in variational problems related to quantum physics can be found in Lieb–Loss [120].

2 Core Kinetic Theory

Classical kinetic theory was born from the heroic efforts of Maxwell and Boltzmann to statistically describe assemblies of particles through their time-dependent distribution function in position and velocity variables: $f = f(t, x, v)$ where, say,

$t \geq 0$, $x \in \Omega_x \subset \mathbb{R}^d$ or $x \in \mathbb{R}^d/\mathbb{Z}^d$, and $v \in \mathbb{R}^d$. If particles are subject to a macroscopic force field $F = F(t, x)$ (possibly induced or partially induced by the particles themselves) and interact through localised binary microscopic "collisions" then the model is the Boltzmann equation

$$\frac{\partial f}{\partial t} + v \cdot \nabla_x f + F \cdot \nabla_v f = Q(f, f) \tag{2.1}$$

and Boltzmann's collision operator Q reads

$$Q(f, f) = \int_{\mathbb{R}^d} \int_{\mathbb{S}^{d-1}} (f' f'_* - f f_*) \, B \, d\sigma \, dv_*, \tag{2.2}$$

where $f' = f(t, x, v')$ and so on, and

$$v' = \frac{v + v_*}{2} + \frac{|v - v_*|}{2} \sigma, \qquad v'_* = \frac{v + v_*}{2} - \frac{|v - v_*|}{2} \sigma, \tag{2.3}$$

$B = B(v - v_*, \sigma)$ only depends on $|v - v_*|$ (relative velocity) and $\frac{v - v_*}{|v - v_*|} \cdot \sigma$ (cosine of the deviation angle). Think of $(v, v_*) \to (v', v'_*)$ as a sudden variation in the velocities of a pair of colliding particles: Conservation of momentum induces $v + v_* = v' + v'_*$, conservation of energy induces $|v|^2 + |v_*|^2 = |v'|^2 + |v'_*|^2$, these $d+1$ conservation laws leave room for $d-1$ parameters for solutions, and that is the role of $\sigma = (v' - v'_*)/|v' - v'_*|$, the direction of the relative postcollisional velocities. Equation (2.1) makes sense only with boundary conditions, but let me skip this issue here, or just think that x lives in $\mathbb{R}^d/\mathbb{Z}^d$ so that there is no boundary in the physical domain.

In one of the greatest artworks of mathematical physics ever, Maxwell and Boltzmann set up the basis of classical kinetic theory in relation to the then-speculative atomistic theory. In particular they showed that

(1) Under an assumption of molecular chaos at initial time, and in a regime in which each particle collides at least a few times per unit of time, (2.1) is a plausible model for the evolution of the density of particles;

(2) If $H(f) = \iint f \log f \, dx \, dv$ then in the absence of force and under appropriate boundary conditions, $dH/dt \leq 0$ (Boltzmann's H Theorem, turning the Second Law of thermodynamics into a plausible *theorem* for this particular but fundamental system). Boltzmann's proof is a classical gem: applying the two basic ways to exchange variables, which are $(v, v') \longleftrightarrow (v_*, v'_*)$ and $(v, v_*, \sigma) \longleftrightarrow (v', v'_*, k)$ $(k = \frac{v - v_*}{|v - v_*|})$, using the additivity of the logarithm, the fact that the transport operator $v \cdot \nabla_x$ preserves all integral functions $\int C(f) \, dv \, dx$, the additivity property of the log and its increasing property,

$$H'(f) \cdot Q(f, f) = \iint \left(\iint (f' f'_* - f f_*) \, B \, dv_* \, d\sigma \right) (\log f + 1) \, dv \, dx$$

$$= \frac{1}{4} \iiiint (f'f'_* - ff_*)\big(\log f + \log f_* - \log f' - \log f'_*\big)$$
$$\times\, B\, dv\, dv_*\, d\sigma\, dx$$
$$= -\frac{1}{4} \iiiint (f'f'_* - ff_*)\big(\log ff_* - \log f'f'_*\big)$$
$$\times\, B\, dv\, dv_*\, d\sigma\, dx \leq 0.$$

From this computation emerges **Boltzmann's dissipation functional** (dissipation of H, or more rigorously its integrand in the x variable)

$$D_B(f) = \frac{1}{4} \iiint (f'f'_* - ff_*)\big(\log f'f'_* - \log ff_*\big)\, B\, dv\, dv_*\, d\sigma. \qquad (2.4)$$

This fundamental functional quantifies the strength of the dissipation process at work in (2.1).

(3) Equilibria for (2.1), in the absence of macroscopic force and except in presence of certain spatial symmetries, are of the form

$$M(x, v) = M_{\rho u T}(v) = \rho\, \frac{e^{-|v-u|^2/(2T)}}{(2\pi T)^{d/2}}, \qquad (2.5)$$

where $\rho \geq 0$, $u \in \mathbb{R}^d$, $T \geq 0$ are constants. In particular these equilibria are space-homogeneous. These Gaussian distributions are called Maxwellians in this context. (There are additional equilibria under certain symmetries, for instance axisymmetric domains with specular reflection of particles at the boundary.)

(4) As $t \to \infty$ solutions of (2.1) converge to equilibrium.

(5) In a suitable regime of many collisions per unit of time, solutions of (2.1) are well approximated by local equilibria $M_{\rho u T}(v)$, where now ρ, u, T are functions of t and x and satisfy certain hydrodynamic equations for density, velocity and temperature.

In this way Maxwell and Boltzmann recovered and discovered some properties of fluids, sometimes consistent with hydrodynamics and sometimes not. The first two striking such discoveries by Maxwell were that the viscosity of a rarefied gas is independent of its density (something that he himself could not believe at first), and that near the boundary a gas may flow from low temperature to high temperature (Maxwell's paradoxical *thermal creep*). By all means the program was a complete success, and a milestone in the discovery of atoms.

Still, even after 150 years and in spite of thousands of works, Boltzmann's equation retains some deep mathematical mysteries. Most famously, we are in want of a large time convergence proof from particle systems to (2.1). (Lanford's theorem, even after half a century of corrections and improvements, only works out for about a fraction of a collision time, which is of the order of $10^{-9}s$ under usual conditions for the air around us.) This and other works suggest that a theory of regular solutions is needed, but as of today the latter is nowhere to be seen, except in particular regimes, for instance very close to equilibrium. Equation (2.1) contains several major difficulties:

- the complexity of the operator Q;
- the degenerate nature of the equation, involving a conservative transport operator $v \cdot \nabla_x$ (or $v \cdot \nabla_x + F \cdot \nabla_v$) and a dissipation, but only in v variable, the collision operator—this is akin to a hypoelliptic or hypocoercive issue;
- the quadratic nature of Q, whose amplitude is proportional to ρ^2 ($\rho = \int f\,dv =$ spatial density), making a priori estimates a challenge;
- the singular nature of the hydrodynamic approximation, in a regime when B becomes very large.

In this dire but exciting situation, it is legitimate to study all difficulties in parallel. A good way to better understand the structure of Q is to focus on the *spatially homogeneous Boltzmann equation*,

$$\frac{\partial f}{\partial t} = Q(f, f), \tag{2.6}$$

and this will be the main focus of this course.

Bibliographical Notes

The founding papers of kinetic theory are those of Maxwell [132] and Boltzmann [28]. Maxwell's creep thermal effect is from [133]. Boltzmann's book [29] is a bright synthesis of the whole theory and had considerable influence, as shown for instance in Perrin's classic book *Les Atomes* [148]. Albert Einstein, Erwin Schrödinger, Max Planck all built from Boltzmann's work to develop their theories of atomistic laws. The story is described in various books of history of science, like Lindley's [122].

Modern survey books on Boltzmann's theory from the past thirty years are those by Cercignani [47, 48], Cercignani–Illner–Pulvirenti [49], Sone [162], and my own review of collisional kinetic theory [180]. The most elaborate version of Lanford's theorem is the book by Bodineau, Saint-Raymond and Gallagher [88]. The recent PhD by Corentin Le Bihan provides a good review of the derivation problem [117].

3 Fisher Information into Kinetic Theory

Fisher information was imported into kinetic theory by Henry P. McKean in a remarkable paper of 1966. His motivation came from the problem raised by Kac on the speed of approach to equilibrium, starting with a one-dimensional caricature of the Boltzmann equation. The idea was sharp and clear: Draw a parallel between the convergence to equilibrium in Boltzmann's theory, and the central limit theorem in classical probability theory and statistics. Repeated interactions tend to push the distribution closer to a Maxwellian (or Gaussian); by writing the solution as a superposition of contributions from all interaction histories, represented by interaction trees, show that most trees are "deep" and thus the corresponding terms are close to Maxwellian.

McKean borrowed from Linnik the use of Fisher's functional I, which had been instrumental in getting some results of explicit convergence for the central limit theorem. He worked on Kac's simplified equation and (a) proved that $I(f)$ is nonincreasing for this model, (b) established the first result of exponential convergence for the distribution function (not just for the solution of the linearised equation) as $t \to \infty$. Note that in dimension 1, the bound on I is already a strong regularity estimate, implying Hölder continuity of f.

Since that time, McKean's program has been extended and refined through a number of tools—functional inequalities, probability metrics, entropy methods, Fourier transform, linearisation... Some notable authors in this program have been Arkeryd, Bobylev, Carlen, Carvalho, Cercignani, Dolera, Desvillettes, Gabetta, Loss, Regazzini, Tanaka and myself. In particular, in 2013 Dolera and Regazzini achieved a very close parallel between Boltzmann equilibration and central limit theorem, and established that the rate of exponential convergence of the spatially homogeneous Boltzmann equation with Maxwell kernel and cutoff (that is, when B only depends on $k \cdot \sigma$ and $\int B\, d\sigma < \infty$) is given precisely by the spectral gap of the linearised equation. Eventually they did not use the Fisher information for that, relying rather on Fourier transform. But in the meantime, Fisher information had proven useful in a number of related problems in kinetic theory. Here are five examples:

(a) *Equilibration for the Fokker–Planck equation:* Consider the basic linear Fokker–Planck equation

$$\frac{\partial f}{\partial t} = \Delta f + \nabla \cdot (f v), \qquad f = f(t, v), \qquad v \in \mathbb{R}^d.$$

Then by the Stam–Gross logarithmic Sobolev inequality

$$-\frac{d}{dt} H_\gamma(f) = I_\gamma(f), \qquad -\frac{d}{dt} I_\gamma(f) \geq 2 I_\gamma(f), \qquad H_\gamma(f) \leq \frac{1}{2} I_\gamma(f),$$

which readily implies the exponential convergence estimates

$$H_\gamma(f(t)) \leq e^{-2t} H_\gamma(f_0), \qquad I_\gamma(f(t)) \leq e^{-2t} I_\gamma(f_0).$$

Using classical inequalities in information theory, this provides a convergence of $f(t)$ to γ like $O(e^{-t})$. Note: That rate of convergence is the same as in the linear theory (going through the spectral analysis of $-\Delta - v \cdot \nabla$ in $L^2(\gamma\, dv)$), but assumptions are much less stringent since the linear theory corresponds to $f/\gamma \in L^2(\gamma)$, or equivalently $\int e^{|v|^2/2} f(v)^2\, dv < \infty$, while here the only requirement is that $f_0 = f_0(v)$ satisfying either $H_\gamma(f_0) < \infty$ or the stronger condition $I_\gamma(f_0) < \infty$; and one may even relax those conditions through a regularisation study.

(b) *Entropy production estimates for Boltzmann's collision operator:* At the end of the nineties, Toscani and I established bounds of the form

$$\forall f \qquad D_B(f) \geq K(B, \varepsilon, f)\, H_\gamma(f)^{1+\varepsilon},$$

where $\varepsilon \in (0, 1)$ can be arbitrarily small ($\varepsilon = 0$ is impossible in general, as shown by counterexamples of Bobylev and Cercignani) and $K(B, \varepsilon, f)$ only depends on B through a polynomial lower bound (like $B \geq a \min(|v - v_*|^\alpha, |v - v_*|^{-\alpha}), a, \alpha > 0$) and on f through regularity bounds (upper control of moments and Sobolev norms of f of large enough order, lower bound on f like $f \geq K_0\, e^{-A_0|v|^{q_0}}$). The result does not use Fisher information but the proof does; it involves a semigroup argument à la Stam, using the Fokker–Planck semigroup $(S_t)_{t\geq 0}$, and goes roughly like this:

$$\begin{aligned} D_B(f) &\gtrsim \int_0^{+\infty} e^{-2\lambda t}\, D_L(S_t f)\, dt \\ &\gtrsim \int_0^{+\infty} e^{-2\lambda t}\, I_\gamma(S_t f)\, dt \\ &\geq \int_0^{+\infty} I_\gamma\big(S_{(1+\lambda)t} f\big)\, dt = \frac{H_\gamma(f)}{1+\lambda}, \end{aligned}$$

where D_L is Landau's dissipation functional, which I will present in detail later in these notes.

(c) *Inhomogeneous entropy estimates:* Around 2000, Desvillettes and I analysed the (hypocoercive) equilibration for the linear inhomogeneous kinetic Fokker–Planck equation with potential, white noise forcing and friction:

$$\frac{\partial f}{\partial t} + v \cdot \nabla_x f - \nabla V(x) \cdot \nabla_v f = \Delta_v f + \nabla_v \cdot (f v) \qquad f = f(t, x, v) \geq 0. \tag{3.1}$$

Then the equilibrium is $f_\infty(x, v) = e^{-V(x)}\gamma(v)$ (assume that V is normalised so that $\int e^{-V} = 1$). We found out that the basic Gronwall-type inequality $dH_\gamma(f)/dt \leq -2H_\gamma(f)$, typical of the spatially homogeneous regime, should be replaced by the more sophisticated system

$$\begin{cases} -\dfrac{d}{dt} H_{f_\infty}(f) \geq H_{\rho\gamma}(f) \qquad \rho = \displaystyle\int f\, dv \\[2ex] \dfrac{d^2}{dt^2} H_{\rho\gamma}(f) \geq I_{e^{-V}}(\rho) - O\Big(\|f - f_\infty\|\, \|f - \rho\gamma\|\Big), \end{cases}$$

leading to long-time estimates on how f approaches f_∞. (Here I am cheating a bit, but the spirit is correct.) This was later extended in my memoir on Hypocoercivity and other works.

(d) *Entropic hypocoercive estimates:* About twenty years ago, I used a mixed (x, v)-Fisher information to establish the first explicit rates of exponential convergence to equilibrium for the kinetic Fokker–Planck equation in spaces of finite entropy : the functional was

$$\widetilde{I}(f) = \int f \left\langle A\nabla_{x,v} f, \nabla_{x,v} f \right\rangle dx\, dv,$$

where $\nabla_{x,v} = (\nabla_x, \nabla_v)$ and

$$A = \begin{pmatrix} aI & bI \\ bI & cI \end{pmatrix}, \qquad a > b > c, \quad b < \sqrt{ac},$$

a, b, c being well-chosen. Coupled with global hypoelliptic regularisation estimates, this approach also led to equilibration results from measure initial data with enough moments. These results were later recovered and improved by Mouhot and collaborators.

(e) *Refined notions of propagation of chaos*: In various works, Carrapataso, Fournier, Hauray, Mischler have shown how to use Fisher information bounds to refine the usual notion of chaos (which says that a sequence of symmetric probability measures μ^N in N variables is close, in weak sense, to a tensor product) into the stronger notion of entropic chaos (which says, in addition, that there is convergence of the entropy). This is a kind of infinite-dimensional interpolation: convergence in weak sense and bound in Fisher sense imply convergence in entropy sense; the HWI inequality allows to do that when the reference measure is log concave.

Still, a more basic development of McKean's paper was almost left aside as a curiosity: When does I decrease along solutions of Boltzmann's equation? The transport flow does not leave I invariant (unlike H) so it seemed natural to focus on the effect of just Q, that is the spatially homogeneous case. McKean seemed to think that the decreasing property of I was a specific feature of dimension 1, but in 1992 Toscani generalised it to the 2-dimensional Boltzmann equation with Maxwell collisions; and in 1998, during my PhD I obtained the generalisation to all dimensions (a question asked by my advisor on the first day of my PhD!). I actually established the following analogues of Stam's inequalities, from which the decreasing property follows at once: If

$$Q^+(f, g) = \int_{\mathbb{R}^d} \int_{\mathbb{S}^{d-1}} f' g'_* b(k\cdot\sigma)\, d\sigma, \qquad \int b(k\cdot\sigma)\, d\sigma = 1, \quad b(k\cdot\sigma) = b(-k\cdot\sigma),$$

then for all distributions f, g with finite energy,

$$I\big(Q^+(f,g)\big) \le \frac{1}{2}\big[I(f)+I(g)\big], \qquad H\big(Q^+(f,g)\big) \le \frac{1}{2}\big[H(f)+H(g)\big].$$

This seemed like a satisfactory answer and (contrary to Toscani) I did not expect I to be decreasing for more general Boltzmann kernels. In the twenty-five years to follow, there was no extension of this result (save for an alternative proof by Matthes and Toscani of a particular calculation in my paper).

Before I turn to the rest of the story, let me mention that McKean's 1966 paper also contained some other questions or conjectures; some were disproved, but others are still standing, concerning either higher order functionals or higher order time derivations.

Another area of research opened in my PhD, on the other hand, was well identified as incomplete, and frustratingly so: the regularity of solutions for "very soft" potentials, that is, when the collision kernel has a strong singularity as $|v - v_*| \to 0$, say like $O(|v - v_*|^\gamma)$ for $\gamma < -2$ (here γ is just a number, has nothing to do with the Gaussian distribution). In dimension 3 this corresponds to power law forces like $O(1/r^s)$ with $2 < s \le 7/3$. The most important motivation for such singularities is the Landau–Coulomb model, a limit case for Boltzmann's equation describing near encounters in plasma physics: then $\gamma = -3$. In such a singular regime, usual notions of weak solutions seemed untractable, and I had constructed a class of "very weak' solutions using a time-integrated formulation and the functional D_B, or its Landau counterpart D_L; they were called H-solutions and the subject of my very first international seminar (Pavia, 1997).

Later it was found that when the kernel is *singular enough* in the angular variable, there is enough time-integrated regularity in the entropy production estimate to reinforce the notion of H-solutions in a more classical notion of weak solutions. It could also be shown, under the same assumption of angular singularity, that there is conditional regularity: If the solutions satisfy some integrability bounds (L^p for p large enough) then they are actually quite regular. But in the following years, nobody could establish whether there was actually regularity, or whether singularity could get in the way, despite enormous work by Golse, Gualdani, Imbert and Vasseur who identified more and more precise blow-up criteria.

Before I go on with my story, let me describe more precisely the taxonomy of collision kernels and their associated equations. This will be the subject of the next section.

Bibliographical Notes

The founding papers are those of Kac [115] and McKean [134]. The most accomplished parallel between central limit theorem and Boltzmann theory is achieved, for spatially homogeneous Maxwell collisions, by Dolera and Regazzini [75], which also includes a review of the field since McKean.

It took enormous effort before one could put Kac's program on a sound footing, estimating the propagation of chaos in such a way that the limit $N \to \infty$ and the limit $t \to \infty$ could commute; this was achieved by Mischler and Mouhot, building on a number of previous authors and ideas [135].

Entropy production estimates for Boltzmann's collision operator are in my works [171], jointly with Toscani, and [181]. Before that, Carlen and Carvalho [38, 39] had obtained the first entropy production estimates relating D and H (but with a much worse dependence), also using Fisher's information. Slightly before Carlen and Carvalho, there had been works by Desvillettes [58, 59] establishing stable entropy production bounds and convergence for the spatially homogeneous Boltzmann equation, albeit not quantitative.

I worked with Desvillettes on inhomogeneous models, conditional to regularity estimates [70, 72]. I wrote several lecture notes on this [184–186] and a memoir on hypocoercivity [187].

The combination of Fisher information bounds and chaos property was examined by Hauray and Mischler [100] and Carrapataso [42], and applied to mean-field limits of particle systems by Fournier, Hauray and Mischler [85] (for the two-dimensional viscous vortex model) and by Carrapataso [42] (for the Boltzmann equation).

Toscani [169] proved decay of the Fisher information for the spatially homogeneous Boltzmann equation with Maxwellian molecules; this was in 1992 and revived the interest for the problem, showing that McKean's 1966 result was not limited to the one-dimensional case. A few years after Toscani, I established Stam's inequalities for the Boltzmann collision kernel in [174], then Matthes and Toscani found an alternative road to a key lemma in [131].

Several types of questions have been called "McKean's conjectures" after McKean's original paper [134]. Some are about successive derivatives of the entropy along the Boltzmann flow: Is H a convex function of time? or even a completely monotone function of time? The latter is known to be false even for the most simple models of Boltzmann equation, but the former might be true for the spatially homogeneous Boltzmann equation. Other conjectures are about the functionals obtained by successive derivation of the entropy and Fisher information along the heat flow: do they have alternate signs, are they optimised by Gaussians under constraint of fixed variance? Some comments are in my review [180, Chapter 4, Section 4.3] and a much more up-to-date discussion, including recent results by Gao, Guo, Yuan, Wang, can be found in Ledoux [119].

I introduced H-solutions in [175] to handle the Cauchy problem for the spatially homogeneous Boltzmann equation with a strong singularity in the relative velocity; the refinement to time-integrated weak solutions was proven as a consequence of the fine study of entropy production which we did with Alexandre, Desvillettes and Wennberg [4]. The control of the growth of moments for these solutions was done by Carlen, Carvalho and Lu [41]. The partial or conditional regularity of these solutions, in the particular Landau–Coulomb case was the subject of a series of works by Golse, Gualdani, Imbert, Vasseur, e.g. [90, 91].

In those works, entropy production is exploited to overcome the strong singularity of the collision kernel. Note that already at the end of the eighties, entropy production had played a key role in the inhomogeneous theory of weak solutions by DiPerna and Lions [73]; but that was to overcome the different problem of the genuinely quadratic nature of the collision operator in the full space-inhomogeneous configuration.

4 Taxonomy of Boltzmann Collisions

Let me recollect notation. If $f : \mathbb{R}^d \to \mathbb{R}_+$ is a probability density, define $Q(f, f)$ as a function of v by

$$Q(f, f) = \int_{\mathbb{R}^d} \int_{\mathbb{S}^d} \left[f(v')f(v'_*) - f(v)f(v_*) \right] B(v - v_*, \sigma)\, dv_*\, d\sigma, \tag{4.1}$$

where v, v_*, v', v'_* all belong in the *collision sphere* of diameter $[v, v_*]$, and $B(v - v_*, \sigma) = B(|v - v_*|, \cos\theta)$ (by abuse of notation I use the same symbol B for both writings), and $\theta \in [0, \pi]$ is the angle between k and σ, the unit vectors directing $v - v_*$ and $v' - v'_*$ respectively; see Fig. 1. The collision kernel is related to the **cross-section** Σ by the formula $B(z, \sigma) = |z|\Sigma(z, \sigma)$.

For intermolecular forces deriving from a repulsive central potential ψ, Maxwell established the formula for the cross-section; this amounts to solving a classical scattering problem with given relative velocity z and impact parameter $p \geq 0$. Working in the referential of center of mass (fixed during the collision), the picture is as in Fig. 2.

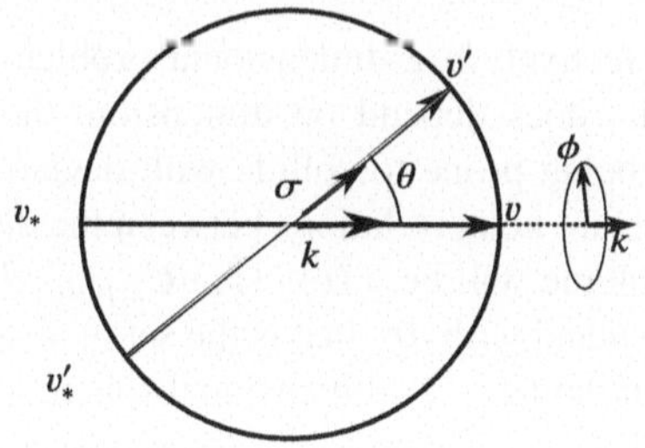

Fig. 1 The collision sphere: a collision changes velocities v, v_* into v', v'_* and *vice versa*; the relative velocity axis is changed from k to σ; and $\theta \in [0, \pi]$ is the deviation angle. It is often convenient to express σ as a combination of k and ϕ, a $(d-2)$-dimensional unit vector in $k^\perp$

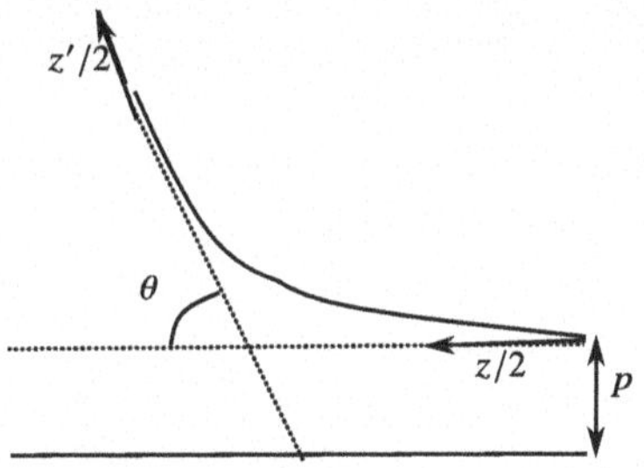

Fig. 2 Maxwell's classical scattering problem: In the referential of the center of mass, with velocity $(v + v_*)/2$, one of the particles arrives from the right with asymptotic relative velocity $z/2 = (v - v_*)/2$ and goes away to the left with asymptotic relative velocity $z'/2 = (v' - v'_*)/2$, the impact parameter is p, the deviation angle is θ

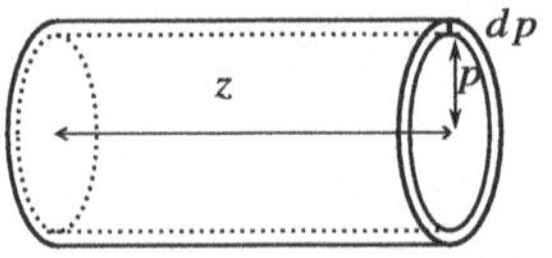

Fig. 3 The cylinder of collisions: the lateral crust of the cylinder is where collision partners will be found with impact parameter p (up to dp) and relative velocity z, during the time interval $[t, t+dt]$

The deviation angle θ is computed from p and z as

$$\theta(p,z) = \pi - 2p \int_{r_0}^{\infty} \frac{dr/r^2}{\sqrt{1 - \frac{p^2}{r^2} - \frac{4\psi(r)}{|z|^2}}}, \tag{4.2}$$

where r_0 is the minimum distance between one particle and the center of mass; its value is the positive root of

$$1 - \frac{p^2}{r_0^2} - \frac{4\psi(r_0)}{|z|^2} = 0. \tag{4.3}$$

The interaction being effectively a 2-dimensional problem, this formula is independent of $d \geq 2$. What does depend on dimension, though, is the volume of particles in phase space being prone to collide with deviation angle θ, per unit of time. Assuming that there are no correlations between the distributions of v and v_*, for given z and p this volume will be $f(v)f(v_*)|C_{z,p,dp}|\,dt$, where $C_{z,p,dp}$ is the cylinder of axis z, radius p and with dp, that is the set of all $tz + [p, p+dp]\phi$, with $0 \leq t \leq 1$, and ϕ lying in the $(d-2)$-dimensional sphere unit orthogonal to z; see Fig. 3.

The infinitesimal volume $|C_{z,p,dp}|$ should be identified with the integral of the collision kernel over the variable ϕ: taking into account $d\sigma = \sin^{d-2}\theta\, d\theta\, d\phi$ (usual polar coordinates around z), this yields

$$B(|z|, \cos\theta)\, \sin^{d-2}\theta\, |\mathbb{S}^{d-2}|\, d\theta = |C_{z,p,dp}| = p^{d-2} |\mathbb{S}^{d-2}|\, |z|\, dp.$$

(One can also avoid ϕ-integration and directly identify $B(|z|, \cos\theta)\, d\sigma$ with the volume of a longitudinal strip in $C_{z,p,dp}$ at angle ϕ, with width $d\phi$, so $B(|z|, \cos\theta)\, \sin^{d-2}\theta\, d\theta\, d\phi = p^{d-2}\, |z|\, dp\, d\phi$.) Anyway we end up with

$$B(|z|, \cos\theta) = \left(\frac{p}{\sin\theta}\right)^{d-2} \left|\frac{dp}{d\theta}\right| |z|. \tag{4.4}$$

Main Example: Inverse Power Laws Assume that the force is an inverse power law of exponent s, so $-\psi'(r) = (s-1)\psi_0/r^s$ for some $s > 1$, or

$$\psi(r) = \frac{\psi_0}{r^{s-1}}.$$

In practice ψ_0 will involve physical constants quantifying the strength of interaction. Then

$$\theta\left(|z|^{-\frac{2}{s-1}} p, z\right) = \theta(p, 1).$$

So for given θ the product $p|z|^{2/(s-1)}$ is independent of $|z|$. It follows that $B(|z|, \cos\theta) = |z|^\gamma B(1, \cos\theta)$ with $\gamma = (s - (2d - 1))/(s - 1)$. Let us denote

$$B(v - v_*, \sigma) = |v - v_*|^\gamma b(\cos\theta), \qquad \gamma = \frac{s - (2d - 1)}{s - 1}. \tag{4.5}$$

As for the function $b(\cos\theta)$, it is nonexplicit and in general involves transcendantal integrals. To study it more explicitly let us assume that the choice of units is such that

$$\psi_0 = \frac{1}{4}.$$

Writing $\theta(p) = \theta(p, 1)$, we have

$$b(\cos\theta) = \left(\frac{p}{\sin\theta(p)}\right)^{d-2} \left|\frac{dp}{d\theta}\right|, \quad \theta(p) = \pi - 2p \int_{r_0}^{\infty} \frac{dr/r^2}{\sqrt{1 - \frac{p^2}{r^2} - \frac{1}{r^{s-1}}}}. \tag{4.6}$$

There are two cases which lead to explicit computation: $s = 2$, $s = 3$. For the particular exponent $s = 2$ the computation can be carried out completely and yields, in dimension 3, **Rutherford's cross-section formula**:

$$B_2(|z|, \cos\theta) = \frac{1}{16} \frac{|z|}{\left(|z| \sin\frac{\theta}{2}\right)^4}. \tag{4.7}$$

More generally, in dimension d,

$$B_2^{(d)}(|z|, \cos\theta) = \frac{|z|}{\left(2|z| \sin\frac{\theta}{2}\right)^{2(d-1)}}. \tag{4.8}$$

For more general s things remain implicit, but at least the behaviour for $\theta \to 0$ can be estimated, though. For this we need to recast the formula for θ in (4.6) as an integral with fixed boundaries. This computation will be important in the sequel, so let us take some care to perform this.

A first possibility is to choose as new variable

$$u = 1 - \frac{p^2}{r^2} - \frac{1}{r^{s-1}},$$

and noting $p/r \simeq \sqrt{1-u}$ as $p \to \infty$ (i.e. $\theta \to 0$), we rewrite

$$\theta = \pi - \int_0^1 \frac{\sqrt{1 + \frac{1}{p^2 r^{s-3}}}}{\left(1 + \frac{2(s-1)\psi_0}{p^2 r^{s-3}}\right)} \frac{du}{\sqrt{u(1-u)}}$$

$$\simeq \frac{2\psi_0(s-2)}{p^{s-1}} \int_0^1 (1-u)^{\frac{s-4}{2}} \frac{du}{\sqrt{u}} \qquad \text{as } p \to \infty. \tag{4.9}$$

Thus

$$p \simeq \left(\frac{C(s)\,\psi_0}{\theta}\right)^{\frac{1}{s-1}} \quad \text{as } \theta \to 0,$$

and eventually

$$b(\cos\theta)\,\sin^{d-2}\theta \sim_{\theta\to 0} \left(\frac{(C(s)\,\psi_0)^{\frac{d-1}{s-1}}}{s-1}\right)\theta^{-(1+\nu)}, \qquad \nu = \frac{d-1}{s-1}. \tag{4.10}$$

Here ν quantifies the degree of nonintegrability of the collision kernel in the θ variable. That formula converges for $s > 2$; as $s \to 2$ the factor $s-2$ vanishes, but the integral (4.9) goes to ∞, so it is indeterminate.

Another change of variable is $y = r_0/r$. Then, using $1 - p^2/r_0^2 = 1/r_0^{s-1}$, we obtain, as $p \to \infty$,

$$\theta(p) = \pi - 2\int_0^1 \frac{dy}{\sqrt{\frac{r_0^2}{p^2} - y^2 - \frac{y^{s-1}}{r_0^{s-3}p^2}}}$$

$$= 2\int_0^1 \left(\frac{1}{\sqrt{1-y^2}} - \frac{1}{\sqrt{\frac{r_0^2}{p^2} - y^2 - \frac{y^{s-1}}{r_0^{s-3}p^2}}}\right)$$

$$\simeq \frac{r_0^2}{p^2}\int_0^1 \frac{1}{(1-y^2)^{3/2}}\left(\frac{1}{r_0^{s-1}} - \frac{y^{s-1}}{r_0^{s-1}}\right) dy$$

$$\simeq \frac{1}{r_0^{s-3}p^2}\int_0^1 \frac{1}{(1-y^2)^{3/2}}(1-y^{s-1})\,dy$$

$$\simeq \left(\int_0^1 \frac{1}{(1-y^2)^{3/2}}(1-y^{s-1})\,dy\right)\frac{1}{p^{s-1}}$$

$$= \left(\frac{\sqrt{\pi}\,\Gamma\left(\frac{s}{2}\right)}{\Gamma\left(\frac{s-1}{2}\right)}\right)\frac{1}{p^{s-1}}.$$

That formula applies to the whole range $s > 1$.

Remarks 4.1

- $\nu \to 0$ as $s \to \infty$, whatever the dimension, and likewise $\gamma \to 1$. This limit corresponds to the hard sphere kernel, which is better expressed in terms of ω, the direction of the centers of mass of colliding particles (with obvious notation, $\omega = (x - x_*)/|x - x_*|$):

$$\widetilde{B}_{HS}(z, \omega) = |z \cdot \omega| \, 1_{k \cdot \omega < 0},$$

The directions σ and ω are related through $|k \cdot \omega| = \sin(\theta/2)$, and the Jacobian is $|d\sigma/d\omega| = (2\sin(\theta/2))^{d-2}$. So eventually, in dimension d,

$$B_{HS}^{(d)}(z, \sigma) = \frac{|z|}{2^{d-2}\sin^{d-3}(\theta/2)}. \tag{4.11}$$

When $d = 2$ the angular dependence of that kernel is proportional to $\sin(\theta/2)$ and when $d \geq 4$ to the singular function $(\sin(\theta/2))^{-(d-3)}$ (the singularity is integrable and may be seen as an artifact of the change of variables $\omega \to \sigma$). Thus in the limit $s \to \infty$, B becomes integrable over $\mathbb{S}^{d-1}$ even though for any finite s it is not. In the particular case $d = 3$, Hard Spheres correspond to a kernel which is just proportional to the relative velocity (*constant cross-section*) $B(z, \sigma) = |z|/2$.
- Power law forces satisfy $\gamma + 2\nu = 1$.
- Fortunately the Boltzmann equation still makes sense if B is nonintegrable, provided that the cross-section for momentum transfer $M = M(z)$ is well defined:

$$M(z) = \int_{\mathbb{S}^{d-1}} B(z, \sigma)\,(1 - k \cdot \sigma)\, d\sigma. \tag{4.12}$$

In fact $M(z)|z|$ is the typical change of momentum in a collision of relative velocity z; by symmetry M only depends on $|z|$ and I shall often write it as $M(|z|)$. Since $1 - k \cdot \sigma \sim \theta^2/2$ as $\theta \to 0$, the finiteness of M requires $\nu < 2$. We shall see in the next section that $\nu = 2$ can be handled as a limit case.
- Coulomb interaction in dimension d is for $s = d - 1$, so $\gamma = -d/(d-2)$, $\nu = (d-1)/(d-2)$. In particular the singularity in velocity for Coulomb interactions is -3 for $d = 3$, -2 for $d = 4$, $-5/3$ for $d = 5$, and converges to -1 as $d \to \infty$. In other words,

  - for $d = 2$, Boltzmann's equation definitely does not make sense for Coulomb potential;
  - for $d = 3$ it is borderline integrable in θ ($\nu = 2$) and also borderline integrable in z ($\gamma = -3$);
  - for $d \geq 4$ it is well defined with γ varying from -2 up to -1 and ν from $3/2$ down to 1.

More generally, as a reference family for collision kernels, we may consider the factorised kernels with exponents γ, ν with the following behaviour as $\theta \to 0$:

$$B(v - v_*, \sigma) = |v - v_*|^\gamma \, b(\cos\theta), \quad b(\cos\theta) \sin^{d-2}\theta \sim_0 \beta_0 \, \theta^{-(1+\nu)}$$
$$\text{for some } \beta_0 > 0. \tag{4.13}$$

This defines a (γ, ν) parameter space, where $0 \le \nu \le 2$ (let us agree that $\nu = 0^-$ means cutoff and $\nu = 2$ is the diffusive limit, explored in the next section) and $\gamma \ge -\min(4, d)$. For $\gamma < -4$ or $\gamma < -d$ there are great mathematical difficulties, and not much physical motivation. On the other hand, on physical grounds there is not much motivation for powers $\gamma > 2$, neither for potential more singular than Coulomb, since the latter's slow decay is already quite a challenge. To summarise, the whole range $0 \le \nu \le 2$ is interesting, but on γ there is a tough mathematical restriction to $\gamma \ge -\min(4, d)$ and a natural physical restriction to $-d/(d-2) \le \gamma \le 2$. For inverse power laws this means: $s \ge 5/3$ if $d = 1$, $s \ge 2$ if $d = 3$, $s \ge d - 1$ if $d \ge 4$.

Decades of research have explored the various regions in this diagram, first putting much emphasis on hard potentials with cutoff. See Fig. 4 for a synthetic presentation of that taxonomy.

Remark 4.2 In the presentation and in Fig. 4 I am simplifying a bit by pretending to forget that sometimes it is the behaviour of B as $|z| \to \infty$ which matters, and sometimes the behaviour as $|z| \to 0$; to be very rigorous we could use a pair of exponents to allow for different behaviours in those regimes, but this is already subtle enough.

Bibliographical Notes

The scattering formulas (4.2) are already in Maxwell's founding paper [132] for $d = 3$. Rutherford derived his cross-section formula (4.7) as part of his epoch-making work on scattering of α particles [155], leading to the discovery of nuclei. It is not difficult to generalise his calculation to any dimension.

The first mathematical study of the Boltzmann equation as an evolution problem for the probability density is due to Carleman [37] and focused on hard spheres under an assumption of spatial homogeneity. Grad initiated an ambitious program starting from the linearised theory and insisting on the integrability of Boltzmann's kernel, which is never satisfied for long-range interactions; so that assumption of integrability is often called "Grad's angular cutoff assumption". The classical study of hard potentials with cutoff ($\gamma > 0$, integrable b) started with Arkeryd [7] (after a preliminary work by Povzner [149]) and was followed by Elmroth, Gustafsson, Wennberg, Desvillettes, Mischler, Mouhot and others. Soft potentials without cutoff ($s < 5$ for $d = 3$) were studied by Arkeryd [8], Goudon [93] and I [175], followed by a number of others. A review can be found in my survey [180]. That was more than twenty years ago and back then, hardly anything was known about very soft potentials, which is when $\gamma < -2$ ($2 \le s < 7/3$ for $d = 3$); but this part of

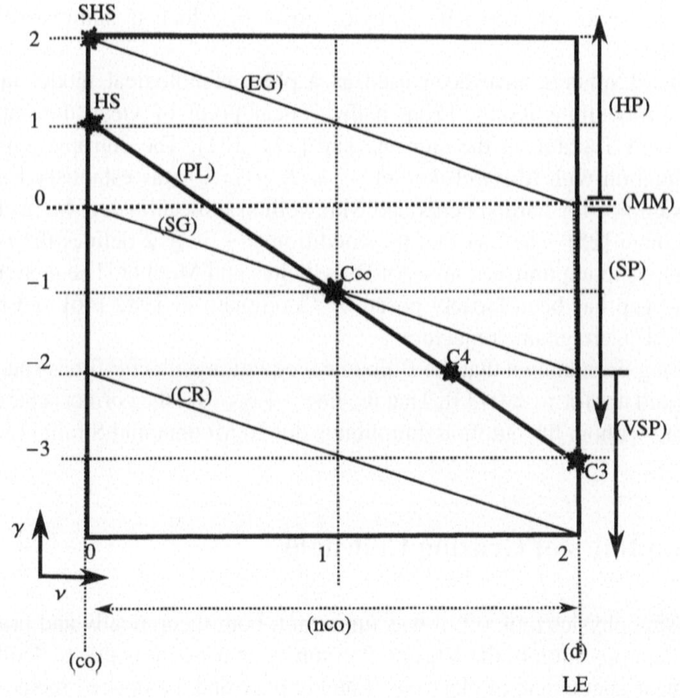

Fig. 4 This diagram presents the landscape of the various regimes for the Boltzmann equation, according to the behaviour of the cross section with respect to relative velocity (γ = power law) and deviation angle (ν = inverse power law singularity). Read as follows. *HS* Hard spheres: $B = \sin^{(d-3)}(\theta/2)|z|$ ($\gamma = 1$, $\nu = 0^-$); *SHS* Super hard spheres: $B = |z|^2$ ($\gamma = 2$, $\nu = 0^-$); *MM* Maxwellian molecules ($\gamma = 0$); *HP* Hard potentials ($\gamma > 0$); *SP* Soft potentials ($\gamma < 0$); *VSP* Very soft potentials ($\gamma < -2$); (*co*) cutoff ($\nu = 0^-$); (*nco*) noncutoff ($0 < \nu < 2$); (*d*) diffusive ($\nu = 2$); LE = Landau equation; (*PL*) (Inverse) Power Laws = the line ($\gamma + 2\nu = 1$). Coulomb interactions are distinguished members of this family; C3 is the Coulomb interaction for $d = 3$ ($\gamma = -3$, $\nu = 2$), C4 for $d = 4$ and so on, accumulating in large dimension at C∞ ($\gamma = -1$, $\nu = 1$); *EG* Entropic gap, i.e. when $D_B(f) \geq K[H(f) - H(M)]$ for all f: above the line ($\gamma + \nu = 2$); *SG* Spectral gap, i.e. when $-\langle Lh, h\rangle_{L^2(M)} \geq K\,\|h-1\|^2_{L^2(M)}$ for all h: above the line ($\gamma + \nu = 0$); *CR* Conditional regularity, i.e. when an a priori estimate in L^p large enough actually implies that f is a priori regular: above the line ($\gamma + \nu = -2$), restricted to $\gamma \geq -d$; in the spatially homogeneous case away from equilibrium, the new estimates based on Fisher information monotonicity allow to prove that above this line (and with some structure conditions covering all natural cases of interest) one can work with smooth solutions, while below this line only H-solutions have been constructed, whose regularity is unknown

the diagram will be the main focus in Sects. 23 and 24, see also the more complete bibliographical notes in these upcoming sections.

Maxwellian molecules ($s = 5$ for $d = 3$), or more generally Maxwellian collision kernels ($\gamma = 0$) were noticed to have special properties, already by Maxwell himself. There is a long history of particular tools in this special case, involving in particular Kac [115], McKean [134] Tanaka [168], Toscani [169] and an enormous series of works by Bobylev, most of which is summarised in his review [23].

Super hard spheres have been used as a phenomenological model in kinetic theory, for modelling issues. From a theoretical point of view, they appear in my work with Toscani on the entropic gap [171, 181]. The entropic gap for the Landau equation with Maxwell kernel ($\nu = 2$, $\gamma = 0$) was established in [179]. Counterexamples for hard spheres and Maxwellian molecules are due to Bobylev and Cercignani [25]. The fact that the condition $\gamma + \nu \geq 2$ defines the region of entropic gap is in unpublished notes of Desvillettes and Mouhot. The mere problem of entropic gap has been loosely posed by Cercignani in 1982 [46] and has long been dubbed "Cercignani conjecture".

It has long been known that the Boltzmann equation with cutoff has spectral gap for $\gamma \geq 0$ and not for $\gamma < 0$ [33]. That the line $\gamma + \nu \geq 0$ is the correct separation for spectral gap without the cutoff assumption is due to Mouhot and Strain [138, 142].

5 Asymptotics of Grazing Collisions

When plasma physics took off, it was important, both theoretically and practically, to model the correction to the Vlasov–Poisson system which is due to "collisions", or rather near encounters of electrons. Landau provided the desired recipe as early as 1936 by working out a suitable converging approximation from the diverging Boltzmann operator. Later, plasma specialists like Balescu and Bogoljubov proposed other, more subtle, collision operators, bypassing Boltzmann's formula and going back to the analysis of Coulomb particle systems; but Landau's approximation is still fine for the vast majority of purposes.

Naturally, we first focus on the Coulomb interaction in dimension $d = 3$: the interaction potential is $\psi(r) = \psi_0/r$; fix $\psi_0 > 0$ such that the Rutherford cross section reads

$$\frac{B_2(|z|, \cos\theta)}{|z|} = \frac{1}{\left(|z| \sin\frac{\theta}{2}\right)^4}.$$

For this kernel B the cross-section for momentum transfer, M, is infinite, making the Boltzmann equation meaningless. This is due to the large tails of the Coulomb potential. A classical remedy is to tame ψ at large distances, replacing ψ by $\psi_\lambda(r) = \psi(r)e^{-r/\lambda}/r$, where λ is Debye's "screening length'. The resulting collision kernel does make sense, the price to pay is that now it is not explicit. A decent approximation is

$$B_\lambda(|z|, \cos\theta) \simeq \frac{|z|}{\left(\left(|z| \sin\frac{\theta}{2}\right)^2 + \frac{1}{\lambda}\right)^2}.$$

Then

$$M_\lambda(|z|) \simeq \frac{8\pi \log \lambda}{|z|^3}\left(1 + O\left(\frac{1}{\lambda}\right)\right).$$

Of course this diverges as $\lambda \to \infty$, but "not much' for practical purpose, since only logarithmically. The approximation prefactor is called the Coulomb logarithm.

In the above, the kernel $B_\lambda/(8\pi \log \lambda)$ is an approximation, so to speak, of $|z|^{-3}\delta''_{\theta=0}$. This can be generalised into a procedure of concentrating the whole influence of B on infinitesimally small deviation angles θ, whose effect dwarves all contributions of $\theta \geq \theta_0$ for any fixed $\theta_0 > 0$. Starting from the nineties, this concentrating procedure was formalised in increasing generality, ending up with the following framework by Alexandre and myself. Handling the borderline singularity $\gamma = -d$ requires a new kernel, which will also turn out to be useful for other purpose.

Definition 5.1 (Compensated Adjoint Boltzmann Kernel) If $B = B(|z|, \cos\theta)$ is a given collision kernel, define $S = S(|z|)$ by

$$S(z) = |\mathbb{S}^{d-2}| \int_0^\pi \left[\frac{1}{\cos^d(\theta/2)}\, B\left(\frac{|z|}{\cos(\theta/2)}, \cos\theta\right) - B(|z|, \cos\theta)\right] \sin^{d-2}\theta\, d\theta. \tag{5.1}$$

The meaning of this kernel is the following: $S * f = Q(f, 1) = \iint (f'_* - f_*)\, B\, dv_*\, d\sigma$. For the next definition, the other meaningful kernel is the cross-section for momentum transfer M defined in (4.12), or equivalently the momentum transfer kernel, $|z|M(|z|)$.

Definition 5.2 (Asymptotics of Grazing Collisions, AGC) A sequence $(B_n)_{n\in\mathbb{N}}$ of collision kernels concentrates on grazing collisions if

(a) The associated compensated adjoint and momentum transfer kernels, $S_n(|z|)$ and $|z|M_n(|z|)$, as measures on $\mathbb{R}^d$, are bounded in total variation on compact sets, uniformly in n;
(b) $\forall \theta_0 > 0$, $S_n^{\theta_0}(|z|)$ and $|z|M_n^{\theta_0}(|z|)$, defined in a similar way with B_n replaced by $B_n\, 1_{\theta\geq\theta_0}$, converge to as $n \to \infty$, locally uniformly in $\mathbb{R}^d \setminus \{0\}$;
(c) There is a measurable $M_\infty : \mathbb{R}_+ \to \mathbb{R}_+$ such that $z\, M_n(|z|) \longrightarrow zM_\infty(|z|)$, locally weakly in the sense of measures.

The short meaning is that $M_n \to M_\infty$ and only $\theta \simeq 0$ count. Recalling the definition of M_n, actually

$$M_n = \int_{\mathbb{S}^{d-1}} (1 - \cos\theta)\, B_n\, d\sigma \simeq \int_{\mathbb{S}^{d-1}} \frac{\theta^2}{2}\, B_n\, d\sigma \xrightarrow[AGC]{} M_\infty.$$

In the sequel of these notes I shall not worry about regularity issues and often omit the index n, using the symbols AGC to stand for this limit procedure.

Proposition 5.3 (Landau Approximation) *Under the asymptotics of grazing collisions,*

(i) $Q(f,f) \xrightarrow[AGC]{} Q_L(f,f)$ *where*

$$Q_L(f,f) = \nabla_v \cdot \left(\int_{\mathbb{R}^d} a(v-v_*) \big[f(v_*) \nabla f(v) - f(v) \nabla f(v_*) \big] \, dv_* \right) \tag{5.2}$$

is Landau's collision operator, and the matrix-valued function $a = a(z)$ *is defined by*

$$a(z) = \Psi(|z|)\, \Pi_{z^\perp}, \qquad \Pi_{z^\perp} = I_d - \frac{z \otimes z}{|z|^2}, \qquad \Psi(|z|) = \frac{|z|^2 M_\infty(|z|)}{4(d-1)}; \tag{5.3}$$

(ii) $D_B(f) \xrightarrow[AGC]{} D_L(f,f)$*, Landau's dissipation functional (H-dissipation, or entropy production, functional), defined by*

$$D_L(f) = \frac{1}{2} \iint_{\mathbb{R}^d \times \mathbb{R}^d} f f_* \Psi(|v-v_*|) \left| \Pi_{k^\perp} \big(\nabla \log f - (\nabla \log f)_* \big) \right|^2 dv\, dv_*, \tag{5.4}$$

with the shorthands $f = f(v)$, $f_* = f(v_*)$, $k = \frac{v-v_*}{|v-v_*|}$.

The proof rests on the useful

Lemma 5.4 (Averages on Subspheres) *Let* A *be a* $d \times d$ *symmetric matrix,* $k \in \mathbb{S}^{d-1}$ *and* $\mathbb{S}^{d-2}_{k^\perp}$ *be the* $(d-2)$ *unit sphere in the plane* $k^\perp$*. Then*

$$\frac{1}{|\mathbb{S}^{d-2}|} \int_{\mathbb{S}^{d-2}_{k^\perp}} \langle A\phi, \phi \rangle \, d\phi = \frac{\mathrm{tr}\,(A \Pi_{k^\perp})}{d-1}. \tag{5.5}$$

As a particular case, for any vector $\xi \in \mathbb{R}^d$,

$$\frac{1}{|\mathbb{S}^{d-2}|} \int_{\mathbb{S}^{d-2}_{k^\perp}} (\xi \cdot \phi)^2 \, d\phi = \frac{|\Pi_{k^\perp} \xi|^2}{d-1}. \tag{5.6}$$

and for any two vectors $\xi, \eta \in R^d$,

$$\frac{1}{|\mathbb{S}^{d-2}|} \int_{\mathbb{S}^{d-2}_{k^\perp}} (\xi \cdot \phi)\,(\eta \cdot \phi) \, d\phi = \frac{\langle \Pi_{k^\perp} \xi, \Pi_{k^\perp} \eta \rangle}{d-1}. \tag{5.7}$$

Proof of Lemma 5.4 $\langle A\phi, \phi \rangle = \langle \Pi_{k^\perp} A \Pi_{k^\perp} \phi, \phi \rangle$ and $\Pi_{k^\perp} A \Pi_{k^\perp}$ is a symmetric endomorphism of $k^\perp$, of the form $\sum \lambda_i e_i \otimes e_i$, where $(e_i)_{1 \le i \le d-1}$ is an orthonormal basis of $k^\perp$. Then $\mathrm{tr}\,(A \Pi_{k^\perp}) = \sum \lambda_i$. So it suffices to check that if e is a unit vector in $k^\perp$, then

$$\frac{1}{|\mathbb{S}^{d-2}|}\int_{\mathbb{S}^{d-2}}(e\cdot\phi)^2\,d\phi = \frac{1}{d-1}. \tag{5.8}$$

This will also imply (5.6) since $\xi\cdot\phi = (\Pi_{k^\perp}\xi)\cdot\phi = |\Pi_{k^\perp}\xi|\,e\cdot\phi$, $e = (\Pi_{k^\perp}\xi)/|\Pi_{k^\perp\xi}|$; and (5.7) will follow also by polarisation.

To prove (5.8), write $\phi = (\cos\chi)e + (\sin\chi)\psi$, where $\psi\in\mathbb{S}^{d-3}_{e^\perp}$ ($e^\perp$ is a subspace of $k^\perp$), so the left-hand side of (5.8) is

$$\frac{\int_0^\pi \cos^2\chi\,\sin^{d-3}\chi\,d\chi}{\int_0^\pi \sin^{d-3}\chi\,d\chi} = \left(\frac{1}{d-2}\right)\frac{\int \sin^{d-1}\chi\,d\chi}{\int\sin^{d-3}\chi\,d\chi}$$
$$= \left(\frac{1}{d-2}\right)\left(\frac{d-2}{d-1}\right) = \frac{1}{d-1},$$

thanks to the classical Wallis integral formula. □

Sketch of Proof of Proposition 5.3 Depending on the needs, (i) can be proven either directly on the operator Q, or from its weak formulation $\int Qh$. I shall go for the latter. Testing Q against a smooth $h(v)$ and using $(v, v_*, \sigma)\leftrightarrow(v', v'_*, k)$,

$$\int Q(f,f)\,h\,dv = \iiint D(f'f'_* - ff_*)\,h = \iiint Bff_*(h'-h)$$
$$= \iint ff_*\left(\int B(h'-h)\,d\sigma\right)dv\,dv_*. \tag{5.9}$$

On the other hand,

$$\int Q_L(f,f)h = -\iint a(v-v_*)(\nabla-\nabla_*)ff_*\cdot\nabla h$$
$$= \iint ff_*(\nabla-\nabla_*)\cdot\big[a(v-v_*)\nabla h\big]\,dv\,dv_*.$$

Next, using

$$(\nabla-\nabla_*)\cdot\Pi_{k^\perp} = -(\nabla-\nabla_*)(k\otimes k) = -2(d-1)\,k,$$

and $\nabla_*\nabla h = 0$, $\Pi_{k^\perp}k = 0$, $(\nabla-\nabla_*)\Psi(|v-v_*|) = 0$, we find

$$(\nabla-\nabla_*)[a(v-v_*)\nabla h] = (\nabla-\nabla_*)\cdot\big[\Psi(|v-v_*|)\,\Pi_{k^\perp}\nabla h\big]$$
$$= \Psi(|v-v_*|)\,(\nabla-\nabla_*)\cdot\Pi_{k^\perp}\nabla h + \Psi(|v-v_*|)\,\Pi_{k^\perp} : \nabla^2 h.$$

So

$$\int Q_L(f,f)h = \iint ff_*\Psi\left(\Pi_{k^\perp} : \nabla^2 h - 2(d-1)\frac{(v-v_*)}{|v-v_*|^2}\cdot\nabla h\right) dv\,dv_*. \tag{5.10}$$

Now for the Boltzmann term. Write

$$v' - v = \frac{v_* - v}{2} + \frac{|v-v_*|}{2}\sigma = \frac{|v-v_*|}{2}(\sigma - k). \tag{5.11}$$

Decompose

$$\sigma = (\cos\theta)k + (\sin\theta)\phi, \qquad \phi \in \mathbb{S}^{d-2}_{k^\perp}, \tag{5.12}$$

$$\sigma - k = (\cos\theta - 1)k + (\sin\theta)\phi = \theta\phi + O(\theta^2). \tag{5.13}$$

So

$$\begin{aligned} h' - h &= \nabla h(v)\cdot(v'-v) + \frac{1}{2}\langle\nabla^2 h(v)\cdot v'-v, v'-v\rangle + o(|v'-v|^2) \\ &= -\frac{|v-v_*|}{2}k\cdot\nabla h\,(1-\cos\theta) + \frac{|v-v_*|}{2}\sin\theta\,(\phi\cdot\nabla h) \\ &\qquad + \frac{|v-v_*|^2}{8}\langle\nabla^2 h\cdot\phi,\phi\rangle\theta^2 + o(|v-v_*|^2\theta^2). \end{aligned}$$

This gives rise to four terms in (5.9), let us consider them separately and check that they match the corresponding integral terms in (5.10):

$$\begin{aligned} &-\int_{\mathbb{S}^{d-1}} \frac{|v-v_*|}{2}k\cdot\nabla h\,(1-\cos\theta)\,B\,d\sigma \\ &= -\frac{|v-v_*|}{2}(k\cdot\nabla h)\int_{\mathbb{S}^{d-1}}(1-\cos\theta)\,B\,d\sigma \\ &\longrightarrow -\frac{|v-v_*|}{2}(k\cdot\nabla h)M_\infty = -\frac{(v-v_*)}{2}\cdot\nabla h\,M_\infty, \end{aligned}$$

$$\frac{|v-v_*|}{2}\int_{\mathbb{S}^{d-1}}\sin\theta\,(\phi\cdot\nabla h)B\,d\sigma = 0 \qquad \text{by symmetry } \phi\to-\phi,$$

$$\begin{aligned} \int_{\mathbb{S}^{d-1}}\frac{|v-v_*|^2}{8}\langle\nabla^2 h\cdot\phi,\phi\rangle\theta^2\,B\,d\sigma &= \frac{|v-v_*|^2}{4(d-1)}\mathrm{tr}\left(\nabla^2 h\,\Pi_{k^\perp}\right)\int_{\mathbb{S}^{d-1}}\frac{\theta^2}{2}\,B\,d\sigma \\ &\longrightarrow \frac{|v-v_*|^2}{4(d-1)}M_\infty(|v-v_*|)\,\mathrm{tr}\,(\nabla^2 h\,\Pi_{k^\perp}), \end{aligned}$$

$$\int o(|v-v_*|^2\theta^2)\, B\, d\sigma \longrightarrow 0.$$

This concludes the proof of (i). As for (ii), first note that

$$\begin{aligned}
D_L(f) &= -\int \nabla\cdot\left(\int \Psi\Pi_{k^\perp}(f_*\nabla f - f(\nabla f)_*)\, dv_*\right)(\log f + 1)\, dv \\
&= \iint ff_*\Psi\Pi_{k^\perp}\left(\frac{\nabla f}{f} - \left(\frac{\nabla f}{f}\right)_*\right)\cdot\frac{\nabla f}{f}\, dv\, dv_* \\
&= \frac{1}{2}\iint ff_*\Psi\Pi_{k^\perp}\left(\frac{\nabla f}{f} - \left(\frac{\nabla f}{f}\right)_*\right)\cdot\left(\frac{\nabla f}{f} - \left(\frac{\nabla f}{f}\right)_*\right)\, dv\, dv_* \\
&= \frac{1}{2}\iint ff_*\Psi\left|\Pi_{k^\perp}\left(\frac{\nabla f}{f} - \left(\frac{\nabla f}{f}\right)_*\right)\right|^2 dv\, dv_*.
\end{aligned}$$

Next the infinitesimal velocity variations

$$v'_* - v_* = v - v' = -\frac{|v_* - v|}{2}\theta\phi + o(|v-v_*|\theta) \tag{5.14}$$

imply, by Taylor expansion of $f(v')$ around $f(v)$ and $f(v'_*)$ around $f(v_*)$,

$$f'f'_* - ff_* = \frac{|v-v_*|}{2}(f_*\nabla f - f(\nabla f)_*)\cdot\theta\phi + o(|v-v_*|\theta).$$

As $b-a\to 0$ we have $(a-b)(\log a - \log b)\simeq (a-b)^2/a$, so

$$\begin{aligned}
&(f'f'_* - ff_*)(\log f'f'_* - \log ff_*) \\
&= \frac{(f'f'_* - ff_*)^2}{ff_*} + o(|v-v_*|^2\theta^2) \\
&= \frac{|v-v_*|^2}{4} ff_*\left|\left[\frac{\nabla f}{f} - \left(\frac{\nabla f}{f}\right)_*\right]\cdot\phi\right|^2\theta^2 + o(|v-v_*|^2\theta^2).
\end{aligned}$$

Thus $D_B(f)$ is

$$\begin{aligned}
&\frac{1}{4}\iiint (f'f'_* - ff_*)(\log f'f'_* - \log ff_*)\, B\, d\sigma\, dv\, dv_* \\
&\simeq \frac{1}{4}\iint \frac{|v-v_*|^2}{4(d-1)} ff_*\left|\Pi_{k^\perp}\left(\frac{\nabla f}{f} - \left(\frac{\nabla f}{f}\right)_*\right)\right|^2 2M_\infty\, dv\, dv_* \\
&= \frac{1}{2}\iint \Psi ff_*\left|\Pi_{k^\perp}\left(\frac{\nabla f}{f} - \left(\frac{\nabla f}{f}\right)_*\right)\right|^2 dv\, dv_* = D_L(f).
\end{aligned}$$

□

The classification of Boltzmann's kernels leads, via the AGC, to a similar taxonomy for Landau's equation: if $\Psi(z) = K|z|^{\gamma+2}$, one talks of hard potentials for $\gamma > 0$, Maxwell kernel for $\gamma = 0$, soft potentials for $\gamma < 0$ and very soft potentials for $\gamma < -2$.

Bibliographical Notes

The founding paper is Landau [116]. Background on plasma physics can be found in Landau–Lipschitz [121] and Delcroix-Bers [55]. Detailed discussions of kinetic models for plasmas are in the books by Bogoljubov [26] and Balescu [17]. The so-called Balescu–Lenard model (which should rather be called Bogoljubov–Balescu) appears in those works; it makes sense even without the Debye cut, but its popularity has remained limited, since its complexity is much higher and its usefulness only appears in very special cases.

Asymptotics of grazing collision was mathematically considered, with increasing generality, by Arsen'ev and Buryak [12], Degond and Lucquin-Desreux [54], Desvillettes [60], Goudon [93], Villani [175], Alexandre and Villani [3]. The more recent of these works prove not only the convergence of the operator, but also the convergence of solutions, with increasing generality, until [3] covers the most general mathematical setting that one may hope for: namely "renormalised" solutions in the sense of DiPerna–Lions, a concept of very weak solutions for inhomogeneous solutions of the Boltzmann equation [73] and Landau equation [125, 173]. In [3] one has also sought the most general conditions on the cross-section, not assuming any factorisation property for instance. Along the way, as there was progress in the understanding of weak solutions for the Boltzmann equation with nonintegrable collision kernels, the AGC served both as a motivation and a consistency check for any new such theory. Background on the physical meaning of the asymptotics is also included in [3].

Many authors have studied the properties of the Landau equation itself. For the purpose of this course, the most relevant works are those handling qualitative properties (smoothness, moments, lower bound, equilibration) in the spatially homogeneous situation: my study of the Maxwellian case ($\gamma = 0$) [176] and my study of hard potentials ($\gamma > 0$) with Desvillettes [67, 68]. Then the special situation $d = 3$, $\gamma = -3$ was considered with special care, but up to recently there were only partial or conditional results [90, 91].

There is another way from the Boltzmann to the Landau equation, distinct from the AGC, which goes via the study of fluctuations in the derivation from particle system; then Ψ is always of Coulomb type (proportional to $1/|z|$ if $d = 3$) and the Boltzmann collision kernel only enters via a multiplicative constant; see Spohn [163] and references therein.

6 The Guillen–Silvestre Theorem

Let us resume the story which I started at the end of Sect. 4. Around 1998 the picture was the following:

- Fisher information was proven decreasing for Maxwellian kernels ($\gamma = 0$) and it seemed (at least to me) the best to hope for;
- the Landau–Coulomb equation ($d = 3, \gamma = -3$) was resisting attempts to study its smoothness or singularity. Actually there was, at first, reason to suspect spontaneous development of singularities. Indeed, for $\Psi(z) = 1/|z|$ we have a special form: writing $\overline{a} = a * f$,

$$\begin{aligned}\frac{\partial f}{\partial t} &= \nabla \cdot (\overline{a}\nabla f - (\nabla \cdot \overline{a}) f) \qquad (6.1)\\ &= \overline{a} : \nabla^2 f - (\nabla^2 : \overline{a}) f \\ &= \overline{a} : \nabla^2 f + 8\pi f^2,\end{aligned}$$

where I used $\nabla^2 : a = -8\pi\delta_0$. If f is smooth, then at best $\overline{a} = a * f$ is positive, smooth, degenerate at infinity, so this equation does not seem more regularising than the model equation

$$\frac{\partial f}{\partial t} = \Delta f + 8\pi f^2, \qquad (6.2)$$

which is known to generally blow up in finite time in L^p for all $p > 3/2$ in dimension $d = 3$, even starting from smooth initial conditions.

So it was tempting to believe that there could be blow-up in such L^p spaces, and still the solution would remain in $L \log L$ (by H-Theorem) or maybe in L^p for some $p < 3/2$; in this way weak solutions would still be well-defined but smooth solutions would break up. This was my initial guess, and that of my PhD advisor as well, as I was starting to work on the Landau equation. But on precisely this topic I remember a heated discussion in Oberwolfach with Sacha Bobylev, who criticized the whole idea of writing the equation in nonconservative form.

It turned out that Bobylev was right (as often): shortly after, numerical simulations arrived—it was the time at which the progress of computing power and numerical analysis opened the path to deterministic schemes for collisional kinetic equations, yielding much more convincing qualitative insight than probabilistic schemes—and they showed clearly that the behaviour of (6.1) and that of (6.2) were completely different. Refining the analysis and eager to test their numerical methods on various qualitative problems, Christophe Buet and Stéphane Cordier observed in those days that (a) solutions of (6.1) do not seem to exhibit any degradation of smoothness, (b) Fisher information seems to be decreasing along solutions of (6.1) even it is far from the Maxwellian case, and (c) numerical schemes accurate enough to capture this monotonicity were also the more stable. This was intriguing and I mentioned this as food for thought in my habilitation dissertation.

And then on this front nothing occurred for 25 years!

But after this quarter-of-Sleeping Beauty period, in 2023 Guillen and Silvestre proved:

- I is nonincreasing along solutions of (6.1) in dimension $d = 3$;
- As a consequence, (6.1) does not exhibit blowup, and in fact as soon as the initial condition $f_0 = f_0(v)$ on $\mathbb{R}^d$ satisfies $I(f_0) < \infty$ one can solve the Cauchy problem with a smooth solution;
- The monotonicity property is true for a whole range of exponents γ, positive or negative (provided that the Cauchy problem makes sense, which is not clear for $\gamma < -3$ when $d = 3$, but that is a different story): they proved it for $|\gamma| \leq \sqrt{19}$, and since then that bound has been slightly improved to $\sqrt{22}$.

Observe that with this theorem at hand the figure of known parameters for the monotonicity of I was now the union of two segments in Fig. 4, one horizontal in the middle, and one vertical on the right, like a $\dashv$ symbol, calling for extension and clarification.

Upon learning of this spectacular result, I first started to dig in the manuscript of Guillen and Silvestre, then I eventually resumed my old papers and worked out on the Boltzmann equation again, trying to improve the known results up to a point where the Guillen–Silvestre theorem would be included. At the same time, symmetrically, Imbert and Silvestre were working to generalise the Guillen–Silvestre approach to the Boltzmann equation. In the Summer of 2024 we joined forces and after a particularly fruitful season we obtained some quite general results covering a large part of the parameter space of collision kernels; this will be the central focus in the rest of these notes. Stated informally, our central result is

Theorem 6.1 *Fisher's information is nonincreasing along solutions of the spatially homogeneous Boltzmann equation, for the vast majority of kernels of interest, including all power law forces between Coulomb and hard spheres, for all dimensions.*

Bibliographical Notes

The description of singularities for the nonlinear heat equation with quadratic nonlinearity is a classical topic, addressed for instance by Zaag [193].

Deterministic numerical models for the Landau equation and the evaluation of the Fisher information were performed by Buet and Cordier [32].

Guillen and Silvestre proved their result in [97]. (Silvestre reminded me that, back in 2009, upon hearing him lecture in IAS Princeton about jump diffusion processes, I encouraged him to start working on the Boltzmann equation. In retrospect my encouragement was an investment with extraordinary payback.) Their method to handle differentiation of the Fisher information is based on a vector field formalism, contrary to my older approach in [174], which is in the spirit of connections. (I prefer the latter but I am obviously biased.) Our joint work with Imbert and Silvestre is in [111].

7 Tensorisation

As it models the effect of binary collisions, the Boltzmann operator takes the form of the marginal of a linear operator in the tensor product $f \otimes f$. Already Maxwell and Boltzmann put this to good use in their discussion of equilibria: The Gaussian arises here because it is the only tensor product with spherical symmetry. Now Guillen and Silvestre observed that this can also be explicitly useful for the computation of the time-evolution of the Fisher information.

Definition 7.1 (Joint Linear Boltzmann Operator) Let $\mathcal{B}$ be the linear operator, acting on distributions $F = F(v, v_*)$ on $\mathbb{R}^{2d}$, by

$$(\mathcal{B}F)(v, v_*) = \int_{\mathbb{S}^{d-1}} \left[F(v', v'_*) - F(v, v_*)\right] B(v - v_*, \sigma)\, d\sigma.$$

Obviously $\mathcal{B}$ still makes sense when applied to a function defined on the collision sphere with diameter $[v, v_*]$. Equally obviously,

$$Q(f, f) = \int_{\mathbb{R}^d} \mathcal{B}(f \otimes f)\, dv_*. \tag{7.1}$$

I shall note $V = (v, v_*)$, $V' = (v', v'_*)$, $z(V) = v - v_*$.

Proposition 7.2 *For any probability distribution f on $\mathbb{R}^d$,*

$$I(f) = \frac{1}{2} I(f \otimes f), \qquad I'(f)\, Q(f, f) = \frac{1}{2} I'(f \otimes f)\, \mathcal{B}(f \otimes f). \tag{7.2}$$

Practical consequence: To show that $dI/dt \le C\, I$ along the Boltzmann equation ($C \in \mathbb{R}$) it is sufficient to prove it along the flow of the linear Boltzmann equation $\partial_t F = \mathcal{B}F$, at least when F has the special form $f \otimes f$.

Remark 7.3 $e^{t\mathcal{B}}$ does not preserve the tensor product structure (it would need $\mathcal{B}(f \otimes f) = Q(f, f) \otimes f + f \otimes Q(f, f)$, a very special property which probably forces either the interaction to be trivial, or f to be Maxwellian). So e^{tQ} and $e^{t\mathcal{B}}$ are not equivalent. This is what makes Property 7.2 interesting.

Proof of Proposition 7.2 Let us write $\boldsymbol{\nabla} = \nabla_V = [\nabla_v, \nabla_{v_*}]$. I use the same letter I for $I(f) = \int f |\nabla \log f|^2$ and $I(F) = \iint F |\boldsymbol{\nabla} \log F|^2$.

The first equality in (7.2) is well-known: If $F = f \otimes f$ then $|\boldsymbol{\nabla} \log F|^2 = |\nabla \log f|^2 + |(\nabla \log f)_*|^2$, so $I(ff_*) = \iint ff_* |\nabla \log f|^2 + \iint ff_* |(\nabla \log f)_*|^2 = 2 \int f |\nabla \log f|^2$.

Next,

$$I'(f)\,Q(f,f) = \int Q(f,f)|\nabla \log f|^2 + 2\int f\nabla \log f \cdot \nabla\left(\frac{Q(f,f)}{f}\right). \tag{7.3}$$

On the other hand,

$$I'(f\otimes f)\,\mathcal{B}(f\otimes f) = \int \mathcal{B}(f\otimes f)\,|\boldsymbol{\nabla} \log f\otimes f|^2\,dV$$
$$+2\int f\otimes f\,\boldsymbol{\nabla}\log(f\otimes f)\cdot \boldsymbol{\nabla}\left(\frac{\mathcal{B}(f\otimes f)}{f\otimes f}\right)dV$$

$$= \iiint (f'f'_* - ff_*)B\left(|\nabla \log f|^2 + |(\nabla \log f)_*|^2\right)d\sigma\,dv\,dv_*$$
$$+2\iint ff_*\nabla \log f\cdot\nabla\left(\frac{1}{ff_*}\int (f'f'_* - ff_*)\,B\,d\sigma\right)dv\,dv_*$$
$$+2\iint ff_*(\nabla\log f)_*\cdot\nabla_*\left(\frac{1}{ff_*}\int (f'f'_* - ff_*)\,B\,d\sigma\right)dv\,dv_*.$$

Obviously $(1/ff_*)\int(f'f'_* - ff_*)\,B\,d\sigma$ is symmetric in (v,v_*) so the last two integrals are identical; and multiplication by $1/f_*$ commutes with ∇, so eventually

$$I'(f\otimes f)\mathcal{B}(f\otimes f)$$
$$= 2\iiint (f'f'_* - ff_*)\,B\,|\nabla\log f|^2\,d\sigma\,dv\,dv_*$$
$$+4\iint ff_*\nabla\log f\cdot\nabla\left(\frac{1}{ff_*}\int(f'f'_* - ff_*)\,B\,d\sigma\right)dv\,dv_*$$
$$= 2\int\left(\iint (f'f'_* - ff_*)\,B\,d\sigma\,dv_*\right)|\nabla\log f|^2\,dv$$
$$+4\iint f\nabla\log f\cdot\nabla\left(\frac{1}{f}\int(f'f'_* - ff_*)B\,d\sigma\right)dv\,dv_*.$$

In the end, this is

$$2\int\left(\iint (f'f'_* - ff_*)\,B\,d\sigma\,dv_*\right)|\nabla\log f|^2\,dv$$
$$+4\int f\nabla\log f\cdot\nabla\left(\frac{1}{f}\iint(f'f'_* - ff_*)\,B\,d\sigma\,dv_*\right)dv,$$

whence the result upon comparison with (7.3). □

Remark 7.4 The proof above would fail for any generic functional of the form $\int A(F, \nabla F)\, dV$. But it works for the particular choice $A(F, \nabla F) = |\nabla F|^2/F$. The same can be said of Boltzmann's entropy: only for $A(F) = F \log F$ can one show an interesting relation between $\int A(F)$ and $\int A(f)$. Actually, it is easy to establish the entropic counterpart of Proposition 7.2:

$$H(f \otimes f) = 2H(f), \qquad H'(f \otimes f)\, \mathcal{B}(f \otimes f) = 2H'(f)\, Q(f, f).$$

Remark 7.5 Let us recover the classical H-Theorem with the help of those identities:

$$\begin{aligned} -H'(F)\,(\mathcal{B}F) &= -\int (\log F + 1)\, \mathcal{B}F = -\iint (F' - F)\, B\, d\sigma\, (\log F + 1)\, dV \\ &= -\iint F(\log F' - \log F)\, B\, d\sigma\, dV = \iint F \log \frac{F}{F'}\, B\, d\sigma\, dV. \end{aligned}$$

Now $C : (a, b) \longmapsto a \log(a/b)$ is jointly convex on $\mathbb{R}^2_+$ and 1-homogeneous, so we conclude that $-H'(F) \cdot \mathcal{B}F \geq 0$ by invoking Jensen's inequality, in the form

$$\int (F - G)\, d\nu = 0 \Longrightarrow \int C(F, G)\, d\nu \geq 0,$$

with $F = F(V)$, $G = G(V')$, $d\nu = B\, d\sigma\, dV$. (The homogeneity makes it unnecessary to impose $\int d\nu = 1$, and an approximation argument also covers the case when $\int d\nu = \infty$.) Now, estimating $I'(F)\,(\mathcal{B}F)$ will turn out to be much more complicated that $H'(F)\,(\mathcal{B}F)$!

Bibliographical Notes

The use of the tensor product structure of the Boltzmann collision operator appears already in Maxwell [132], Boltzmann [29], or in the treatment of Boltzmann's dissipation functional by Toscani and I [171, 181], but Guillen and Silvestre seem to be the first to have explicitly shown its full power [97].

McKean [134] makes some remarks about the properties of those particular nonlinearities appearing in H and I; actually in information theory one often starts *a priori* from desired identities under tensor product, to justify the introduction of H. [53].

8 Playing Γ Calculus with Boltzmann and Fisher

A cornerstone of the study of diffusion processes, everywhere present in the theory of log Sobolev inequalities, is the structure of the infinitesimal commutator between the diffusion under scrutiny and a local nonlinearity, typically the integrand of an integral functional of interest. It captures the "genuinely dissipative part' in some sense. As the most classical example, starting from the quadratic functional $\int f^2$, given a linear operator L, define

$$\Gamma(f,f) = \Gamma_1(f,f) = \frac{1}{2}\big(Lf^2 - 2fLf\big). \tag{8.1}$$

Then

$$-\frac{d}{2\,dt}\bigg|_{t=0} \int f^2 = \int \Gamma(f,f),$$

but Γ is a local quantity, which may be technically useful, and computationally more elementary, than its integral version. Notice, if $Lf = \Delta f + \xi \cdot \nabla f$, for any vector field ξ, then $\Gamma(f,f) = |\nabla f|^2$. In certain circles, Γ is called the *carré du champ*. Then Γ can be extended into a bilinear operator $\Gamma(f,g)$, and the procedure can be iterated, replacing fg by $\Gamma(f,g)$:

$$\Gamma_2(f,f) = \frac{1}{2}\big(L\Gamma(f,f) - 2\Gamma(f,Lf)\big), \tag{8.2}$$

whence

$$-\frac{d}{2\,dt}\bigg|_{t=0} \int \Gamma(f,f) = \int \Gamma_2(f,f).$$

Remark 8.1 As another warmup, let us rewrite Remark 7.5 in that way. Introduce the functional $\Gamma_{H,\mathcal{B}}$ by

$$\Gamma_{H,\mathcal{B}}(F) = \mathcal{B}(F\log F) - (\log F + 1)\,\mathcal{B}F$$

and then compute

$$\begin{aligned}
\Gamma_{H,\mathcal{B}}(F) &= \mathcal{B}(F\log F) - (\log F + 1)\,\mathcal{B}F \\
&= \int \big(F'\log F' - F\log F\big)\,B\,d\sigma - (\log F + 1)\int (F' - F)\,B\,d\sigma \\
&= \int \left(F'\log\frac{F'}{F} - F' + F\right) B\,d\sigma \\
&= F\int \left(\frac{F'}{F}\log\frac{F'}{F} - \frac{F'}{F} + 1\right) B\,d\sigma \geq 0,
\end{aligned}$$

since $x\log x - x + 1 \geq 0$ for all x. This implies the H Theorem. Note that this computation is in a way simpler than the usual one, and even Remark 7.5, since (a) integration in V was not used, (b) the pre-post-collisional change of variables is not explicitly used, it enters only via $\int \mathcal{B}G = 0$ after the computation is performed, (c) neither is Jensen's inequality. Of course all three ingredients are implicit and useful for the H-Theorem itself, but in the Γ computation the algebraic structure for the monotonicity of H is reduced to its core.

In this section I shall consider as nonlinearity the integrand of Fisher's information, in the form $\mathcal{F}(F, \mathbf{\nabla} F) = F|\mathbf{\nabla} \log F|^2$, and take $\mathcal{B}$ as linear operator. As before, $\mathbf{\nabla} = [\nabla, \nabla_*]$.

Definition 8.2 (Gamma Calculus for Fisher and Boltzmann) If B is a Boltzmann kernel and $\mathcal{B}$ the associated joint linear operator as in Definition 7.1, let $\Gamma_{I,\mathcal{B}}$ be defined by

$$\Gamma_{I,\mathcal{B}}(F) = \mathcal{B}\big(F|\mathbf{\nabla} \log F|^2\big) - \left[(\mathcal{B}F)|\mathbf{\nabla} \log F|^2 + 2\mathbf{\nabla} F \cdot \mathbf{\nabla}\left(\frac{\mathcal{B}F}{F}\right)\right]. \tag{8.3}$$

(No factor $1/2$ here since $\mathcal{F}$ is 1-homogeneous in F.) By construction, using $\int \mathcal{B}G = 0$,

$$\int \Gamma_{I,\mathcal{B}}(F)\, dV = -I'(F)(\mathcal{B}F). \tag{8.4}$$

So in view of Proposition 7.2, any positivity estimate on $\Gamma_{I,\mathcal{B}}$ will imply a decay for I along the Boltzmann equation; and will even more precise, as it will allow for multiplication by a function of V, for whatever purpose.

We go on with the computation of $\Gamma_{I,\mathcal{B}}(F)$. Obviously it is needed to compute $\mathbf{\nabla}\mathcal{B}F$. I start with three computational lemmas, using the same formalism as in older work of mine on Fisher information and Boltzmann equation.

Lemma 8.3 $\mathbf{\nabla}(F') = \mathbb{A}_{\sigma k}(\mathbf{\nabla} F)'$ *where, for any* k, σ *in* $\mathbb{S}^{d-1}$,

$$\mathbb{A}_{\sigma k} = \frac{1}{2}\begin{bmatrix} I + k \otimes \sigma & I - k \otimes \sigma \\ I - k \otimes \sigma & I + k \otimes \sigma \end{bmatrix},$$

and by convention $k \otimes \sigma(x) = (x \cdot \sigma)k$.

For the next one, recall that $B = B(|z|, \cos\theta)$.

Lemma 8.4 $\mathbf{\nabla} B = [\nabla B, -\nabla B]$, *where*

$$\nabla B = \left(\frac{\partial B}{\partial |z|}\right) k + \frac{1}{|z|}\left(\frac{\partial B}{\partial(\cos\theta)}\right)\Pi_{k^\perp}\sigma.$$

Lemma 8.5 *For any differentiable functions* $g : \mathbb{R}^d \to \mathbb{R}$, $b : (-1, 1) \to \mathbb{R}$, *any* $k \in \mathbb{S}^{d-1}$,

$$\int_{\mathbb{S}^{d-1}} b'(k \cdot \sigma) g(\sigma) \big(\Pi_{k^\perp}\sigma\big)\, d\sigma = \int b(k \cdot \sigma) M_{\sigma k} \nabla g(\sigma)\, d\sigma,$$

where $M_{\sigma k}(x) = (k \cdot \sigma)x - (x \cdot k)\sigma$.

(I will write indifferently $M_{\sigma k}(x)$ or $M_{\sigma k}x$.) The proofs of these lemmas will be given after the main results of this section, namely the computation of $\nabla \mathcal{B}F$ and that of $\Gamma_{I,\mathcal{B}}(F)$.

Proposition 8.6

$$\nabla \mathcal{B}F = \int_{\mathbb{S}^{d-1}} \left[\mathbb{G}_{\sigma k}(\nabla F)' - \nabla F\right] B\, d\sigma + \left(\int_{\mathbb{S}^{d-1}} (F' - F)\,\frac{\partial B}{\partial |z|}\, d\sigma\right)[k, -k],$$

where

$$\mathbb{G}_{\sigma k} = \frac{1}{2}\begin{bmatrix} I + P_{\sigma k} & I - P_{\sigma k} \\ I - P_{\sigma k} & I + P_{\sigma k} \end{bmatrix} \tag{8.5}$$

and $P_{\sigma k} = k \cdot \sigma + k \otimes \sigma - \sigma \otimes k$, *or explicitly*

$$P_{\sigma k}(x) = (k \cdot \sigma)x + (\sigma \cdot x)k - (x \cdot k)\sigma. \tag{8.6}$$

Theorem 8.7

$$\Gamma_{I,\mathcal{B}}(F) = \int_{S^{d-1}} F'\Big\langle \mathbb{C}_{\sigma k}\big[\nabla \log F, (\nabla \log F)'\big], \big[\nabla \log F, (\nabla \log F)'\big]\Big\rangle B\, d\sigma$$
$$- 2\left(\int (F' - F)\,\frac{\partial B}{\partial |z|}\, d\sigma\right) k \cdot z(\nabla \log F), \tag{8.7}$$

where $z(v, v_*) = v - v_*$ *and for any* k, σ *in* $\mathbb{S}^{d-1}$, $\mathbb{C}_{\sigma k}$ *is the* $(4d) \times (4d)$ *linear operator defined by*

$$\mathbb{C}_{\sigma k} = \begin{bmatrix} I & \mathbb{G}_{\sigma k} \\ \mathbb{G}_{k\sigma} & I \end{bmatrix} = \begin{bmatrix} \begin{pmatrix} I & 0 \\ 0 & I \end{pmatrix} & \frac{1}{2}\begin{pmatrix} I + P_{\sigma k} & I - P_{\sigma k} \\ I - P_{\sigma k} & I + P_{\sigma k} \end{pmatrix} \\ \frac{1}{2}\begin{pmatrix} I + P_{k\sigma} & I - P_{k\sigma} \\ I - P_{k\sigma} & I + P_{k\sigma} \end{pmatrix} & \begin{pmatrix} I & 0 \\ 0 & I \end{pmatrix} \end{bmatrix}.$$

In particular, with the notation

$$\xi = -\nabla \log f, \quad z = v - v_*, \quad k = \frac{v - v_*}{|v - v_*|},$$

$$|y - x|^2_{k,\sigma} = |x|^2 + |y|^2 - 2(P_{k\sigma}x) \cdot y,$$

we have the following formula when $F = f \otimes f$:

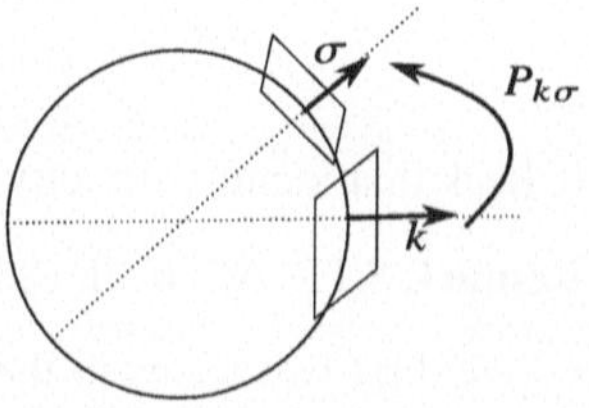

Fig. 5 The operator $P_{k\sigma}$ sends k onto σ, and the tangent plane $k^\perp$ onto the tangent plane $\sigma^\perp$, thereby allowing to compare vectors tangent to the sphere at k and σ respectively

$$\begin{aligned}\Gamma_{I,\mathcal{B}}(f\otimes f) &= \frac{1}{2}\int f'f'_*\left|(\xi'+\xi'_*)-(\xi+\xi_*)\right|^2 B\,d\sigma \\ &+\frac{1}{2}\int f'f'_*\left|(\xi'-\xi'_*)-(\xi-\xi_*)\right|^2_{k,\sigma} B\,d\sigma \\ &+2\int (f'f'_*-ff_*)\,k\cdot(\xi-\xi_*)\,\frac{\partial B}{\partial|z|}\,d\sigma.\end{aligned} \tag{8.8}$$

Corollary 8.8 *$\Gamma_{I,\mathcal{B}}(f\otimes f)\geq 0$ if B does not depend on $|z|$ (Maxwellian kernel), in particular I is nonincreasing along the spatially homogeneous Boltzmann equation with such a kernel.*

Some words of explanation about the notation $|y-x|^2_{k,\sigma}$. One way to think of $P_{k\sigma}$ is a recipe to map $k^\perp = T_k\mathbb{S}^{d-1}$ onto $\sigma^\perp = T_\sigma\mathbb{S}^{d-1}$ (a connection, in pedantic words, although much simpler than the Riemannian one; see Fig. 5). This allows to compare vectors which are tangent to the sphere (e.g. gradients) defined at k and σ respectively. Note carefully that when I write $|y-x|^2_{k,\sigma}$, there is no such thing as $y-x$, it is the whole expression $|y-x|^2_{k,\sigma}$ which makes sense.

Example 8.9 When $d=2$, then under the identification $\mathbb{S}^1 = 2\pi\mathbb{R}/\mathbb{Z}$, $|y-x|^2_{k,\sigma}$ is just the usual squared norm $|y-x|^2$ in $\mathbb{R}$.

Example 8.10 When $d=3$, we may use spherical coordinates (α,ϕ) (with the notation α rather than θ to avoid confusion with the deviation angle); if k has coordinates (α,ϕ) and σ has coordinates (α',ϕ'), then

$$\begin{aligned}|\nabla f(\sigma)-\nabla f(k)|^2_{k,\sigma} &= (\partial_\alpha f)^2+(\partial_\phi f)^2+(\partial_\alpha f)'^2+(\partial_\phi f)'^2 \\ -2\cos(\phi-\phi')(\partial_\alpha f)(\partial_\alpha f)' &-2\big(\sin\alpha\sin\alpha'+\cos\alpha\cos\alpha'\cos(\phi-\phi')\big)(\partial_\phi f)(\partial_\phi f)' \\ &+2\sin(\phi'-\phi)\big(\cos\alpha'(\partial_\alpha f)(\partial_\phi f)'-\cos\alpha(\partial_\alpha f)'(\partial_\phi f)\big).\end{aligned}$$

So already in dimension 3 this explicit formula becomes cumbersome and it will be more convenient, in practice, to use the abstract, intrinsic expression.

In the end of this section, I shall provide a list of useful properties for the operators $P_{\sigma k}$ (Proposition 8.12). For the moment let me note that

$$\left|(v' - v'_*) - (v - v_*)\right|^2_{k,\sigma} = \left(|v' - v'_*| - |v - v_*|\right)^2. \tag{8.9}$$

(Check that identity directly or apply Proposition 8.12.)

Remark 8.11 All in all, (8.8) has a rather neat formal structure:

- The first two terms in the right-hand side of (8.8) are related to the two conservation laws of elastic collisions: conservation of momentum and conservation of energy. Indeed, they both take the form $\int f' f'_* \, C_i(\xi, \xi_*, \xi', \xi'_*)\, d\sigma$, $i = 1, 2$, where $C_1 = |(\xi' + \xi'_*) - (\xi + \xi_*)|^2$ and $C_2 = |(\xi' - \xi'_*) - (\xi - \xi_*)|^2_{k,\sigma}$ (the latter expression depends also on k, σ) and conservation of momentum reads $C_1(v, v_*, v', v'_*) = 0$ while in view of (8.9) conservation of energy reads $C_2(v, v_*, v', v'_*) = 0$. In particular, if f were Maxwellian, then $\xi = av + b$ for some $a > 0$, $b \in \mathbb{R}^d$, and the integrands in those integral expressions would both vanish.
- The last term in the right-hand side of (8.8) combines the non-Maxwellianity of f (that is, by how much $f' f'_* - f f_*$ is nonzero) and the non-Maxwellianity of B (that is, by how much B depends on $|z|$).

Now let us prove the basic computational lemmas.

Proof of Lemma 8.3 By definition,

$$F' = F\left(\frac{v + v_*}{2} + \frac{|v - v_*|}{2}\sigma, \frac{v + v_*}{2} - \frac{|v - v_*|}{2}\sigma\right),$$

so with $\nabla = \nabla_v$, $\nabla_* = \nabla_{v_*}$ we have

$$\nabla(F') = \frac{1}{2}\big(\nabla F + k(\sigma \cdot \nabla F)\big)(V') + \frac{1}{2}\big(\nabla_* F - k(\sigma \cdot \nabla_* F)\big)(V'),$$

and symmetrically

$$\nabla_*(F') = \frac{1}{2}\big(\nabla F - k(\sigma \cdot \nabla F)\big)(V') + \frac{1}{2}\big(\nabla_* F + k(\sigma \cdot \nabla_* F)\big)(V').$$

□

Proof of Lemma 8.4 By assumption,

$$B = B\left(|v - v_*|, \frac{v - v_*}{|v - v_*|} \cdot \sigma\right),$$

so writing this as a function of two variables, $B(|z|, \cos\theta)$, and using chain rule,

$$\nabla B = \frac{v - v_*}{|v - v_*|}\frac{\partial B}{\partial |z|} + \frac{1}{|v - v_*|}\left(\sigma - \left(\frac{v - v_*}{|v - v_*|} \cdot \sigma\right)\frac{v - v_*}{|v - v_*|}\right)\frac{\partial B}{\partial(\cos\theta)},$$

and $\nabla_* B = -\nabla B$. □

Proof of Lemma 8.5 First, let $u : \mathbb{R}^d \to \mathbb{R}$ and $\varphi = \varphi(r)$ concentrating on $r = 1$ in such a way that $\int \varphi(|x|)\, dx = 1$. By polar change of coordinates $x = r\sigma$,

$$\begin{aligned}\int \nabla u(x)\, \varphi(|x|)\, dx &= \iint \nabla u(r\sigma)\, \varphi(r)\, r^{d-1}\, dr\, d\sigma \\ &\simeq \left(\int \nabla u(\sigma)\, d\sigma\right)\left(\int \varphi(r)\, r^{d-1}\, dr\right).\end{aligned}$$

On the other hand, $\nabla(u\varphi) = \nabla u(x)\, \varphi(r) + u\sigma\varphi'(r)$, so

$$\begin{aligned}\int \nabla u(x)\, \varphi(|x|)\, dx &= -\iint u(r\sigma)\sigma\varphi'(r)\, r^{d-1}\, dr\, d\sigma \\ &\simeq -\left(\int u(\sigma)\, \sigma\, d\sigma\right)\left(\int \varphi'(r)\, r^{d-1}\, dr\right) \\ &= \left(\int u(\sigma)\, \sigma\, d\sigma\right)(d-1)\left(\int \varphi(r)\, r^{d-2}\, dr\right) \\ &\simeq \left(\int u(\sigma)\, \sigma\, d\sigma\right)(d-1)\left(\int \varphi(r)\, r^{d-1}\, dr\right),\end{aligned}$$

where I used again integration by parts and the concentration of φ on $(r = 1)$. The conclusion of this first step is

$$\int_{\mathbb{S}^{d-1}} \nabla u(\sigma)\, d\sigma = (d-1) \int_{\mathbb{S}^{d-1}} u(\sigma)\, \sigma\, d\sigma. \tag{8.10}$$

This is true for any $u : \mathbb{R}^d \to \mathbb{R}$, so also for any vector-valued function $\xi : \mathbb{R}^d \to \mathbb{R}^d$, so

$$\int_{\mathbb{S}^{d-1}} \nabla \xi(\sigma)\, d\sigma = (d-1) \int_{\mathbb{S}^{d-1}} \xi(\sigma) \otimes \sigma\, d\sigma. \tag{8.11}$$

As corollaries, here are some formulas of integration by parts on the sphere: For any functions u, v and vector field ξ defined in a neigborhood of $\mathbb{S}^{d-1}$, for any $k \in \mathbb{S}^{d-1}$,

$$\begin{aligned}\int_{\mathbb{S}^{d-1}} u(\sigma)\, \nabla v(\sigma)\, d\sigma = -\int_{\mathbb{S}^{d-1}} \nabla u(\sigma)\, v(\sigma)\, d\sigma + (d-1) \\ \times \int_{\mathbb{S}^{d-1}} u(\sigma)\, v(\sigma)\, \sigma\, d\sigma;\end{aligned} \tag{8.12}$$

$$\begin{aligned}\int_{\mathbb{S}^{d-1}} u(\sigma)\, [k \cdot \nabla v(\sigma)]\, d\sigma = -\int_{\mathbb{S}^{d-1}} k \cdot \nabla u(\sigma)\, v(\sigma)\, d\sigma + (d-1) \\ \times \int_{\mathbb{S}^{d-1}} u(\sigma)\, v(\sigma)\, (k \cdot \sigma)\, d\sigma\end{aligned} \tag{8.13}$$

$$\int_{\mathbb{S}^{d-1}} \xi(\sigma)\,[k\cdot\nabla v(\sigma)]\,d\sigma = -\int_{\mathbb{S}^{d-1}} [k\cdot\nabla\xi(\sigma)]\,v(\sigma)\,d\sigma + (d-1)\int_{\mathbb{S}^{d-1}} \xi(\sigma)\,v(\sigma)\,(k\cdot\sigma)\,d\sigma. \tag{8.14}$$

(Recall $(k\cdot\nabla u)_i = (k\cdot\nabla)u_i$ by convention.) The next preliminary is the elementary formula

$$\Pi_{k^\perp}\sigma = (\Pi_{\sigma^\perp}k\cdot k)\sigma - (k\cdot\sigma)\Pi_{\sigma^\perp}k, \tag{8.15}$$

indeed the right-hand side is $[k-(k\cdot\sigma)\sigma]\cdot k\sigma - (k\cdot\sigma)(k-(k\cdot\sigma)\sigma) = \sigma - (k\cdot\sigma)^2\sigma - (k\cdot\sigma)k + (k\cdot\sigma)^2\sigma = \sigma - (\sigma\cdot k)k$.

It follows from (8.15) that

$$\int_{\mathbb{S}^{d-1}} b'(k\cdot\sigma)\Pi_{k^\perp}\sigma\, g(\sigma)\,d\sigma = \int b'(k\cdot\sigma)\big(\Pi_{\sigma^\perp}k\cdot k\big)\,\sigma\, g(\sigma)\,d\sigma - \int (k\cdot\sigma)\,b'(k\cdot\sigma)\,\Pi_{\sigma^\perp}k\, g(\sigma)\,d\sigma. \tag{8.16}$$

To transform the first integral on the right hand side, choose $\xi(x) = \frac{x}{|x|}g(x)$, $v(x) = b(k\cdot\frac{x}{|x|})$, apply (8.14) and $(k\cdot\nabla g)\sigma = (\sigma\otimes\nabla g(\sigma))k$ to obtain

$$-\int \big(\Pi_{\sigma^\perp}k\, g(\sigma) + (\sigma\otimes\nabla g(\sigma)\big)\,k\,b(k\cdot\sigma)\,d\sigma + (d-1) \times \int g(\sigma)b(k\cdot\sigma)\,(k\cdot\sigma)\,\sigma\,d\sigma. \tag{8.17}$$

The second integral on the right hand side of (8.16) is handled by setting $b_1(s) = sb(s)$, so $sb'(s) = b_1'(s) - b(s)$, and letting $v(x) = b_1(k\cdot\frac{x}{|x|})$, so $\nabla v(x) = b_1'(k\cdot\frac{x}{|x|})\Pi_{\sigma^\perp}k$; then letting $u(x) = g(x)$ and applying (8.13):

$$\begin{aligned}
\int (k\cdot\sigma)b'(k\cdot\sigma)\,\Pi_{\sigma^\perp}k\, g(\sigma)\,d\sigma &= \int \big[b_1'(k\cdot\sigma) - b(k\cdot\sigma)\big]\Pi_{\sigma^\perp}k\, g(\sigma)\,d\sigma \\
= -\int \nabla g(\sigma)\,b_1(k\cdot\sigma)\,d\sigma &+ (d-1)\int b_1(k\cdot\sigma)\,g(\sigma)\,\sigma\,d\sigma - \int b(k\cdot\sigma)\Pi_{\sigma^\perp}k\, g(\sigma)\,d\sigma \\
= -\int (k\cdot\sigma)b(k\cdot\sigma)\,\nabla g(\sigma)\,d\sigma &+ (d-1)\int b(k\cdot\sigma)(k\cdot\sigma)\,g(\sigma)\,\sigma\,d\sigma \\
&- \int b(k\cdot\sigma)\Pi_{\sigma^\perp}k\, g(\sigma)\,d\sigma.
\end{aligned}$$

Subtracting this from (8.17), one sees that (8.16) is equal to

$$\int_{\mathbb{S}^{d-1}} \big[(k\cdot\sigma)\nabla g(\sigma) - (k\cdot\nabla g(\sigma))\sigma\big]\, b(k\cdot\sigma)\, d\sigma,$$

as announced in Lemma 8.5. □

Proof of Proposition 8.6 By Lemma 8.3,

$$\begin{aligned}
\boldsymbol{\nabla}(\mathcal{B}F) &= \boldsymbol{\nabla}\int_{\mathbb{S}^{d-1}} (F'-F)\, B\, d\sigma \\
&= \int \big[\boldsymbol{\nabla}(F') - \boldsymbol{\nabla}F\big]\, B\, d\sigma + \int (F'-F)\, \boldsymbol{\nabla}B\, d\sigma \\
&= \int \big[\mathbb{A}_{\sigma k}(\boldsymbol{\nabla}F)' - \boldsymbol{\nabla}F\big]\, B\, d\sigma + \int (F'-F)\, [\nabla B, -\nabla B]\, d\sigma.
\end{aligned}$$

By Lemmas 8.4 and 8.5,

$$\begin{aligned}
\int (F'-F)\,\nabla B\, d\sigma &= \int (F'-F)\left(\frac{\partial B}{\partial |z|}\, k + \frac{1}{|z|}\frac{\partial B}{\partial(\cos\theta)}\, \Pi_{k^\perp}\sigma\right) d\sigma \\
&= \left(\int (F'-F)\frac{\partial B}{\partial |z|}\, d\sigma\right) k + \frac{1}{|z|}\int M_{\sigma k}\nabla_\sigma (F'-F)\, B\, d\sigma.
\end{aligned}$$

The second term involves $\nabla_\sigma(F') = (|v-v_*|/2)(\nabla F - \nabla_* F)'$ (here σ lives in $\mathbb{R}^d$), so

$$\begin{aligned}
\int (F'-F)\,\nabla B\, d\sigma &= \left(\int (F'-F)\frac{\partial B}{\partial |z|}\, d\sigma\right) k \\
&\quad + \frac{1}{2}\int \big[M_{\sigma k}(\nabla F)' - M_{\sigma k}(\nabla_* F)'\big]\, B\, d\sigma.
\end{aligned}$$

All in all,

$$\begin{aligned}
\nabla(\mathcal{B}F) = \int \Big[&\frac{1}{2}(\nabla F)' + \frac{1}{2}\sigma\cdot(\nabla F)' k + \frac{1}{2}(\nabla_* F)' - \frac{1}{2}\sigma\cdot(\nabla_* F)' k \\
&+ \frac{1}{2}(k\cdot\sigma)(\nabla F)' - \frac{1}{2}k\cdot(\nabla F)'\sigma \\
-\frac{1}{2}(k\cdot\sigma)(\nabla_* F)' &+ \frac{1}{2}k\cdot(\nabla_* F)\,\sigma - \nabla F\Big]\, B\, d\sigma + \left(\int (F'-F)\frac{\partial B}{\partial |z|}\, d\sigma\right) k,
\end{aligned}$$

and by symmetry there is a similar formula for $\nabla_*(\mathcal{B}F)$, whence the result. □

Proof of Theorem 8.7 Let us compute one after the other the various terms appearing in $\Gamma_{I,\mathcal{B}}$:

$$\mathcal{B}(F|\nabla \log F|^2) = \int \Big[(F|\nabla \log F|^2)' - (F|\nabla \log F|^2) \Big] \, B \, d\sigma;$$
$$(\mathcal{B}F)|\nabla \log F|^2 = \int \Big(F'|\nabla \log F|^2 - F|\nabla \log F|^2 \Big) \, B \, d\sigma;$$

$$2F\nabla \log F \cdot \nabla \left(\frac{\mathcal{B}F}{F} \right) = 2\nabla \log F \cdot \nabla \mathcal{B}F - 2F\nabla \log F \cdot \left(\mathcal{B}F \, \frac{\nabla F}{F^2} \right)$$
$$= 2 \int \nabla \log F \cdot \big[\mathbb{G}_{\sigma k}(\nabla F)' - \nabla F \big] \, B \, d\sigma - 2 \int \nabla \log F \cdot (F' - F) \frac{\nabla F}{F} \, B \, d\sigma$$
$$+ 2 \left(\int (F' - F) \frac{\partial B}{\partial |z|} \, d\sigma \right) [k, -k] \cdot \nabla \log F$$
$$= 2 \int F' \nabla \log F \cdot \mathbb{G}_{\sigma k}(\nabla \log F)' B \, d\sigma - 2 \int F' |\nabla \log F|^2 B \, d\sigma$$
$$+ 2 \left(\int (F' - F) \, \frac{\partial B}{\partial |z|} \, d\sigma \right) [k, -k] \cdot \nabla \log F.$$

All in all,

$$\Gamma_{I,\mathcal{B}}(F) = \int F' \Big(|\nabla \log F|^2 + |(\nabla \log F)'|^2 - 2(\nabla \log F) \cdot \mathbb{G}_{\sigma k}(\nabla \log F)' \Big) \, B \, d\sigma$$
$$- 2 \left(\int (F' - F) \, \frac{\partial B}{\partial |z|} \, d\sigma \right) [k, -k] \cdot \nabla \log F,$$

which is the same as (8.7) upon noting that $(P_{\sigma k})^* = P_{k\sigma}$ (Proposition 8.12(iii) below).

Now let us turn to (8.8). Plug $F = f \otimes f = f f_*$ into (8.7). Then

$$\big[\nabla \log F, (\nabla \log F) \big] = -\big[\xi, \xi_*, \xi', \xi'_* \big].$$

Then the integrand in the integral expression of Γ will be $f' f'_* B$ multiplied by

$$|\xi|^2 + |\xi_*|^2 + |\xi'|^2 + |\xi'_*|^2 - \big\langle (I + P_{k\sigma})\xi, \xi' \big\rangle - \big\langle (I - P_{k\sigma})\xi, \xi'_* \big\rangle$$
$$- \big\langle (I - P_{k\sigma})\xi_*, \xi' \big\rangle - \big\langle (I + P_{k\sigma})\xi_*, \xi'_* \big\rangle$$
$$= |\xi|^2 + |\xi_*|^2 + |\xi'|^2 + |\xi'_*|^2 - \big(\xi' \cdot \xi + \xi \cdot \xi'_* + \xi_* \cdot \xi' + \xi_* \cdot \xi'_* \big) - P_{k\sigma}\xi \cdot \xi' + P_{k\sigma}\xi \cdot \xi'_*$$
$$+ P_{k\sigma}\xi_* \cdot \xi' - P_{k\sigma}\xi_* \cdot \xi'_*$$
$$= \frac{1}{2} \Big| (\xi' + \xi'_*) - (\xi + \xi_*) \Big|^2 + \frac{1}{2} \Big(|\xi' - \xi'_*|^2 + |\xi - \xi_*|^2 - 2P_{k\sigma}(\xi - \xi_*) \cdot (\xi' - \xi'_*) \Big),$$

which yields the desired expression. □

I end up this section with the promised list of properties for $P_{\sigma k}$. (Actually for future use it will be more convenient to swap σ and k.)

Proposition 8.12 *For any k, σ in $\mathbb{S}^{d-1}$, let $P_{k\sigma}(x) = (k \cdot \sigma)x + (x \cdot k)\sigma - (\sigma \cdot x)k$, or equivalently $P_{k\sigma} = (k \cdot \sigma)I + \sigma \otimes k - k \otimes \sigma$. Then*

(0) $P_{(-k)\sigma} = P_{k(-\sigma)} = -P_{k\sigma}$;
(i) $P_{kk} = \mathrm{Id}$;
(ii) $P_{k\sigma}k = \sigma, \quad P_{k\sigma}(k^\perp) \subset \sigma^\perp$;
(iii) $P^*_{k\sigma} = P_{\sigma k}$;
(iv) $(P_{k\sigma}\sigma) \cdot k = 2(k \cdot \sigma)^2 - 1 = (P_{\sigma k}k) \cdot \sigma$;
(v) *If $k \neq \pm\sigma$ then the restriction of $P_{k\sigma}$ to* $\mathrm{Vect}(k, \sigma)$*, the plane generated by k and σ, is the rotation sending k onto σ, and its restriction to* $\mathrm{Vect}(k, \sigma)^\perp$ *is the multiplication by $k \cdot \sigma$;*
Moreover, for all $x \in \mathbb{R}^d$,
(vi) *If $\Pi^\perp_{\{k,\sigma\}}$ stands for the orthogonal projection on* $\mathrm{Vect}(k, \sigma)^\perp$*, then*

$$\begin{aligned}
|x|^2 - |P_{k\sigma}x|^2 &= \left[1 - (k \cdot \sigma)^2\right]\left|\Pi^\perp_{\{k,\sigma\}}x\right|^2 \\
&= \left(1 - (k \cdot \sigma)^2\right)|x|^2 - (x \cdot k)^2 - (x \cdot \sigma)^2 \\
&\quad + 2(k \cdot \sigma)(x \cdot k)(x \cdot \sigma) \\
&= |\Pi_{k^\perp}x|^2 - |P_{k\sigma}\Pi_{k^\perp}x|^2,
\end{aligned}$$

(vii) $|P_{k\sigma}x| \le |x|$ *with equality only if $x \in$* $\mathrm{Vect}(k, \sigma)$ *or $k = \pm\sigma$;*
(viii) $P_{k\sigma}x = 0$ *if and only if (k, σ, x) are pairwise orthogonal;*
(ix) $\sigma \cdot (P_{k\sigma}x) = k \cdot x$;
(x) *In addition, for all $x, y \in \mathbb{R}^d$ and with the notation $|y - x|^2_{k,\sigma} = |x|^2 + |y|^2 - 2(P_{k\sigma}x) \cdot y$,*
$|y - x|^2_{k,\sigma} = \left|\Pi_{\sigma^\perp}y - \Pi_{k^\perp}x\right|^2_{k,\sigma} + (y \cdot \sigma - x \cdot k)^2$,
$|y - x|^2_{k,\sigma} = (|x|^2 - |P_{k\sigma}x|^2) + |y - P_{k\sigma}x|^2 \ge |y - P_{k\sigma}x|^2$,
$|y - P_{k\sigma}x|^2 = \left|\Pi_{\sigma^\perp}y - P_{k\sigma}\Pi_{k^\perp}x\right|^2 + (y \cdot \sigma - x \cdot k)^2$,
$|y - x|^2_{k,\sigma} = \left|\Pi_{\sigma^\perp}y - P_{k\sigma}\Pi_{k^\perp}x\right|^2 + (|x|^2 - |P_{k\sigma}x|^2) + (y \cdot \sigma - x \cdot k)^2$,
$|y - x|^2_{k,\sigma} = |x - y|^2_{\sigma,k}$.

Proof of Proposition 8.12 (0), (i), (ii), (iii), (iv) are immediate.

To prove (v), let $k^\perp$ and $\sigma^\perp$ be obtained from k and σ by the action of the rotation of angle $\pi/2$ in $P = \mathrm{Vect}(k, \sigma)$. (Choose an arbitrary orientation of P.) Then $x = x_P + x_\perp$, where x_P is the projection of x onto P and $x_\perp$ on its orthogonal, so that $x_\perp \perp k, \sigma, k^\perp$. Then we have the orthogonal decomposition

$$x = x_\perp + (x \cdot k)k + (x \cdot k^\perp)k^\perp, \qquad |x|^2 = |x_\perp|^2 + (x \cdot k)^2 + (x \cdot k^\perp)^2,$$

and

$$\begin{aligned}
P_{k\sigma}x &= P_{k\sigma}x_\perp + (x\cdot k)P_{k\sigma}k + (x\cdot k^\perp)P_{k\sigma}k^\perp \\
&= (k\cdot\sigma)x_\perp + (x\cdot k)\sigma + (x\cdot k^\perp)P_{k\sigma}k^\perp \\
&= (k\cdot\sigma)x_\perp + (x\cdot k)\sigma + (x\cdot k^\perp)\big[(k\cdot\sigma)k^\perp - (\sigma\cdot k^\perp)k\big] \\
&= (k\cdot\sigma)x_\perp + (x\cdot k)\sigma + (x\cdot k^\perp)\sigma^\perp.
\end{aligned}$$

This implies (v), the first equality of (vi), (vii) and (viii). As for the second equality of (vi), write

$$\begin{aligned}
|x|^2 - |P_{k\sigma}x|^2 = |x|^2 - \Big(&(k\cdot\sigma)^2|x|^2 + (k\cdot x)^2 + (\sigma\cdot x)^2 + 2(k\cdot\sigma)(x\cdot k)(\sigma\cdot x) \\
&- 2(k\cdot\sigma)(k\cdot x)(\sigma\cdot x) - 2(k\cdot\sigma)(\sigma\cdot x)(x\cdot k)\Big) \\
= |x|^2 - \Big(&(k\cdot\sigma)^2|x|^2 + (k\cdot x)^2 + (\sigma\cdot x)^2 - 2(k\cdot\sigma)(\sigma\cdot x)(x\cdot k)\Big).
\end{aligned}$$

Then to prove the third (and last) equality of (vi), insert the decomposition $x = \Pi_{k^\perp}x + (k\cdot x)k$ into the previous formula for $|x|^2 - |P_{k\sigma}x|^2$ and see that it simplifies into $(1-(k\cdot\sigma)^2)|\Pi_{k^\perp}x|^2 - ((\Pi_{k^\perp}x)\cdot\sigma)^2$, which is the same as $|\Pi_{k^\perp}x|^2 - |P_{k\sigma}\Pi_{k^\perp}x|^2$.

Identity (ix) is immediate. It remains to prove (x). Start from

$$y = \Pi_{\sigma^\perp}y + (\sigma\cdot y)\sigma, \qquad x = \Pi_{k^\perp}x + (k\cdot x)k, \tag{8.18}$$

and deduce

$$\begin{aligned}
|y-x|^2_{k,\sigma} &= |x|^2 + |y|^2 - 2P_{k\sigma}x\cdot y \\
= |\Pi_{k^\perp}x|^2 &+ (x\cdot k)^2 + |\Pi_{\sigma^\perp}y|^2 + (y\cdot\sigma)^2 - 2(P_{k\sigma}\Pi_{k^\perp}x)\cdot\Pi_{\sigma^\perp}y - 2(P_{k\sigma}k)\cdot(x\cdot k)\Pi_{\sigma^\perp}y \\
&- 2P_{k\sigma}(\Pi_{k^\perp}x)\cdot(y\cdot\sigma)\sigma - 2P_{k\sigma}k\cdot(x\cdot k)(y\cdot\sigma)\sigma.
\end{aligned} \tag{8.19}$$

Now,

$$(P_{k\sigma}k)\cdot(x\cdot k)\Pi_{\sigma^\perp}y = (x\cdot k)\sigma\cdot\Pi_{\sigma^\perp}y = 0,$$

$$P_{k\sigma}(\Pi_{k^\perp}x)\cdot(y\cdot\sigma)\sigma = (y\cdot\sigma)(\Pi_{k^\perp}x)\cdot P_{\sigma k}\sigma = (y\cdot\sigma)(\Pi_{k^\perp}x)\cdot k = 0,$$

$$P_{k\sigma}k\cdot(x\cdot k)(y\cdot\sigma)\sigma = (x\cdot k)(y\cdot\sigma).$$

All of this reduces (8.19) to

$$|\Pi_{k^\perp}x|^2 + |\Pi_{\sigma^\perp}y|^2 - 2(P_{k\sigma}\Pi_{k^\perp}x)\cdot\Pi_{\sigma^\perp}y + (x\cdot k)^2 + (y\cdot\sigma)^2 - 2(x\cdot k)(y\cdot\sigma)$$

$$= \left|\Pi_{k^\perp} x - \Pi_{\sigma^\perp} y\right|^2_{k,\sigma} + (x \cdot k - y \cdot \sigma)^2,$$

which yields the first part of (x).

The second part of (x) is immediate from algebra and (vii).

For the third part of (x), using (8.18) again and (ii),

$$\begin{aligned}
|y - P_{k\sigma} x|^2 &= \left|\left(\Pi_{\sigma^\perp} y + (y \cdot \sigma)\sigma\right) - P_{k\sigma}\left(\Pi_{k^\perp} x + (x \cdot k)k\right)\right|^2 \\
&= \left|\left(\Pi_{\sigma^\perp} y - P_{k\sigma} \Pi_{k^\perp} x\right) + (y \cdot \sigma - x \cdot k)\sigma\right|^2 \\
&= \left|\Pi_{\sigma^\perp} y - P_{k\sigma} \Pi_{k^\perp} x\right|^2 + (y \cdot \sigma - x \cdot k)^2.
\end{aligned}$$

Then, starting again from

$$|y - P_{k\sigma} x|^2 = |y - x|^2_{k,\sigma} - \left(|x|^2 - |P_{k\sigma} x|^2\right)$$

and likewise

$$\left|\Pi_{\sigma^\perp} y - P_{k\sigma} \Pi_{k^\perp} x\right|^2 = \left|\Pi_{\sigma^\perp} y - \Pi_{k^\perp} x\right|^2_{k,\sigma} - \left(|\Pi_{k^\perp} x|^2 - |P_{k\sigma} \Pi_{k^\perp} x|^2\right),$$

we see that the first and third parts of (x) are equivalent provided that

$$|x|^2 - |P_{k\sigma} x|^2 = |\Pi_{k^\perp} x|^2 - \left|P_{k\sigma} \Pi_{k^\perp} x\right|^2. \tag{8.20}$$

Here is a direct proof of (8.20): Write $x = \Pi_{k^\perp} x + (x \cdot k)k$, then

$$\begin{aligned}
|x|^2 - |P_{k\sigma} x|^2 &= |\Pi_{k^\perp} x|^2 + (x \cdot k)^2 - |P_{k\sigma} \Pi_{k^\perp} x|^2 - |(x \cdot k) P_{k\sigma} k|^2 \\
&\quad - 2(k \cdot \sigma)(P_{k\sigma} \Pi_{k^\perp} x) \cdot (P_{k\sigma} k) \\
&= |\Pi_{k^\perp} x|^2 + (x \cdot k)^2 - |P_{k\sigma} \Pi_{k^\perp} x|^2 - (x \cdot k)^2 \\
&\quad - 2(k \cdot \sigma)(P_{k\sigma} \Pi_{k^\perp} x) \cdot \sigma \\
&= |\Pi_{k^\perp} x|^2 - |P_{k\sigma} \Pi_{k^\perp} x|^2 - 2(k \cdot \sigma)(\Pi_{k^\perp} x \cdot P_{\sigma k} \sigma)
\end{aligned}$$

and the last term is 0 since $P_{\sigma k} \sigma = k$. This concludes the proof of (8.20) and the but-to-last bit of (x) as well. The final bit of (x) follows at once from (iii). □

Bibliographical Notes

The *carré du champ itéré* Γ_2 appears in Bakry–Émery [15] and many further works by these authors and the research community which has developed from there. As for the *carré du champ* itself, it is present in innumerable works.

In the context of kinetic theory, the local nature of Γ it is exploited in [181] to establish an entropic gap.

Lemmas 8.3–8.5 and Proposition 8.6 are reformulations of my older work on the Fisher information for Boltzmann equation [174], while Corollary 8.8 captures the main result of that work. The proof of Lemma 8.5 given here is more intrinsic than in [174], but there is certainly an even more synthetic formulation in the language of intrinsic differential geometry. Of course the main novelty in this section is the more precise calculation leading to Theorem 8.7, allowing B to depend on $|z|$.

Expressions like $(F+G)|\nabla \log F - \nabla \log G|^2$, underlying the estimates in this section, also play a key role in my older estimates, with Toscani, of Boltzmann's dissipation functional [171, 181]. This coincidence is not surprising: As noticed in [171, Section 8], for Maxwellian kernels the dissipation of D_B by the Fokker–Planck equation coincides, by commutation of the flows, with the dissipation of I by the Boltzmann equation.

9 Qualitative Discussion

The intricate nature of $\Gamma_{I,\mathcal{B}}$ is the result of the interaction of three "geometries': tensorisation, collision spheres, and Fisher information. Integration, Proposition 7.2 and Theorem 8.7 yield

$$-I'(f)\cdot Q(f,f) = \frac{1}{2}\int \Gamma_{I,\mathcal{B}}(f\otimes f)\,dv\,dv_* = \text{(I)} + \text{(II)} + \text{(III)},$$

where

$$\text{(I)} = \frac{1}{4}\iiint f'f'_*\big|(\xi'+\xi'_*) - (\xi+\xi_*)\big|^2\,B\,d\sigma\,dv\,dv_* \tag{9.1}$$

$$\text{(II)} = \frac{1}{4}\iiint f'f'_*\big|(\xi'-\xi'_*) - (\xi-\xi_*)\big|^2_{k,\sigma}\,B\,d\sigma\,dv\,dv_* \tag{9.2}$$

$$\begin{aligned}\text{(III)} &= \iiint (f'f'_* - ff_*)\,k\cdot(\xi-\xi_*)\,\frac{\partial B}{\partial |z|}\,d\sigma\,dv\,dv_* \qquad (9.3)\\ &= \frac{1}{2}\iiint (f'f'_* - ff_*)\big[k\cdot(\xi-\xi_*) - \sigma\cdot(\xi'-\xi'_*)\big]\,\frac{\partial B}{\partial |z|}\,d\sigma\,dv\,dv_*.\end{aligned}$$

By symmetry, in (I) and (II) one may replace $f'f'_*$ by ff_* or $(f'f'_*+ff_*)/2$. In a way (I) is associated with momentum conservation ($((v'+v'_*)-(v+v_*)=0)$, (II) with energy conservation, keeping collisions on the sphere ($|v'-v'_*|=|v-v_*|$ implies $|(v'-v'_*)-(v-v_*)|^2_{k,\sigma}=0$).

Now I will choose to write (II) with ff_* in the integrand, and decompose, according to Proposition 8.12(x),

$$|y-x|_{k,\sigma}^2 = \left|\Pi_{\sigma^\perp} y - P_{k\sigma}\Pi_{k^\perp} x\right|^2 + \left(|x|^2 - |P_{k\sigma}x|^2\right) + (y\cdot\sigma - x\cdot k)^2$$

with $y = (\xi' - \xi'_*)$ and $x = (\xi - \xi_*)$. This yields

$$(\mathrm{II}) = (\mathrm{II})_1 + (\mathrm{II})_2 + (\mathrm{II})_3,$$

where

$$(\mathrm{II})_1 = \frac{1}{4}\iiint ff_*\left|\Pi_{\sigma^\perp}(\xi'-\xi'_*) - P_{k\sigma}\Pi_{k^\perp}(\xi-\xi_*)\right|^2 B\,d\sigma\,dv\,dv_* \qquad (9.4)$$

$$(\mathrm{II})_2 = \frac{1}{4}\iiint ff_*\left(|\xi-\xi_*|^2 - |P_{k\sigma}(\xi-\xi_*)|^2\right) B\,d\sigma\,dv\,dv_* \qquad (9.5)$$

$$(\mathrm{II})_3 = \frac{1}{4}\iiint ff_*\left[(\xi'-\xi'_*)\cdot\sigma - (\xi-\xi_*)\cdot k\right]^2 B\,d\sigma\,dv\,dv_*. \qquad (9.6)$$

To summarise: four nonnegative terms, (I), $(\mathrm{II})_1$, $(\mathrm{II})_2$, $(\mathrm{II})_3$, and one unsigned term (III). The monotonicity property (or not) of I will depend on the possibility to control the unsigned term by the nonnegative ones. Among the latter, $(\mathrm{II})_2$ may be simplified using the following lemma.

Lemma 9.1 *If $X \in \mathbb{R}^d$ and $k \in \mathbb{S}^{d-1}$ are given, then*

$$\int_{\mathbb{S}^{d-1}} \left(|X|^2 - |P_{k\sigma}X|^2\right) B\,d\sigma = \left(\frac{d-2}{d-1}\right)\left(\int_{\mathbb{S}^{d-1}} [1-(k\cdot\sigma)^2]\,B\,d\sigma\right)|\Pi_{k^\perp}X|^2.$$

Remark 9.2 $|\int[1-(k\cdot\sigma)^2]\,B\,d\sigma \le 2\int(1-k\cdot\sigma)\,B\,d\sigma = 2M(z)$ is finite even when B is nonintegrable.

Proof of Lemma 9.1 Write $\cos\theta = k\cdot\sigma$. By Proposition 8.12(vi),

$$\begin{aligned}|X|^2 - |P_{k\sigma}X|^2 &= \left[1-(k\cdot\sigma)^2\right]|X|^2 - (k\cdot X)^2 - (\sigma\cdot X)^2 \\ &\quad + 2(k\cdot X)(\sigma\cdot X)(k\cdot\sigma) \\ &= \sin^2\theta|X|^2 - (k\cdot X)^2 - (\sigma\cdot X)^2 + 2\cos\theta(k\cdot X)(\sigma\cdot X).\end{aligned}$$

Decompose $\sigma = (\cos\theta)k + (\sin\theta)\phi$, $\phi \in \mathbb{S}^{d-2}_{k^\perp}$. Then

$$\begin{aligned}&|X|^2 - |P_{k\sigma}X|^2 \\ &= \sin^2\theta|X|^2 - (k\cdot X)^2 - \cos^2\theta(k\cdot X)^2 - \sin^2\theta(\phi\cdot X)^2 - 2(\cos\theta)(\sin\theta)(k\cdot X)(\phi\cdot X) \\ &\qquad + 2\cos^2\theta(k\cdot X)^2 + 2\cos\theta\sin\theta(k\cdot X)(\phi\cdot X).\end{aligned}$$

By Lemma 5.4,

$$\frac{1}{|\mathbb{S}^{d-2}|}\int_{S^{d-2}_{k\perp}}\left(|X|^2-|P_{k\sigma}X|^2\right)d\phi$$

$$= \sin^2\theta|X|^2-(k\cdot X)^2-\cos^2\theta(k\cdot X)^2-\frac{\sin^2\theta}{d-1}|\Pi_{k^\perp}X|^2+2\cos^2\theta(k\cdot X)^2.$$

(Note that $\int\phi\cdot X\,d\phi=0$ by symmetry.) All in all, using again $|X|^2=(k\cdot X)^2+|\Pi_{k^\perp}X|^2$,

$$\frac{1}{|\mathbb{S}^{d-2}|}\int_{S^{d-2}_{k\perp}}\left(|X|^2-|P_{k\sigma}X|^2\right)d\phi$$

$$= (k\cdot X)^2\left(\sin^2\theta-1+\cos^2\theta\right)+|\Pi_{k^\perp}X|^2\left(\sin^2\theta-\frac{\sin^2\theta}{d-1}\right)$$

$$= \left(\frac{d-2}{d-1}\right)\sin^2\theta\,|\Pi_{k^\perp}X|^2,$$

and the result follows. □

Lemma 9.1 implies the following simplified formula for $(\mathrm{II})_2$:

$$(\mathrm{II})_2=\frac{1}{4}\left(\frac{d-2}{d-1}\right)\iint ff_*\,\Sigma(|v-v_*|)\left|\Pi_{k^\perp}\left(\frac{\nabla f}{f}-\left(\frac{\nabla f}{f}\right)_*\right)\right|^2dv\,dv_*, \tag{9.7}$$

where

$$\Sigma(|z|)=\int_{\mathbb{S}^{d-1}}\sin^2\theta\,B(z,\sigma)\,d\sigma=\int_{\mathbb{S}^{d-1}}[1-(k\cdot\sigma)^2]\,B(z,\sigma)\,d\sigma. \tag{9.8}$$

This formula resembles Landau's dissipation functional, and suggests that a better intuition for the various terms in $-I'(f)\,Q(f,f)$ will be obtained by first going to the asymptotics of grazing collisions. By Proposition 5.3 this will also amount to considering the problem of monotonicity of Fisher information along the spatially homogeneous Landau equation. In the Proposition below, $\|M\|^2_{\mathrm{HS}}=\mathrm{tr}\,(M^*M)$ will stand for the Hilbert–Schmidt square norm of a matrix M (nothing to do with Hard Spheres...).

Proposition 9.3 (Asymptotics of Grazing Collisions for the Fisher Information Dissipation) *With the notation*

$$\xi=-\nabla\log f,\qquad A=-\nabla^2\log f,\qquad \Psi=\frac{M_\infty(v-v_*)\,|v-v_*|^2}{4(d-1)}$$

as in Proposition 5.3, one has the following limits in the AGC:

$$\text{(I)} \xrightarrow[AGC]{} \frac{1}{2} \iint f f_* \Psi \, \|(A - A_*)\Pi_{k^\perp}\|_{\mathrm{HS}}^2 \, dv \, dv_*$$

$$\text{(II)}_1 \xrightarrow[AGC]{} \frac{1}{2} \iint f f_* \Psi \left\| \left((A + A_*) - 2k \cdot \frac{(\xi - \xi_*)}{|v - v_*|} I \right) \Pi_{k^\perp} \right\|_{\mathrm{HS}}^2 dv \, dv_*$$

$$\text{(II)}_2 \xrightarrow[AGC]{} 2(d-2) \iint f f_* \frac{\Psi}{|v - v_*|^2} \left| \Pi_{k^\perp}(\xi - \xi_*) \right|^2 dv \, dv_*$$

$$\text{(II)}_3 \xrightarrow[AGC]{} \frac{1}{2} \iint f f_* \Psi \left| \Pi_{k^\perp} \left((A + A_*)k + 2 \frac{\xi - \xi_*}{|v - v_*|} \right) \right|^2 dv \, dv_*$$

$$\text{(III)} \xrightarrow[AGC]{} - \iint f f_* \left(\Psi' - \frac{2\Psi}{|v - v_*|} \right) \times \left\langle \Pi_{k^\perp}(\xi - \xi_*), \Pi_{k^\perp} \left[(A + A_*)k + 2 \frac{(\xi - \xi_*)}{|v - v_*|} \right] \right\rangle dv \, dv_*.$$

(*where* Ψ' *is just the usual derivative of the function* $\Psi(r)$ *applied to* $r = |v - v_*|$).

Before giving the proof, here are some comments on these various terms. At several places, one computes $\|M\Pi_{k^\perp}\|_{\mathrm{HS}}^2 = \mathrm{tr}\,(\Pi_{k^\perp} M^* M \Pi_{k^\perp})$, in particular with $M = A + A_*$; note, this is larger than $\|\Pi_{k^\perp} M \Pi_{k^\perp}\|_{HS}^2$, the Hilbert–Schmidt norm of the operator induced on $k^\perp$ by M (because $A + A_*$ acting on $k^\perp$ in general has a component in the k direction). Terms (I) and (II)_1 take the form of an integrated squared log Hessian: roughly speaking they look like $\int \|(\nabla^2 - \nabla_*^2) \log(f f_*)\|_{\mathrm{HS}}^2 f f_* \, dv \, dv_*$ and $\int \|(\nabla^2 + \nabla_*^2) \log(f f_*)\|_{\mathrm{HS}}^2 f f_* \, dv \, dv_*$, respectively; some kind of (weighted) tensorised version of the familiar term $\int \|\nabla^2 \log f / f_\infty\|_{\mathrm{HS}}^2 f$ which appears in the theory of log Sobolev inequalities. And pursuing this analogy, the term (II)_2, which looks something like $(d-2) \int |(\nabla - \nabla_*) \log(f f_*)|^2 f f_* \, dv \, dv_*$, is akin to a *curvature* term, something like a tensorised version of $(d-2) \int f |\nabla \log f|^2 \, dv$, which is exactly the first order term popping out in Bakry–Émery's famous computation of the derivative of the Fisher information along the Fokker–Planck equation. (Note that $d - 2$ is exactly the value of the Ricci curvature on the sphere, up to multiplication by the metric of course.) Next, to understand (II)_3, consider the following: If $f f_*$ were intrinsincally defined on the collision sphere (not on the whole space), then $(\xi - \xi_*) \cdot k = 0$ would hold throughout the whole sphere, and differentiation would yield

$$\Pi_{k^\perp}(\nabla - \nabla_*)\big((\xi - \xi_*) \cdot k\big) = 0,$$

but this is the same as

$$\Pi_{k^\perp}\left((A + A_*)k + 2\frac{\xi - \xi_*}{|v - v_*|}\right) = 0.$$

In other words, $(\mathrm{II})_3$ is a kind of *extrinsic term* measuring the extent to which ff_* fails to be preserved along the collision sphere. Finally (III) is the only "bad" term in the monotonicity property, and in it one can appreciate the effect of the non-Maxwellian nature of the interaction (if Ψ is a power law $|z|^{\gamma+2}$ then $|\Psi' - 2\Psi/|z||$ is just $|\gamma||z|^{\gamma+1}$, hence vanishes only for $\gamma = 0$), and the amount of variation of ff_* along the collision spheres (recall that a tensor product invariant along collision spheres is a Maxwellian equilibrium).

This intuition behind the various terms in (8.7) is the main outcome of this section. But also we can be a straightforward derivation of a suboptimal version of Guillen–Silvestre theorem.

Theorem 9.4 (A Rough Version of the Guillen–Silvestre Theorem) *Assume* $|r\Psi'(r)/\Psi(r) - 2| \le 2\sqrt{d-2}$*; then the Fisher information is monotonous along solutions of the spatially homogeneous Landau equation with function* Ψ.

Proof of Theorem 9.4 Let $\overline{\gamma}$ be an upper bound for $|r\Psi'(r)/\Psi(r) - 2|$. Then from the asymptotic expressions in Proposition 9.3

$$\begin{aligned}
|(\mathrm{III})| &\le \overline{\gamma} \iint ff_* \frac{\Psi(|v - v_*|)}{|v - v_*|} \left|\Pi_{k^\perp}(\xi - \xi_*)\right| \\
&\quad \times \left|\Pi_{k^\perp}\left[(A + A_*)k + 2\frac{\xi - \xi_*}{|v - v_*|}\right]\right| dv\, dv_* \\
&\le \frac{1}{2} \iint ff_* \Psi \left|\Pi_{k^\perp}\left[(A + A_*)k + 2\frac{\xi - \xi_*}{|v - v_*|}\right]\right|^2 dv\, dv_* \\
&\quad + \frac{(\overline{\gamma})^2}{2} \iint ff_* \frac{\Psi}{|v - v_*|^2} \left|\Pi_{k^\perp}(\xi - \xi_*)\right|^2 dv\, dv_* \\
&= (\mathrm{II})_3 + \frac{(\overline{\gamma})^2}{4(d-2)} (\mathrm{II})_2.
\end{aligned}$$

So if $\overline{\gamma} \le 2\sqrt{d-2}$, this is bounded above by $(\mathrm{II})_3 + (\mathrm{II})_2$, and decay follows. □

Remark 9.5 In dimensions 6 and higher this covers already all mathematical and physical cases in which one would reasonably be interested. If one restricts to potentials decaying faster than d-dimensional Coulomb, as I chose to do, this even covers them for $d = 4, 5$. But it misses the range $\gamma = [-3, -2)$ in dimension $d = 3$ (which is crucial since it contains the true Landau–Coulomb equation) and it says nothing about $d = 2$. It is clear from the proof that to improve this theorem one will need to use the Hessian terms.

It remains to sketch the proof of Proposition 9.3, which is a tricky exercise in Taylor expansions, based on repeated use of (5.14).

Sketch of Proof of Proposition 9.3 First with (I). From (5.11) and (5.13),

$$\xi' - \xi = \nabla\xi(v)(v' - v) + o(\theta|v - v_*|) = \frac{|v - v_*|}{2}\nabla\xi(v)\cdot\theta\phi + o(\theta|v - v_*|)$$

and likewise

$$\xi'_* - \xi_* = -\frac{|v - v_*|}{2}\nabla\xi(v_*)\cdot\theta\phi + o(\theta|v - v_*|),$$

so that

$$\left|(\xi' + \xi'_*) - (\xi + \xi_*)\right|^2 = \frac{|v - v_*|^2}{4}\left|(A - A_*)\phi\right|^2\theta^2 + o(\theta^2|v - v_*|^2).$$

In the AGC, using Lemma 5.4,

$$\int \left|(\xi' + \xi'_*) - (\xi + \xi_*)\right|^2 B\, d\sigma \longrightarrow \frac{|v - v_*|^2}{4}\frac{1}{d-1}\mathrm{tr}\left[(A - A_*)\Pi_{k^\perp}\right]^2(2M_\infty),$$

thus

$$\frac{1}{4}\int \left|(\xi' + \xi'_*) - (\xi + \xi_*)\right|^2 B\, d\sigma \longrightarrow \frac{1}{2}\Psi(|v - v_*|)\left\|(A - A_*)\Pi_{k^\perp}\right\|_{\mathrm{HS}}^2.$$

Now for (II). From

$$\xi' - \xi'_* = (\xi - \xi_*) + \frac{|v - v_*|}{2}(A + A_*)\theta\phi + o(\theta|v - v_*|)$$

one obtains, recalling (5.13),

$$\begin{aligned}
&(\xi' - \xi'_*) - P_{k\sigma}(\xi - \xi_*) \\
&\qquad = (1 - k\cdot\sigma)(\xi - \xi_*) - k\cdot(\xi - \xi_*)\sigma - \sigma\cdot(\xi - \xi_*)k \\
&\qquad\quad + \frac{|v - v_*|}{2}(A + A_*)\theta\phi + o(\theta|v - v_*|) \\
&\qquad = (1 - k\cdot\sigma)(\xi - \xi_*) - \left[k\cdot(\xi - \xi_*)(\sigma - k) + k\cdot(\xi - \xi_*)k\right] \\
&\qquad\quad + \left[(\sigma - k)\cdot(\xi - \xi_*)k + k\cdot(\xi - \xi_*)k\right] \\
&\qquad\quad + \frac{|v - v_*|}{2}(A + A_*)\theta\phi + o(\theta|v - v_*|) \\
&= -k\cdot(\xi - \xi_*)\theta\phi + \theta\phi\cdot(\xi - \xi_*)k + \frac{|v - v_*|}{2}(A + A_*)\phi\theta + o(\theta|v - v_*|).
\end{aligned} \tag{9.9}$$

On the other hand, using (5.12),

$$\begin{aligned}
&\sigma \cdot (\xi' - \xi'_*) - k \cdot (\xi - \xi_*) \\
&= \big[(\cos\theta)k + (\sin\theta)\phi\big] \cdot \Big[(\xi - \xi_*) + \frac{|v - v_*|}{2}(A + A_*)\theta\phi + o(\theta|v - v_*|)\Big] - k \cdot (\xi - \xi_*) \\
&= -(1 - \cos\theta)\, k \cdot (\xi - \xi_*) + \theta\phi \cdot (\xi - \xi_*) \\
&\quad + \frac{|v - v_*|}{2}\Big(\theta^2 \langle (A + A_*)\phi, \phi\rangle + \theta \langle (A + A_*)\phi, k\rangle\Big) + o(\theta|v - v_*|) \\
&= \frac{|v - v_*|}{2}\theta \left(\langle (A + A_*)\phi, k\rangle + 2\left\langle \phi, \frac{\xi - \xi_*}{|v - v_*|}\right\rangle \right) + o(\theta|v - v_*|),
\end{aligned} \tag{9.10}$$

combining this with $\sigma = k + O(\theta)$ it follows that

$$\begin{aligned}
\big[\sigma \cdot (\xi' - \xi'_*) - k \cdot (\xi - \xi_*)\big]\sigma &= \frac{|v - v_*|}{2}\theta \\
&\times \left(\langle (A + A_*)\phi, k\rangle + 2\left\langle \phi, \frac{\xi - \xi_*}{|v - v_*|}\right\rangle \right) k + o(\theta|v - v_*|).
\end{aligned}$$

But the left-hand side is also $\Pi_\sigma(\xi' - \xi'_*) - P_{k\sigma}\Pi_k(\xi - \xi_*)$. Subtracting this expression from (9.9) one obtains

$$\begin{aligned}
&\Pi_{\sigma^\perp}(\xi' - \xi'_*) - P_{k\sigma}\Pi_{k^\perp}(\xi - \xi_*) \\
&= \frac{|v - v_*|}{2}\theta \left[\left((A + A_*) - 2k \cdot \frac{(\xi - \xi_*)}{|v - v_*|} I \right) \phi - \langle (A + A_*)\phi, k\rangle k \right] + o(\theta|v - v_*|).
\end{aligned} \tag{9.11}$$

Since $\phi \cdot k = 0$,

$$\begin{aligned}
&\left| \Pi_{\sigma^\perp}(\xi' - \xi'_*) - P_{k\sigma}\Pi_{k^\perp}(\xi - \xi_*) \right|^2 \\
&= \frac{|v - v_*|^2\theta^2}{4} \left(\left| \left((A + A_*) - 2k \cdot \frac{(\xi - \xi_*)}{|v - v_*|} I \right) \phi \right|^2 + \langle (A + A_*)\phi, k\rangle^2 \right) \\
&\quad + o(|v - v_*|\theta).
\end{aligned} \tag{9.12}$$

Let $C = (A + A_*) - 2k \cdot (\xi - \xi_*)/|v - v_*|$. Note that $\langle (A + A_*)\phi, k\rangle = \langle C\phi, k\rangle$. Applying Lemma 5.4 to (9.12),

$$\frac{1}{|\mathbb{S}^{d-2}|} \int_{S^{d-2}_{k^\perp}} \left| \Pi_{\sigma^\perp}(\xi' - \xi'_*) - P_{k\sigma}\Pi_{k^\perp}(\xi - \xi_*) \right|^2 d\phi$$

$$= \frac{1}{d-1} \|\Pi_{k^\perp} C \Pi_{k^\perp}\|_{\mathrm{HS}}^2 + \|\Pi_{k^\perp}(Ck)\|^2 + o(\theta^2|v-v_*|^2)$$

$$= \frac{1}{d-1} \|\Pi_{k^\perp} C\|_{\mathrm{HS}}^2 + o(\theta^2|v-v_*|^2).$$

From this the asymptotic behaviour of $(\mathrm{II})_1$ follows at once.

Handling $(\mathrm{II})_2$: just start from (9.8), note that in the AGC, $\sin^2\theta = (1+\cos\theta)(1-\cos\theta) \simeq 2(1-\cos\theta)$, and apply Lemma 9.1.

Now for $(\mathrm{II})_3$: Starting again from (9.10),

$$\Big[\sigma\cdot(\xi'-\xi'_*) - k\cdot(\xi-\xi_*)\Big]^2 = \frac{|v-v_*|^2}{4}\theta^2 \left[\left((A+A_*)k + 2\frac{\xi-\xi_*}{|v-v_*|}\right)\cdot\phi\right]^2$$
$$+ o(\theta^2|v-v_*|^2),$$

so by Lemma 5.4 again,

$$\frac{1}{|\mathbb{S}^{d-2}|}\int_{\mathbb{S}^{d-2}_{k\perp}} \big[\sigma\cdot(\xi'-\xi'_*) - k\cdot(\xi-\xi_*)\big]^2\, d\phi$$
$$= \frac{|v-v_*|^2\theta^2}{4(d-1)} \left|\Pi_{k^\perp}\left((A+A_*)k + 2\frac{\xi-\xi_*}{|v-v_*|}\right)\right|^2 + o(\theta^2|v-v_*|^2),$$

and the asymptotics of $(\mathrm{II})_3$ follows.

Finally (III). Combine (9.10) with

$$f'f'_* - ff_* = \frac{|v-v_*|^2}{2}\big(f_*\nabla f - f(\nabla f)_*\big)\cdot\theta\phi + o(|v-v_*|\theta)$$
$$= \frac{|v-v_*|^2}{2}(\nabla-\nabla_*)(ff_*)\cdot\theta\phi + o(|v-v_*|\theta),$$

to get

$$\frac{1}{|\mathbb{S}^{d-2}|}\int_{\mathbb{S}^{d-2}_{k\perp}} (f'f'_* - ff_*)\big[k\cdot(\xi-\xi_*) - \sigma\cdot(\xi'-\xi'_*)\big]\, d\phi + o(|v-v_*|^2\theta^2)$$

$$= \frac{|v-v_*|^2}{4|\mathbb{S}^{d-2}|}\int_{\mathbb{S}^{d-2}_{k\perp}} \big((\nabla-\nabla_*)(ff_*)\cdot\phi\big)\left(\left[(A+A_*)k + 2\frac{(\xi-\xi_*)}{|v-v_*|}\right]\cdot\phi\right) d\phi$$
$$+ o(|v-v_*|^2\theta^2)$$

$$= \frac{|v-v_*|^2}{4(d-1)}\Pi_{k^\perp}\big((\nabla-\nabla_*)(ff_*)\big)\cdot\Pi_{k^\perp}\left[(A+A_*)k + 2\frac{(\xi-\xi_*)}{|v-v_*|}\right]$$
$$+ o(|v-v_*|^2\theta^2)$$

$$= \frac{|v-v_*|^2}{4(d-1)} ff_* \Pi_{k^\perp}\big((\nabla - \nabla_*)(\log ff_*)\big) \cdot \Pi_{k^\perp}\left[(A+A_*)k + 2\frac{(\xi-\xi_*)}{|v-v_*|}\right]$$
$$+ o(|v-v_*|^2\theta^2).$$

To conclude, it only remains to relate $\partial_{|z|}B$ and $\partial_{|z|}\Psi$ in the AGC. For this,

$$\lim_{AGC} \frac{|z|^2}{4(d-1)} \int_{\mathbb{S}^{d-1}} \left(\frac{\partial B}{\partial |z|}\right)\frac{\theta^2}{2}\,d\sigma = \frac{|z|^2}{4(d-1)}\frac{\partial}{\partial |z|}\left(\lim_{AGC}\int_{\mathbb{S}^{d-1}} B\frac{\theta^2}{2}\,d\sigma\right)$$
$$= |z|^2 \frac{\partial}{\partial |z|}\left(\frac{\Psi(|z|)}{|z|^2}\right) = \Psi'(|z|) - 2\frac{\Psi(|z|)}{|z|}.$$

□

Bibliographical Notes

I prepared the conjunction of AGC and Fisher information monotonicity for the purpose of this course. Precursors are the study of the AGC *per se* [3], the relation between Boltzmann and Landau dissipation functionals [171, 181], and of course the Guillen–Silvestre theorem [97].

10 Reduction to the Sphere

Section 8 has revealed how to compute $I'(f)\,Q(f,f)$. In this section we shall see that the "bad" term is controlled by the "good" terms as soon as a certain functional inequality holds for even functions on the sphere. I shall start with an intermediate result.

Proposition 10.1 *If, for all probability distributions f on $\mathbb{R}^d$,*

$$\iiint \frac{(f'f'_* - ff_*)^2}{ff_* + f'f'_*}\,\frac{1}{B}\left(\frac{\partial B}{\partial |z|}\right)^2 d\sigma\,dv\,dv_*$$
$$\leq \frac{1}{2}\iiint ff_* \Big|\Pi_{\sigma^\perp}(\xi' - \xi'_*) - \Pi_{k^\perp}(\xi - \xi_*)\Big|^2_{k,\sigma} B\,d\sigma\,dv\,dv_*, \tag{10.1}$$

then I is nonincreasing along solutions of the spatially homogeneous Boltzmann equation with kernel B.

Proof of Proposition 10.1 Start again from the results of Sects. 8 and 9; the issue is to control (III) in (9.3) by (I) + (II) in (9.1)–(9.2). It turns out that (II) will suffice. (I do not know whether a more general result can be obtained by exploiting (I).) By Cauchy–Schwarz,

$$
\begin{aligned}
|(\mathrm{III})| &\le \frac{1}{2}\left(\iiint (f'f'_* + ff_*)\big[\sigma\cdot(\xi'-\xi'_*) - k\cdot(\xi-\xi_*)\big]^2\, B\, d\sigma\, dv\, dv_*\right)^{1/2} \\
&\quad \left(\iiint \frac{(f'f'_* - ff_*)^2}{ff_* + f'f'_*}\,\frac{1}{B}\left(\frac{\partial B}{\partial |z|}\right)^2 d\sigma\, dv\, dv_*\right)^{1/2} \\
&= \frac{1}{2}(8(\mathrm{II})_3)^{1/2}\left(\iiint \frac{(f'f'_* - ff_*)^2}{ff_* + f'f'_*}\,\frac{1}{B}\left(\frac{\partial B}{\partial |z|}\right)^2 d\sigma\, dv\, dv_*\right)^{1/2} \\
&\le (\mathrm{II})_3 + \frac{1}{2}\iiint \frac{(f'f'_* - ff_*)^2}{ff_* + f'f'_*}\,\frac{1}{B}\left(\frac{\partial B}{\partial |z|}\right)^2 d\sigma\, dv\, dv_*.
\end{aligned}
$$

By applying (10.1),

$$
|(\mathrm{III})| \le (\mathrm{II})_3 + \frac{1}{4}\iiint ff_*\Big|\Pi_{\sigma^\perp}(\xi'-\xi'_*) - \Pi_{k^\perp}(\xi-\xi_*)\Big|^2_{k,\sigma} B\, d\sigma\, dv\, dv_*.
$$

By the but-to-last formula of Proposition 8.12(x) and the last bit of Proposition 8.12(vi),

$$
\begin{aligned}
|\Pi_{\sigma^\perp}(\xi'-\xi'_*) - \Pi_{k^\perp}(\xi-\xi_*)|^2_{k,\sigma} &= |\Pi_{\sigma^\perp}(\xi'-\xi'_*) - P_{k\sigma}\Pi_{k^\perp}(\xi-\xi_*)|^2 \\
&\quad + \left(|\xi-\xi_*|^2 - |P_{k\sigma}(\xi-\xi_*)|^2\right).
\end{aligned}
$$

Thus the latter integral coincides with $(\mathrm{II})_1 + (\mathrm{II})_2$. Thus $|(\mathrm{III})| \le (\mathrm{II})$, so $\Gamma_{I,\mathcal{B}} \ge 0$ and the proof is complete. □

To go from (10.1) to a more reduced integral criterion on the sphere, let us perform a change of variables

$$
(v, v_*, \sigma) \longrightarrow \left(\frac{v+v_*}{2}, |v-v_*|, \frac{v-v_*}{|v-v_*|}, \sigma\right) = (c, r, k, \sigma).
$$

Then

$$
dv\, dv_*\, d\sigma = |\mathbb{S}^{d-1}|\, r^{d-1}\, dc\, dr\, dk\, d\sigma.
$$

For each c, r write

$$
F_{c,r}(\sigma) = f\left(c + \frac{r}{2}\sigma\right) f\left(c - \frac{r}{2}\sigma\right),
$$

note that $F_{c,r}$ is an even function of σ, in the sense that $F_{c,r}(-\sigma) = F_{c,r}(\sigma)$. Then the left hand side in (10.1) is

$$|\mathbb{S}^{d-1}| \iiiint \frac{[F_{c,r}(\sigma) - F_{c,r}(k)]^2}{F_{c,r}(k) + F_{c,r}(\sigma)} \frac{1}{B} \left(\frac{\partial B}{\partial |z|}\right)^2 (r, k \cdot \sigma)\, r^{d-1}\, dc\, dr\, dk\, d\sigma. \tag{10.2}$$

On the other hand, treating σ as an independent variable in $\mathbb{R}^d$,

$$\nabla_\sigma \log F_{c,r}(\sigma) = \frac{r}{2}\left[(\nabla \log f)\left(c + \frac{r}{2}\sigma\right) - (\nabla \log f)\left(c - \frac{r}{2}\sigma\right)\right] = -\frac{r}{2}(\xi' - \xi'_*).$$

So the integral in the right-hand side of (10.1) is

$$\frac{|\mathbb{S}^{d-1}|}{2} \iiiint F_{c,r}(k) \frac{4}{r^2} \left|\Pi_{\sigma^\perp} \nabla \log F_{c,r}(\sigma) - \Pi_{k^\perp} \nabla \log F_{c,r}(k)\right|^2_{k,\sigma} \times B(r, k \cdot \sigma)\, r^{d-1}\, dc\, dr\, dk\, d\sigma. \tag{10.3}$$

Requiring the integrand in (10.2) to be bounded above by the integrand in (10.3) *for all* c, r, we are left with the following criterion: For all r, c,

$$\iint_{\mathbb{S}^{d-1} \times \mathbb{S}^{d-1}} \frac{[F(c,r)(\sigma) - F_{c,r}(k)]^2}{F_{c,r}(k) + F_{c,r}(\sigma)} \frac{1}{B} \left(\frac{\partial B}{\partial |z|}\right)^2 (r, k \cdot \sigma)\, dk\, d\sigma$$
$$\leq \frac{2}{r^2} \iint F_{c,r}(k) \left|\Pi_{\sigma^\perp} \nabla \log F_{c,r}(\sigma) - \Pi_{k^\perp} \nabla \log F_{c,r}(k)\right|^2_{k,\sigma} B(r, k \cdot \sigma)\, r^{d-1}\, dk\, d\sigma.$$

The integral in the right-hand side involves the gradient $\nabla F(\sigma)$ only via its projection on the tangent space $\sigma^\perp$; so we might as well assume that $F_{c,r}$ is defined intrinsically on the sphere. The inequality is also translation-invariant in c. Eventually we arrive at the sufficient compact condition: For all even functions $F : \mathbb{S}^{d-1} \to \mathbb{R}_+$,

$$\iint_{\mathbb{S}^{d-1} \times \mathbb{S}^{d-1}} \frac{[F(\sigma) - F(k)]^2}{F(k) + F(\sigma)} \frac{1}{B} \left(\frac{\partial B}{\partial |z|}\right)^2 (r, k \cdot \sigma)\, dk\, d\sigma$$
$$\leq \frac{2}{r^2} \iint_{\mathbb{S}^{d-1} \times \mathbb{S}^{d-1}} F(k) \left|\nabla \log F(\sigma) - \nabla \log F(k)\right|^2_{k,\sigma} B(r, k \cdot \sigma)\, r^{d-1}\, dk\, d\sigma.$$

To turn this into a more tractable criterion, we may use two simplifications:

- Assume that $|\partial_{|z|} B(r, k \cdot \sigma)| \leq \overline{\gamma}(r) B(r, k \cdot \sigma)/r$; then $\overline{\gamma}(r)$ is something like the maximum singularity/growth admissible in r; and of course, we may simplify even further by taking $\overline{\gamma}$ to be constant;
- Replace the nonlinearity $(b - a)^2/(a + b)$ by $(\sqrt{b} - \sqrt{a})^2$, in view of

$$\frac{(F(\sigma) - F(k))^2}{F(k) + F(\sigma)} \leq 2\left(\sqrt{F(\sigma)} - \sqrt{F(k)}\right)^2.$$

Let us summarise all the above discussion in the next

Theorem 10.2 *Let $B = B(|z|, \cos\theta)$ be a collision kernel. Let*

$$\overline{\gamma}(B, r) = \sup_{-1\leq\cos\theta\leq 1} \left(\frac{|z|}{B}\frac{\partial B}{\partial |z|}\right)(r, \cos\theta), \qquad \overline{\gamma}(B) = \sup_{r\geq 0} \overline{\gamma}(B, r) \tag{10.4}$$

Let us write

$$\beta_r(\cos\theta) = B(r, \cos\theta),$$

and say that $\beta : [-1, 1] \to \mathbb{R}_+$ is a section of B if there is $r \geq 0$ such that $\beta(\cos\theta) = B(r, \cos\theta)$ for all θ. Then a sufficient condition for I to be nonincreasing along solutions of the Boltzmann equation with kernel B is the following: For all even $F : \mathbb{S}^{d-1} \to \mathbb{R}_+$, for all $r \geq 0$,

$$\begin{aligned}&\iint_{\mathbb{S}^{d-1}\times\mathbb{S}^{d-1}} \frac{(F(\sigma) - F(k))^2}{F(k) + F(\sigma)}\, \beta_r(k\cdot\sigma)\, dk\, d\sigma \\ &\leq \frac{2}{\overline{\gamma}(B, r)^2} \iint_{\mathbb{S}^{d-1}\times\mathbb{S}^{d-1}} F(k) \left|\nabla \log F(\sigma) - \nabla \log F(k)\right|^2_{k,\sigma} \beta_r(k\cdot\sigma)\, dk\, d\sigma.\end{aligned} \tag{10.5}$$

This is true in particular if: For all even $F : \mathbb{S}^{d-1} \to \mathbb{R}_+$, for all $r \geq 0$,

$$\begin{aligned}&\iint_{\mathbb{S}^{d-1}\times\mathbb{S}^{d-1}} \left(\sqrt{F}(\sigma) - \sqrt{F}(k)\right)^2 \beta_r(k\cdot\sigma)\, dk\, d\sigma \\ &\leq \frac{1}{\overline{\gamma}(B, r)^2} \iint_{\mathbb{S}^{d-1}\times\mathbb{S}^{d-1}} F(k) \left|\nabla \log F(\sigma) - \nabla \log F(k)\right|^2_{k,\sigma} \beta_r(k\cdot\sigma)\, dk\, d\sigma.\end{aligned} \tag{10.6}$$

The latter is true in particular if: For all even $F : \mathbb{S}^{d-1} \to \mathbb{R}_+$, for all sections β of B,

$$\begin{aligned}&\iint_{\mathbb{S}^{d-1}\times\mathbb{S}^{d-1}} \left(\sqrt{F}(\sigma) - \sqrt{F}(k)\right)^2 \beta(k\cdot\sigma)\, dk\, d\sigma \\ &\leq \frac{1}{\overline{\gamma}(B)^2} \iint_{\mathbb{S}^{d-1}\times\mathbb{S}^{d-1}} F(k) \left|\nabla \log F(\sigma) - \nabla \log F(k)\right|^2_{k,\sigma} \beta(k\cdot\sigma)\, dk\, d\sigma.\end{aligned} \tag{10.7}$$

Remark 10.3 Inequalities (10.5)–(10.7) look familiar—control of an integral difference by an integral gradient, like in Poincaré or log Sobolev—and at the same time weird—a nonlocal nonlinearity with gradients is unusual, and we have a coupling by β instead of a reference measure. Most importantly and possibly most disturbingly, the *value* of the optimal constant in these inequalities will determine the range of exponents covered by the theorem.

In the sequel I will focus on the simplest inequality (10.7), and the question is to find conditions under which it may hold, or not. But first of all, in view of the turn taken by the discussion, it will be good to devote some time to equations on the sphere.

Bibliographical Notes

Theorem 10.2 is proven by Imbert, Silvestre and myself [111].

11 Games of Boltzmann and Laplace on the Sphere

Let $\beta = \beta(\cos\theta) \geq 0$ be an angular collision kernel, satisfying the minimum integrability condition

$$\mu_\beta = \int_{\mathbb{S}^{d-1}} (1 - k \cdot \sigma)\, \beta(k \cdot \sigma)\, d\sigma < \infty. \tag{11.1}$$

For $f : \mathbb{S}^{d-1} \to \mathbb{R}$, define

$$L_\beta f(k) = \int_{\mathbb{S}^{d-1}} [f(\sigma) - f(k)]\, \beta(k \cdot \sigma)\, d\sigma. \tag{11.2}$$

This is a linear Boltzmann operator on $L^1(\mathbb{S}^{d-1})$, a kind of diffusion on the sphere. It is actually the same object that I denoted by $\mathcal{B}$ in Sect. 7 but let us keep notation different since the latter was acting on functions of (v, v_*) and the current one on functions of just one variable. The operator L_β acts on all $L^p(\mathbb{S}^{d-1})$, in particular on L^2. Thanks to the exchange $k \leftrightarrow \sigma$,

$$\begin{aligned}\langle L_\beta f, g\rangle_{L^2(\mathbb{S}^{d-1})} &= \iint [f(\sigma) - f(k)]\, g(k)\, \beta(k \cdot \sigma)\, d\sigma\, dk \\ &= -\frac{1}{2} \iint [f(\sigma) - f(k)]\, [g(\sigma) - g(k)]\, \beta(k \cdot \sigma)\, d\sigma\, dk.\end{aligned}$$

So L_β is self-adjoint and nonpositive,

$$-\langle L_\beta f, f\rangle_{L^2} = \frac{1}{2} \iint_{\mathbb{S}^{d-1} \times \mathbb{S}^{d-1}} [f(\sigma) - f(k)]^2\, \beta(k \cdot \sigma)\, dk\, d\sigma. \tag{11.3}$$

Further, along e^{tL_β} all convex integrals decay: if C is any differentiable convex function, bounded below by a linear function, then $C(f) = \int C(f)\, d\sigma$ satisfies

$$C(f)\cdot L_\beta f = \iint_{\mathbb{S}^{d-1}\times\mathbb{S}^{d-1}} C'(f(k))[f(\sigma)-f(k)]\,\beta(k\cdot\sigma)\,dk\,d\sigma$$
$$= -\frac{1}{2}\iint_{\mathbb{S}^{d-1}\times\mathbb{S}^{d-1}} \Big(C'(f(\sigma))-C'(f(k))\Big)\big(f(\sigma)-f(k)\big) \le 0.$$

Of course the most famous dissipation on the sphere is the heat equation, whose generator, the Laplace–Beltrami operator Δ, may be introduced in several ways. For instance, Δf at x is the usual Laplacian of $f(x/|x|)$, evaluated at x. Or $\Delta = \sum_{ij}(x_i\partial_j - x_j\partial_i)^2/2$ (an average of "pure" second derivatives along rotations in the planes (i,j), thus yielding Δ in any dimension from Δ in dimension 1). In any case, the explicit expression in coordinates is

$$\Delta f = \sum_{ij}(\delta_{ij}-x_ix_j)\frac{\partial^2 f}{\partial x_i\,\partial x_j} - (d-1)\,x\cdot\nabla f$$
$$= \operatorname{tr}(\Pi_{x^\perp}\nabla^2 f) - (d-1)\,x\cdot\nabla f,$$

where ∇^2 and ∇ are usual differential operators in $\mathbb{R}^d$, and f is extended outside $\mathbb{S}^{d-1}$ in any way. Then Δ is also self-adjoint and

$$-\int_{\mathbb{S}^{d-1}}(\Delta f)g\,d\sigma = \int_{\mathbb{S}^{d-1}}\nabla f\cdot\nabla g\,d\sigma.$$

In the sequel, ∇ and ∇^2 will be the intrinsic gradient and Hessian on the sphere: when f is 0-homogeneous ($f = f(x/|x|)$) they coincide with the usual expressions in $\mathbb{R}^d$; when defined in an intrinsic way, $\nabla f(\sigma)$ is always orthogonal to σ and $\nabla^2 f$ is a symmetric operator on $\sigma^\perp$.

The usual Γ procedure yields

Proposition 11.1 *If Δ is the Laplace operator on $\mathbb{S}^{d-1}$ then* (8.1) *and* (8.2) *reduce to the following expressions, for any $f:\mathbb{S}^{d-1}\to\mathbb{R}$:*

$$\Gamma(f) = \Gamma(f,f) = |\nabla f|^2,$$

which will be denoted also $\Gamma_1(f)$ or $\Gamma_1(f,f)$; and

$$\Gamma_2(f) = \Gamma_2(f,f) = \Delta\frac{|\nabla f|^2}{2} - \nabla f\cdot\nabla\Delta f = \|\nabla^2 f\|_{\mathrm{HS}}^2 + (d-2)|\nabla f|^2$$

(a particular case of Bochner's formula).

Note carefully, the expressions above only apply to a function defined on $\mathbb{S}^{d-1}$; if f is defined on, say, $\mathbb{R}^d$ then for instance $\nabla f(\sigma)$ should be replaced by $\Pi_{\sigma^\perp}\nabla f(\sigma)$, the orthogonal component of the gradient.

It turns out that Δ is part of the Boltzmann family on the sphere; not surprisingly, its limit in the asymptotics of grazing collisions. Let us convene that a sequence $(\beta_n)_{n\in\mathbb{N}}$ concentrates on grazing collisions if $\mu_n = \int(1-k\cdot\sigma)\beta_n(k\cdot\sigma)\,d\sigma$ has a finite limit $\mu > 0$, while for any $\theta_0 > 0$, $\int 1_{k\cdot\sigma\le\cos\theta_0}\beta_n(k\cdot\sigma)\,d\sigma \longrightarrow 0$. (Here k is any fixed unit vector.) Then

Proposition 11.2

(i) For any two kernels β_0 and β_1, $[L_{\beta_0}, L_{\beta_1}] = 0$;
(ii) In the AGC, if $\int(1-k\cdot\sigma)\beta(k\cdot\sigma) \longrightarrow \mu > 0$, then

$$L_\beta \longrightarrow \left(\frac{\mu}{d-1}\right)\Delta;$$

(iii) For any kernel β, $[L_\beta, \Delta] = 0$.

Proof of Proposition 11.2 We start with (ii), which follows the same pattern as the AGC for Boltzmann, using $k\cdot\sigma = \cos\theta$, and the identities (5.12) and (5.13). As $\theta \simeq 0$,

$$\begin{aligned} f(\sigma) - f(k) &\simeq (\sigma - k)\cdot\nabla f(k) + \frac{1}{2}\nabla^2 f(k)\cdot(\sigma-k,\sigma-k) \\ &= \left(-\frac{\theta^2}{2}k + \theta\phi\right)\cdot\nabla f(k) + \frac{1}{2}\langle\nabla^2 f(k)\cdot\phi,\phi\rangle\theta^2 + o(\theta^2). \end{aligned}$$

So from Lemma 5.4,

$$\begin{aligned} &\int [f(\sigma) - f(k)]\,\beta(k\cdot\sigma)\,d\sigma \\ &= -k\cdot\nabla f(k)\int\frac{\theta^2}{2}\,\beta(k\cdot\sigma)\,d\sigma + \left(\int\phi\,\beta(k\cdot\sigma)\,d\sigma\right)\cdot\nabla f(k) \\ &\quad + \int\langle\nabla^2 f(k)\cdot\phi,\phi\rangle\,\frac{\theta^2}{2}\,\beta(k\cdot\sigma)\,d\sigma + o(1) \\ &= -k\cdot\nabla f(k)\,\mu + 0 + \frac{1}{d-1}\mathrm{tr}\left(\nabla^2 f(k)\,\Pi_{k^\perp}\right)\mu + o(1). \end{aligned}$$

Of course (iii) will result from (i) and (ii). So it only remains to check (i). Let us assume that $\int\beta_0$, $\int\beta_1$ are finite; the general case will follow by approximation. Then, for any $\ell\in\mathbb{S}^{d-1}$,

$$\begin{aligned} L_{\beta_1}L_{\beta_0}f(\ell) &= \int\Big[(L_{\beta_0}f)(k) - (L_{\beta_0}f)(\ell)\Big]\,\beta_1(\ell\cdot k)\,dk \\ &= \iint f(\sigma)\,\beta_0(k\cdot\sigma)\,\beta_1(\ell\cdot k)\,dk\,d\sigma - \iint f(k)\,\beta_0(k\cdot\sigma)\,\beta_1(\ell\cdot k)\,dk\,d\sigma \end{aligned}$$

$$-\iint f(\sigma)\,\beta_0(\ell\cdot\sigma)\,\beta_1(\ell\cdot k)\,dk\,d\sigma + \iint f(\ell)\,\beta_0(\ell\cdot\sigma)\,\beta_1(\ell\cdot k)\,dk\,d\sigma.$$

Let us try to exchange β_0 and β_1 in the latter expression made up of four integrals.

The last of the four integrals is the easiest: exchanging k and σ,

$$\begin{aligned} f(\ell)\iint \beta_0(\ell\cdot\sigma)\,\beta_1(\ell\cdot k)\,dk\,d\sigma &= f(\ell)\iint \beta_0(\ell\cdot k)\,\beta_1(\ell\cdot\sigma)\,dk\,d\sigma \\ &= f(\ell)\iint \beta_1(\ell\cdot\sigma)\,\beta_0(\ell\cdot k)\,dk\,d\sigma. \end{aligned}$$

Now turn to the first of the four integrals. Here is the trick (which I learnt from Sacha Bobylev) to use in such a situation: For any fixed σ we may find a change of variables $k \to \widetilde{k}$ by an isometry of $\mathbb{S}^{d-1}$ which exchanges σ and ℓ. Then

$$\begin{aligned} \int \beta_0(k\cdot\sigma)\,\beta_1(\ell\cdot k)\,dk &= \int \beta_0(\widetilde{k}\cdot\sigma)\,\beta_1(\ell\cdot\widetilde{k})\,dk \\ &= \int \beta_0(k\cdot\ell)\,\beta_1(\sigma\cdot k)\,dk, \end{aligned}$$

whence

$$\iint f(\sigma)\,\beta_0(k\cdot\sigma)\,\beta_1(\ell\cdot k)\,dk\,d\sigma = \iint f(\sigma)\,\beta_1(k\cdot\sigma)\,\beta_0(\ell\cdot k).$$

Finally the same trick will apply to *the sum* of the two central integrals:

$$\iint f(k)\,\beta_0(k\cdot\sigma)\,\beta_1(\ell\cdot k)\,dk\,d\sigma \qquad + \iint f(\sigma)\,\beta_0(\ell\cdot\sigma)\,\beta_1(\ell\cdot k)\,dk\,d\sigma$$

$$= \iint f(k)\,\beta_0(\ell\cdot\sigma)\,\beta_1(\ell\cdot k)\,dk\,d\sigma \qquad + \iint f(\sigma)\,\beta_0(\ell\cdot\sigma)\,\beta_1(\sigma\cdot k)\,dk\,d\sigma$$

(change of variable: σ, exchange: k & ℓ) (c.o.v. k, exch. ℓ & σ)

$$= \iint f(\sigma)\,\beta_0(\ell\cdot k)\,\beta_1(\ell\cdot\sigma)\,dk\,d\sigma \qquad + \iint f(k)\,\beta_0(\ell\cdot k)\,\beta_1(\sigma\cdot k)\,dk\,d\sigma.$$

(exchanging k and σ) (idem)

□

Now let us go on with the Γ formalism for L_β and Δ:

Definition 11.3 (Gamma Commutators Involving Linear Boltzmann) Given a Boltzmann kernel β, let

$$\Gamma_\beta(f,f) = \frac{1}{2}\Big(L_\beta(f^2) - 2f\,L_\beta f\Big);$$

$$\Gamma_{1,\beta}(f,f) = \frac{1}{2}\Big[L_\beta\,\Gamma_1(f,f) - 2\Gamma_1(f,L_\beta f)\Big].$$

It turns out that by commutation of L_β and Δ, it does not matter in which order we let the two operators act: also

$$\Gamma_{1,\beta}(f,f) = \frac{1}{2}\Big[\Delta\,\Gamma_\beta(f,f) - 2\Gamma_\beta(f,\Delta f)\Big].$$

Explicit computations yield

$$\Gamma_\beta(f,f)(k) = \frac{1}{2}\int_{\mathbb{S}^{d-1}} (f(\sigma) - f(k))^2\,\beta(k\cdot\sigma)\,d\sigma \tag{11.4}$$

$$\Gamma_{1,\beta}(f,f)(k) = \frac{1}{2}\int_{\mathbb{S}^{d-1}} \big|\nabla f(\sigma) - \nabla f(k)\big|^2_{k,\sigma}\,\beta(k\cdot\sigma)\,d\sigma, \tag{11.5}$$

where as before $|y-x|^2_{k,\sigma} = |x|^2 + |y|^2 - 2(P_{k\sigma}x)\cdot y$ and $P_{k\sigma}$ is defined as in Proposition 8.12. Notice: since $\nabla f(\sigma)\cdot\sigma = 0$, here $M_{\sigma k}\nabla f(\sigma) = P_{\sigma k}\nabla f(\sigma)$.

Eventually there is also a nice formula for ∇L_β:

$$\nabla L_\beta(f) = \int_{\mathbb{S}^{d-1}} \big[P_{\sigma k}\nabla f(\sigma) - \nabla f(k)\big]\,\beta(k\cdot\sigma)\,d\sigma. \tag{11.6}$$

Further, in the AGC, as a consequence of the commutation,

$$\Gamma_{1,\beta}(f,f) \xrightarrow[AGC]{} \frac{\mu}{d-1}\,\Gamma_2(f,f). \tag{11.7}$$

And of course, we may also study the evolution of the Fisher information I under the Boltzmann flow and define as in (8.3)

$$\Gamma_{I,L_\beta}(f) = L_\beta\big(f|\nabla\log f|^2\big) - \left[(L_\beta f)|\nabla\log f|^2 + 2\nabla f\cdot\nabla\left(\frac{L_\beta f}{f}\right)\right], \tag{11.8}$$

so that

$$-\,I'(f)\cdot L_\beta f = \int_{\mathbb{S}^{d-1}} \Gamma_{I,L_\beta}(f); \tag{11.9}$$

and we find, as in Sect. 8 and with the natural simplifications,

$$\Gamma_{I,L_\beta}(f)\,(k) = \int_{\mathbb{S}^{d-1}} f(\sigma)\big|\nabla\log f(\sigma) - \nabla\log f(k)\big|^2_{k,\sigma}\,\beta(k\cdot\sigma)\,d\sigma \tag{11.10}$$

a limit case being

$$\Gamma_{I,\Delta}(f) = 2f\,\Gamma_2(\log f). \tag{11.11}$$

A final subject we should explore is, quite naturally in this linear context, the spectrum. This will deserve a section on its own.

Bibliographical Notes

Linear Boltzmann equations are classical in neutronics [47]; but also, interacting particle systems are naturally described by a linear Boltzmann equation in a phase space for several particles, shaped by the conservation laws. Kac [115] took the list of this approach by letting his "particles" be described by a point in a high-dimensional sphere. But in this section the sphere may be low-dimensional, and as we have seen the linear Boltzmann equation on this space can help understanding the nonlinear Boltzmann equation.

I am not aware of any reference where the subject of this section is explicitly treated, but at least part of it must be folklore. I learnt the important trick for the commutation of the linear Boltzmann equations in Bobylev's survey paper on the Boltzmann equation with Maxwellian kernels [23], he used this to prove the commutation between (nonlinear) spatially homogeneous Maxwellian Boltzmann and Fokker–Planck semigroups. This is a particular case of much more general theorems used in representation theory and spectral analysis.

12 Apparition Spectrale

Eigenfunctions of the Laplace–Beltrami operator constitute an orthonormal basis of $L^2(\mathbb{S}^{d-1})$, generalising to higher dimensions the decomposition of $L^2(\mathbb{S}^1)$ in Fourier series. These functions are the spherical harmonics on $\mathbb{S}^{d-1}$ and they have been studied and classified for more than two centuries, especially on $\mathbb{S}^2$. There are several ways to introduce them, one is that spherical harmonics are restrictions to $\mathbb{S}^{d-1}$ of homogeneous polynomials. By definition, a polynomial $P : \mathbb{R}^d \to \mathbb{R}$ is ℓ-homogeneous ($\ell \in \mathbb{N}_0$) if $P(\lambda x) = \lambda^\ell P(x)$ for all $x \in \mathbb{R}^d$ and $\lambda > 0$. Then the restriction of P to $\mathbb{S}^{d-1}$ satisfies

$$\Delta P = \lambda P \qquad \lambda = \ell(\ell + d - 2).$$

(Think that $\Delta|x|^\ell = \ell(\ell + d - 2)$, even though that is a mnemonic rather than an example, since the restriction of $|x|^\ell$ is trivial.) So the eigenfunctions of $-\Delta$ are $0, d-1, 2d, 3(d+1)$, etc.

Spherical harmonics have "explicit' expressions in terms of special functions: as we see in Wikipedia (why not), they can be written $Y^{(d)}_{\ell,m}(\theta,\phi)$, where $\ell \in \mathbb{N}_0$, m is a $(d-2)$-index, $m = (m_1, \ldots, m_{d-2})$, $|m_1| \le m_2 \le \ldots \le m_{d-2} \le \ell$, and x is represented by its $(d-1)$-dimensional spherical coordinates: $\theta \in [0, \pi]$, $\phi \in \mathbb{S}^{d-2}$ through $x = (\cos\theta)e + (\sin\theta)\phi$, e fixed arbitrarily, $\phi \in S^{d-2}_{e^\perp}$, and recursively $\phi = \phi_{d-2}$, $\phi_j = (\cos\psi_j)e_j + (\sin\psi_j)\phi_{j-1}$ with e_j fixed arbitrarily and $\phi_{j-1} \in \mathbb{S}^{j-1}_{e_j^\perp}$, $0 \le \psi_j \le \pi$, and in the end ϕ is identified with its coordinates $(\psi_1, \dots, \psi_{d-2})$. Then, in complex notation,

$$Y_{\ell,m}(\theta,\phi) = \frac{e^{im_1\psi_1}}{\sqrt{2\pi}} \overline{P}^{m_1,(2)}_{m_2}(\psi_2)\overline{P}^{m_2,(3)}_{m_3}(\psi_3)\dots \overline{P}^{m_{d-2},(d-1)}_{\ell}(\theta),$$

where

$$P_L^{\ell,(j)}(\psi) = \sqrt{\left(\frac{2L+j-1}{2}\right)\frac{(L+\ell+j-2)!}{(L-\ell)!}}\,(\sin^{\frac{2-j}{2}}\psi)\, P_{L+\frac{j-2}{2}}^{-\left(\ell+\frac{j-2}{2}\right)}(\cos\psi),$$

and P^μ_λ is Legendre's special function, one of the solutions of the second-order real singular differential linear equation

$$(1-x^2)y''(x) - 2xy'(x) + \left(\lambda(\lambda+1) - \frac{\mu^2}{1-x^2}\right)y = 0$$

(itself a particular instance of the hypergeometric function); then the associated eigenvalue is $\lambda_\ell^{(d)} = \ell(\ell+d-2)$. (Phew!)

I will not need these explicit formulae, and for the sequel the spherical harmonics of given degree ℓ may be ordered in an arbitrary way. But the following facts about spherical harmonics $Y^{(d)}_{\ell,m}$ in dimension $d \ge 2$ will be useful. (The case $\ell = 0$ is omitted: obviously, the eigenspace is just made of constant functions. Note, I write $\mathbb{N} = \{1, 2, 3, \dots\}$.)

Proposition 12.1 (Spherical Harmonics on $\mathbb{S}^{d-1}$)

(i) For each $\ell \in \mathbb{N}$, the spherical harmonics $(Y^{(d)}_{\ell,m})_{1\le m\le N(d,\ell)}$ constitute an orthonormal basis of H_ℓ, the eigenspace of $-\Delta$ associated with the eigenvalue $\lambda_\ell = \ell(\ell+d-2)$, with multiplicity

$$N(d,\ell) = \left(\frac{2\ell+d-2}{\ell}\right)\binom{\ell+d-3}{\ell-1};$$

(ii) For any $\ell \in \mathbb{N}$ there is a unique harmonic ℓ-homogeneous polynomial $P_\ell = P^{(d)}_\ell$ (sometimes denoted ${}_{d-1}P^0_\ell$) on $[-1,1]$ such that for all k, σ in $\mathbb{S}^{d-1}$,

$$P_\ell(k\cdot\sigma) = \frac{|\mathbb{S}^{d-1}|}{N(d,\ell)} \sum_{m=1}^{N(d,\ell)} Y_{\ell,m}(k)Y_{\ell,m}(\sigma).$$

This is the addition formula, and P_ℓ is the Legendre polynomial of order ℓ in dimension d. Legendre polynomials contain most of the relevant information about spherical harmonics. In particular, the space of spherical harmonics of order ℓ which admit an axis of symmetry $e \in \mathbb{S}^{d-1}$ is a one-dimensional space generated by $\sigma \longmapsto P_\ell(\sigma \cdot e)$. Similarly, there are vectors $(\eta_m)_{1\le m\le N(d,\ell)}$ such that for any $Y \in H_\ell$ there are coefficients $(a_m)_{1\le m\le N(d,\ell)}$ such that one may write $Y(\sigma) = \sum_{m=1}^{N(d,\ell)} a_m\, P_\ell(\sigma \cdot \eta_m)$.

(iii) *Legendre polynomials $P_\ell^{(d)}$ are even for even ℓ and odd for odd ℓ; they satisfy $P_\ell(1) = 1$ and $|P_\ell(x)| \le 1$ for all $x \in [-1, 1]$. Moreover they satisfy the following four important identities:*

- *Rodrigues formula:*

$$P_\ell(x) = \frac{(-1)^\ell}{2^\ell \left(\ell + \frac{d-3}{2}\right)_\ell}(1-x^2)^{-\left(\frac{d-3}{2}\right)} \frac{d}{dx}^{\ell} (1-x^2)^{\ell+\frac{d-3}{2}},$$

  where $a_\ell = a(a-1)\dots(a-\ell+1)$.
- *Hypergeometric differential equation:*

$$(1-x^2)P_\ell''(x) - (d-1)x P_\ell'(x) + \ell(\ell+d-2)\, P_\ell(x) = 0. \qquad (12.1)$$

- *Recursion formula: $P_0(x) = 1$, $P_1(x) = x$, and*

$$(\ell+d-2)P_{\ell+1}(x) - (2\ell+d-2)x P_\ell(x) + \ell P_{\ell-1}(x) = 0.$$

- *Representation by axisymmetric averaging:*

$$P_\ell^{(2)}(x) = \Re(x + i\sqrt{1-x^2})^\ell \qquad (12.2)$$

  and, for $d \ge 3$,

$$P_\ell(x) = \frac{|\mathbb{S}^{d-3}|}{|\mathbb{S}^{d-2}|}\int_{-1}^{1} \left(x + i\, s\sqrt{1-x^2}\right)^\ell (1-s^2)^{(d-4)/2}\, ds. \qquad (12.3)$$

Example 12.2 For $d = 2$, an element in $\mathbb{S}^1$ can be identified to an angle $\theta \in \mathbb{R}/(2\pi\mathbb{Z})$; then Δ is the usual operator $d^2/d\theta^2$, the eigenspace H_ℓ is 2-dimensional for any $\ell \in \mathbb{N}$ and an orthonormal basis is $(\pi^{-1/2}\cos(n\theta), \pi^{-1/2}\sin(n\theta))$; then the addition formula turns into

$$P_\ell(\cos(\theta' - \theta)) = \cos(\ell\theta)\cos(\ell\theta') + \sin(\ell\theta)\sin(\ell\theta'),$$

which by the usual addition formula for trigonometric functions is just $\cos(\ell(\theta' - \theta))$; so P_ℓ is just the Tchebyshev polynomial defined by the identity

$$P_\ell(\cos\alpha) = \cos(\ell\alpha).$$

In complex notation, we immediately see that $P_\ell(\cos\alpha) = \Re e^{i\ell\alpha} = \Re(\cos\alpha + \sqrt{1-\cos^2\alpha})^\ell$, which amounts to (12.2).

Remark 12.3 Formula (12.3) shows that, in some sense, all P_ℓ can be obtained from the 2-dimensional case; indeed, this formula amounts to writing a harmonic polynomial in two variables, say x_1, x_2, and symmetrize it around the axis e_1. Note that the integral in (12.3) is real by symmetry $y \to -y$ and the operation applied to the polynomial is indeed an averaging, since

$$\int_{-1}^{1} \sqrt{1-s^2}^{\,d-4}\, ds = \int_0^\pi (\sin\theta)^{d-3}\, d\theta = \frac{|\mathbb{S}^{d-2}|}{|\mathbb{S}^{d-3}|}.$$

Legendre polynomials have since long been used to describe the spectrum of the linearised Boltzmann equation on $\mathbb{R}^d$. But they will also be useful to describe the spectrum of the linear Boltzmann equation on $\mathbb{S}^{d-1}$.

Theorem 12.4 (Spectrum of the Linear Boltzmann Equation) *If a collisional kernel β is given on $\mathbb{S}^{d-1}$, then the linear operator $-L_\beta$ has eigenvalues $(\lambda_\ell)_{\ell\in\mathbb{N}}$, and for each $\ell \in \mathbb{N}$, H_ℓ is an invariant subspace of dimension $N(d,\ell)$ associated with ν_ℓ, where*

$$\begin{aligned} \nu_\ell &= \int_{\mathbb{S}^{d-1}} \left[1 - P_\ell^{(d)}(k\cdot\sigma)\right] \beta(k\cdot\sigma)\, d\sigma \qquad (12.4)\\ &= |\mathbb{S}^{d-2}| \int_0^\pi \left[1 - P_\ell^{(d)}(\cos\theta)\right] \beta(\cos\theta)\, \sin^{d-2}\theta\, d\theta. \end{aligned}$$

Remark 12.5 It may be that some eigenvalues ν_ℓ coincide, depending on the choice of β; so the multiplicity may not be exactly $N(d,\ell)$.

Proof of Theorem 12.4 Let me skip the technical issue of domain of definition (handled by many authors in the much more tricky case of $\mathbb{R}^d$) and jump directly to the calculation of the spectrum. Since Δ and L_β commute, we know that the basis of spherical harmonics $(Y_{\ell,m})$ is a basis of diagonalisation for L_β. The question is to find the associated eigenvalues

$$\nu(Y) = -\langle L_\beta Y, Y\rangle = \frac{1}{2}\iint \left[Y(\sigma) - Y(k)\right]^2 \beta(k\cdot\sigma)\, dk\, d\sigma.$$

A particular choice for Y is $Y(k) = c\, P_\ell(k\cdot\eta)$ for some $\eta \in \mathbb{S}^{d-1}$, where c is a normalisation constant. If η and ζ are two such unit vectors, again using an isometry of $\mathbb{S}^{d-1}$ which exchanges η and ζ, we find

$$\frac{1}{2}\iint \left[P_\ell(\sigma\cdot\eta) - Y(k\cdot\eta)\right]^2 \beta(k\cdot\sigma)\, dk\, d\sigma$$

$$= \frac{1}{2} \iint \left[P_\ell(\sigma \cdot \zeta) - Y(k \cdot \zeta)\right]^2 \beta(k \cdot \sigma)\, dk\, d\sigma;$$

in other words the eigenvalue ν associated with $P_\ell(k \cdot \eta)$ is independent of the choice of η. Since all spherical harmonics Y are linear combinations of such elementary functions, we deduce that the eigenvalue $\nu(Y)$ is independent of Y and only depends on ℓ. It remains to compute this number.

Consider the elementary Boltzmann equation operator L^θ where β is concentrated at a single deviation angle θ:

$$L^\theta f(k) = \int_{\mathbb{S}^{d-1}} [f(\sigma) - f(k)]\, 1_{k \cdot \sigma = \cos\theta}\, d\sigma.$$

In this way

$$L_\beta = \int_0^\pi L^\theta\, \beta(\cos\theta)\, \sin^{d-2}\theta\, d\theta. \tag{12.5}$$

Each $\sigma \in \mathbb{S}^{d-1}$ such that $k \cdot \sigma = \cos\theta$ is the endpoint of a unit-speed geodesic $\gamma : [0, \theta] \to \mathbb{S}^{d-1}$ such that $\gamma(0) = k$, $\dot\gamma(0) = \phi \in \mathbb{S}^{d-2}_{k\perp}$. In other words, $\sigma = \exp_k(\theta\phi)$. Thus

$$L^\theta f(k) = \int_{\mathbb{S}^{d-2}_{k\perp}} \Big[f\big(\exp_k \theta\phi\big) - f(k)\Big]\, d\phi.$$

Again L^θ commutes with Δ, and the eigenvalues are given by the numbers

$$\nu_\ell(\theta) = -\langle L_\theta Y, Y\rangle,$$

where Y is any normalised spherical harmonic in H_ℓ. Then

$$\begin{aligned}
\nu_\ell(\theta) &= \int_{\mathbb{S}^{d-1}} Y(k) \left(\int_{\mathbb{S}^{d-2}_{k\perp}} \left[Y(k) - Y(\exp_k \theta\phi)\right] d\phi \right) dk \\
&= \int_{\mathbb{S}^{d-1}} Y(k)^2\, |\mathbb{S}^{d-2}_{k\perp}|\, dk - \int_{\mathbb{S}^{d-1}} \int_{\mathbb{S}^{d-2}_{k\perp}} Y(k)\, Y(\exp_k \theta\phi)\, d\phi\, dk \\
&= |\mathbb{S}^{d-2}| - \int_{\mathbb{S}^{d-1}} \int_{\mathbb{S}^{d-2}_{k\perp}} Y(k)\, Y(\exp_k \theta\phi)\, d\phi\, dk,
\end{aligned}$$

and as a consequence $\iint Y(k) Y(\exp_k \theta\phi)\, d\phi\, dk$ does not depend on the choice of Y.

By the addition formula of Proposition 12.1 (ii), for all k and $\sigma = \exp_k \theta\phi$,

$$P_\ell^{(d)}(k \cdot \sigma) = \frac{|\mathbb{S}^{d-1}|}{N(d,\ell)} \sum_{m=1}^{N(d,\ell)} Y_{\ell,m}(k)\, Y_{\ell,m}(\sigma),$$

so upon integration $d\phi\, dk$ and substitution $k \cdot \sigma = \cos\theta$,

$$P_\ell^{(d)}(\cos\theta)\, |\mathbb{S}^{d-2}|\, |\mathbb{S}^{d-1}| = \frac{|\mathbb{S}^{d-1}|}{N(d,\ell)} \sum_{m=1}^{N(d,\ell)} \iint Y_{\ell,m}(k)\, Y_{\ell,m}(\exp_k \theta\phi)\, d\phi\, dk.$$

On the right hand side, all terms in the sum are equal to $|\mathbb{S}^{d-2}| - \nu_\ell(\theta)$. So

$$P_\ell^{(d)}(\cos\theta)\, |\mathbb{S}^{d-2}|\, |\mathbb{S}^{d-1}| = |\mathbb{S}^{d-1}|\, \big(|\mathbb{S}^{d-2}| - \nu_\ell(\theta)\big).$$

All in all,

$$\nu_\ell(\theta) = |\mathbb{S}^{d-2}|\, \big(1 - P_\ell^{(d)}(\cos\theta)\big). \qquad (12.6)$$

Then the conclusion follows from (12.5) and its corollary $\nu_\ell(\beta) = \int_0^\pi \nu_\ell(\theta)\, \beta(\cos\theta)\, \sin^{d-2}\theta\, d\theta$. □

Bibliographical Notes

About spherical harmonics, among innumerable references (the majority of which insist on dimension $d = 3$) I recommend the exceptionally clear lecture notes by Frye and Efthimiou [87] considering general dimensions.

The application of special functions and symmetries to the computation of the spectrum of the Boltzmann equation with Maxwellian kernels goes back at least to Wang Chang and Uhlenbeck in the sixties [190]; general background is in the reference books like Cercignani [47]. More recently, noting that the relevant material is scattered in a number of places, Dolera [74] provides a self-contained presentation and review. The eigenvalues are also given by integral expressions involving the kernel and Legendre polynomials.

13 Reduction to the Sphere, Reformulated

Using the language of Sect. 11 it is possible to rewrite the criterion from Sect. 10 in a more synthetic way. For simplicity I will only consider product collision kernels: $B(v - v_*, \sigma) = \Phi(|v - v_*|)\, \beta(\cos\theta)$; it is not difficult to adapt the statements to more general collision kernels. The associated Landau function Ψ is proportional to $|v - v_*|^2 \Phi(|v - v_*|)$. Theorem 10.2, in its simplest version, can now be rewritten

Theorem 13.1 (Criterion for Decay of the Fisher Information) *Consider the spatially homogeneous Boltzmann equation with kernel* $B(v - v_*, \sigma) = \Phi(|v -$

$v_*|)\,\beta(\cos\theta)$, or the Landau equation with function $\Psi(|v - v_*|)$ proportional to $|v - v_*|^2\Phi(|v - v_*|)$. Let

$$\overline{\gamma} = \sup_{r>0} \frac{r\,|\Phi'(r)|}{\Phi(r)} = \sup_{r>0} \left| \frac{r\Psi'(r)}{\Psi(r)} - 2 \right|.$$

Then, the Fisher information I is nonincreasing along solutions of equation, as soon as

$$\overline{\gamma} \leq \sqrt{K_*},$$

where K_* is the best constant in the following inequality, required for all $F : \mathbb{S}^{d-1}/\{\pm I\} \to \mathbb{R}_+$,

- *For Boltzmann:*

$$\int_{\mathbb{S}^{d-1}} \Gamma_\beta(\sqrt{F}) \leq \frac{1}{K_*} \int_{\mathbb{S}^{d-1}} F\,\Gamma_{1,\beta}(\log F), \tag{13.1}$$

  where the expressions of Γ_β and $\Gamma_{1,\beta}$ are in Definition 11.3;
- *For Landau:*

$$\int_{\mathbb{S}^{d-1}} \Gamma_1(\sqrt{F}) \leq \frac{1}{K_*} \int_{\mathbb{S}^{d-1}} F\,\Gamma_2(\log F), \tag{13.2}$$

  where the expressions of Γ_1 and Γ_2 are in Proposition 11.1.

The constant K_* a priori depends on whether we are considering Boltzmann or Landau, and a priori depends on β. Here I is the identity on $\mathbb{S}^{d-1}$. Note that in the definition of F, taking quotient by $\{\pm I\}$ is the same as requiring F to be even. In differential geometry, $\mathbb{S}^{d-1}/\{\pm I\}$ is called the projective space in dimension d (that is, the set of lines in $\mathbb{R}^d$) and denoted $\mathbb{RP}^{d-1}$.

Obviously these criteria can be refined as in Sect. 10, considering non-factorised kernels B, but also replacing the square roots by the more precise nonlinearity there. The advantage of the formulation above is that it is compact and synthetic. Clearly (13.2) is the AGC of (13.1). The remarkable achievement of Guillen–Silvestre is the explicit reduction of the decay of I to an optimal functional inequality of logarithmic Sobolev type. Before going more deeply in this issue, let us consider the linearisation.

14 Linearisation

Let $M(v) = (2\pi)^{-d/2}e^{-|v|^2/2}$. Linearising the Boltzmann equation around M consists in writing $f = M(1 + \varepsilon h)$ and letting $\varepsilon \to 0$, then $Q(f, f) \simeq \varepsilon \mathcal{L}_B h$,

where

$$\mathcal{L}_B h = \frac{1}{M}\big[Q(M, Mh) + Q(Mh, M)\big]$$
$$= \int \big[h(v') + h(v'_*) - h(v) - h(v_*)\big]\, M(v_*)\, B(v - v_*, \sigma)\, dv_*\, d\sigma.$$

Of course the linearised Boltzmann equation is $\partial_t h = \mathcal{L}_B h$. Without loss of generality I assume $\int h(v)\, dv = 0$, $\int h(v)\, v\, dv = 0$, $\int h(v)\, |v|^2\, dv = 0$. Then as $\varepsilon \to 0$,

$$\begin{cases} H(M(1+\varepsilon h)) \simeq H(M) + \dfrac{\varepsilon^2}{2}\|h\|^2_{L^2(M)}, \\ I(M(1+\varepsilon h)) \simeq I(M) + \varepsilon^2 \|\nabla h\|^2_{L^2(M)}, \\ \Gamma_1((\log(1+\varepsilon M)) \simeq \varepsilon^2 \Gamma_1(h). \end{cases}$$

So Theorem 13.1 admits the following linearised version involving the Dirichlet form

$$\mathcal{I}(h) = \int_{\mathbb{R}^d} |\nabla h(v)|^2\, M(v)\, dv.$$

Theorem 14.1 (Decay of the Dirichlet Form Along Linearised Boltzmann) *With the same notation as in Theorem 13.1, for $\mathcal{I}$ to be nonincreasing along either the spatially homogeneous linearised Boltzmann or linearised Landau equation, it is sufficient that $\overline{\gamma} \leq \sqrt{4P_*}$, where P_* is the best constant in the following inequality: for all even $h : \mathbb{S}^{d-1}/\{\pm I\} \to \mathbb{R}$ with $\int h\, dv = 0$,*

- *For Boltzmann:*

$$\int_{\mathbb{S}^{d-1}} \Gamma_\beta(h) \leq \frac{1}{P_*} \int_{\mathbb{S}^{d-1}} \Gamma_{1,\beta}(h).$$

- *For Landau:*

$$\int_{\mathbb{S}^{d-1}} \Gamma_1(h) \leq \frac{1}{P_*} \int_{\mathbb{S}^{d-1}} \Gamma_2(h).$$

Remark 14.2 The range of γ may possibly be further improved by restricting to functions h satisfying not only $\int h = 0$, but also $\int h\, v\, dv = 0$ and $\int h|v|^2\, dv = 0$.

The constant P_* a priori depends on whether we are considering Boltzmann or Landau, and a priori depends on β. For the Landau case, it cannot be larger than λ_1, which is thus the best uniform lower bound we may hope for. As we shall see later, that best possible bound does hold true: $P_* \geq \lambda_1$, and in general there is equality. Before that we need some reminders.

Bibliographical Notes

The relation between entropic inequalities and Poincaré (spectral gap) inequalities is very well known, see e.g. [16, 188]. In the special case $\gamma = 0$ (Maxwellian molecules), the monotonicity of I follows from the commutation property with Δ; apart from that special case the general result of Theorem 14.1 is new as far as I know.

15 Reminders on Diffusion Processes

Some reminders from the theory of diffusion of processes. Since Bakry and Émery, the theory has been built upon the concept of **curvature-dimension condition** $\mathrm{CD}(K, N)$: Ricci curvature bounded below by K, dimension bounded above by N. Expressed in terms of the Laplace–Beltrami operator, this condition, set on a Riemannian manifold (M, g), reads

$$\forall f : M \to \mathbb{R}, \qquad \Gamma_2(f) \geq \frac{(\Delta f)^2}{N} + K\Gamma_1(f).$$

(Here $\Gamma_1(f) = \Gamma_1(f, f)$, etc.) For instance the sphere satisfies $\mathrm{CD}(d-2, d-1)$:

$$\forall f : \mathbb{S}^{d-1} \to \mathbb{R}, \qquad \Gamma_2(f) \geq \frac{(\Delta f)^2}{d-1} + (d-2)\,\Gamma_1(f). \tag{15.1}$$

Keeping only the latter term, this obviously yields

$$\int f\Gamma_2(\log f) \geq (d-2)\int f\Gamma_1(\log f) = (d-2)I(f). \tag{15.2}$$

Along the heat flow, $dH/dt = -I$, $dI/dt = -2\int f\Gamma_2(\log f)$; so (15.2) is the same as

$$-\frac{d}{dt}I(e^{t\Delta}f) \geq -2(d-2)\frac{d}{dt}H(e^{t\Delta}f) \geq 0.$$

Upon integrating from $0 \to \infty$ *à la* Stam, we discover

$$H(f) \leq \frac{1}{2(d-2)}I(f), \tag{15.3}$$

that is, the log Sobolev inequality with constant $d-2$. This is good, but not optimal. On the one hand, in dimension 2 the unit sphere $\mathbb{S}^1$ has no curvature, but still satisfies log Sobolev, with constant 1. On the other hand, linearisation of (15.3) yields a Poincaré inequality with constant $d-2$, but we know from the discussion of spherical harmonics that the spectral gap for $-\Delta$ on $\mathbb{S}^{d-1}$ is $d-1$, not $d-2$. It turns out that $d-1$ is indeed optimal, not only for Poincaré but also for log Sobolev: For all probability densities f on $\mathbb{S}^{d-1}$, for all functions h on $\mathbb{S}^{d-1}$ with $\int h = 0$,

$$\int h^2 \leq \frac{1}{d-1} \int \Gamma_1(h), \tag{15.4}$$

$$H(f) \leq \frac{1}{2(d-1)} I(f), \tag{15.5}$$

$$\int f\,\Gamma_1(\log f) \leq \frac{1}{d-1} \int f\,\Gamma_2(\log f). \tag{15.6}$$

To summarise: There are three functional inequalities of interest for us:

- Optimal Poincaré (spectral gap): For all functions h,

$$\int_M h^2 - \left(\int_M h\right)^2 \leq \frac{1}{\lambda_1(M)} \int_M |\nabla h|^2 \tag{15.7}$$

- Optimal log Sobolev inequality: For all probability densities f,

$$\int_M f \log f \leq \frac{1}{2L(M)} \int_M f|\nabla \log f|^2 \tag{15.8}$$

- Optimal differential log Sobolev inequality, or optimal differential Bakry–Émery inequality: For all probability densities f,

$$\int_M f|\nabla \log f|^2 \leq \frac{1}{L_*(M)} \int f\Gamma_2(\log f); \tag{15.9}$$

and always

$$L_* \leq L \leq \lambda_1,$$

while

$$L_*(\mathbb{S}^{d-1}) = L(\mathbb{S}^{d-1}) = \lambda_1(\mathbb{S}^{d-1}) = d-1.$$

When this is applied to Theorems 13.1 and 14.1, taking into account $\int \Gamma_1(\sqrt{F}) = (1/4)I(F)$, this immediately yields a range of exponents for decay of the Fisher information along the Landau equation: $K_* = 4L_* = 4(d-1)$, $P_* = \lambda_1 = d-1$, and in both cases $\overline{\gamma} \leq 2\sqrt{d-1}$. This is already good, and better than the crude bounds in Theorem 9.4. But it still fails to cover the most important case, physically speaking, $\gamma = -3$ in dimension $d = 3$, in fact with this bound $\overline{\gamma}$ can only go up to $2\sqrt{2}$, which is < 3 (not by much, but that is life!).

However there is a crucial information which was not yet exploited: Since F is assumed even in the criteria of Theorems 13.1 and 14.1, this can only improve the constants; in fact we are looking for optimal constants on $\mathbb{RP}^{d-1}$, rather than $\mathbb{S}^{d-1}$. For the spectral gap, we already know the answer: considering even functions amounts to rule out spaces H_ℓ with odd ℓ, so

$$\lambda_1(\mathbb{RP}^{d-1}) = \lambda_2(\mathbb{S}^{d-1}) = 2d \tag{15.10}$$

(much better than $d - 1$). Also, for $d = 2$ one can work out completely

$$L_*(\mathbb{RP}^1) = L(\mathbb{RP}^1) = 4 \quad (= \lambda_1(\mathbb{RP}^1)).$$

For higher dimensions the constants L and L_* are not as good: In fact

$$d \geq 3 \Longrightarrow \qquad L_*(\mathbb{RP}^{d-1}) \leq L(\mathbb{RP}^{d-1}) < 2d.$$

However, the excellent spectral gap already implies at least an improved bound on L: Indeed, a theorem by Rothaus shows that if M satisfies $\mathrm{CD}(K, N)$ and $\lambda_1 > KN/(N-1)$ then $L(M) \geq [K(1-1/N) + (4/N)\lambda_1]/(1+1/N)^2$; in our case this yields

$$L(\mathbb{RP}^{d-1}) \geq (d-1)\left(\frac{d+2}{d}\right)^2. \tag{15.11}$$

(Note that it does capture the optimal constant 4 when $d = 2$.) In dimension 3 it yields an excellent bound $60/9 \simeq 5.55$ for L. While this gives no lower bound for L_*, it suggests at least that the latter is also improved. In fact, one can establish

$$5.5 \leq L_*(\mathbb{RP}^2) < 5.74. \tag{15.12}$$

In higher dimensions, the bound gets worse and worse, but still one has at least that $2d = \lambda_1(\mathbb{RP}^{d-1}) \geq L_*(\mathbb{RP}^{d-1}) \geq L_*(\mathbb{S}^{d-1}) = d - 1$.

Now we are ready to rework Theorems 13.1 and 14.1, and obtain

Theorem 15.1 (Guillen–Silvestre Theorem, Full Version) *Assume that the function Ψ appearing in Landau's collision operator satisfies $\overline{\gamma} = \sup |r\Psi'(r)/\Psi(r) - 2| \leq 4$. Then*

(i) Fisher's information can only go down with t along the spatially homogeneous Landau equation;
(ii) The Dirichlet form $\int |\nabla h|^2 M$ can only go down with t along the spatially homogeneous linearised Landau equation.

Remark 15.2 The choice of the constant 4 in Theorem 15.1 is because the Landau equation theory assumes at least $\gamma \geq -4$.

Proof of Theorem 15.1 For (i): In view of Theorem 13.1, suffices to check $L_*(\mathbb{RP}^{d-1}) \geq 4$. This is postponed to Sect. 20: There I shall prove that $L_*(\mathbb{RP}^1) = 4$, $L_*(\mathbb{RP}^2) \geq 4.5$ (a bit short of the better bounds 4.75 and 5.5 obtained respectively by Guillen–Silvestre and by Ji, but quite sufficient for this theorem), and for higher dimensions $L_*(\mathbb{RP}^{d-1}) \geq 6d/(d+1)$ (quite bad in large dimension since actually $L_* \geq d-1$, but also quite sufficient for this theorem).

Then, (ii) is simpler: It follows directly from (15.10) and Theorem 14.1; actually any $\overline{\gamma} \leq \sqrt{8d}$ will do. □

Now what about the Boltzmann equation, that is, what to do of the inequalities appearing in the first half of Theorems 13.1 and 14.1? These ones do not fall within the range of usual functional inequalities, but they are related as the preceding chapters demonstrate. The next chapter will be devoted to the linearised case: a spectral gap problem. The subsequent four chapters will consider the nonlinear inequality.

Bibliographical Notes

There are two lines of thought behind curvature-dimension bounds: operators and geodesics. The first one goes via functional inequalities involving Δ, as in the works of Bakry, Ledoux and others. The other one involves the behaviour of geodesics and the way they distort measure. This dichotomy can be seen as an Eulerian/Lagrangian duality; see [188, Chapter 14]. The second approach has been at the core of the synthetic theory of curvature-dimension bounds [188, Part III]. In the smooth setting, the Bochner formula relates the two approaches, which are therefore equivalent in a smooth context.

For functional inequalities in log Sobolev style, Bakry–Gentil–Ledoux [16] is a standard rich introduction to the subject; the differential Bakry–Émery inequality with optimal constants is Eq. (5.7.5) in that book. Another excellent introduction is the collective work [6], in french.

Inequality (15.4) for CD$(d-2, d-1)$ manifolds is due to Lichérowicz, while (15.5) and (15.6) are part of the Bakry–Émery theory. The optimal inequality (15.5) is discussed in Ledoux [118], who develops a unified approach to Poincaré and log Sobolev inequalities.

Michel Ledoux has kindly provided references on the problem of constants on $\mathbb{RP}^{d-1}$, with the help of Fabrice Baudoin and Dominique Bakry. The theorem by Rothaus is from [154]. Saloff-Coste [156] considers the log Sobolev and Poincaré constants on $\mathbb{RP}^{d-1}$ and notices that the log Sobolev and Poincaré inequality differ (Theorem 9.2 therein), and that the ratio should go down to $1/2$ as $d \to \infty$ (Remark 3), so that in large dimension the sphere and projective space would have nearly similar L and L_* constants. Details do not seem to have been written down before. A lecture by Fontenas [80] addresses this and related topics. Sehyun Ji worked on this issue and presented preliminary results in the summer school Mathemata (July 2024), then announced [113] a lower bound $L_*(\mathbb{RP}^{d-1}) \geq d + 3 - 1/(d-1)$.

16 Spectral Gap

This section addresses the problem of monotonicity for the Dirichlet form I along the spatially homogeneous linearised Boltzmann equation; that is, the linearised version of the initial quest.

Consider the following commutative diagram, where h is a function on $\mathbb{S}^{d-1}$ with zero average.

$$\begin{array}{ccc} \displaystyle\frac{1}{2}\int_{\mathbb{S}^{d-1}} h^2 & \xrightarrow{-\frac{d}{dt}\big|_{L_\beta}} & \displaystyle\int_{\mathbb{S}^{d-1}} \Gamma_\beta(h) \\ \big\downarrow{\scriptstyle -\frac{d}{dt}|_\Delta} & & \big\downarrow{\scriptstyle -\frac{d}{dt}|_\Delta} \\ \displaystyle\int_{\mathbb{S}^{d-1}} \Gamma_1(h) & \xrightarrow{-\frac{d}{dt}\big|_\Delta} & \displaystyle 2\int_{\mathbb{S}^{d-1}} \Gamma_{1,\beta}(h) \end{array}$$

Knowing that $-\frac{d}{dt}\big|_\Delta \int h^2 \geq \lambda_1 \int \Gamma_1(h)$ does not imply in itself that $\frac{d}{dt}\big|_{L_\beta} \frac{d}{dt}\big|_\Delta \frac{1}{2}\int h^2 \geq -\lambda_1 \frac{d}{dt}\big|_{L_\beta} \int \Gamma_1(h)$; it is actually the other way round. But these are symmetric operators with a common basis of eigenfunctions (spherical harmonics e_k), so we may write (λ_k) and (ν_k) for the eigenvalues of $-\Delta$ and $-L_\beta$, with repetition, and if $h = \sum x_k\, e_k$, assuming $\int h = 0$,

$$\frac{1}{2}\int h^2 = \frac{1}{2}\sum_{k\geq 1} x_k^2, \qquad \int \Gamma_1(h) = \sum_{k\geq 1} \lambda_k x_k^2$$

$$\int \Gamma_\beta(h) = \sum_{k\geq 1} \nu_k x_k^2 \qquad 2\int \Gamma_{1,\beta}(h) = 2\sum_{k\geq 1} \lambda_k \nu_k x_k^2.$$

Then

$$2\int \Gamma_{1,\beta}(h) \geq 2\lambda_1 \sum_{k\geq 1} \nu_k x_k^2 = 2\lambda_1 \int \Gamma_\beta(h).$$

Let P_* be as in Theorem 14.1 (first forget about the parity requirement); then the above computation shows that $P_* \geq \lambda_1$, whatever β, and also in the AGC. It is also immediate from those formulae that λ_1 is achieved for $x = e_1$, except if $\nu_1 = 0$ (which could arise from a peculiar choice of β; recall formula (12.4)). When only even functions are considered, as in the statement of Theorem 14.1, the discussion is the same with λ_1 and ν_1 replaced by λ_2 and ν_2.

The above discussion is summarised into the

Proposition 16.1 *Whatever the collision kernel* β*, it holds* $P_* \geq \lambda_1(\mathbb{RP}^{d-1}) = \lambda_2(\mathbb{S}^{d-1})$*, with equality as soon as* $\nu_2(\beta) \neq 0$.

Remark 16.2 It is enlightening to work out directly the one-dimensional case. When $d = 2$, identify $\mathbb{S}^1$ with $\mathbb{R}/(2\pi\mathbb{Z})$, write $\dot{g}$ for the usual derivative of g, then the problem becomes

$$\iint \left(g(\alpha') - g(\alpha)\right)^2 \beta(\cos(\alpha - \alpha'))\, d\alpha\, d\alpha' \leq$$
$$\frac{1}{P_*} \iint \left(\dot{g}(\alpha') - \dot{g}(\alpha)\right)^2 \beta(\cos(\alpha - \alpha'))\, d\alpha\, d\alpha'. \tag{16.1}$$

Let us prove (16.1) by elementary means. Upon first sight, the optimal constant P_* will depend on β. But rewrite everything in Fourier series. Writing by abuse of notation $\widehat{\beta}$ for $\widehat{\beta \circ \cos}$,

$$\beta(\alpha) = \sum_{k\in\mathbb{Z}} \widehat{\beta}(k) \cos(k\alpha), \qquad g(\alpha) = \sum_{k\in\mathbb{Z}} \widehat{g}(k)\, e^{ik\alpha},$$

where

$$g(\alpha) = \frac{1}{2\pi} \int_0^{2\pi} g(t)\, e^{-i\,kt}\, dt.$$

By approximation we may assume that β is integrable. Then, using standard formulas of Fourier transform,

$$\frac{1}{2} \iint \left(g(\alpha') - g(\alpha)\right)^2 \beta(\cos(\alpha' - \alpha))\, d\alpha\, d\alpha'$$
$$= \int g(\alpha)^2\, d\alpha \int \beta(\cos\alpha'))\, d\alpha'' - 2 \int g(\alpha)\, g(\alpha')\, \beta(\cos(\alpha' - \alpha))\, d\alpha\, d\alpha'$$
$$= \|g\|_{L^2}^2 \left(\int \beta\right) - \int g(g * \beta)$$
$$= 2\pi^2 \Big[\Big(\sum_k |\widehat{g}(k)|^2\Big)\widehat{\beta}(0) - \Re \sum_k \widehat{g}(k)(\widehat{g}(k)\widehat{\beta}(k))^*\Big]$$
$$= 2\pi^2 \Big[\sum_k |\widehat{g}(k)|^2 (\widehat{\beta}(0) - \Re\widehat{\beta}(k))\Big]$$
$$= 2\pi^2 \Big[\sum_k |\widehat{g}(k)|^2 (\widehat{\beta}(0) - \widehat{\beta}(k))\Big].$$

So (16.1) is equivalent to

$$\sum_{k\in\mathbb{Z}} |\widehat{g}(k)|^2 \big(\widehat{\beta}(0) - \widehat{\beta}(k)\big) \le \frac{1}{P_*} \sum_{k\in\mathbb{Z}} |k|^2 |\widehat{g}(k)|^2 \big(\widehat{\beta}(0) - \widehat{\beta}(k)\big). \tag{16.2}$$

The coefficients $\widehat{\beta}(0) - \widehat{\beta}(k)$ are nonnegative since β is, and vanish for $k = 0$. So (16.2) holds with $P_* = 1$. Now if we recall the additional condition that g is even, i.e. $g(\alpha + \pi) = g(\alpha)$, only even coefficients k remain, and $\widehat{g}(\pm 1) = 0$, so the constant P_* improves from 1 to 4. This handles the one-dimensional case. □

The Boltzmann analogue of Theorem 15.1 (ii) now comes as an easy consequence of the previous estimates:

Theorem 16.3 (Decay of the Dirichlet Form Along Linearised Boltzmann) *With the notation* (10.4), *let* $B = B(|v - v_*|, \cos\theta)$ *be a Boltzmann kernel satisfying* $\overline{\gamma}(B) \le 4$. *Then the Dirichlet form* $\int |\nabla h|^2 M$ *can only go down with* t *along the spatially homogeneous linearised Boltzmann equation.*

Proof of Theorem 16.3 Combining Theorem 14.1 with Proposition 16.1 yields the decay for $\overline{\gamma}(B) \le \sqrt{4\lambda_2(\mathbb{S}^{d-1})} = \sqrt{8d}$, which is always at least 4. □

17 Local Monotonicity Criterion via Curvature

Now we are back to the study of Criterion 13.1 for the nonlinear Boltzmann equation. Recall the expression of Γ_β and $\Gamma_{1,\beta}$ from (11.4)–(11.5). By analogy with the diffusive case, we may call inequality (13.1) a β-nonlocal differential log Sobolev inequality. As in Sect. 9, let us separate $\Gamma_{1,\beta}$ into two terms, one involving variations of the gradient and the other being "curvature-like", pointwise in the gradient:

$$\begin{aligned}\Gamma_{1,\beta}(f) &= \frac{1}{2}\int \big|\nabla f(\sigma) - P_{k\sigma}\nabla f(k)\big|^2 \beta(k\cdot\sigma)\, d\sigma \\ &\quad + \frac{1}{2}\int \big(|\nabla f(k)|^2 - |P_{k\sigma}\nabla f(k)|^2\big)\, \beta(k\cdot\sigma)\, d\sigma \\ &= \frac{1}{2}\int \big|\nabla f(\sigma) - P_{k\sigma}\nabla f(k)\big|^2 \beta(k\cdot\sigma)\, d\sigma + (d-2)\,\Sigma(\beta)\,|\nabla f(k)|^2,\end{aligned} \tag{17.1}$$

where I used the second identity in Proposition 8.12(x) and Lemma 9.1, and

$$\Sigma(\beta) = \frac{1}{2(d-1)}\int \big[1 - (k\cdot\sigma)^2\big]\, \beta(k\cdot\sigma)\, d\sigma. \tag{17.2}$$

In particular,

$$\Gamma_{1,\beta}(f) \ge (d-2)\,\Sigma(\beta)\,|\nabla f|^2.$$

The first lowest hanging fruit from Criterion 13.1 is to rely on just the curvature term:

$$\int_{\mathbb{S}^{d-1}} F\,\Gamma_{1,\beta}(\log F) \geq (d-2)\,\Sigma(\beta) \int_{\mathbb{S}^{d-1}} F(k)|\nabla \log F(k)|^2\,dk \qquad (17.3)$$
$$= 4(d-2)\,\Sigma(\beta) \int_{\mathbb{S}^{d-1}} |\nabla\sqrt{F}|^2.$$

Remarkably, Σ turns out the be *exactly* the optimal constant relating $\int \Gamma_\beta(g)$ and $\int \Gamma_1(g)$.

Proposition 17.1 *For any even function $g : \mathbb{S}^{d-1} \to \mathbb{R}$,*

$$\int_{\mathbb{S}^{d-1}} \Gamma_\beta(g) \leq \Sigma(\beta) \int_{\mathbb{S}^{d-1}} \Gamma_1(g), \qquad (17.4)$$

and the constant $\Sigma(\beta)$ is optimal (lowest).

Before proving Proposition 17.1, I will state some striking consequences. First, combined with (17.3), Proposition 17.1 implies a universal (independent of β) bound:

Corollary 17.2 (Curvature-based Nonlocal Differential Log Sobolev) *For any even function $F : \mathbb{S}^{d-1} \to \mathbb{R}_+$, for any $\beta : [-1,1] \to \mathbb{R}_+$ with $\int \theta^2 \beta(\cos\theta)\sin^{d-2}\theta\,d\theta < \infty$,*

$$\int_{\mathbb{S}^{d-1}} \Gamma_{1,\beta}(\log F) \geq 4(d-2) \int_{\mathbb{S}^{d-1}} \Gamma_\beta(\sqrt{F}). \qquad (17.5)$$

A very simple decay criterion follows:

Theorem 17.3 (Curvature-induced Decay of I for the Boltzmann Equation) *If the collision kernel $B = B(|v - v_*|, \cos\theta)$ satisfies*

$$\sup_{r>0,\ 0\leq\theta\leq\pi} \left\{ \frac{r}{B} \left| \frac{\partial B}{\partial r} \right| (r, \cos\theta) \right\} \leq 2\sqrt{d-2},$$

then I is nonincreasing along solutions of the spatially homogeneous Boltzmann equation.

That covers for instance $|\gamma| \leq 2$ in any dimension $d \geq 3$.

Proof of Theorem 17.3 As in Corollary 17.2, for any section $\beta_r(\cos\theta) = B(r, \cos\theta)$, we have

$$\int_{\mathbb{S}^{d-1}} F\Gamma_{1,\beta_r}(\log F) \geq (d-2)\Sigma(\beta_r) \int_{\mathbb{S}^{d-1}} F|\nabla \log F|^2$$

$$= 4(d-2)\,\Sigma(\beta_r)\int_{\mathbb{S}^{d-1}} \Gamma_1(\sqrt{F})$$

$$\geq 4(d-2)\int_{\mathbb{S}^{d-1}} \Gamma_{\beta_r}(\sqrt{F}),$$

and Theorem 13.1 (or actually the slightly generalised version coming from Theorem 10.2 when B is not necessarily in product form and (10.4) is used) applies with $\overline{\gamma}(B,r) = \sqrt{4(d-2)}$. □

It remains to prove Proposition 17.1. I shall provide two arguments. The first one carries some intuition of Poincaré inequalities but fails to achieve the sharp bound. The second one carries no intuition at all, but provides the sharp bound.

Proof of Proposition 17.1, Nonsharp Bound Here I will prove

$$\frac{1}{2}\iint \left[g(\sigma)-g(k)\right]^2 \beta(k\cdot\sigma)\,dk\,d\sigma \leq \frac{1}{2(d-1)}\left(|\mathbb{S}^{d-2}|\int_0^\pi \theta^2\beta(\cos\theta)\,\sin^{d-2}\theta\,d\theta\right)\int_{\mathbb{S}^{d-1}} |\nabla g|^2. \tag{17.6}$$

Replacing β by $[\beta(\cos\theta)+\beta(-\cos\theta)]1_{\cos\theta\geq 0}$ does not change the integral since $g(-\sigma) = g(\sigma)$. So we may assume that β is supported in $\theta \in [0, \pi/2]$. Then (17.6) differs from (17.4) by a factor at most $\sup(\theta^2/\sin^2\theta) = \pi^2/4$.

By linearity it is enough to prove (17.6) when β is concentrated on some fixed angle θ. Then the estimate becomes

$$\iint \left[g(\sigma)-g(k)\right]^2 1_{k\cdot\sigma=\cos\theta}\,dk\,d\sigma \leq \frac{\theta^2}{d-1}\int_{\mathbb{S}^{d-1}} |\nabla g|^2.$$

Given k, each σ such that $k\cdot\sigma = \cos\theta$ is the endpoint of a geodesic $[0,\theta]\to\mathbb{S}^{d-1}$ coming from k with initial velocity $\phi \in S_\perp^{d-2}$, the $(d-2)$ orthonormal sphere centered at k and orthogonal to it. In other words, ϕ lies in the unitary tangent space to $\mathbb{S}^{d-1}$ at k. So the estimate to prove really is

$$\iint_{U\mathbb{S}^{d-1}} \left[g(\exp_k\theta\phi)-g(k)\right]^2 d\phi\,d\sigma \leq \frac{\theta^2}{d-1}\int_{\mathbb{S}^{d-1}} |\nabla g|^2,$$

where $U\mathbb{S}^{d-1}$ is the unitary tangent bundle. For each k and ϕ we write $\gamma_{k,\phi}(t) = \exp_k(t\phi)$, and

$$g(\exp_k(\theta\phi)) - g(k) = \int_0^\theta \nabla g(\gamma_{k,\phi}(t))\cdot\dot{\gamma}_{k,\phi}(t)\,dt$$

so

$$\Big(g(\exp_k(\theta\phi)) - f(k)\Big)^2 = \left(\int_0^\theta \nabla g(\gamma_{k,\phi}(t)) \cdot \dot\gamma_{k,\phi}(t)\,dt\right)^2$$
$$\leq \theta \int_0^\theta \big(\nabla g(\gamma_{k,\phi}(t)) \cdot \dot\gamma_{k,\phi}(t)\big)^2\,dt.$$

Upon integration and using the fact that for each t the exponential map is measure preserving from the unitary tangent bundle to itself,

$$\iint_{U\mathbb{S}^{d-1}} \Big(g(\exp_k(\theta\phi)) - g(k)\Big)^2\,dk\,d\phi$$
$$\leq \theta \int_0^\theta \iint_{U\mathbb{S}^{d-1}} \big(\nabla g(\gamma_{k,\phi}(t)) \cdot \dot\gamma_{k,\phi}(t)\big)^2\,dt\,dk\,d\phi$$
$$= \theta \int_0^\theta \iint_{U\mathbb{S}^{d-1}} \big(\nabla g(x) \cdot v\big)^2\,dt\,dx\,dv$$
$$= \theta^2 \iint_{U\mathbb{S}^{d-1}} \big(\nabla g(x) \cdot v\big)^2\,dx\,dv$$
$$= \frac{\theta^2}{d-1} \int_{\mathbb{S}^{d-1}} |\nabla g(x)|^2\,dx.$$

□

Proof of Proposition 17.1, Sharp Bound Now the goal is

$$\frac{1}{2}\iint \big[g(\sigma) - g(k)\big]^2 \beta(k\cdot\sigma)\,dk\,d\sigma$$
$$\leq \frac{1}{2(d-1)}\left(|\mathbb{S}^{d-2}| \int_0^\pi \sin^2\theta\,\beta(\cos\theta)\,\sin^{d-2}\theta\,d\theta\right)\int_{\mathbb{S}^{d-1}} |\nabla g|^2. \qquad (17.7)$$

In other words,

$$\langle -L_\beta g, g\rangle \leq \Sigma(\beta)\,\langle -\Delta g, g\rangle.$$

It suffices to check this for all even spherical harmonics; in other words, that for all $\ell \in 2\mathbb{N}$,

$$\nu_\ell(\beta) \leq \Sigma(\beta)\,\lambda_\ell^{(d)},$$

where $\nu_\ell(\beta)$ is the eigenvalue of $-L_\beta$ in H_ℓ, and $\lambda_\ell^{(d)}(\beta)$ the one of $-\Delta$. This is the same as

$$\int_{\mathbb{S}^{d-1}} \big[1 - P_\ell(k\cdot\sigma)\big]\,\beta(k\cdot\sigma)\,d\sigma \leq \frac{\ell(\ell+d-2)}{2(d-1)} \int_{\mathbb{S}^{d-1}} \big[1 - (k\cdot\sigma)^2\big]\,\beta(k\cdot\sigma)\,d\sigma.$$

The proof is concluded by Lemma 17.4 below. □

Lemma 17.4 *For all $d \in \mathbb{N}$, for all $\ell \in 2\mathbb{N}$,*

$$\sup_{-1 \le x \le 1} \left(\frac{1 - P_\ell^{(d)}(x)}{1 - x^2} \right) \le \frac{\ell(\ell + d - 2)}{2(d - 1)}, \tag{17.8}$$

with equality for $\ell = 2$ (for all x) and for $x \to \pm 1$ (for all ℓ).

Proof of Lemma 17.4 Throughout the proof, superscripts (d) are implicit. First some reformulations: Since $\lambda_\ell = \ell(\ell + d - 2)$ and $P_2(x) = (dx^2 - 1)/(d - 1)$, the inequality amounts to

$$\ell \text{ even} \implies \forall x, \qquad \frac{1 - P_\ell(x)}{\lambda_\ell} \le \frac{1 - P_2(x)}{\lambda_2}.$$

Since $P_\ell(\pm 1) = 1$ and $P_\ell'(\pm 1) = \pm \lambda_\ell$ from (12.1), the inequality also says that P_ℓ remains above the parabola which is tangent to its graph at $x = \pm 1$.

Now for the proof of the Lemma itself. For pedagogical reasons, first consider $d = 2$. Then $\lambda_2 = 4$, $\lambda_{2n} = 4n^2$, and from the interpretation of P_ℓ (Example 12.2), (17.8) is the same as

$$\forall n \in \mathbb{N}, \qquad \frac{1 - \cos(2n\alpha)}{1 - \cos^2 \alpha} \le 2n^2.$$

Equivalently, since $1 - \cos(2n\alpha) = 2\sin^2(n\alpha)$ and $1 - \cos^2 \alpha = \sin^2 \alpha$, the inequality amounts to just

$$|\sin(n\alpha)| \le n \sin \alpha, \tag{17.9}$$

which is geometrically obvious and can also be proven by induction using $|\sin(n\alpha + \alpha)| \le |\cos(n\alpha)| \sin \alpha + |\sin(n\alpha)||\cos \alpha| \le \sin \alpha + |\sin(n\alpha)|$.

To handle the general case, recall that Formula 12.3 reduces the general case to the 2-dimensional situation. In view of that formula and Remark 12.3,

$$1 - P_\ell(x) = \frac{|\mathbb{S}^{d-3}|}{|\mathbb{S}^{d-2}|} \int_{-1}^{1} \left[1 - (x + is\sqrt{1 - x^2})^\ell\right] (1 - s^2)^{\frac{d-4}{2}} \, ds. \tag{17.10}$$

We turn to estimate the real part of the integrand (the imaginary part vanishes by symmetry). Given s and x both $[-1, 1]$, let $\rho = \sqrt{x^2 + s^2(1 - x^2)}$, $\cos \alpha = x/\rho$, $\sin \alpha = s\sqrt{1 - x^2}/\rho$, so that $x + is\sqrt{1 - x^2} = \rho e^{i\alpha}$.
Claim:

$$\forall n \in \mathbb{N}_0,\ \rho \in [0, 1],\ \alpha \in [0, \pi] \qquad \Re\left[1 - (\rho e^{i\alpha})^{2n}\right] \le n(1 - \rho^2) + 2n^2 \rho^2 \sin^2 \alpha. \tag{17.11}$$

The proof of (17.11) is by induction again. For $n = 0$ the identity is obvious. If true at step n, then

$$\begin{aligned}
&\mathfrak{R}\big[1 - (\rho e^{i\alpha})^{2n+2}\big]\\
&\quad = \mathfrak{R}\big[1 - (\rho e^{i\alpha})^{2n}\big] + \mathfrak{R}\big[(\rho e^{i\alpha})^{2n} - (\rho e^{i\alpha})^{2n+2)}\big]\\
&\quad = \mathfrak{R}\big[1 - (\rho e^{i\alpha})^{2n}\big] + \rho^{2n}(1-\rho^2)\cos(2n\alpha) + \rho^{2n+2}[\cos 2n\alpha - \cos(2n+2)\alpha]\\
&\quad = \mathfrak{R}\big[1 - (\rho e^{i\alpha})^{2n}\big] + \rho^{2n}(1-\rho^2)\cos(2n\alpha) - 2\rho^{2n+2}\sin(2n+1)\alpha\,\sin\alpha.
\end{aligned}$$

Using both the induction hypothesis and the bound (17.9), the above is bounded by

$$\begin{aligned}
&\big(n(1-\rho^2) + 2n^2\rho^2\sin^2\alpha\big) + (1-\rho^2) + 2(2n+1)\rho^2\sin^2\alpha \le (n+1)(1-\rho^2)\\
&\quad + 2(n+1)^2\rho^2\sin^2\alpha,
\end{aligned}$$

which propagates the induction hypothesis and proves (17.11). So

$$\begin{aligned}
\mathfrak{R}\big[1 - (x + is\sqrt{1-x^2})^{2n}\big] &\le \big[n(1-s)^2 + 2n^2s^2\big](1-x^2)\\
&\le \big[n(1-2n)(1-s^2) + 2n^2\big](1-x^2).
\end{aligned}$$

Plugging this back in (17.10), we get

$$\begin{aligned}
\frac{1-P_\ell(x)}{1-x^2} &\le \frac{|\mathbb{S}^{d-3}|}{|\mathbb{S}^{d-2}|}\int_{-1}^{1}\big[n(1-2n)(1-s^2)^{\frac{d-2}{2}} + 2n^2(1-s^2)^{\frac{d-4}{2}}\big]\,ds\\
&= \frac{|\mathbb{S}^{d-3}|}{|\mathbb{S}^{d-2}|}\left(n(1-2n)\frac{|\mathbb{S}^{d}|}{|\mathbb{S}^{d-2}|} + 2n^2\frac{|\mathbb{S}^{d-2}|}{|\mathbb{S}^{d-3}|}\right)\\
&= n(1-2n)\frac{|\mathbb{S}^{d-3}|\,|\mathbb{S}^{d}|}{|\mathbb{S}^{d-2}|\,|\mathbb{S}^{d-1}|} + 2n^2\\
&= n(1-2n)\left(\frac{\int_0^\pi \sin^{d-1}\psi\,d\psi}{\int_0^\pi \sin^{d-3}\psi\,d\psi}\right) + 2n^2\\
&= n(1-2n)\left(\frac{d-2}{d-1}\right) + 2n^2 = \frac{2n(n+d-2)}{d-1},
\end{aligned}$$

as required, after application of the Wallis integrals again. □

Bibliographical Notes

Poincaré estimates in positive curvature are classical and appear e.g. in [188, Chapter 18], but the setting here allows for much more explicit constants, as Proposition 17.1 shows. All this section (except the nonoptimal argument for Proposition 17.1) is taken from [111].

18 (Counter)examples

Theorem 17.3 provides a general answer, even if nonoptimal, in dimension $d \geq 3$, but says nothing of dimension 2. This may seem awkward as dimension 2 is usually simpler for the theory of the Boltzmann equation, and as a matter of fact the monotonicity property of the Fisher information for Maxwellian kernels was established in dimension 2 several years before higher dimensions.

It turns out that this is in the order of things. The general criterion established in Sect. 10 fails for dimension 2, at least for certain classes of singular kernels. At the opposite side, when the collision kernel is constant in the angular variable (which is arguably the most regular that can be!) then the criterion holds even in dimension 2, and this will allow for a first positive result for non-Maxwell kernels in that dimension.

Theorem 18.1

(i) Consider a measure β on $[0, \pi]$ defined by

$$\beta(d\theta) = \sum_{i=0}^{N} \beta_i \, \delta_{\theta_i}, \quad \theta_i = \frac{i\pi}{N}, \; N \in \mathbb{N}, \; \beta_i \geq 0, \; \sum_{i=1}^{N-1} \beta_i > 0.$$

Then

$$\inf \left\{ \frac{\int_{\mathbb{S}^1} F\,\Gamma_{1,\beta}(\log F)}{\int_{\mathbb{S}^1} \Gamma_\beta(\sqrt{F})}; \quad F : \mathbb{S}^1/\{\pm I\} \to \mathbb{R}_+ \right\} = 0. \tag{18.1}$$

(ii) If β is constant on $\mathbb{S}^{d-1}$, then for any $d \geq 2$,

$$\inf \left\{ \frac{\int_{\mathbb{S}^{d-1}} F\,\Gamma_{1,\beta}(\log F)}{\int_{\mathbb{S}^{d-1}} \Gamma_\beta(\sqrt{F})}; \quad F : \mathbb{S}^{d-1}/\{\pm I\} \to \mathbb{R}_+ \right\} \geq 4d. \tag{18.2}$$

(iii) For the symmetrised hard spheres kernel in dimension 2,

$$\inf \left\{ \frac{\int_{\mathbb{S}^1} F\,\Gamma_{1,\beta}(\log F)}{\int_{\mathbb{S}^1} \Gamma_\beta(\sqrt{F})}; \quad F : \mathbb{S}^1/\{\pm I\} \to \mathbb{R}_+ \right\} \geq 4\sqrt{2}. \tag{18.3}$$

In particular, I is nonincreasing along solutions of the 2-dimensional spatially homogeneous Boltzmann equation for hard spheres.

Remarks 18.2 1. In Part (i), δ_0 and δ_π play no role in either Γ_β or $\Gamma_{1,\beta}$ (δ_0 obviously corresponds to no deviation in the collision, and so does δ_π when applied to π-periodic functions). So we may work directly with $\theta_1, \ldots \theta_{N-1}$. By setting some coefficients β_i equal to 0, we see that the theorem covers any kernel with finite support in $\mathbb{Q}\pi$. Extension to angles which are irrational to π is a more difficult problem. Another question left open by this theorem is whether one can use this negative result to construct $f = f(v)$ on $\mathbb{R}^2$ such that $I'(f) \cdot Q(f, f) > 0$, say for $B = |v - v_*|\, \beta$. Theorem 19.1 below will show that Part (i) would be false if β would not vanish.

(ii) Part (ii) implies that for $B(v - v_*, \sigma) = |v - v_*|^\gamma$, $|\gamma| \leq 2\sqrt{d}$, I is nonincreasing along solutions of the spatially homogeneous Boltzmann equation. But only the case $\gamma = 1$, $d = 3$ corresponds to a classical interaction considered in physics, namely three-dimensional hard spheres (already covered by Theorem 17.3). Recall that in dimension 2 the hard spheres kernel vanishes for $\theta = 0$ and in dimensions 4 and higher it has an integrable singularity as $\theta \to 0$. Still (iii) tells us that hard spheres in dimension 2 can be handled, this will be done by using the obvious perturbation lemma below.

Lemma 18.3 (Perturbation Lemma for the Integral Criterion) *If β_0 and β are two collision kernels, and $K_*(\beta_0)$, $K_*(\beta)$ are the respective associated optimal constants in the criterion from Theorem 13.1. If there are constants $m, M > 0$ such that*

$$\forall \theta \in [0, \pi], \qquad m\, \beta_0(\cos\theta) \leq \beta(\cos\theta) \leq M\, \beta_0(\cos\theta),$$

Then

$$K_*(\beta) \geq \frac{m}{M}\, K_*(\beta_0).$$

Proof of Theorem 18.1 (i) Fix $0 < \delta < 1/2$ and in $\mathbb{R}/\pi\mathbb{Z}$ let $I_i = [(i-\delta)\pi/N, (i+\delta)\pi/N]$, $1 \leq i \leq N$, so that the intervals I_i are disjoint neighbourhoods of θ_i. Pick up two smooth functions $\mathbb{R}/\pi\mathbb{Z} \to \mathbb{R}_+$, h and ψ, such that

- h is constant on each I_i and takes distinct integer values on all these intervals;
- ψ is π/N-periodic, nonzero and supported over the intervals I_i.

Then for $A > 0$ let $F = (1+A\psi)h$. Since ψ is π/N-periodic, so is $\log(1+A\psi)^{\cdot}$, so

$$(\log F)^{\cdot}(\alpha + \theta_i) - (\log F)^{\cdot}(\alpha) = (\log h)^{\cdot}(\alpha + \theta_i) - (\log h)^{\cdot}(\alpha)$$

and thus

$$\int F\Gamma_{1,\beta}(\log F) = \frac{1}{2}\sum_i \beta_i \int (1+A\psi(\alpha))h(\alpha)\big[(\log h)^{\cdot}(\alpha+\theta_i) - (\log h)^{\cdot}(\alpha)\big]^2\, d\alpha.$$

On the support of ψ, h is constant, and so is $h(\cdot + \theta_i)$; thus $\psi[(\log h)^{\cdot}(\alpha + \theta_i) - (\log h)^{\cdot}(\alpha)]^2 = 0$. In the end,

$$\int F\Gamma_{1,\beta}(\log F) = \frac{1}{2}\sum_i \beta_i \int h(\alpha)\big[(\log h)^{\cdot}(\alpha + \theta_i) - (\log h)^{\cdot}(\alpha)\big]^2 \, d\alpha$$
$$= \int h\Gamma_{1,\beta}(h).$$

This is independent of A and ψ.

On the other hand,

$$\Gamma_\beta(\sqrt{F}) \geq \sum_{i,j} \beta_i \int_{I_j} \big[\sqrt{F(\alpha + \theta_i)} - \sqrt{F(\alpha)}\big]^2 \, d\alpha$$
$$= A\left(\sum_i \beta_i\right)\left(\sqrt{\psi(I_j)} - \sqrt{\psi(I_i)}\right)^2 \left(\sum_j \int_{I_j} h\right)^2.$$

□

Proof of Theorem 18.1 (ii) Take $\beta \equiv 1$. Then

$$\int F\Gamma_{1,\beta}(\log F) = \frac{1}{2}\iint F(k)\big|\nabla \log F(\sigma) - \nabla \log F(k)\big|^2_{k,\sigma} \, dk \, d\sigma$$
$$= \frac{1}{2}\iint F(k)|\nabla \log F(k)|^2 \, dk \, d\sigma$$
$$+ \frac{1}{2}\iint F(k)|\nabla \log F(\sigma)|^2 \, dk \, d\sigma$$
$$- \iint \nabla F(k) P_{\sigma k} \nabla \log F(\sigma) \, dk \, d\sigma$$
$$= \frac{|\mathbb{S}^{d-1}|}{2}\int F|\nabla \log F|^2 + \left(\int_{\mathbb{S}^{d-1}} F\right)\left(\int_{\mathbb{S}^{d-1}} |\nabla\sqrt{F}|^2\right) + 0,$$

since $\int P_{\sigma k}\nabla \log F(\sigma) \, d\sigma = 0$ by Lemma 8.5 (recall $M_{\sigma k}\nabla f = P_{\sigma k}\nabla f$ if $\nabla f(\sigma) \cdot \sigma = 0$). So

$$\int F\Gamma_{1,\beta}(\log F) \geq 2|\mathbb{S}^{d-1}| \int |\nabla\sqrt{F}|^2. \tag{18.4}$$

On the other hand,

$$\int \Gamma_\beta(\sqrt{F}) = \frac{1}{2}\iint \left(\sqrt{F(\sigma)} - \sqrt{F(k)}\right)^2 dk \, d\sigma$$

$$
\begin{aligned}
&= |\mathbb{S}^{d-1}| \int F - \left(\int \sqrt{F}\right)^2 \\
&= |\mathbb{S}^{d-1}|^2 \left[\frac{1}{|\mathbb{S}^{d-1}|} \int F - \left(\frac{1}{|\mathbb{S}^{d-1}|} \int \sqrt{F}\right)^2\right] \\
&\leq \frac{|\mathbb{S}^{d-1}|^2}{\lambda_1} \left(\frac{1}{|\mathbb{S}^{d-1}|} \int_{\mathbb{S}^{d-1}} |\nabla \sqrt{F}|^2\right),
\end{aligned}
$$

where $\lambda_1 = \lambda_1(\mathbb{RP}^{d-1}) = 2d$. The conclusion follows. □

Proof of Theorem 18.1 (iii) For $d = 2$ the hard spheres kernel $b(\cos\theta)$ varies like $\sin(\theta/2)$. Its symmetric version $[b(\cos\theta) + b(-\cos\theta)]/2$ like $[\sin(\theta/2) + \cos(\theta/2)]/2$, which takes values in $[1/2, \sqrt{2}/2]$. So we may apply Lemma 18.3 with $\beta_0 = 1$ and $M/m = \sqrt{2}$, leading to (18.3). □

Bibliographical Notes

Counterexample (i) was shown to me by Luis Silvestre (for $N = 2$). Estimates (i)–(ii)–(iii) are published in [111].

The perturbation Lemma 18.3 is obvious, but it is natural, in this context of log Sobolev type inequalities, to link it to the Holley–Stroock perturbation lemma [106].

19 Local-integral Monotonicity Criterion via Positivity

Combining the example of constant kernel in Sect. 18, a comparison argument and the Poincaré inequality from Sect. 17, is enough to show that $K_* > 0$, thereby ruling out the situation of counterexample Theorem 18.1(i) and proving the following monotonicity theorem.

Theorem 19.1 (Positivity-induced Decay of I for the Boltzmann Equation) *If the collision kernel $B = B(|v - v_*|, \cos\theta)$ is such that for all $r > 0$ and (any) $k \in \mathbb{S}^{d-1}$,*

$$
\sup_{0\leq\theta\leq\pi} \left\{\frac{r}{B}\left|\frac{\partial B}{\partial r}\right|(r, \cos\theta)\right\} \leq \left(\frac{4(d-1)|\mathbb{S}^{d-1}| \left(\min_{0\leq\theta\leq\pi} B(r, \cos\theta)\right)}{\int_{\mathbb{S}^{d-1}} [1 - (k\cdot\sigma)^2] B(r, k\cdot\sigma)\, d\sigma}\right)^{1/2},
$$

then I is nonincreasing along solutions of the spatially homogeneous Boltzmann equation. Such is the case if $B = |v - v_*|^\gamma \beta(\cos\theta)$ with

$$|\gamma| \leq \left(\frac{4(d-1)|\mathbb{S}^{d-1}|(\min\beta)}{|\mathbb{S}^{d-2}| \int_0^\pi \sin^d\theta\, \beta(\cos\theta)\, d\theta} \right)^{1/2}. \tag{19.1}$$

Proof of Theorem 19.1 Let $\beta_r(\cos\theta) = \beta(r, \cos\theta)$ and $m_r = \min(\beta_r)$. On the one hand, by (18.4),

$$\int F\Gamma_{1,\beta_r}(\log F) \geq \int F\Gamma_{1,m_r}(\log F) \geq 2|\mathbb{S}^{d-1}|m_r \int |\nabla\sqrt{F}|^2,$$

on the other hand, from Proposition 17.1,

$$\int \Gamma_{\beta_r}(\sqrt{F}) \leq \Sigma(\beta_r) \int |\nabla\sqrt{F}|^2.$$

So for all even $F : \mathbb{S}^{d-1} \to \mathbb{R}_+$,

$$K_*(\beta_r) \geq \frac{2|\mathbb{S}^{d-1}|\, m_r}{\Sigma(\beta_r)}$$

and the conclusion follows from Theorem 10.2 and the definitions of m_r, $\Sigma(\beta_r)$.

□

While this approach provides a considerable list of kernels, even in dimension 2, for which the Fisher information is decaying, it is not powerful enough to handle all classical inverse-law potentials.

Example 19.2 Assume $d = 2$ and inverse power law forces like $1/r^s$. The collision kernel takes the form $B = |v - v_*|^\gamma b(\cos\theta)$ with $\gamma = (s-3)/(s-1)$. Maxwellian molecules correspond to $s = 3$. This kernel is bounded from below, so the right-hand side of (19.1) is a strictly positive number. As s varies, $\gamma = \gamma(s)$ varies continuously, but also the right-hand side of (19.1). By continuity there is an interval around $s = 3$ in which the inequality (19.1) remains satisfied. Numerics can tell us exactly which is this range, but we already see that this method covers some of the classical inverse power law forces in dimension 2, which the curvature-based approach was unable to tackle.

Example 19.3 Let us now consider the case $s = 2$, still in dimension $d = 2$. Then $\gamma = -1$ and $b(\cos\theta)$ is proportional to $1/\sin^2(\theta/2)$. (This is the 2-dimensional version of Rutherford's cross-section formula.) An explicit calculation shows that the right-hand side in (19.1) is $1/\sqrt{2}$, which is not enough to squeeze $|\gamma| = 1$ in the left-hand side. So Theorem 19.1 is not general enough to cover $s = 2$ in dimension $d = 2$.

20 Exploiting Gradient Variations I: The Diffusive Case

In this and the next two sections the question is how to exploit the variation of gradients in $\Gamma_{1,\beta}$, that is mainly the term $(\mathrm{II})_1$ in (9.4). In the diffusive case ($\nu = 2$) this reduces to a Hessian term. Exploiting this Hessian is a classical topic: it corresponds to the difference between $\mathrm{CD}(K,\infty)$ and $\mathrm{CD}(K,N)$ bounds. It will be the subject of the present section. Unfortunately (or fortunately for later work), most if not all the recipes of the diffusive case will turn out difficult to adapt to the Boltzmann situation, as will be reviewed in Sect. 21. However, in Sect. 22 a strategy will be presented, which is based on the reduction to the diffusive case and suffices to handle all the classical situations which are not covered by Theorem 17.3.

20.1 Dimension 2: The Circle

Let us start with $d = 2$, then there is no curvature, and Theorem 10.2 is about functions of one variable θ in $\mathbb{S}^1$ or $\mathbb{S}^1/\{\pm I\}$. In this diffusive context there is no risk of confusion with post-collisional velocities, so I shall use primes for regular derivatives in $\mathbb{R}$. The problem is the differential Bakry–Émery inequality

$$\int_{\mathbb{S}^1} [(\sqrt{f})']^2 \leq \frac{1}{K_*} \int_{\mathbb{S}^1} f[(\log f)'']^2. \tag{20.1}$$

At this point there is a fascinating calculus combining nonlinearities and derivatives in dimension 1, which was discovered and rediscovered several times, in particular by McKean:

$$\begin{aligned}\int f[(\log f)'']^2 &= \int f\left(\frac{f''}{f} - \frac{f'^2}{f^2}\right)^2 \\ &= \int \frac{(f'')^2}{f} - 2\int \frac{f'' f'^2}{f^2} + \int \frac{f'^4}{f^3},\end{aligned}$$

and by integration by parts

$$\int \frac{f'' f'^2}{f^2} = \frac{2}{3}\int \frac{f'^4}{f^3}. \tag{20.2}$$

So that

$$\int f[(\log f)'']^2 = \int \frac{f''^2}{f} - \frac{1}{3}\int \frac{f'^4}{f^3}, \tag{20.3}$$

an algebraic relation involving three quantities which all are natural "higher order" generalisations of the Fisher information.

Remark 20.1 Choosing $\alpha \in (0, 1)$ we find

$$\frac{1}{\alpha}\frac{((f^\alpha)'')^2}{f^{2\alpha-1}} = \frac{f''^2}{f} + (\alpha-1)^2\frac{f'^4}{f^3} + 2(\alpha-1)\frac{f''f'^2}{f^2},$$

so by (20.2)

$$\frac{1}{\alpha}\int\frac{(f^\alpha)''^2}{f^{2\alpha-1}} = \int\frac{f''^2}{ff} + \left[(\alpha-1)^2 + \frac{4}{3}(\alpha-1)\right]\int\frac{f'^4}{f^3}.$$

In particular, the nonnegativity of the left hand side implies

$$\int\frac{f'^4}{f^3} \le \frac{1}{(1-\alpha)(\alpha+\frac{1}{3})}\int\frac{f''^2}{f},$$

which is optimal for $\alpha = 1/3$ and yields

$$\int_{\mathbb{S}^1}\frac{f'^4}{f^3} \le \frac{9}{4}\int_{\mathbb{S}^1}\frac{f''^2}{f}. \tag{20.4}$$

Plugging this back in (20.3) further yields

$$\int f[(\log f)'']^2 \ge \frac{1}{4}\int\frac{f''^2}{f} \ge \frac{1}{9}\int\frac{f'^4}{f^3}.$$

Back to the proof of (20.1), write $g = \sqrt{f}$, the goal is

$$\int g'^2 \le \frac{1}{K_*}\int g^2(\log g^2)''^2. \tag{20.5}$$

But

$$\int g^2(\log g^2)''^2 = 4\int g^2\left(\frac{g''}{g} - \frac{g'^2}{g^2}\right)^2$$
$$= 4\int g''^2 - 8\int\frac{g''g'^2}{g} + 4\int\frac{g'^4}{g^2}.$$

Now the relevant integration by parts reads

$$\int\frac{g''g'^2}{g} = \frac{1}{3}\int\frac{g'^4}{g^2},$$

so

$$\int f(\log f)''^2 = 4\int g''^2 + \frac{4}{3}\int \frac{g'^4}{g^2} = 4\int (\sqrt{f})''^2 + \frac{1}{12}\int \frac{f'^4}{f^3}. \tag{20.6}$$

Combining this with the usual Poincaré inequality on $\mathbb{S}^1$ yields

$$\int f\Gamma_2(\log f) = \int f(\log f)''^2 \geq 4\int (\sqrt{f})''^2 \geq 4\int (\sqrt{f})'^2 = I(f),$$

which yields $K_* \geq 1$ and also implies the log Sobolev inequality with constant 1. Combining this, instead, with the better Poincaré inequality for even functions (that is, when f is π-periodic) yields

$$\int f\Gamma_2(\log f) \geq 4\int (\sqrt{f})''^2 \geq 16\int (\sqrt{f})'^2 = 4I(f),$$

which also implies log Sobolev with constant 4.

To summarise:

$$L_*(\mathbb{RP}^1) = L(\mathbb{RP}^1) = \lambda_1(\mathbb{RP}^1) = 4,$$

and this implies the bound announced in the proof of Theorem 15.1.

20.2 Dimensions 3 and Higher

For $d \geq 3$ there are various ways to combine the Hessian and curvature estimates in order to estimate the best constant K_* in the inequality

$$\int \Gamma_1(\sqrt{f}) \leq \frac{1}{K_*}\int f\Gamma_2(\log f). \tag{20.7}$$

Here I will combine the optimal spectral gap with a change of nonlinearity. So the chain of inequalities will be

$$\int \Gamma_1(\sqrt{f}) \leq \frac{1}{\lambda_1}\int \Gamma_2(\sqrt{f}) \leq \frac{1}{\widetilde{K}\lambda_1}\int f\Gamma_2(\log f). \tag{20.8}$$

Note that $\int \Gamma_1(\sqrt{f}) = (1/4)I(f)$, so this implies log Sobolev with constant $\widetilde{K}\lambda_1/4$. Since $L_* \leq \lambda_1$, $\widetilde{K}$ can be at most 4.

Now

$$\int \Gamma_2(\log f) = \int f\|\nabla^2 \log f\|_{\mathrm{HS}}^2 + (d-2)\int f|\nabla \log f|^2,$$

$$\int \Gamma_2(\sqrt{f}) = \int \|\nabla^2 \sqrt{f}\|_{HS}^2 + (d-2) \int |\nabla \sqrt{f}|^2,$$

and since $\int f |\nabla \log f|^2 = 4 \int |\nabla \sqrt{f}|^2$, it will be sufficient to show

$$\int \|\nabla^2 \sqrt{f}\|_{HS}^2 \leq \frac{1}{\widetilde{K}} \int f \|\nabla^2 \log f\|_{HS}^2 \qquad (20.9)$$

for some $\widetilde{K} \leq 4$.

For this let us play with nonlinear changes of variables: writing $\varphi \circ f = \varphi(f)$, the three basic such formulas are

$$\chi'(f)\Delta f = \Delta \chi(f) - \chi''(f)\Gamma_1(f), \qquad (20.10)$$

$$\Gamma_1(\varphi(f)) = (\varphi')^2(f)\,\Gamma_1(f), \qquad (20.11)$$

$$\Gamma_2(\varphi(f)) = (\varphi')^2(f)\,\Gamma_2(f) + 2\varphi'(f)\,\varphi''(f)\Gamma\Gamma(f) + (\varphi'')^2(f)\,\Gamma_1(f)^2, \qquad (20.12)$$

where I write indifferently $\Gamma_i(f) = \Gamma_i(f, f)$, $\Gamma_1 = \Gamma$, and

$$\Gamma\Gamma(f) = \frac{1}{2}\Gamma\big(f, \Gamma(f, f)\big)$$

(morally this is $\langle \nabla^2 f \cdot \nabla f, \nabla f \rangle$).

Dominique Bakry has been virtuose in showing the power of these local identities, combined with a fourth, integral, formula:

$$\int \psi(f)(\Delta f)^2 = \int \Big[\psi(f)\,\Gamma_2(f) + 3\psi'(f)\,\Gamma\Gamma(f) + \psi''(f)\,\Gamma(f)^2\Big]^2. \qquad (20.13)$$

(Notice that $\psi(f) = 1/f$ in dimension 1 reduces to (20.2).) Let us apply the above formula (20.12) with $f = g^2$ and $\varphi(g) = \log g$:

$$\Gamma_2(\log g) = \frac{1}{g^2}\Gamma_2(g) - \frac{2}{g^3}\Gamma\Gamma(g) + \frac{1}{g^4}\Gamma(g)^2;$$

or equivalently

$$f\Gamma_2(\log f) = 4g^2\Gamma_2(\log g) = 4\Gamma_2(g) - 8\frac{\Gamma\Gamma(g)}{g} + 4\frac{\Gamma(g)^2}{g^2};$$

using $\Gamma\Gamma(\log f) = 8\,\Gamma\Gamma(\log g) = 8\,(\Gamma\Gamma(g)/g^3 - \Gamma(g)^2/g^4)$ and $\Gamma(\log f)^2 = 16\,\Gamma(\log g)^2$, we obtain

$$\int \Gamma_2(g) = \frac{1}{4}\int f\Gamma_2(\log f) + \frac{1}{4}\int f\Gamma\Gamma(\log f) + \frac{1}{16}\int f\Gamma(\log f)^2. \qquad (20.14)$$

From (20.13),

$$\int f(\Delta \log f)^2 = \int f\Gamma_2(\log f) + 3\int f\Gamma\Gamma(\log f) + \int f\Gamma(\log f)^2,$$

so

$$\int f\Gamma\Gamma(\log f) = \frac{1}{3}\left(\int f\big[(\Delta \log f)^2 - \Gamma_2(\log f)\big]\right) - \frac{1}{3}\int f\Gamma(\log f)^2.$$

Plugging this back in (20.14),

$$\begin{aligned}\int \Gamma_2(\sqrt{f}) = \frac{1}{4}\int f\Gamma_2(\log f) + \frac{1}{12}\left(\int f\big[(\Delta \log f)^2 - \Gamma_2(\log f)\big]\right)\\ - \frac{1}{12}\int f\Gamma(\log f)^2 + \frac{1}{16}\int f\Gamma(\log f)^2;\end{aligned}$$

that is

$$\begin{aligned}\int \Gamma_2(\sqrt{f}) = \frac{1}{4}\int f\Gamma_2(\log f) + \frac{1}{12}\left(\int f\big[(\Delta \log f)^2 - \Gamma_2(\log f)\big]\right)\\ - \frac{1}{48}\int f\Gamma(\log f)^2. \qquad (20.15)\end{aligned}$$

The CD$(d-2, d-1)$ criterion yields

$$\Gamma_2(u) \geq \frac{(\Delta u)^2}{d-1} + (d-2)\,\Gamma(u),$$

or

$$(\Delta u)^2 - \Gamma_2(u) \leq (d-2)[\Gamma_2(u) - (d-1)\Gamma(u)].$$

Plugging this back in (20.15) yields

$$\begin{aligned}\int \Gamma_2(\sqrt{f}) &\leq \left(\frac{1}{4} + \frac{d-2}{12}\right)\int f\Gamma_2(\log f) - \frac{(d-2)(d-1)}{12}\int f\Gamma(\log f)\\ &\qquad - \frac{1}{48}\int f\Gamma(\log f)^2\\ &= \left(\frac{d+1}{12}\right)\int f\Gamma_2(\log f) - \frac{(d-2)(d-1)}{12}\int f\Gamma(\log f)\\ &\qquad - \frac{1}{48}\int f\Gamma(\log f)^2.\end{aligned}$$

This shows that

$$\widetilde{K} \geq \frac{12}{d+1}, \qquad \forall d \geq 2.$$

Combining this in (20.8) with $\lambda_1(\mathbb{RP}^{d-1}) = 2d$ yields for the optimal constant in (20.7)

$$K_* \geq \frac{24\,d}{d+1} \tag{20.16}$$

As d becomes large this is a poor estimate but for $d \leq 6$ it improves on the obvious bound $L_*(\mathbb{RP}^{d-1}) \geq L_*(\mathbb{S}^{d-1})$, and for $d = 2$ it yields again $K_* = 16$ (optimal), for $d = 3$, $K_* \geq 18$ (not bad) and it is anyway bigger than 16 for any d, completing the proof of Theorem 15.1.

Bibliographical Notes

For $d = 2$ the key integration by parts (20.2) and its use appear in many places but the oldest reference that I am aware of is McKean [134].

Formulas (20.11)–(20.13) can be found in various places of the literature on Γ_2 and logarithmic Sobolev inequalities; e.g. Bakry [14]. These are also Lemmas 5.4.3 and 5.5.5 in [6].

For higher dimensions there are many works about the estimate of the log Sobolev and differential Bakry–Émery constants [16] on $\mathbb{S}^{d-1}$, but estimates about best constants on $\mathbb{RP}^{d-1}$ are much more rare: Fore increasingly precise results, see Saloff-Coste [156], Fontenas [80], Guillen–Silvestre [97] and Ji [113].

21 Exploiting Gradient Variations II: Failed Attempts

How to adapt or extend the good theory and tools from the diffusive case, to more general Boltzmann equation, possibly of fractional diffusion nature? Reviewed here are several attempts which did not succeed, but may possibly be partly saved.

21.1 Decomposing the Inequality

The desired inequality

$$\int \Gamma_\beta(\sqrt{F}) \leq \frac{1}{K_*} \int F\Gamma_{1,\beta}(\log F)$$

involves at the same time a change of derivation order (1 more on the right hand side) and of nonlinearity ($\sqrt{\ }$ vs log). If a direct approach is elusive, one can try and handle one change at a time, decomposing in any of the following two ways:

(a) $K_* \geq \overline{K}\widetilde{K}$, where $\overline{K}$, $\widetilde{K}$ are best in

$$\int \Gamma_\beta(\sqrt{F}) \leq \frac{1}{\overline{K}} \int \Gamma_{1,\beta}(\sqrt{F}) \tag{21.1}$$

$$\int \Gamma_{1,\beta}(\sqrt{F}) \leq \frac{1}{\widetilde{K}} \int F\,\Gamma_{1,\beta}(\log F). \tag{21.2}$$

(b) $K_* \geq \dot{K}\widehat{K}$, where $\dot{K}$, $\widehat{K}$ are best in

$$\int \Gamma_\beta(\sqrt{F}) \leq \frac{1}{\dot{K}} D_\beta(F) \tag{21.3}$$

$$D_\beta(F) \leq \frac{1}{\widehat{K}} \int F\,\Gamma_{1,\beta}(\log F). \tag{21.4}$$

Here

$$D_\beta(F) = \frac{1}{2} \iint \big(F(\sigma) - F(k)\big)\big(\log F(\sigma) - \log F(k)\big)\,\beta(k\cdot\sigma)\,dk\,d\sigma$$

is the dissipation of Boltzmann's H functional along the operator L_β. In all these inequalities, F is assumed even.

It turns out that two of the above constants are easy to estimate. From the linear discussion we recall that $\overline{K} \geq \lambda_1(\mathbb{RP}^{d-1}) = 2d$ (with equality in general). And from the elementary inequality $4(\sqrt{x} - \sqrt{y})^2 \leq (x - y)\log(x/y)$ we have $\dot{K} \geq 4$ (with equality in general). This reduces the problem to the estimates of either $\widetilde{K}$ (as I have done in the diffusive case) or $\widehat{K}$ (which in the diffusive case is just as tricky as estimating K_*).

21.2 Nonlocal Γ_2 Changes of Variables

Nonlinear changes of variables, both for Γ and Γ_2, proved powerful in the previous section and it is natural to try and repeat them in the nonlocal context of the Boltzmann collision operator. A seemingly good news is that all four formulas (20.10)–(20.13) have nonlocal analogues. The analogue of (20.10) is

$$\chi'(f)L_\beta f = L_\beta(\chi(f)) - \frac{1}{2}\int \Big(\chi(f(\sigma)) - \chi(f(k)) - \chi'(f(k))[f(\sigma) - f(k)]\Big)\beta(k\cdot\sigma)\,d\sigma, \tag{21.5}$$

note that inside the large brackets the expression looks like $\chi''(f)[f(\sigma) - f(k)]^2$. The analogue of (20.11) is the rather tautological

$$\big(\varphi(f(\sigma)) - \varphi(f(k))\big)^2 = \left(\frac{\varphi(f(\sigma)) - \varphi(f(k))}{f(\sigma) - f(k)}\right)^2 \big(f(\sigma) - f(k)\big)^2. \tag{21.6}$$

The analogue of (20.12) is the less obvious

$$\begin{aligned} \Big|\Phi'(f(\sigma))\nabla f(\sigma) - \Phi'(f(k))\nabla f(k)\Big|^2_{k,\sigma} &= \Phi'(f(k))^2 \left|\nabla f(\sigma) - \nabla f(k)\right|^2_{k,\sigma} \\ &\quad + 2\Phi'(f(k))\big[\Phi'(f(\sigma)) - \Phi'(f(k))\big]\nabla f(\sigma)\cdot\big[\nabla f(\sigma) - P_{k\sigma}\nabla f(k)\big] \\ &\quad + \big[\Phi'(f(\sigma)) - \Phi'(f(k))\big]^2 |\nabla f(\sigma)|^2. \end{aligned} \tag{21.7}$$

Finally the analogue of (20.13) is

$$\begin{aligned} \int \psi(f)\, L_\beta f\, \Delta f &= \frac{1}{2}\iint \psi(f(k)) \left|\nabla f(\sigma) - \nabla f(k)\right|^2_{k,\sigma} \beta(k\cdot\sigma)\, dk\, d\sigma \\ &\quad + \frac{1}{2}\iint \Big[2\psi'(f(\sigma))\big[f(\sigma) - f(k)\big]^2 + \big(\psi(f(\sigma)) - \psi(f(k))\big)\Big] \\ &\qquad \left(\frac{|\nabla f(\sigma)|^2}{2} - \frac{|\nabla f(k)|^2}{2}\right) \beta(k\cdot\sigma)\, dk\, d\sigma \\ &\quad + \frac{1}{2}\iint \big[\psi'(f(\sigma)) - \psi'(f(k))\big]\big(f(\sigma) - f(k)\big)\, |\nabla f(k)|^2\, \beta(k\cdot\sigma)\, dk\, d\sigma. \end{aligned} \tag{21.8}$$

At first sight this looks promising, and also one can hope to exploit

$$\int (\Delta g)(L_\beta g) = \int \Gamma_{1,\beta}(g)^2.$$

But notice that expressions which were similar in the local case, in particular in (20.12) and (20.13), are now different, as in (21.7) and (21.8). So arranging the nonlocal formulas together necessarily comes at a cost in the constants. For a classical problem of partial differential equation estimate this might not be a big issue, but here the value of constants does matter, and the problem is rather tight! In the end I did not manage to put this strategy up and alive, but who knows.

21.3 Semigroup One Step Beyond

What about going for a semigroup argument to estimate

$$I'_{\beta}(F) = \int F \left|\nabla \log F(\sigma) - \nabla \log F(k)\right|^2_{k,\sigma} \beta(k\cdot\sigma)\, dk\, d\sigma$$

from its dissipation via the heat equation on $\mathbb{S}^1$?

By commutation of semigroups, this amounts to estimate the dissipation of

$$J(F) = \int F(\log F)''^2$$

via the Boltzmann semigroup on $\mathbb{S}^1$. (The Boltzmann-dissipation of the heat-dissipation of Fisher's information is the same as the heat-dissipation of the Boltzmann-dissipation of Fisher's information.)

Let us start with $d = 2$. Similarly to what was done in Sect. 8, let us define

$$\Gamma_{J,\beta}(F) = \left(F(\log F)''^2\right) - (Lf)(\log f)''^2 - 2f(\log f)''\left(\frac{Lf}{f}\right)''$$

in the hope to find a good leading nonnegative term. It turns out that there is a good algebra and the result is neat:

$$\Gamma_{J,\beta}(F)(x) = \int F(y)\Big(\big[\ell''(y)-\ell''(x)\big]^2 - 2\ell''(x)\big(\ell'(y)-\ell'(x)\big)^2\Big)\, \beta(\cos(x-y))\, dy,$$

where $\ell = \log f$. But then, what to do? It is normal that log concavity improves the estimates, but on $\mathbb{S}^1$ functions cannot be log concave everywhere. Is there a nice inequality involving this functional? It seems, we have reduced a tricky problem to another one. Furthermore, in view of the counterexample in Sect. 18 this can turn out nicely only if β has some regularity, and so far nothing has been used. Should one look for a clever integration by parts involving β? Again, who knows.

21.4 Interpolation

Consider the family of inverse power laws interactions, as in Sect. 4. In dimension 3, we are able to handle the "right" end $\nu = 2$ (diffusive case, Sect. 20); and also the "left" end $\nu = 0$ (hard spheres) thanks to either Theorems 17.3 or 18.1(ii). It would be nice if we could interpolate between both ends in the inequality

$$\int \Gamma_{\beta}(\sqrt{F}) \le \frac{1}{K_*} \int F\Gamma_{1,\beta}(\log F).$$

and cover all power laws with exponents $s \in (2, \infty)$. More generally, once the covered area comprises $\gamma = 0$, $\nu = 0$, $\nu = 2$ (at least part of a "H" in the diagram), it is tempting to hope that interpolation fills the whole region.

But obstacles are not easy to overcome. The functionals involved are homogeneous, but not convex. If working with inverse power laws, the formulas for the collision kernels are certainly analytic, but implicit and indirect. Estimating the complex parts in an attempt of the three line lemma (complex interpolation) is not obvious. So I do not know if the idea can be saved.

Bibliographical Notes

A pedagogical introduction to the formulas of changes of variables in diffusion processes can be found in the collective work [6]. A pedagogical introduction to classical interpolation theory, together with many references, can be found in Lunardi [130].

In the context of the Boltzmann equation, nonlocal formulas for Γ_1 under change of functions were introduced by Alexandre [1] to solve the problem of "renormalising" the Boltzmann equation without cutoff; then the two of us simplified and improved the formulas and estimates [2–4]. See also Sect. 23.4 below.

22 Exploiting Gradient Variations III: Combinations of Heat Kernels

A powerful principle in harmonic analysis consists in representing functions and operators as combinations of heat kernels. In this way one may perform regularisation and interpolation while take advantage of the good constants and algebra coming from diffusion. This is obviously a core principle of harmonic analysis. Here is a basic example:

$$\int_0^\infty (e^{t\Delta} - 1)\,\frac{dt}{t^{1+\alpha}} = -c_\alpha(-\Delta)^\alpha \tag{22.1}$$

for some positive constant $c_\alpha = c(\alpha, d) > 0$, $0 < \alpha < 1$; so nice estimates on $e^{t\Delta}$ may imply nice estimates for the fractional diffusion $(-\Delta)^\alpha$. This point of view will lead to a more precise estimate than Theorem 19.1, which was by comparison to a constant kernel.

To motivate the method in our Boltzmann context, assume that $\beta = \mathcal{K}_t$ is the heat kernel at some time $t > 0$ on $\mathbb{S}^{d-1}$: that is

$$e^{t\Delta}F(k) = \int_{\mathbb{S}^{d-1}} F(\sigma)\mathcal{K}_t(k\cdot\sigma)\,d\sigma.$$

(Do not be mistaken: t here is no longer the time variable in the Boltzmann equation, it is just a parameter.) The regularising property of the heat equation implies that $\mathcal{K}_t$

is smooth for any $t > 0$. Obviously $\int_{\mathbb{S}^{d-1}} \mathcal{K}_t(k\cdot\sigma)\,d\sigma = 1$ and the linear Boltzmann operator associated with β is

$$L_{\mathcal{K}_t} F = e^{t\Delta} F - F.$$

Applying successively (11.9), (11.10), Fubini, the last bit of Proposition 8.12(x), and (11.5),

$$\begin{aligned}
&- I'(F) \cdot L_{\mathcal{K}_t} F \\
&= \int_{\mathbb{S}^{d-1}} \Gamma_{I,L_\beta}(F) \qquad (22.2)\\
&= \iint_{\mathbb{S}^{d-1}\times\mathbb{S}^{d-1}} F(\sigma)\,\big|\nabla \log F(\sigma) - \nabla \log F(k)\big|^2_{k,\sigma}\, \mathcal{K}_t(k\cdot\sigma)\,d\sigma\,dk \\
&= \int_{\mathbb{S}^{d-1}} F(\sigma)\left(\int_{\mathbb{S}^{d-1}} \big|\nabla \log F(\sigma) - \nabla \log F(k)\big|^2_{k,\sigma} \mathcal{K}_t(k\cdot\sigma)\,dk\right) d\sigma \\
&= \int_{\mathbb{S}^{d-1}} F(\sigma)\left(\int_{\mathbb{S}^{d-1}} \big|\nabla \log F(k) - \nabla \log F(\sigma)\big|^2_{\sigma,k} \mathcal{K}_t(k\cdot\sigma)\,dk\right) d\sigma \\
&= 2\int F\,\Gamma_{1,\mathcal{K}_t}(\log F).
\end{aligned}$$

But by convexity of I,

$$- I'(F) \cdot L_{\mathcal{K}_t} F = -I'(F)\cdot\big(e^{t\Delta}F - F\big) \geq I(F) - I(e^{t\Delta}F). \qquad (22.3)$$

Also

$$- I'(F)\cdot \Delta F = \int F\Gamma_2(\log F) \geq 2L_* I(F),$$

where L_* is the optimal constant in the differential Bakry–Émery inequality, as in (15.9), so by Gronwall (along the heat equation),

$$I(e^{t\Delta}F) \leq e^{-2L_* t} I(F).$$

Plugging this back in (22.3) yields

$$\int_{\mathbb{S}^{d-1}} F\,\Gamma_{1,\mathcal{K}_t}(\log F) \geq \left(\frac{1 - e^{-2L_* t}}{2}\right) I(F). \qquad (22.4)$$

On the other hand, using successively the definition of Γ_β (Definition 11.3), the symmetry of $\mathcal{K}_t$ and the dissipation of L^2 norm,

$$\begin{aligned}\int \Gamma_{\mathcal{K}_t}(\sqrt{F}) &= -\int \sqrt{F} L_{\mathcal{K}_t} \sqrt{F} \\ &= \int F - \int \sqrt{F} e^{t\Delta} \sqrt{F} \\ &= \int F - \int (e^{(t/2)\Delta} \sqrt{F})^2 \\ &= \int_0^t \left(-\frac{d}{ds} \int (e^{(s/2)\Delta} \sqrt{F})^2 \right) ds \\ &= \frac{1}{2} \int_0^t 2 \int |\nabla e^{(s/2)\Delta} \sqrt{F}|^2 \, ds. \end{aligned} \tag{22.5}$$

From

$$\frac{d}{dt}\bigg|_{\partial_t u = \Delta u} \int |\nabla u|^2 = -2 \int |\nabla^2 u|^2 \leq -2\lambda_1 \int |\nabla u|^2$$

(no confusion: λ_1 here is the spectral gap for $-\Delta$, not for L_β), we have

$$\int |\nabla e^{(s/2)\Delta} \sqrt{F}|^2 \leq e^{-\lambda_1 s} \int |\nabla \sqrt{F}|^2 = \frac{e^{-\lambda_1 s}}{4} I(F).$$

Plugging this back in (22.5),

$$\int_{\mathbb{S}^{d-1}} \Gamma_{\mathcal{K}_t}(\sqrt{F}) \leq \left(\frac{1 - e^{-\lambda_1 t}}{4\lambda_1} \right) I(F). \tag{22.6}$$

The combination of (22.4) and (22.6) yields

$$\int_{\mathbb{S}^{d-1}} \Gamma_{\mathcal{K}_t}(\sqrt{F}) \leq \frac{1}{2\lambda_1} \left(\frac{1 - e^{-\lambda_1 t}}{1 - e^{-2L_* t}} \right) \int_{\mathbb{S}^{d-1}} F \, \Gamma_{1,\mathcal{K}_t}(\log F),$$

in other words

$$\beta = \mathcal{K}_t \Longrightarrow \quad K_* \geq 2\lambda_1 \left(\frac{1 - e^{-2L_* t}}{1 - e^{-\lambda_1 t}} \right).$$

As t varies from 0 to ∞ this interpolates in a monotone way between $4L_*(\mathbb{RP}^{d-1})$ (consistent with the fact that $L_{\mathcal{K}_t}$ for small t is approximately t times Δ) and $2\lambda_1(\mathbb{RP}^{d-1}) = 4d$ (consistent with (18.2) and the fact that $L_{\mathcal{K}_t}$ for large t is approximately the averaging operation on the sphere). It follows that the constant K_* is never below the minimum of those two numbers.

Remark 22.1 At the moment it is not known exactly which is the optimal value L_*. In these notes I have proven that $L_* \geq 6d/(d+1)$. Sehyun Ji has shown that

$$L_*(\mathbb{RP}^d) \geq d + 3 - \frac{1}{d-1}. \tag{22.7}$$

(Compare with the bound by Rothaus (15.11), which is $L \geq d + 3 - 4/d^2$.) In any case it shows that

$$\min(4L_*, 2\lambda_1) = 2\lambda_1 = 4d,$$

and ironically it is now λ_1 the limiting factor in our inequalities.

Let us summarise with the following statement:

Proposition 22.2 (Functional Inequality for Heat Kernels) *For any $t > 0$, for all even functions $F : \mathbb{S}^{d-1} \to \mathbb{R}_+$,*

$$\int_{\mathbb{S}^{d-1}} F\Gamma_{1,\mathcal{K}_t}(\log F) \geq 4d \int_{\mathbb{S}^{d-1}} \Gamma_{\mathcal{K}_t}(\sqrt{F}).$$

Once that striking property $K_*(\beta) \geq 4d$ has been established for any $\beta = \mathcal{K}_t$, it readily extends to any linear combination, discrete or continuous, of such kernels, say

$$\beta(k \cdot \sigma) = \int_0^\infty \mathcal{K}_t(k \cdot \sigma)\, \lambda(dt)$$

for some (nonnegative) measure λ on $\mathbb{R}_+$. This means that the Boltzmann operator defined by the kernel β is

$$L_\beta f(k) = -\int_0^\infty \int_{\mathbb{S}^{d-1}} \big[f(k) - \mathcal{K}_t(k \cdot \sigma) f(\sigma)\big]\, \lambda(dt)\, d\sigma$$

or

$$L_\beta = -\int_0^\infty (I - e^{t\Delta})\, \lambda(dt). \tag{22.8}$$

Warning: A possible source of confusion is that the kernel of an operator L is usually defined via the formula $Lf(x) = \int k(x,y) f(y)\, dy$; but in the context of Boltzmann equation, the formula is $Lf(x) = \int k(x,y)[f(y) - f(x)]\, dy$. This is just a terminology issue. For a kernel of infinite mass, the second convention still makes sense. In any case, I will write informally (22.8) as

$$\beta = \int_0^\infty \mathcal{K}_t\, \lambda(dt). \tag{22.9}$$

The above discussion is summarised by the following estimate (Compare Corollary 17.2):

Corollary 22.3 (Regularity-improved Nonlocal Differential Log Sobolev) *If* $\beta : [-1,1] \to \mathbb{R}_+$ *satisfies* $\int [1 - k \cdot \sigma]\, \beta(k \cdot \sigma)\, d\sigma < \infty$ *on* $\mathbb{S}^{d-1}$ *and admits the integral heat kernel representation*

$$\beta(k \cdot \sigma) = \int_0^\infty \mathcal{K}_t(k \cdot \sigma)\, \lambda(dt)$$

for some measure λ *on* $\mathbb{R}_+$*, then for all even functions* $F : \mathbb{S}^{d-1} \to \mathbb{R}_+$*,*

$$\int_{\mathbb{S}^{d-1}} F \Gamma_{1,\beta}(\log F) \geq 4d \int_{\mathbb{S}^{d-1}} \Gamma_\beta(\sqrt{F}).$$

In other words, (13.1) *holds with* $K_* \geq 4d$.

Asking for the heat kernel representation is the same as asking for L_β to be of the form $-g(-\Delta)$, where $g(\omega) = \int_0^\infty (I - e^{-\omega t})\, \lambda(dt)$. So g is the Laplace transform of λ, up to a constant term. A famous criterion by Bernstein says that this amounts for g to be nonnegative and completely monotonous, that is, $(-1)^n g^{(n)} \geq 0$ for all $n \in \mathbb{N}$. Such functions g are called *Bernstein functions*, and Bernstein's representation theorem precisely says that a completely monotonous function $g : \mathbb{R}_+ \to \mathbb{R}_+$ can be written

$$g(\omega) = a\omega + b + \int_0^\infty (1 - e^{-\omega t})\, \lambda(dt)$$

for some Borel measure λ with $\int \min(1, t)\, \lambda(dt) < \infty$. Note that accordingly

$$-g(-\Delta) = a\Delta - bI + \int_0^\infty (e^{t\Delta} - I)\, \lambda(dt)$$

is the combination of the usual diffusion, a constant, and a possibly non-cutoff linear Boltzmann operator. In the case under discussion here, only the latter part is relevant, so we may impose $g(0) = 0$, and $g(\omega)/\omega \to 0$ (sublinearity) to impose $a = b = 0$.

Now the following theorem is a direct consequence of Corollary 22.3 and Theorem 13.1:

Theorem 22.4 *Let* $B(v - v_*, \sigma) = |v - v_*|^\gamma \beta(k \cdot \sigma)$ *be a collision kernel in dimension* d*. Assume that the angular kernel* β *is associated with* $-g(-\Delta)$ *for some* g *which is nonnegative, completely monotone, sublinear, such that* $g(0) = 0$*. Further assume that* $|\gamma| \leq 2\sqrt{d}$*. Then* I *is nonincreasing along solutions of the spatially homogeneous Boltzmann equation with kernel* B*.*

Example 22.5 The main cases of application for this theorem are fractional powers of the Laplacian, choosing $g(\omega) = \omega^{\nu/2}$ for some $\nu \in (0, 2)$. It is known that such a kernel has the same angular singularity as the Boltzmann kernel for inverse power forces like r^{-s} if $s = 1 + (d-1)/\nu$. Since $2\sqrt{d} \geq \min(d, 4)$ for all $d \geq 2$, in this way we have the monotonicity property for a natural family of collision kernels covering all values of γ, ν of interest for the Boltzmann equation (recall Fig. 4).

The latter example, as well as motivation from physics, naturally lead to ask whether these power-law induced kernels fall in the range of the theorem. But as recalled in Sect. 4, these kernels are tricky and defined in an indirect, implicit way. They are in general distinct from fractional heat kernels, and so far nobody knows if they can be represented by heat kernels. But we may use Lemma 18.3 to compare such a kernel to a proxy, for instance the kernel of the fractional diffusion with the same singularity, or any other combination of heat kernels.

The following theorem, stated here for collision kernels which are not necessarily in product form, summarises all this.

Theorem 22.6 (Curvature-dimension Induced Decay Estimates via Heat Kernel Representation) *Let $B = B(|v - v_*|, \cos\theta)$ be a collision kernel in dimension d. Assume that for each $r > 0$ there are $m_r, M_r > 0$, an angular collision kernel $\beta_{0,r} = \beta_{0,r}(\cos\theta)$ and a measure λ_r on $\mathbb{R}_+$ such that*

$$\beta_{0,r} = \int \mathcal{K}_t\, \lambda_r(dt), \qquad \int_0^\infty (1 - e^{-t\omega})\, \lambda_r(dt) < \infty, \qquad \int \beta_{0,r}(k\cdot\sigma)(1 - k\cdot\sigma) < +\infty,$$

and for all $r > 0$,

$$m_r \leq \frac{\beta_r}{\beta_{0,r}} \leq M_r,$$

and

$$\sup_{0 \leq \theta \leq \pi} \left\{ \frac{r}{B} \left| \frac{\partial B}{\partial r} \right| (r, \cos\theta) \right\} \leq 2\sqrt{d\, \frac{m_r}{M_r}}.$$

Then I is nonincreasing along solutions of the spatially homogeneous Boltzmann equation with kernel B.

Remark 22.7 This theorem uses the rather sharp bound by Ji on $L_*(\mathbb{RP}^{d-1})$, namely (22.7). But if one applies the cruder bound proven in Sect. 20.2, namely (20.16), one obtains in the end a similar result only with d replaced by $\min(d, 4)$ and as of today that is sufficient for applications to the Boltzmann equation we may think of.

Let us see how to apply this Theorem to inverse s-power laws in the most important cases left out by Theorem 17.3, namely

- $d = 2$ and $-2 \leq \gamma < 1$ ($0 < \nu \leq 1$, or $2 \leq s < \infty$);
- $d = 3$ and $-3 \leq \gamma < -2$ ($3/2 \leq \nu < 2$, or $2 < s < 7/3$).

There is also a little bit which is outside the "natural" mathematical range but still makes physical sense, in principle:

- $d = 2$ and $-3 < \gamma < -2$ ($1 < \nu < 2$, or $3/2 < s < 2$).

In each case, we may work out numerically the angular kernel of the Boltzmann equation for various values of s and compare them to the angular kernel of the fractional diffusion with the same singularity, hoping that the two remain mutually bounded. While this is not a purely mathematical proof, the procedure is quite safe: we are here numerically comparing functions which are well-defined through special functions and other classical tools of numerical analysis, so errors remain under control. Dimension 2 is handled by Fourier series, dimension 3 by classical spherical harmonics, which are numerically addressed in available software. Working along these lines and trying empirically some variations, one obtains:

- In dimension $d = 2$, the symmetrised Boltzmann kernel for s-power law forces stays close to the associated fractional Laplace kernel, with ratio M/m always less than $\sqrt{2}$ (which is obtained for the limit of hard spheres). This yields $\overline{\gamma} \geq 2 \cdot 2^{1/4} > 2$: already enough to get the range $[-2, 2]$.
- Still in dimension 2, refining the analysis, one can see that the criterion applies up to get $\gamma > -2.7$, but fails to get all the way to the physical limit $\gamma = -3$. But by empirically changing the weight function to

$$\lambda(dt) = \Big[1 + 2(\nu - 1)^2 (1 - \exp(-2t))\Big]\, t^{-(1+2\nu)}\, dt,$$

  one obtains $\overline{\gamma} > 3.3$ all throughout the range $s \in [3/2, 2]$. So this covers all cases which make physical sense.
- In dimension $d = 3$, the numerically observed M/m ratio approaches 1.6 for $\nu = 1$. This yields $\overline{\gamma} > \sqrt{7.5}$, not quite sufficient for very singular kernels. But by empirically changing the weight function to

$$\lambda(dt) = \frac{t^{-(1+2\nu)}\, dt}{\sqrt{1 + (2-\nu)t}}$$

  one obtains ratios not larger than 1.1; then this yields $\overline{\gamma} \geq \sqrt{12/1.1}$ which is > 3 (not by much!). Another more sophisticated weighting, also obtained by trial and error, is

$$\lambda(dt) = \left[1 - \min\left(\frac{13}{8} - \frac{3}{2}s, \frac{2}{5}\right)\big(1 - \exp(-2t)\big)\right] t^{1-2\nu}\, dt,$$

  which yields estimates $\overline{\gamma} \geq 4.3$ all throughout the range $\gamma \in (-3, -2]$, thus a more comfortable margin, robust to numeric errors.

This concludes our investigation of the monotonicity of the Fisher information. All in all, we discovered that the three basic methods presented here (curvature, positivity, Hessian variations), taken together, are enough to establish the monotonicity of the Fisher information for the vast majority of collision kernels of interest, including all power law interactions between Coulomb and hard spheres, in all dimensions.

Remark 22.8 If one does not want to restrict to potentials decaying at least as fast than Coulomb, then there is actually one little bit still not covered: $\gamma \in (-3, -2\sqrt{2}]$ in dimension $d = 4$. There is no particular motivation for that range, except mathematical consistency.

Bibliographical Notes

Heat kernel approximation is developed at large in harmonic analysis [86, 165], and combinations of simple elementary kernels are used in the theory of stochastic processes with the concept of "subordination". Apart from these elements of context, I am not aware of any precedent in the context of Boltzmann equation of the technique developed in this section, whose basic principle was discovered in a discussion between Luis Silvestre and Cyril Imbert in the Mathemata summer school in Crete (July 2024), explored later by the three of us in [111].

Bernstein proved his representation theorem in [22] and there is a modern textbook by Schilling, Song and Vondraček [157].

The proximity of the heat kernel for power law forces and the fractional Laplace operator was noticed by a number of authors, starting at least with Desvillettes, and quantitatively exploited for regularity issues by Alexandre, Desvillettes, Wennberg and myself [4]. Using the ingredients in the latter work, one easily proves, for instance, that $-\langle L_\beta h, h\rangle \geq K\langle(-\Delta)^{\nu/2}h, h\rangle$ when β has the form (4.10).

Numerics quoted in this section was done by Silvestre [160] and the results published in [111].

23 Regularity of Solutions for Singular Collision Kernels

Regularity theory for the spatially homogeneous Boltzmann equation classically rests on four types of estimates:

- moments
- integrability
- smoothness
- positivity (lower bounds)

While this is more or less the natural order for the four estimates, they may be combined or related in many ways, and also intertwined with the equilibration problem. It turns out that the Fisher information estimate unlocks all the remaining blanks in the theory of soft potentials, at least in the region of "conditional regularity" $\gamma + \nu \geq -2$, $\gamma \geq -d$.

I shall go through all four types of estimates, not searching for exhaustivity or optimality—a complete exposition of the state of the art would easily fill up a 500-page book. I will focus on the particular case of factorised collision kernels (4.13); this is for simplicity, as more general forms could be handled, and only for very soft potentials (with a high singularity in the relative velocity, like $\gamma < -2$), whose regularity is most tricky, and which is precisely the compartment of the theory which was unlocked by the new Fisher information estimates. I will systematically consider first the Landau equation and then the Boltzmann equation, and most of the time only sketch the proofs, pointing to the existing literature for more information. I will also skip the uniqueness issue, which in this field always follows in practice from good a priori estimates.

Besides the assumption of very soft potential, I shall limit myself to the regime of "conditional regularity". So the precise bounds on the singularities in relative velocity and angular variables will be

- For Landau's equation,

$$-4 < \gamma < -2, \qquad \gamma \geq -d; \tag{23.1}$$

- For Boltzmann's equation,

$$-2 < \gamma + \nu < 0, \qquad \nu < 2, \qquad \gamma \geq -d \tag{23.2}$$

(of which (23.1) is obviously a limit case)

The initial datum will be denoted by f_0: by assumption it is a probability density; and its initial energy will be $E_0 = (1/2) \int f_0(v)|v|^2\, dv$.

23.1 Integrability

Assume that the initial datum f_0 has finite Fisher information and the monotonicity property holds. Then solutions $f(t) = f(t, \cdot)$ satisfy the a priori estimate $I(f(t)) \leq I(f_0)$ for all $t \geq 0$. By Sobolev embedding

$$I(f) = 4\|\nabla\sqrt{f}\|^2_{L^2(\mathbb{R}^d)} \geq K(d)\|\sqrt{f}\|^2_{L^{p^\star}(\mathbb{R}^d)} = K\|f\|_{L^{p^\star/2}(\mathbb{R}^d)}$$

with $p^\star = 2d/(d-2)$ for $d \geq 3$ and any $p^\star < \infty$ for $d = 2$. So,

$$\sup_{t\geq 0} \|f(t)\|_{L^{p_0}(\mathbb{R}^d)} \leq C(f_0, d) \quad p_0 = \frac{d}{d-2} \text{ if } d \geq 3, \quad \forall p_0 < \infty \text{ if } d = 2. \tag{23.3}$$

This is already a major information, as it rules out, in dimension $d = 3$, a behaviour of the type of the quadratic heat equation $\partial_t f = \Delta f + f^2$ (which generically blows up in L^p for all $p > d/2$). Furthermore, by Hardy–Littlewood–Sobolev inequality,

$$\iint_{\mathbb{R}^d\times\mathbb{R}^d} \frac{f(v)f(v_*)}{|v-v_*|^\alpha}\,dv\,dv_* \le C(d,\alpha)\|f\|^2_{L^q(\mathbb{R}^d)}$$

as soon as $\alpha = 2d(1-1/q)$ and $0 < \alpha < d$. Choosing q as close to q_0 as these conditions allow, yields the second key a priori estimate

$$\sup_{t\ge 0}\int_{\mathbb{R}^d\times\mathbb{R}^d} \frac{f(t,v)\,f(t,v_*)}{|v-v_*|^{\alpha_0}}\,dv\,dv_* \le C(d,\alpha_0,f_0) \qquad \forall\alpha_0,\ 0<\alpha_0<\min(4,d). \tag{23.4}$$

23.2 Weak Solutions

Various notions of weak solutions have been introduced and used in the context of the Boltzmann and Landau equations. But the above integrability bounds allow to use the most natural one, already considered by Maxwell seventy years before Sobolev and Schwartz would formalize the notion of distributions: For any smooth test function $\varphi = \varphi(v)$,

$$\frac{d}{dt}\int_{\mathbb{R}^d} f\varphi = \int_{\mathbb{R}^d} Q(f,f)\,\varphi.$$

Let us check that indeed the right-hand side makes perfect sense under the integrability bounds of Sect. 23.1.

For Landau, as already noted in Sect. 5 (Proposition 5.3),

$$\begin{aligned}\int Q_L(f,f)\,\varphi &= \iint ff_*\,a(v-v_*):\nabla^2\varphi\,dv\,dv_* \\ &\qquad + 2\iint ff_*\,b(v-v_*)\cdot\nabla\varphi\,dv\,dv_* \\ &= \iint ff_*\Big(a(v-v_*):\Big(\frac{\nabla^2\varphi+(\nabla^2\varphi)_*}{2}\Big) \\ &\qquad + b(v-v_*)\cdot\big(\nabla\varphi-(\nabla\varphi)_*\big)\Big)\,dv\,dv_*,\end{aligned} \tag{23.5}$$

where symmetrisation $v \leftrightarrow v_*$ was used. Now $a = O(|v-v_*|^{\gamma+2})$ and $b = O(|v-v_*|^{\gamma+1})$ so if φ is smooth the integrand in the above integral is $O(|v-v_*|^{\gamma+2})$. If $0 \ge \gamma \ge -2$, the integral would converge just by the mass and energy estimates, but for $\gamma < -2$ this is no longer true. (This is where "very soft" potentials begin.) But the regularity estimate (23.4) ensures the convergence of the integral as soon as $-d < \gamma + 2 < 0$.

Now for Boltzmann:

$$\int Q(f,f)\varphi = \frac{1}{4}\iiint f f_*(\varphi' + \varphi'_* - \varphi - \varphi_*)\, B\, d\sigma\, dv\, dv_*.$$

Then if φ is smooth, for any $k \in \mathbb{S}^{d-1}$

$$\int_{\mathbb{S}^{d-2}_{k\perp}} (\varphi' + \varphi'_* - \varphi - \varphi_*)\, d\phi = O\big(|v - v_*|^2 \theta^2\big),$$

so

$$\int_{\mathbb{S}^{d-1}} (\varphi' + \varphi'_* - \varphi - \varphi_*)\, B\, d\sigma = O\left(|v - v_*|^2 \int B(1 - k\cdot\sigma)\, d\sigma\right)$$
$$= O(|v - v_*|^{\gamma+2}),$$

and again the integral converges thanks to (23.4).

23.3 Moments

Moments are the first concern in any mathematical implementation of kinetic theory, so it is worth spending some energy on them.

Let $\langle v\rangle = \sqrt{1 + |v|^2}$, so that

$$\nabla\langle v\rangle = \frac{v}{\langle v\rangle}, \qquad \partial_{ij}\langle v\rangle = \frac{1}{\langle v\rangle}\left(\delta_{ij} - \frac{v_i v_j}{\langle v\rangle^2}\right),$$

and for $s > 0$ let $\varphi_s = \langle v\rangle^s$, so that

$$\nabla\varphi_s = s\, v\langle v\rangle^{s-2}, \qquad \partial_{ij}\varphi_s = s\langle v\rangle^{s-2}\left(\delta_{ij} - \frac{v_i v_j}{\langle v\rangle^2}\right) + s(s-1)\langle v\rangle^{s-4} v_i v_j. \tag{23.6}$$

Let $s > 2$ be a real number: Bounding the moment of order s, $M_s = \int f|v|^s$, is the same as bounding $\int f\varphi_s$. For that we may use the weak formulation. Recall that the weighted Lebesgue spaces, L^p_s, are in interpolation: $L^p_s = [L^{p_0}_{s_0}, L^{p_1}_{s_1}]_\theta$ if $1/p = (1-\theta)/p_0 + \theta/p_1$, $s = (1-\theta)s_0 + \theta s_1$. I will denote by C various constants changing from time to time and only depending on the parameters involved.

First attempt for Landau: Using (23.5) and replacing a, b by their expression in terms of Ψ,

$$\frac{d}{dt}\int f\varphi_s = \iint ff_* A_s(v, v_*)\, dv\, dv_*,$$

where

$$A_s(v, v_*) = \Psi(|v-v_*|)\left(\Pi_{k^\perp} : \left(\frac{\nabla^2\varphi_s + (\nabla^2\varphi_s)_*}{2}\right)\right.$$
$$\left. -(d-1)\frac{v-v_*}{|v-v_*|^2}\cdot\left(\nabla\varphi_s - (\nabla\varphi_s)_*\right)\right).$$

To fix ideas, assume that $\Psi(|z|) = |z|^{\gamma+2}$. Let us separate four cases:

(a) $|v| = O(1)$, $|v_*| = O(1)$. Noting that $|\nabla\varphi_s(v) - \nabla\varphi_s(v_*)| = O(|v-v_*|)$, one easily has $|A_s(v, v_*)| = O(|v-v_*|^{\gamma+2})$.

(b) $|v-v_*| \gg 1$ and either $|v|/|v_*| \gg 1$ or $|v|/|v_*| \ll 1$. By symmetry, consider for instance the first situation. Then

$$\begin{aligned} A_s(v, v_*) &= |v|^{\gamma+2}\Big(s\frac{(d-1)}{2}|v|^{s-2} - s(d-1)|v|^{s-2} + o(|v|^{s-2})\Big) \\ &= |v|^{\gamma+2}\left(-\frac{s(d-1)}{2}|v|^{s-2} + o(|v|^{s-2})\right) \\ &\le -\frac{(d-1)}{4}|v|^{s+\gamma} \qquad \text{for } |v| \text{ large enough} \end{aligned}$$

(c) $|v-v_*| \gg 1$ and $|v|, |v_*|$ comparable (and thus both large). Then, assuming for instance $|v-v_*| \ge |v|/2$,

$$|A_s(v, v_*)| \le C|v|^{\gamma+2}\langle v\rangle^{s-2} \le C\langle v\rangle^{s+\gamma} \le C\langle v\rangle^{s+\gamma-2}\langle v_*\rangle^2.$$

(d) $|v-v_*| = O(1)$ and $|v|, |v_*|$ both very large. Then $|v|, |v_*|$ are of course comparable and

$$A_s(v, v_*) \le C|v-v_*|^{\gamma+2}(|v|^{s-2} + |v_*|^{s-2}) \le C|v-v_*|^{\gamma+2}\langle v\rangle^{\frac{s}{2}-1}\langle v_*\rangle^{\frac{s}{2}-1}.$$

We end up with

$$\begin{aligned} A_s(v, v_*) \le &-K_s(\langle v\rangle^{s+\gamma} + \langle v_*\rangle^{s+\gamma}) + C(\langle v\rangle^{s+\gamma-2}\langle v_*\rangle^2 + \langle v\rangle^2\langle v_*\rangle^{s+\gamma-2}) \\ &+ C|v-v_*|^{\gamma+2}\big(1 + \langle v\rangle^{\frac{s}{2}-1}\langle v_*\rangle^{\frac{s}{2}-1}\big). \end{aligned}$$

Integrating this against $ff_* \, dv \, dv_*$ yields (possibly changing the constants), after using again the Hardy–Littlewood–Sobolev inequality, and the Young inequality for the last term,

$$\iint ff_* \, A_s(v, v_*) \, dv \, dv_* \leq -K_s \|f\|_{L^1_{s+\gamma}} + C_s \|f\|_{L^1_{s+\gamma-2}} (1 + 2E_0)$$
$$+ C_s \left\| f \langle v \rangle^{\frac{s-2}{2}} \right\|^2_{L^q}, \qquad (23.7)$$

where

$$E_0 = \frac{1}{2} \int f |v|^2 \, dv, \qquad \frac{1}{q} = 1 + \frac{\gamma + 2}{2d}.$$

The best we can hope is that the first term in the right-hand side of (23.7) be dominant in the sense of controlling the contribution of high velocities. Not only will this imply that M_s remains locally bounded, but also, since that first term comes with a negative sign, it will show that actually dM_s/dt remains $O(1)$, hence a linear bound in time, for any s. After that, by interpolation with higher order moments it is possible to lower the exponent at leisure. But for that one first has to control the tricky last term in (23.7). Let us interpolate to bound that term, which is actually

$$\|f\|^2_{L^1_{\frac{s-2}{2}}} \leq \|f\|^{2a}_{L^{p_0}} \|f\|^{2b}_{L^1_2} \|f\|^{2c}_{L^1_\sigma},$$

where $1/p_0 = 1 - 2/d$ (if $d \geq 3$), and $a, b, c \in [0, 1]$, $\sigma > 0$ satisfy

$$a + b + c = 1, \qquad \frac{1}{q} = \frac{a}{p_0} + b_c, \qquad \frac{s-2}{2} = 2b + \sigma c.$$

The idea here is that both $\|f\|_{L^{p_0}}$ and $\|f\|_{L^1_2}$ are bounded uniformly, so the last bit of (23.7) will be bounded by a multiple of $\|f\|_{L^1_\sigma}$ and we wish σ to be as small as possible. A bit of playing shows that our best choice is $a = -(\gamma + 2)/4$, $c = 1/2$, $b = (\gamma + 4)/4$, then $\sigma = s - \gamma - 6$. Grand conclusion:

$$\frac{dM_s}{dt} \leq -K M_{s+\gamma} + C(1 + M_{s-\gamma-6}).$$

But $s - \gamma - 6 < s + \gamma$ only for $\gamma > -3$, so this misses an important case. Try again!

The way out will come from another weak formulation, and the use of the entropy estimate. First note that

$$\frac{d}{dt} \int f \varphi \, dv = \int \nabla \cdot \left(\int a(\nabla - \nabla_*) ff_* \, dv_* \right) \varphi \, dv$$
$$= - \iint a(\nabla - \nabla_*) ff_* \nabla \varphi \, dv \, dv_*$$

$$= -\frac{1}{2}\iint a(\nabla - \nabla_*) f f_* \big(\nabla\varphi - (\nabla\varphi)_*\big)\, dv\, dv_*.$$

Here $a = a(v - v_*) = \Psi(|v - v_*|)\Pi_{(v-v_*)^\perp}$. On the other hand Landau's dissipation functional (or entropy production functional) is

$$\begin{aligned} D_L(f) &= \frac{1}{2}\iint a f f_* (\nabla - \nabla_*) \log f f_* (\nabla - \nabla_*) \log f f_*\, dv\, dv_* \\ &= 2\iint a(\nabla - \nabla_*)\sqrt{f f_*}(\nabla - \nabla_*)\sqrt{f f_*}\, dv\, dv_*. \end{aligned}$$

The point is that here the estimate, contrary to the Fisher information bound, becomes stronger when γ becomes more negative and a more singular. There is a cost: along the flow, $D_L(f)$ is bounded only after time-integration. But for the control of moments this will not be a problem. Now, by Cauchy–Schwarz,

$$\begin{aligned} \frac{d}{dt}\int f\varphi &= -\iint a\sqrt{f f_*}\,(\nabla - \nabla_*)\sqrt{f f_*}\big(\nabla\varphi - (\nabla\varphi)_*\big)\, dv\, dv_* \\ &\leq C\, D_L(f)^{1/2}\left(\iint f f_* a\big(\nabla\varphi - (\nabla\varphi)_*\big)\big(\nabla\varphi - (\nabla\varphi)_*\big)\, dv\, dv_*\right)^{1/2} \\ &= C\, D_L(f)^{1/2}\iint f f_* \Psi(v - v_*)\left|\Pi_{(v-v_*)^\perp}\big[\nabla\varphi - (\nabla\varphi)_*\big]\right|^2 dv\, dv_*. \end{aligned}$$

Observe that if φ is smooth and in $W^{2,\infty}$ (second derivatives in L^∞), then the integrand in the last integral is $f f_*$ multiplied by $O(|v - v_*|^{\gamma+4})$ and so bounded. Let us do the estimate now for φ_s in place of φ. Let us decompose $\Psi(z)$ into $\Psi(z)1_{|z|\leq\lambda} + \Psi(z)1_{|z|>\lambda}$ (small relative velocities, large relative velocities) where λ should be fixed large enough. Then with obvious notation

$$\frac{d}{dt}\int f\varphi_s = \int Q_L^{>\lambda}\varphi_s + \int Q_L^{<\lambda}\varphi_s.$$

For the first one, the estimate of the beginning of this section is now more favourable, since the last bit of (23.7), coming from (a), will not be present:

$$\int Q_L^{>\lambda}\varphi_s \leq -K_s M_{s+\gamma} + C_s. \tag{23.8}$$

For the second one, the estimate just above yields

$$\int Q_L^{<\lambda}\varphi \leq C\lambda^{\gamma+2} D_L(f)^{1/2} M_{s-2}.$$

By interpolation and using $\gamma \geq -4$,

$$\int Q_L^{<\lambda} \varphi \leq C\lambda^{\gamma+2} D_L(f)^{1/2} M_s^{1/2} M_{s+\gamma}^{1/2}$$

$$\leq \varepsilon M_{s+\gamma} + C_\varepsilon D_L(f) M_s + C.$$

Choosing ε small enough and combining this with (23.8) yields

$$\frac{d}{dt} M_s \leq -K M_{s+\gamma} + C D_L(f) M_s + C. \tag{23.9}$$

This inequality, first obtained by Carlen, Carvalho and Lu, implies an excellent bound on M_s. Indeed, by time-integration, it yields

$$M_s(t) + K \int_0^t M_{s+\gamma}(\tau)\, d\tau \leq M_s(0) + C(1+t) + \int_0^t D_L(f(\tau)) M_s(\tau)\, d\tau.$$

And then by Gronwall's lemma

$$M_s(t) + K \int_0^t M_{s+\gamma}(\tau)\, d\tau \leq C(1+t) \exp\left(\int_0^\infty D_L(f(\tau))\, d\tau\right). \tag{23.10}$$

This proves in particular

$$M_s(t) \leq C(s, \gamma, f_0)(1+t),$$

where $C(s, f_0)$ depends on f_0 only through E_0, $H(f_0)$ and $M_s(f_0)$.

Now the same strategy for Boltzmann. For this we need to use two weak formulations:

$$\int Q(f, f)\varphi = \frac{1}{2} \iint f f_* (\varphi' + \varphi'_* - \varphi - \varphi_*)\, dv\, dv_*\, B(v - v_*, \sigma)\, d\sigma$$

(useful for large relative velocities) and

$$\int Q(f, f)\varphi = -\frac{1}{4} \iint (f' f'_* - f f_*)(\varphi' + \varphi'_* - \varphi - \varphi_*)\, dv\, dv_*\, B(v - v_*, \sigma)\, d\sigma$$

(useful for small relative velocities). For large relative velocities we need a good estimate of $\int (\varphi' + \varphi'_* - \varphi - \varphi_*)\, d\sigma$; for small relative velocities it will be a good estimate of $(\varphi + \varphi'_* - \varphi - \varphi_*)$, when $\varphi(v) = \varphi_s(v) = \langle v \rangle^s$, with $s > 2$.

Let's go. Write

$$z = v - v_*, \qquad c = \frac{v + v_*}{2}, \qquad y = \frac{z}{|c|}, \qquad k = \frac{z}{|z|} = \frac{y}{|y|}, \qquad e = \frac{c}{|c|}.$$

Then

$$v = c + \frac{z}{2}, \qquad v_* = c - \frac{z}{2}, \qquad v' = c + \frac{|z|}{2}\sigma, \qquad v'_* = c - \frac{|z|}{2}\sigma.$$

Then

$$|v|^2 = |c|^2\left(1 + e\cdot y + \frac{|y|^2}{4}\right)$$

$$|v_*|^2 = |c|^2\left(1 - e\cdot y + \frac{|y|^2}{4}\right)$$

$$|v'|^2 = |c|^2\left(1 + (e\cdot\sigma)|y| + \frac{|y|^2}{4}\right)$$

$$|v'_*|^2 = |c|^2\left(1 - (e\cdot\sigma)|y| + \frac{|y|^2}{4}\right).$$

For $\varepsilon \ll 1$,

$$(1+\varepsilon)^{s/2} = 1 + \frac{s}{2}\varepsilon + \frac{s}{2}\left(\frac{s}{2}-1\right)\frac{\varepsilon^2}{2} + \frac{s}{2}\left(\frac{s}{2}-1\right)\left(\frac{s}{2}-2\right)\frac{\varepsilon^3}{6} + O(\varepsilon^4).$$

It follows

$$|v'|^s + |v'_*|^s - |v|^s - |v_*|^s = |c|^s\left(\frac{s}{2}\left(\frac{s}{2}-1\right)\left[(e\cdot\sigma)^2|y|^2 - (e\cdot y)^2\right] + O(|y|^4)\right). \tag{23.11}$$

Again, let us separate small relative velocities $|v - v_*|^{\leq\lambda}$ and larger relative velocities $|v - v_*| \geq \lambda$, informally written $|v - v_*| = O(1)$ and $|v - v_*| \gg 1$, separate the collision kernel, with obvious notation, as $Q^{\leq\lambda} + Q^{>\lambda}$, and apply two different weak formulations for these two parts:

$$\begin{aligned}\int Q(f,f)\varphi_s = \frac{1}{2}\iint ff_*\Big(\int &\left(\varphi_s(v') + \varphi_s(v'_*) - \varphi_s(v) - \varphi_s(v_*)\right)\\ &B^{>\lambda}(v - v_*,\sigma)\,d\sigma\Big)\,dv\,dv_*\\ - \frac{1}{4}\iiint (f'f'_* - ff_*)&\left(\varphi_s(v') + \varphi_s(v'_*) - \varphi_s(v) - \varphi_s(v_*)\right)\\ &B^{\leq\lambda}(v - v_*,\sigma)\,d\sigma\,dv\,dv_*.\end{aligned} \tag{23.12}$$

- For $|v - v_*| \gg 1$ and $|v|, |v_*|$ incomparable, say $|v| \gg |v_*|$, one has

$$v' \simeq \frac{v + |v|\sigma}{2} = |v|\left(\frac{k+\sigma}{2}\right), \qquad v'_* \simeq |v|\left(\frac{k-\sigma}{2}\right),$$

so

$$
\begin{aligned}
&\int \big(\varphi_s(v') + \varphi_s(v'_*) - \varphi_s(v) - \varphi_s(v_*)\big)\, B^{>\lambda}(v - v_*, \sigma)\, d\sigma \\
&\quad \simeq -|v|^{s+\gamma} \int_{\mathbb{S}^{d-1}} \left(1 - \left|\frac{k+\sigma}{2}\right|^s - \left|\frac{k-\sigma}{2}\right|^s\right) b(k\cdot\sigma)\, d\sigma \\
&\quad = -|v|^{s+\gamma}|\mathbb{S}^{d-2}| \int_0^\pi \left[1 - \left(\frac{1+\cos\theta}{2}\right)^{s/2} - \left(\frac{1-\cos\theta}{2}\right)^{s/2}\right] \\
&\qquad \times b(\cos\theta)\, \sin^{d-2}\theta\, d\theta \\
&\quad = -|v|^{s+\gamma}|\mathbb{S}^{d-2}| \int_0^\pi \left[1 - (\cos(\theta/2))^s - (\sin(\theta/2))^s\right] b(\cos\theta)\, \sin^{d-2}\theta\, d\theta \\
&\quad \le -K(s,d)|v|^{s+\gamma} \int_0^\pi \theta^2 b(\cos\theta)\, \sin^{d-2}\theta\, d\theta.
\end{aligned}
$$

- For $|v - v_*| \gg 1$ and $|v|$, $|v_*|$ comparable, from (23.11)

$$
\begin{aligned}
&\left|\int \big(\varphi_s(v') + \varphi_s(v'_*) - \varphi_s(v) - \varphi_s(v_*)\big) B^{>\lambda}(v - v_*, \sigma)\, d\sigma\right| \\
&\quad \le C|v - v_*|^{\gamma+2}(1 + |v|^{s-2} + |v_*|^{s-2}) \\
&\qquad \times \left(\int \sin^2\theta\, b(\cos\theta)\, \sin^{d-2}\theta\, d\theta\right) \\
&\quad \le C'(1+|v|)^{s+\gamma-2}(1+|v_*|^2).
\end{aligned}
$$

Combining both estimates,

$$
\begin{aligned}
&\iint ff_* \int \big(\varphi_s(v') + \varphi_s(v'_*) - \varphi_s(v) - \varphi_s(v_*)\big)\, B^{>\lambda}(v - v_*, \sigma)\, d\sigma\, dv\, dv_* \\
&\qquad \le -K(s,d,\gamma,b) \int f\varphi_{s+\gamma} + C(s,d,\gamma,b,E_0) \int f\varphi_{s+\gamma-2}.
\end{aligned}
$$

Now for smaller relative velocities:

$$
\begin{aligned}
\left|\int Q^{<\lambda}(f,f)\,\varphi_s\right| &\le \frac{1}{4}\left|\iiint (f'f'_* - ff_*)\Big[(\varphi_s)' + (\varphi_s)'_* - \varphi - (\varphi)_*\Big]\right. \\
&\quad \times B\, dv\, dv_*\, d\sigma\Big| \qquad (23.13) \\
&\le \frac{1}{4}\left(\iiint \left(\sqrt{f'f'_*} - \sqrt{ff_*}\right)^2 B\, dv\, dv_*\, d\sigma\right)^{1/2}
\end{aligned}
$$

$$\times \left(\left(\sqrt{f'f'_*} + \sqrt{ff_*} \right)^2 \left[(\varphi_s)' + (\varphi_s)'_* - \varphi - (\varphi)_* \right]^2 B \, dv \, dv_* \, d\sigma \right)^{1/2}$$

$$\leq 2D_B(f)^{1/2} \left(\iint ff_* C(B, \varphi_s) \, dv \, dv_* \right)^{1/2}$$

where

$$C(B, \varphi_s) = \int \left[(\varphi_s)' + (\varphi_s)'_* - \varphi_s - (\varphi_s)_* \right]^2 B \, d\sigma.$$

Playing with (23.11), we see that the term within square brackets in the integrand above is $O(|v - v_*|^2 \theta)(1 + |v|^{s-2} + |v_*|^{s-2})$. When $|v|, |v_*| = O(1)$ the square is $O(|v - v_*|^4 \theta^2)$. When $|v|, |v_*| \gg 1$, then necessarily $|v|, |v_*|$ are comparable to each other and the square is bounded by $C|v - v_*|^4 \theta^2 |v|^{s-2} |v_*|^{s-2}$. All in all,

$$\begin{aligned} &\left| \iint ff_* C(B, \varphi_s) \, dv \, dv_* \right| \\ &\leq C(s, d, \gamma) \left(\int \theta^2 b(\cos\theta) \sin^{d-2}\theta \, d\theta \right) \\ &\quad \times \iint ff_* |v - v_*|^{\gamma+4} 1_{|v - v_*| \leq \lambda} \langle v \rangle^{s-2} \langle v_* \rangle^{s-2} \, dv \, dv_* \\ &\leq C \left(\int f \varphi_{s-2} \right)^2. \end{aligned}$$

Plugging back in (23.13) and interpolating L^1_{s-2} between L^1_{s-4} and L^1_s yields

$$\begin{aligned} \left| \int Q^{<\lambda}(f, f) \varphi_s \right| &\leq C D_B(f)^{1/2} \|f\|^{1/2}_{L^1_s} \|f\|^{1/2}_{L^1_{s-4}} \\ &\leq \varepsilon \|f\|_{L^1_{s-4}} + C D_B(f) \|f\|_{L^1_s}. \end{aligned}$$

Putting the pieces together, we obtain (23.9) and it follows that $M_s(t)$ remains $O(1+t)$ globally in time.

Once that conclusion is obtained, by interpolation it follows readily that: For any s and any $\varepsilon > 0$, there is $s_0 = s/\varepsilon$ such that if $M_{s_0}(f_0) < \infty$ then $M_s(t)$ remains $O((1+t)^\varepsilon)$ for all $t \geq 0$, and the constant can be made explicit in terms of $s, d, \gamma, b, \varepsilon, M_{s/\varepsilon}(f_0), H(f_0)$. To write it in a synthetic form:

$$f_0 \in L^1_\infty(\mathbb{R}^d) \cap L \log L(\mathbb{R}^d) \Longrightarrow \qquad \|f\|_{L^1_\infty} = O((1+t)^{0}). \tag{23.14}$$

This excellent control on the moments allows to truncate all large velocities, which is a key step in all subsequent developments. The assumption of finite moments of arbitrarily large order is usually quite reasonable in the context of kinetic theory.

23.4 Higher Integrability

At this stage large velocities are under control and the Landau equation is a quasilinear parabolic equation with diffusion matrix $\overline{a} = a * f$. The symmetric matrix $a(z)$ is degenerate in the direction z and has eigenvalues roughly $|z|^{\gamma+2}$ and 0. When averaged over f, which is not concentrated on a subspace (due to the L^{p_0} bound or just the bound on $H(f)$), one finds

$$\overline{a}(v) \geq K_r |v|^{\gamma} \Pi_v + K_{\perp} |v|^{\gamma+2} \Pi_{v\perp}, \qquad \overline{a}(v) \leq C|v|^{\gamma+2}. \tag{23.15}$$

The gap between exponents γ and $\gamma + 2$ could be a problem, but the moment estimates are enough to handle this. Very classically, let us look for energy-type estimates: if $\Phi_p(f) = f^p$, $p > 1$, then from $\partial_t f = \nabla \cdot (\overline{a}\nabla f - \overline{b} f)$, $\overline{b} = \nabla \cdot \overline{a}$, results

$$\begin{aligned} \partial_t \Phi_p(f) &= \Phi_p'(f)\nabla \cdot (\overline{a}\nabla f - \overline{b} f) \\ &= \nabla \cdot (\overline{a}\nabla \Phi_p(f) - \overline{b}\Phi(f)) - \Phi_p''(f)\overline{a}\nabla f \nabla f \\ &\quad + (\nabla \cdot \overline{b})\big[f\Phi_p'(f) - \Phi_p(f)\big]. \end{aligned} \tag{23.16}$$

Writing $\overline{c} = \nabla \cdot \overline{b}$ and integrating, yields the a priori estimate

$$\frac{d}{dt}\int \Phi_p(f) = -\int \Phi_p''(f)\overline{a}\nabla f \nabla f + \int \overline{c}\big[f\Phi_p'(f) - \Phi_p(f)\big].$$

Then

$$\Phi_p''(f) = p(p-1)f^{p-2}, \quad \Phi_p'(f) = pf^{p-1}, \qquad f\Phi_p'(f) - \Phi_p(f) = (p-1)f^p.$$

So

$$\begin{aligned} \frac{d}{dt}\int \Phi_p(f) &= -p(p-1)\int f^{p-2}\overline{a}\nabla f \nabla f - (p-1)\int \overline{c} f^p \\ &= -\frac{p(p-1)}{(p/2)^2}\int \overline{a}\nabla \Phi_{p/2}(f)\nabla \Phi_{p/2}(f) - (p-1)\int \overline{c} f^p. \end{aligned} \tag{23.17}$$

The first term on the right-hand side is handled through the positivity estimate (23.15), an easy commutator and Sobolev embedding:

$$\begin{aligned}\int \left|\sqrt{\overline{a}}\nabla\Phi_{p/2}(f)\right|^2 &\geq K(d,\gamma,E_0,H(f_0)) \int \langle v\rangle^\gamma \, |\nabla\Phi_{p/2}(f)|^2 \, dv \\ &\geq K \left\|\nabla(\langle v\rangle^{\gamma/2} f^{p/2})\right\|_{L^2}^2 - C \int f^p \langle v\rangle^{\gamma-2}\, dv \\ &\geq K \|f^{p/2}\|_{L^{2^\star}_{\gamma/p}} - C\|f\|^2_{L^p_{(\gamma-2)/p}} \qquad (2^\star = (2d)/(d-2)) \\ &= K\|f\|_{L^{\lambda p}_{\gamma/p}} - C\|f\|^p_{L^p_{(\gamma-2)/p}}, \qquad \lambda = \frac{d}{d-2}.\end{aligned}$$

(The case $d = 2$ should be treated separately, any λ will do.)

Let us turn to the second term on the right hand side of (23.17). First, if $a(z) = |z|^{\gamma+2}\Pi_{z^\perp}$, then $b(z) = -(d-1)|z|^\gamma z$, $c(z) = -(d-1)(\gamma+d)|z|^\gamma$, to be understood as a multiple of the Dirac mass δ_0 in the limit case $d = -\gamma$ (for $d = 2, 3$). From these bounds and Hardy–Littlewood–Sobolev inequality, this second term is controlled by

$$C(\gamma+d) \iint |v - v_*|^\gamma f(v_*) f(v)^p \, dv\, dv_* \leq C\|f\|_{L^{p_0}} \|f^p\|_{L^r}, \tag{23.18}$$

where

$$2 + \frac{\gamma}{d} = \frac{1}{p_0} + \frac{1}{r},$$

and it is required that $-(d+2) < \gamma < -2$, which is implied by (23.1). In the limit case $\gamma = -d$, the left-hand side of (23.18) should be replaced by $C\int f^{p+1}$ and the bound still holds true by Lebesgue interpolation. Then, by interpolation again,

$$\|f^p\|_{L^r} = \|f\|^p_{L^{pr}} \leq \left(\|f\|^p_{L^{\lambda p}_{\gamma/p}}\right)^{r/\lambda} \|f\|_{L^1_s}^{p(1-(r/\lambda))}$$

for $s = -(\gamma r)/(\lambda - r)$ provided of course that $r < \lambda$, i.e. $1/r < (d-2)/d$, i.e. $\gamma > -4$ (as assumed). All in all, we have identified $s > 0$ and $\theta \in (0,1)$ such that

$$\frac{d}{dt}\|f\|^p_{L^p} \leq -K\|f\|^p_{L^{\lambda p}_{\gamma/p}} + C\|f\|^p_{L^p_{(\gamma-2)/p}} + C(\|f\|^p_{L^{\lambda p}_{\gamma/p}})^\theta \|f\|^{1-\theta}_{L^1_s}.$$

It follows, through Young's inequality and interpolation again,

$$\frac{d}{dt}\|f\|^p_{L^p} \leq -K\|f\|^p_{L^{\lambda p}_{\gamma/p}} + C\|f\|_{L^1_s}. \tag{23.19}$$

Recall that $\|f_0\|^{L^{p_0}}$ is uniformly bounded, that $\|f\|_{L^1_s}$ is $O((1+t)^\varepsilon)$ for any given ε if f has finite moments of sufficiently high order. Further note that, by interpolation again, (23.19) implies

$$\frac{d}{dt}\|f\|^p_{L^p} \le -K(\|f\|^p_{L^p})^{\lambda'} + C\|f\|_{L^1_{s'}}. \tag{23.20}$$

for any $\lambda' < \lambda$, and $s' = s'(s, \lambda, p, \gamma)$. Choosing $\lambda' > 1$ implies, by a classical reasoning, the short-time appearance of all L^p norms of f for $p = \lambda' p_0$, $(\lambda')^2 p_0$, etc. up to any $(\lambda')^\ell p_0$, $\ell \in \mathbb{N}$, and the bounds will be controlled in time like powers of the moments, which in turn are controlled like small powers of t. To summarise: Given any $p \ge 1$ we have the a priori bound

$$\forall t \ge 0 \qquad \|f(t)\|_{L^p} \le C\left(\frac{1}{t^{\alpha_p}} + t^\varepsilon\right),$$

where α_p depends on $p, d, \gamma, \varepsilon$ is as small as required, and C depends on p, d, γ, b, E_0, $H(f_0)$, $I(f_0)$, ε and $\|f_0\|_{L^1_s}$ for some $s = s(\varepsilon, \gamma, p, d)$. In short, all L^p norms appear instantly and grow slowly in large time.

Let us repeat the same scheme for the Boltzmann equation. This will work out if the kernel is *singular enough* in the angular variable. First a Γ formula. From the elementary identity

$$-pf^{p-1}(f'f'_*-ff_*)+[(f')^p f'_*-f^p f_*]=f'_*[(f')^p-f^p-pf^{p-1}(f'-f)] \\ -(p-1)(f'_*-f_*)f^p$$

(for any four numbers $f, f_*, f', f'_* \ge 0$) one finds

$$Q(f, f^p) - pf^{p-1}Q(f, f) = \iint f'_*\Gamma_{\Phi_p}(f, f')\, B\, dv_*\, d\sigma - (p-1)\mathcal{S} * f, \tag{23.21}$$

where $Q(g, f) = \iint (g'_* f' - g_* f)\, B\, dv_*\, d\sigma$,

$$\Gamma_{\Phi_p}(f, f') = (f')^p - f^p - pf^{p-1}(f' - f)$$

and

$$\mathcal{S} * f = \iint (f'_* - f_*) B(v - v_*, \sigma)\, dv_*\, d\sigma.$$

Changing variable $v'_* \to v_*$, for fixed σ, shows that $\mathcal{S}$ is well-defined and $|\mathcal{S}(z)| \le C|v - v_*|^\gamma$ for $\gamma > -d$, while in the limit case $\gamma = -d$, $\mathcal{S}$ is proportional to the Dirac mass δ_0. Also the elementary convexity inequality

$$x^p - 1 - p(x-1) \ge K_p(x^{p/2} - 1)^2$$

yields

$$\Gamma_{\Phi_p}(f, f') \ge K_p\big[(f')^{p/2} - f^{p/2}\big]^2.$$

So integrating (23.21) in $\mathbb{R}^d$ yields the a priori estimate

$$\begin{aligned}
\frac{d}{dt}\int f^p &= -\iiint f'_*\Gamma_{\Phi_p}(f,f')\,B\,dv\,dv_*\,d\sigma + (p-1)\int f^p(\mathcal{S}*f)\,dv \\
&= -\iiint f_*\Gamma_{\Phi_p}(f',f)\,B\,dv\,dv_*\,d\sigma + (p-1)\int f^p(\mathcal{S}*f)\,dv \\
&\le -K\iiint f_*\big[f^{p/2}(v')-f^{p/2}(v)\big]^2\,B(v-v_*,\sigma)\,dv\,dv_*\,d\sigma \\
&\quad + C\iint f^p f_*|v-v_*|^\gamma\,dv\,dv_* \\
&\le -K\iiint f_*\langle v-v_*\rangle^\gamma\big[f^{p/2}(v')-f^{p/2}(v)\big]^2\,b(k\cdot\sigma)\,dv\,dv_*\,d\sigma \\
&\quad + C\iint f^p f_*|v-v_*|^\gamma\,dv\,dv_*,
\end{aligned}$$

where as usual $k=(v-v_*)/|v-v_*|$.

Let us focus on

$$\mathcal{D}(f,F)=\iiint f_*\langle v-v_*\rangle^\gamma\big[F(v')-F(v)\big]^2\,b(k\cdot\sigma)\,dv\,dv_*\,d\sigma.$$

The factor $\langle v-v_*\rangle^\gamma$ may be very small compared to $\langle v\rangle^\gamma$ as $|v_*|\to\infty$; so it will be convenient to truncate large velocities v_* in the crudest possible way:

$$\langle v-v_*\rangle^\gamma \ge K\,R^\gamma\,\langle v\rangle^\gamma\,1_{|v_*|\le R}.$$

It will also be convenient to truncate angles $\theta\ge\pi/2$; so let $b_0(\cos\theta)=b(\cos\theta)1_{1\le\theta\le\pi/2}$. Then

$$\begin{aligned}
\mathcal{D}(f,F) &\ge K(R)\iiint (f_*1_{|v_*|\le R})\langle v\rangle^\gamma\big[F(v')-F(v)\big]^2\,b(k\cdot\sigma)\,dv\,dv_*\,d\sigma \\
&\ge \frac{K(R)}{2}\iiint (f_*1_{|v_*|\le R})\big[F(v')\langle v'\rangle^\gamma - F(v)\langle v\rangle^\gamma\big]^2\,b_0(k\cdot\sigma)\,dv\,dv_*\,d\sigma \qquad (23.22) \\
&\quad - K(R)\iiint (f_*1_{|v_*|\le R})\big[\langle v'\rangle^{\gamma/2}-\langle v\rangle^{\gamma/2}\big]^2 \\
&\qquad\qquad \times F(v')^2\,b_0(k\cdot\sigma)\,dv\,dv_*\,d\sigma. \qquad (23.23)
\end{aligned}$$

Then $(\langle v'\rangle^{\gamma/2}-\langle v\rangle^{\gamma/2})^2\le C|v-v_*|^2\theta^2\le C|v'-v_*|^2\theta^2$ (because $\theta\le\pi/2$), so the last integral is crudely bounded above by

$$C \iiint f_*(1 + |v_*|^2)(1 + |v'|^2)\theta^2 b_0(k \cdot \sigma) F(v')^2 \, dv \, dv_* \, d\sigma.$$

For given v_*, σ the change of variables $v \to v'$ has bounded Jacobian, and transforms θ into $\theta/2$; so the bound above may be controlled by a multiple of

$$\iiint f_*(1 + |v_*|^2)(1 + |v|^2) \, b_0(\cos(2\theta)) \, \theta^2 F(v)^2 \, dv \, dv_* \, d\sigma,$$

which in turn is bounded above by

$$C \left(\int f_* |v_*|^2 \, dv_* \right) \left(\int_{\mathbb{S}^{d-1}} \theta^2 \, b(\cos\theta) \, \sin^{d-2}\theta \, d\theta \right) \int F^2 \langle v \rangle^2 \, dv.$$

Now for the other part, (23.22): Here we have reduced to a Maxwellian kernel. A Fourier integral representation going back to Bobylev shows that

$$\begin{aligned} &\iiint h(v_*) \big[G(v') - G(v) \big]^2 \, b_0(k \cdot \sigma) \, dv \, dv_* \, d\sigma \\ &= \int_{\mathbb{R}^d} \int_{\mathbb{S}^{d-1}} b(k \cdot \sigma) \Big[\widehat{h}(0) |\widehat{G}(\xi)|^2 + \widehat{h}(0) |\widehat{G}(\xi^+)|^2 - \widehat{h}(\xi^-) \widehat{G}(\xi^+) \overline{\widehat{G}(\xi)} \\ &\qquad\qquad - \overline{\widehat{h}(\xi^-) \widehat{G}(\xi^+)} \widehat{G}(\xi) \Big] \, d\xi \, d\sigma \end{aligned}$$

where $\overline{z}$ stands for complex conjugate of z, and

$$\widehat{f}(\xi) = \int_{\mathbb{R}^d} e^{-2i\pi \xi \cdot v} f(v) \, dv, \qquad \xi^+ = \frac{\xi + |\xi|\sigma}{2}, \qquad \xi^- = \frac{\xi - |\xi|\sigma}{2}.$$

It follows, with $h(v) = f(v) 1_{|v| \le R}$ and $G(v) = F(v) \langle v \rangle^{\gamma/2}$, that (23.22) is bounded below by

$$K(R) \iint \big(\widehat{h}(0) - |\widehat{h}(\xi^-)| \big) \big(|\widehat{G}(\xi^+)|^2 + |\widehat{G}(\xi)|^2 \big) \, d\xi \, d\sigma.$$

From the bound on $H(f)$ and E_0 follows a non-concentration estimate on f and thus on h (independently of R), so that

$$\widehat{h}(0) - |\widehat{h}(\xi^-)| \ge K \min(|\xi^-|^2, 1),$$

and for R large enough, fixed, the constant K will be bounded bellow. Eventually (23.22) is bounded below by

$$K \int \min(|\xi^-|^2, 1) b\left(\frac{\xi}{|\xi|} \cdot \sigma\right) |\widehat{G}(\xi)|^2 \, d\xi \, d\sigma \geq K \int_{\mathbb{R}^d} |\widehat{G}(\xi)|^2 |\xi|^\nu \, d\xi = \|G\|^2_{\dot{H}^{\nu/2}}.$$

All in all,

$$\mathcal{D}(f, F) \geq K \|F\|^2_{H^{\nu/2}_{\gamma/2}} - C \|F\|^2_{L^2_1},$$

where K, C only depend on $E(f_0)$ and $H(f_0)$.

With this at hand, we may reason as in the proof of higher integrability for the Landau equation, using the Sobolev embedding

$$\|f\|_{L^{2\lambda}} \leq C \|f\|_{H^{\nu/2}},$$

where now

$$\lambda = \frac{d}{d - \nu}.$$

Eventually,

$$\frac{d}{dt} \int f^p \leq -K \|f^{p/2}\|^2_{H^{\nu/2}_{\gamma/p}} + C \|f\|^{\theta p}_{L^p} + C \|f\|_{L^1_s}$$

for some $\theta \in (0, 1)$ and s large enough, as in (23.19), but now the condition $r < \lambda$ becomes

$$\gamma + \nu > -2,$$

as assumed in (23.2). The end of the reasoning is similar: We find that all L^p norms appear instantaneously like inverse power of t, and these bounds are all $O(t^\varepsilon)$ as $t \to \infty$, for ε arbitrarily small, provided that f_0 has sufficiently many finite moments, and again the bounds and exponents are computable.

23.5 Smoothness

The previous subsection was about improving integrability, this one is about improving regularity, working with L^2-type spaces now. Again, start with the Landau equation.

The first issue is to evaluate the regularity of the diffusion matrix $\overline{a} = a * f$. As $|a| \leq C|z|^{\gamma+2}$,

$$\|a * f\|_{L^\infty} \leq C \, \|f\|_{L^q}$$

with $\gamma + 2 + d = d/q$. So, as soon as $\gamma + 2 + d > 0$ (which is implied by (23.1)), choosing q large enough yields an L^∞ bound on $\overline{a}$, with a norm proportional to $\|f\|_{L^q}$, which is known to be $O(t^\varepsilon)$ with ε arbitrarily small, if enough moments are finite. The same reasoning holds for $\nabla\overline{a}$, as soon as $\gamma + 1 + d > 0$. As for $\nabla^2\overline{a}$, as $\gamma + d$ may be equal to 0, we only get a bound on $\|(\nabla^2 a) * f\|_{L^p}$, locally, for p arbitrarily large. The conclusion is $\overline{a} \in W^{2,p}$, locally, for all $p \geq 1$, with a local bound that may grow polynomially, and in particular $\overline{a} \in C^{1,\alpha}$ for some $\alpha > 0$ and $\overline{c} \in L^p$ locally, for all $p > 1$. In the sequel, I shall systematically evaluate $\overline{a}$ in Sobolev norms weighted by a negative power of the velocity—anyway large velocities are controlled by the moment bounds.

As for the lower bound, nonconcentration implies a lower bound like $\overline{a}(z) \geq K\langle z\rangle^\gamma$.

Now one can go through higher regularity with the classical method of commuting the equation both with the quadratic nonlinearity and the derivation:

- Start from $\partial_t f = \nabla \cdot (\overline{a}\nabla f - \overline{b} f)$, $\overline{a} = a * f$, $\overline{b} = b * f = (\nabla \cdot a) * f$.
- Then pick up an index k, and differentiate with respect to v_k, it follows

$$\partial_t(\partial_k f) = \nabla \cdot \left(\overline{a}\nabla\partial_k f - \overline{b}\,\partial_k f\right) + \partial_i\left(\partial_k\overline{a}_{ij}\,\partial_j f\right) - \partial_i\left(\partial_k\overline{b}_i f\right),$$

  with the convention of summation over repeated indices.
- Then multiply by $\partial_k f$ and work out the commutator in Γ style, to get

$$\begin{aligned}\partial_t\left(\frac{(\partial_k f)^2}{2}\right) &= \partial_k f\,\partial_t f \\ &= \nabla\cdot\left(\overline{a}\nabla\frac{(\partial_k f)^2}{2} - \overline{b}\frac{(\partial_k f)^2}{2}\right) - \overline{a}\nabla\partial_k f\nabla\partial_k f - \overline{c}\frac{(\partial_k f)^2}{2} \\ &\qquad + \partial_k f\,\partial_i\left(\partial_k\overline{a}_{ij}\partial_j f\right) - \partial_k f\,\partial_i\left(\partial_k\overline{b}_i f\right).\end{aligned}$$

  The key term is of course $-\overline{a}\nabla\partial_k f\nabla\partial_k f$ which is like a $\Gamma_2(f, f)$ term. After that, the last three terms will be treated as error terms. Upon integrating the previous expression over $\mathbb{R}^d$, summing over k and integrating by parts, comes

$$\begin{aligned}\frac{d}{dt}\int\frac{|\nabla f|^2}{2} &= -\int\overline{a}\nabla\nabla f : \nabla\nabla f - \int\overline{c}\frac{|\nabla f|^2}{2} \\ &\qquad - \int\partial_{ik} f\,\partial_k\overline{a}_{ij}\,\partial_j f + \int\partial_{ik} f\,\partial_k\overline{b}_i f. \qquad (23.24)\end{aligned}$$

  Then

$$\int\overline{a}\nabla\nabla f\nabla\nabla f \geq K\int\langle v\rangle^\gamma|\nabla^2 f|^2.$$

Also

$$\left|\int \partial_{ik} f \, \partial_k \overline{a}_{ij} \, \partial_j f\right| \leq \left(\int \langle v\rangle^{\gamma} |\nabla^2 f|^2\right)^{1/2} \left(\in \langle v\rangle^{-\gamma} |\nabla \overline{a}|^2 |\nabla f|^2\right)^{1/2}$$

$$\leq \varepsilon \int \langle v\rangle^{\gamma} |\nabla^2 f|^2 + C \int \langle v\rangle^{-\gamma} |\nabla \overline{a}|^2 |\nabla f|^2,$$

where ε is as small as desired. Similarly,

$$\left|\int \partial_{ik} f \, \partial_k \overline{a}_{ij} \, \partial_j f\right| \leq \varepsilon \int \langle v\rangle^{\gamma} |\nabla^2 f|^2 + C \int \langle v\rangle^{-\gamma} |\nabla \overline{b}|^2 |\nabla f|^2.$$

To summarise, at this stage

$$\begin{aligned} \frac{d}{dt} \int |\nabla f|^2 \leq & -K \int \langle v\rangle^{\gamma} |\nabla^2 f|^2 \\ & + C \left(\int \langle v\rangle^{-\gamma} |\nabla \overline{a}|^2 |\nabla f|^2 \right. \\ & \left. + \int \langle v\rangle^{-\gamma} |\nabla \overline{b}|^2 |\nabla f|^2 + \int |\overline{c}| |\nabla f|^2 \right). \end{aligned} \tag{23.25}$$

To get the wheel of the estimate go round, one needs to dominate the sum of the three integrals within brackets, at the end of (23.25), by a little bit of $\int \langle v\rangle^{\gamma} |\nabla^2 f|^2$. There are two issues to do this:

- powers of $\langle v\rangle$ do not match (the diffusion is weak at large velocities), but this will be handled by moment estimates;
- $\overline{a}$, $\overline{b}$, $\overline{c}$ are just partly regular and their regularity has to rest partly on the regularity of f, not just a, b, c (this will be particularly true for higher derivatives). For instance in the important case $d = 3$, $\gamma = -3$, we have $|\overline{c}| = f$, which at this stage of the proof is estimated in all L^p spaces but not in L^∞. So the power of f inside the integral will be higher than 2, and Young inequality will be needed to work with indices $2p, 2p'$ with $p \sim \infty$, $p' \sim 1$. For this strategy to work out, we shall use the information that f lies in all weighted L^p spaces. So this will be an interpolation between $W^{k,p}_s$ spaces (k derivatives in L^p and s moments), of the type

$$\|f\|_{W^{1,p}_\sigma} \leq C \|f\|^{\theta}_{W^{2,2}_{\gamma/2}} \|f\|^{1-\theta}_{L^r_s}. \tag{23.26}$$

Taking r, s large enough will make the exponent θ lower. In fact the compatibility condition is

$$-1+\frac{d}{p}=\theta\left(-2+\frac{d}{2}\right)+(1-\theta)\frac{d}{r}.$$

So the condition to find suitable r, s (large enough) for (23.26) to hold is $0 < \theta < 1$, where

$$\theta=\frac{1-\frac{d}{p}}{2-\frac{d}{2}}.$$

As for the dependence of $\overline{a}, \overline{b}, \overline{c}$ in f, it is as follows: One can find σ such that for any p there are r, s such that for all f,

$$\|\overline{a}\|_{W^{2,p}_{-\sigma}}+\|\overline{b}\|_{W^{1,p}_{-\sigma}}+\|\overline{c}\|_{L^p_{-\sigma}}\leq C\|f\|_{L^r_s}. \tag{23.27}$$

So, if $p > 1$ and $p' = p/(p-1)$,

$$\begin{aligned}\int \langle v\rangle^{-\gamma}|\nabla\overline{a}|^2|\nabla f|^2 &\leq C\|\nabla\overline{a}\|^2_{L^{2p}_m}\|\nabla f\|^2_{L^{2p'}_{-\gamma/2}}\\ &\leq C\|f\|^2_{L^{r_1}_{s_1}}\|f\|^{2\theta}_{W^{2,2}_{\gamma/2}}\|f\|^{2(1-\theta)}_{L^{r_2}_{s_2}}.\end{aligned}$$

and this works for r_1, r_2, s_1, s_2 large enough if

$$0<\frac{1-\frac{d}{2p'}}{2-\frac{d}{2}}<1.$$

Obviously this is true for p large enough, so that p' is close to 1.

In the same way, one shows that

$$\int \langle v\rangle^{-\gamma}|\nabla\overline{b}|^2|\nabla f|^2+\int|\overline{c}||\nabla f|^2\leq C\|f\|^{2\theta}_{W^{2,2}_{\gamma/2}}\|f\|^{2(2-\theta)}_{L^r_s}.$$

Also note that

$$\|f\|^2_{W^{2,2}_{\gamma/2}}\leq\int\langle v\rangle^{\gamma}|\nabla^2 f|^2+C\|f\|^2_{L^2}.$$

Then fixing all parameters as above, (23.25) becomes

$$\frac{d}{dt}\int\frac{|\nabla f|^2}{2}\leq -K\|f\|^2_{W^{2,2}_{\gamma/2}}+C\|f\|^4_{L^p_s}. \tag{23.28}$$

This proves an a priori bound on $\int|\nabla f|^2$, which becomes immediately finite and remains so for all times, with a bound like $O(t^\varepsilon)$ in large time, for arbitrarily small ε.

Then go for higher derivatives. Differentiating the equation at higher order, working out the commutators, the degenerate coercivity and the error terms, one arrives at

$$\frac{d}{dt}\int|\nabla^k f|^2 \leq -K\int\langle v\rangle^\gamma|\nabla^{k+1}f|^2 + C\sum_{|r|\leq k,|r|+|\ell|\leq k+1}\int\langle v\rangle^{-\gamma}|\nabla^\ell\overline{a}|^2|\nabla^r f|^2$$
$$+C\sum_{|r|+|\ell|\leq k+1}\int\langle v\rangle^{-\gamma}|\nabla^\ell\overline{b}|^2|\nabla^r f|^2 + C\sum_{|r|+|\ell|\leq k}\langle v\rangle^{-\gamma}|\nabla^\ell\overline{c}|^2|\nabla^r f|^2,$$

and if σ is given we may find s such that

$$\|\overline{a}\|_{W^{j+2,p}_{-\sigma}} + \|\overline{b}\|_{W^{j+1,p}_{-\sigma}} + \|\overline{c}\|_{W^{j,p}_{-\sigma}} \leq C\|f\|_{W^{j,2}_s},$$

and the relevant interpolation now is

$$\|f\|_{W^{\ell,p}_\sigma} \leq C\|f\|^\theta_{W^{L,2}_{\gamma/2}}\|f\|^{1-\theta}_{L^r_s},$$

where

$$-\ell+\frac{d}{p} = \theta\left(-L+\frac{d}{2}\right) + (1-\theta)\frac{d}{r}.$$

Then error terms are bounded like, for instance,

$$\|\nabla^\ell\overline{b}\|^2_{L^{2p}_m}\|\nabla^r f\|^2_{L^{2p'}_{-\gamma/2}} \leq C\|f\|^2_{W^{\ell-1,2p}_{m+\sigma}}\|f\|^2_{W^{r,2p'}_{-\gamma/2}}$$
$$\leq C\left(\|f\|^{\theta_1}_{W^{k+1,2}_{-\gamma/2}}\|f\|^{1-\theta_1}_{L^m_\sigma}\right)^2\left(\|f\|^{\theta_2}_{W^{k+1,2}_{-\gamma/2}}\|f\|^{1-\theta_2}_{L^m_\sigma}\right)^2,$$

and

$$\theta_1+\theta_2 = \frac{(\ell-1)-\frac{d}{2p}+\left(r-\frac{d}{2p'}\right)}{k+1-\frac{d}{2}} \leq \frac{k-\frac{d}{2}}{k+1-\frac{d}{2}}.$$

All in all,

$$\frac{d}{dt}\int|\nabla^k f|^2 \leq -K\int\langle v\rangle^\gamma|\nabla^{k+1}f|^2 + C\|f\|^A_{L^m_s},$$

for some $K > 0$ and m, s, A large enough, and there is $\theta \in (0, 1)$ such that

$$\frac{d}{dt}\int |\nabla^k f|^2 \le -K\left(\int |\nabla^k f|^2\right)^{1/\theta} + C\|f\|^A_{L^m_s},$$

which proves both the immediate appearance of Sobolev norms and their good control in large time since all weighted L^m norms are $O(t^\varepsilon)$ for $t \to \infty$, with ε as small as desired (if large enough moments are finite). By Sobolev embedding, this also implies a control of all derivatives like inverse power laws (rapid decay).

More precisely: For arbitrarily large $k, \kappa > 0$, for arbitrarily small $\varepsilon > 0$, there are $\alpha > 0$ (controlled from below) and $s > 0$ (controlled from above) such that

$$\forall t > 0, \qquad |\nabla^k f(t, v)| \le \frac{C\big(t^{-\alpha} + (1+t)^\varepsilon\big)}{\langle v\rangle^\kappa}, \tag{23.29}$$

and C depends in a computable way on E_0, $H(f_0)$, $I(f_0)$ and $\|f_0\|_{L^1_s}$.

For Boltzmann, the strategy to get higher regularity is exactly the same, but now it rests on the properties of the bilinear Boltzmann operator $Q(g, f) = \iint (g'_* f' - g_* f) B\, dv_*\, d\sigma$, where (without loss of generality) b is supported in $(0 \le \theta \le \pi/2)$. Since $Q(g, \cdot)$ acts like a variant of a fractional diffusion of order ν, one has the estimates

$$\|Q(g, f)\|_{L^p_\sigma} \le C\|g\|_{L^q_s}\|f\|_{W^{\overline{\nu},\overline{p}}},$$

where $\overline{\nu}$ is arbitrarily close to ν and $\overline{p}$ arbitrarily close to p. The coercivity estimate remains the same,

$$\int Q(g, f) f \ge K\|f\|^2_{H^{\nu/2}_{-\gamma/2}} - C\|f\|^2_{L^m_s},$$

and going to higher derivatives is easy thanks to the bilinear identity

$$\partial_\ell Q(g, f) = Q(\partial_\ell g, f) + Q(g, \partial_\ell f).$$

Then all ingredients are here to obtain exactly the same rapid decay estimates (23.29).

23.6 What If the Initial Fisher Information Is Infinite?

Bounds and rapid decay of the distribution function are quite classical assumptions, as is the requirement of finite entropy. But the requirement of finite Fisher information, while statistically meaningful, may seem unsatisfactory. It turns out however that this can also be dispensed with, thanks to the regularising property of the equation. The scheme of proof is the following:

- Control the solution in L^p norms for a short time, as a quasilinear diffusive equation with quadratic nonlinearity, and deduce regularisation during that short time;
- Deduce that the Fisher information becomes finite in short time; then it will be nonincreasing, and we are back to the situation treated before.

As a warmup, let us consider the model case

$$\partial_t f = \langle v\rangle^\gamma \Delta f + f^2 \tag{23.30}$$

which does not have the same nice conservation properties as the Landau equation, but (for $d = 3$, $\gamma = -3$) satisfies the same estimates for diffusivity and nonlinearity. The usual estimate yields

$$\frac{d}{dt}\|f\|_{L^p}^p \leq -K\|f^{p/2}\|_{H^1_{\gamma/p}}^2 + C\|f\|_{L^{p+1}}^{p+1} \tag{23.31}$$

$$\leq -K\|f\|_{L^{\lambda p}_{\gamma/p}}^p + C\|f\|_{L^{p+1}}^{p+1}, \tag{23.32}$$

with $\lambda = d/(d-2) > 1$. Assume $p + 1 < \lambda p$, ie $p > 1/(\lambda - 1) = (d/2) - 1$. Then one can interpolate L^{p+1} between L^p and $L^{\lambda p}$, and a little bit of high order moments to compensate for the negative weight in the $L^{\lambda p}$ norm. This yields

$$\|f\|_{L^{p+1}} \leq C\|f\|_{L^p}^\theta \|f\|_{L^{\lambda p}_{\gamma/p}}^{1-\overline{\theta}} \|f\|_{L^1_s}^{\overline{\theta}-\theta}, \tag{23.33}$$

where $\overline{\theta} > \theta$ (arbitrarily close to θ) and $s > 0$ (large enough), and

$$\frac{1}{p+1} = \frac{\theta}{p} + \frac{1-\overline{\theta}}{\lambda p} + (\overline{\theta} - \theta).$$

This can be achieved for any

$$1 > \theta > \frac{\frac{p}{p+1} - \frac{1}{\lambda}}{1 - \frac{1}{\lambda}}.$$

From (23.32) and (23.33) follows

$$\frac{d}{dt}\|f\|_{L^p}^p \leq -K\|f\|_{L^{\lambda p}_{\gamma/p}}^p + C\|f\|_{L^p}^r + C\|f\|_{L^1_s}^A, \tag{23.34}$$

for any

$$r > \frac{p}{1 - \frac{\lambda-1}{(p+1)(\lambda-1)-1}},$$

provided that the exponent of $\|f\|_{L^{\lambda p}_{\gamma/p}}$ in the positive term is smaller than the exponent in the negative term, i.e. $(p+1)(1-\theta) < p$, which means $p > d/2$. Obviously $r > p$, so (23.34) does not prevent blowup in finite time—as expected. But it does imply a uniform control of $\|f\|_{L^p}$ in short time. This in turn implies a time-integrated control of $\|f\|_{L^{\lambda p}_{\gamma/p}}$ in short time, and by interpolation with moment bounds, a time-integrated control of $\|f\|_{L^q}$ in short time for any $q < \lambda p$. Then we can repeat the reasoning and obtain a uniform control of f in L^q, on a shorter time interval. And by induction, this works out for all L^p norms, and thus for all L^p_s norms, for arbitrary p and s. (The control deteriorates as p and s become larger, and the time interval becomes shorter and shorter, but the procedure can be repeated an arbitrary number of times to get arbitrarily large smoothness in short time.) Then we can repeat the procedure for higher regularity. As a conclusion of this step, solutions of (23.30) satisfy bounds of rapid decay (23.29), at least for short time $t \leq t_0 = t_0(k, \kappa, E_0, H(f_0), \|f_0\|_{L^1_s}, \|f_0\|_{L^{q_0}})$, where q_0 is any Lebesgue exponent satisfying $q_0 > d/2$.

As a second step, this implies the finiteness of the Fisher information for small t, thanks to an estimate of Toscani and myself (see Bibliographical Notes):

$$I(f) \leq C(s,d)\|f\|_{H^2_s}, \qquad \text{for any } s > \frac{d}{2}. \tag{23.35}$$

Now working out the same scheme for the Boltzmann equation with very soft potentials yields

$$\frac{d}{dt}\|f\|^p_{L^p} \leq -K\|f^{p/2}\|^2_{H^{\nu/2}_{\gamma/p}} + C\int |(S*f)|f^p \tag{23.36}$$

where $|S(v)| \leq C|v|^\gamma$. On the one hand,

$$\|f^{p/2}\|^2_{H^{\nu/2}_{\gamma/p}} \geq \|f\|^p_{L^{\lambda p}_{\gamma/p}}, \qquad \lambda = \frac{d}{d-\nu}. \tag{23.37}$$

On the other hand, we can find $\sigma > 0$ such that

$$\|S*f\|_{L^r_{-\sigma}} \leq C\|f\|^{\theta_1}_{L^q}\|f\|^{1-\theta_1}_{L^1_s}, \qquad \gamma + d + \frac{d}{r} = \frac{\theta_1 d}{q} + (1-\theta_1)d.$$

So, with $1/r + 1/r' = 1$,

$$\int |S*f|f^p \leq \|S*f\|_{L^r_{-\sigma}}\|f^p\|_{L^{r'}_\sigma}$$

$$\leq C\|f\|_{L^q_s}^{\theta_1}\|f\|_{L^{pr'}_s}^{p}$$

$$\leq C\|f\|_{L^{\lambda p}_{\gamma/p}}^{\alpha}\|f\|_{L^p}^{\beta}\|f\|_{L^1_s}^{A},$$

where

$$\frac{1}{q} = \frac{\eta}{\lambda p} + \frac{1-\overline{\eta}}{p} + (\overline{\eta} - \eta) \qquad (0 < \eta < \overline{\eta} < 1),$$

$$\frac{1}{pr'} = \frac{\zeta}{\lambda p} + \frac{1-\overline{\zeta}}{p} + (\overline{\zeta} - \zeta) \qquad (0 < \zeta < \overline{\zeta} < 1),$$

$$\alpha = \eta\theta_1 + \zeta p.$$

Using moments of high order we may choose

$$\begin{cases} \theta_1 \simeq 1 \\ \frac{\gamma}{d} + 1 + \frac{1}{r} \simeq \frac{1}{q} \\ \frac{1}{q} \simeq \frac{\eta}{\lambda p} + \frac{1-\eta}{p} \\ \frac{1}{r'} \simeq \frac{\zeta}{\lambda} + (1-\eta) \\ \frac{1}{r'} + \frac{1}{r} = 1 \\ \alpha \simeq \eta + \zeta p. \end{cases}$$

This leads to

$$\eta \simeq \frac{1 - p\left(1 + \frac{\gamma}{d} + \frac{1}{r}\right)}{1 - \frac{1}{\lambda}}$$

$$\zeta \simeq \frac{1 - \frac{1}{r'}}{1 - \frac{1}{\lambda}}$$

$$\alpha \simeq \eta + \zeta p \simeq \frac{1 - p\left(\frac{\gamma+d}{d}\right)}{1 - \frac{1}{\lambda}}.$$

If $\alpha < p$ this leads again to a differential inequality of the form

$$\frac{d}{dt}\|f\|_{L^p}^p \leq -K\|f\|_{L^p_{\gamma/p}}^p + C\|f\|_{L^p}^r + C\|f\|_{L^1_s}^A$$

for some $r > p$. The condition $\alpha < p$ can always be satisfied if

$$p > \frac{1}{2 + \frac{\gamma}{d} - \frac{1}{\lambda}} = \frac{d}{d + \gamma + \nu}.$$

We assumed $\gamma + \nu > -2$, so this estimate is at worse like $f_0 \in L^{d/(d-2)}$ (or f_0 in all L^p for $d = 2$). Note that this is precisely the integrability estimate implied by the Fisher information estimate! To summarise:

Using short-time regularisation, we may weaken the assumption of finite Fisher information into the Lebesgue integrability assumption

$$f_0 \in L^{q_0} \qquad \text{for some } q_0 > \frac{d}{d + \gamma + \nu}, \tag{23.38}$$

an assumption which is always weaker than the assumption of finite Fisher information. This implies short-time regularisation, and then by (23.35) again, Fisher information becomes instantly finite, and we are back into the regime of the previous subsection.

To summarise: Given any $k, \kappa > 0$ and $\varepsilon > 0$, and any q_0 as in (23.38) we may find $\alpha > 0$ and $s > 0$ such that (23.29) holds true for all times, where the constant C only depends on $k, \kappa, \varepsilon, q_0$, the energy E_0, the information $H(f_0)$, $\|f_0\|_{L^{q_0}}$ and $\|f_0\|_{L^1_s}$.

23.7 *Lower Bound*

Already Carleman, almost hundred years ago, understood the interest of a lower bound estimate on solutions of the Boltzmann equation. This is interesting to control the omnipresent log of the distribution function, but also has intrinsic interest as in any diffusion process. Recall that lower bounds of Gaussian nature also appear in Nash's proof of the continuity of solutions of diffusion equations with discontinuous coefficients.

It turns out that the spatially homogeneous Boltzmann equation always leads to a Gaussian lower bound, as in diffusion processes:

$$f(t, v) \geq K(t)\, e^{-A(t)|v|^2},$$

where K, A are positive for $t > 0$ and remain controlled as $t \to \infty$, as $O(t^\varepsilon)$ (through the smoothness bounds). Here I shall not provide such a strong bound but only sketch the proof of a weaker estimate

$$f(t, v) \geq K(t) e^{-A(t)|v|^q},$$

where $\log K(t)$ and $A(t)$ are both $O(t^\varepsilon)$ as $t \to \infty$. This will be the opportunity to touch the phenomenon of spreading which is the key in proving lower bounds. I shall use the following maximum principle:

If $f = f(t, v)$ and $\varphi = \varphi(t, v)$ are smooth and defined on $[0, t_0) \times \mathbb{R}^d$ with respective initial data f_0 and φ_0,

$$\begin{cases} f_0(v) \geq \varphi_0(v) & \text{for all } v \in \mathbb{R}^d \\ \dfrac{\partial f}{\partial t} = Q(f, f) & \\ \dfrac{\partial \varphi}{\partial t} \leq Q(f, \varphi) & \text{for all } t \in (0, t_0),\ v \in \mathbb{R}^d, \end{cases}$$

then $f(t, v) \geq \varphi(t, v)$ for all $(t, v) \in (0, t_0) \times \mathbb{R}^d$.

First consider the Landau equation. By smoothness and energy bound, there is a ball $B(v_0, r)$ with $|v_0| \leq R$ and $r > 0$ controlled from below, such that f is bounded below by $K > 0$ on $[0, t_0) \times B(v_0, r)$. Let's work only on that time interval: If we get a lower bound proportional to $\exp(-\alpha|v - v_0|^q)$ it will easily follow a lower bound independent of the choice of v_0. Let $\varphi(t) = \exp(-\alpha(t)|v - v_0|^q \chi(v - v_0) - B(t))$, where $\alpha(t) \geq 0$ and $B(t)$ large enough will be chosen later, $\chi(v)$ is a cutoff function with value 0 in $|v| \leq r/2$ and 1 for $|v| \geq r$. Let us further assume that $\alpha(0) = +\infty$. Then for $B(0)$ large enough, $\varphi_0 \leq f_0$ in all of $\mathbb{R}^d$. Consider f solution of the Landau equation, in the form

$$\frac{\partial f}{\partial t} = (a * f) : \nabla^2 f - (c * f) f,$$

the goal is

$$\frac{\partial \varphi}{\partial t} \geq (a * f) : \nabla^2 \varphi - (c * f) \varphi. \tag{23.39}$$

But

$$\frac{\partial \varphi}{\partial t} = \Big(- \dot{\alpha}(t)|v - v_0|^q \chi(v - v_0) - \dot{B}(t) \Big) \varphi,$$

$$(a * f) : \nabla^2 \varphi \geq \Big(K \langle v \rangle^\gamma \alpha(t)^2 |v - v_0|^{2(q-1)} \chi(v - v_0)^2 - C \Big) \varphi,$$

$$(c * f) \varphi \leq C \varphi,$$

where C is chosen large enough. (There are also lower powers of $|v - v_0|$ and derivatives of χ, but for large $|v|$ these terms are negligible and everything is controlled.) Choosing $B(t) = Bt$ for B large enough, $\alpha(t) = \alpha/t^2$ for α small enough, and $2(q - 1) + \gamma = q$, i.e. $q = 2 - \gamma$ solves the problem and yields

$$f(t, v) \geq K e^{-A|v|^{2-\gamma}}$$

for $t_0/2 < t < t_0$ and for all $v \in \mathbb{R}^d$.

It is tempting to believe that the same strategy will work out for the noncutoff Boltzmann equation, but that is not so simple. As a distorted fractional diffusion of order ν, one would hope that $Q(f, e^{-\alpha|v|^q})$ is like $|v|^r e^{-\alpha|v|^q}$ for some power r, but then that power would certainly go to 0 as $\nu \to 0$, and that term would never dominate the negative term proportional to $|v|^q e^{-\alpha|v|^q}$ coming from the time derivative of φ when α depends on t.

Instead, we can go through spreading, with an iterative scheme which I will only sketch. Assume that f is bounded below on $[0, t_0)$ by K on $B(0, R)$ and prove that it will be bounded below by βK on $B(0, \lambda R)$ for some $\lambda > 1$, and keep track of the dependence of β, λ on R. Choose

$$\varphi(t, v) = e^{-Bt} \chi\left(\frac{|v|^2}{2R(t)}\right).$$

Then

$$\dot{\varphi}(t) = -B\varphi - e^{-Bt} \frac{v}{R(t)} \cdot \nabla\chi\left(\frac{|v|^2}{2R(t)}\right),$$

$$\begin{aligned} Q(f, \varphi) &= \iint f'_*(\varphi' - \varphi)\, B\, dv_*\, d\sigma - (S * f)\varphi \\ &\geq K R^\gamma \iint f'_*(\varphi' - \varphi)\, dv_*\, d\sigma - C\varphi. \end{aligned}$$

And so on. Working out the dependencies of the constants will eventually lead to an inequality that can be repeated to get the appearance of a lower bound of stretched exponential form.

Bibliographical Notes

The generic blowup of solutions of the quadratic heat equation in $L^p(\mathbb{R}^d)$ for $d > p/2$ was established by Weissler [191], see also Quittner and Souplet [152, Remark 16.2 (iii)]. (Thanks to Hatem Zaag for providing these references.)

"Very" weak solutions for very soft potentials were introduced in my article [175] and improved in our joint work [4], in the region of "conditional regularity", into a more classical notion of time-integrated weak solutions. It is only with the new estimates on Fisher information that one can strengthen the notion of weak solution, not having necessarily to time-integrate.

For interpolation I already mentioned the treatise by Lunardi [130]. In the context of the spatially homogeneous Boltzmann equation, interpolation techniques go back at least to Gustafsson [99] in L^p spaces. From the mid-nineties on, interpolation between weighted Sobolev spaces was heavily used in papers by Desvillettes, Mouhot and myself, see e.g. [67, 70, 72, 143] and many references since then.

Moment estimates have been the first crucial step in the study of the Boltzmann equation, first for hard potentials with cutoff, by Povzner [149], Arkeryd [7], Elmroth [77], Desvillettes [61], Wennberg [192] and others. A key observation by Desvillettes was that for hard potentials there is appearance of moments: Even if the initial datum is not well localised, suffices to have one finite moment of order higher than 2, to get all moments uniformly bounded. The ultimate results are those of Mischler and Wennberg [136] (see also Lu [128, 129]): Under the mere assumption of finite energy of the initial datum, there is instantaneous appearance and uniform boundedness of all moments, and there is also existence and uniqueness. But this "instant localisation" is a specific feature of hard potentials: In the regime of Maxwellian or soft potentials there are no such results and immediate appearance of moments does not hold; at best there is propagation. When the collision kernel is Maxwellian, moments equations can be closed in some sense and there is more structure, allowing precise results, see Ikenberry and Truesdell [107], Bobylev [23, 24]. Propagation estimates for soft potentials were studied by various authors, e.g. Arkeryd [9] or Desvillettes [61]. For very soft potentials the first estimates appeared in my PhD Thesis [175, 178]; there I was already juggling between the two kinds of estimates (with two different symmetrisation formulas) according to the behaviour in $|v - v_*|$, as in Sect. 23.3. Then came the fine work by Carlen, Carvalho and Lu [41] who showed how to combine these ideas to get good large-time moment estimates; inequality (23.9) is established there (for the Boltzmann equation with very soft potentials).

Gronwall's lemma, in the precise form used for (23.10), states that if

$$u(t) \leq \phi(t) + \int_0^t \lambda(\tau) u(\tau)\, d\tau,$$

then

$$u(t) \leq \phi(0) \exp\left(\int_0^t \lambda(\tau)\, d\tau\right) + \int_0^t \exp\left(\int_\tau^t \lambda(s)\, ds\right) \frac{d\phi}{d\tau}\, d\tau.$$

To prove it, check that

$$\varphi(t) \exp\left(-\int_0^t \lambda(\tau)\, d\tau\right) + \left(\int_0^t \lambda(\tau) u(\tau)\, d\tau\right) \exp\left(-\int_0^t \lambda(\tau)\, d\tau\right) \\ - \int_0^t \exp\left(-\int_0^\tau \lambda(s)\, ds\right) \frac{d\varphi}{d\tau}\, d\tau$$

is a nonincreasing function of t.

Integrability and regularity were studied in the context of the Boltzmann equation with hard potentials first, and mainly under a cutoff assumption [99]. Lions [124] first observed that the gain part of the collision operator should act like a dual Radon transform and that the regularity gained in this way would allow theorems of propagation of both regularity and singularity. This important result was revisited by various authors, e.g. Bouchut and Desvillettes [30], or Mouhot and myself [143]; in this last work constructive "hands-on" estimates of propagation of regularity are deduced. As for propagation of singularities, see for instance Boudin and Desvillettes [31] for inhomogeneous solutions of small mass.

But for the Boltzmann equation without cutoff, there is instant regularisation, whenever there is a good theory of Cauchy problem. This was discovered by Desvillettes for the Kac model [62], and then further generalised to several particular instances of the spatially homogeneous Boltzmann equation by Desvillettes [63, 64] and his student Proutière [150]. Desvillettes and I also worked out the regularisation for the Landau equation with hard potentials [67]. The study of Boltzmann's entropy production functional motivated the development of efficient tools: after some precursors [126, 177] came the work by Alexandre, Desvillettes, Wennberg and myself [4], which presented the truncation recipe, the Fourier representation in Bobylev style, the optimal fractional Sobolev estimate which were at the core of Sects. 23.4 and 23.5. The combination of these tools in a Moser-type regularity iteration was sketched in my Peccot lecture notes [182] and provides a much more efficient approach to the regularisation for the Boltzmann equation without cutoff. From there a long list of works started to systematically cover situations of interest; see for instance Desvillettes–Mouhot [66] (also establishing for the first time uniqueness of the solution for the noncutoff Boltzmann with soft potentials), Alexandre–Morimoto–Ukai–Xu–Yang [5], Chen–He [51], Chen–Desvillettes-He [52], Fournier–Guérin [82], Fournier–Mouhot [84], He [101]. Besides the complexity due to the structure of the Boltzmann operator, this method is very much in the flavour of the classical regularity theory by Nash, DeGiorgi and Moser on divergence form parabolic equations: See Mouhot's course in the same volume [141], yielding much more details and a whole historical perspective.

By the way, it is interesting to note that Nash's original proof of his celebrated continuity theorem is based on Boltzmann's entropy, and there is a mathematical filiation with the study of the Boltzmann equation here: Indeed, after the death of Carleman, Lennard Carleson (future Abel Prize winner, like Nash) was in charge of completing and editing Carleman's last papers on the subject, and thus became acquainted with the underlying mathematical physics. So when Nash consulted him, Carleson pointed out to him the interest of Boltzmann's H functional, which in those days was hardly known in the community of mathematical analysts (this was told to me by Carleson himself).

Regularity always follows, as a rule, from regularity estimates through usual tools; the main subtlety is if one wants to prove uniqueness of the weak solution in presence of smooth solutions; see [67] for the spatially homogeneous Landau equation with hard potentials, and Fournier [81], Fournier–Guérin [83] for soft or very soft potentials.

While the large majority of regularity estimates were based on time-derivatives of nonlinear functionals, as in the papers by Nash and Moser, Silvestre introduced a genuinely different method to prove regularity estimates, based on Harnack inequalities for nonlocal equations [159]. See also Chaker–Silvestre [50] for another approach to the regularity induced by the entropy production functional.

All of these results have been, so far, unaccessible to very soft potentials, because of the lack of an adequate L^p bound; there were only partial or conditional results [90–92]. The new Fisher information estimate precisely fills this gap.

Short-time estimates for the spatially homogeneous Landau equation (with Coulomb potential in $d = 3$) were first established by Arsen'ev and Peskov [13] in the seventies. The inequality (23.35) appears in [172] and rests on another inequality by Lions and myself [127], controlling (among other estimates) $\sqrt{a}$ in $W^{1,4}$ by a in $W^{2,2}$. The strategy of short-time regularisation to get the finite Fisher information is due to Imbert, Silvestre and myself [111]. Short-time estimates for the Boltzmann equation with very soft potentials were established with full details in weighted L^∞ spaces by Henderson, Snelson and Tarfulea [104, 105], while the same authors [102] and Snelson and Solomon [161] treated the Landau equation. Since the a priori estimates hold in L^p spaces as soon as $p > d/(d + \gamma + \nu)$, it is natural to expect that this is a natural space to achieve short-time regularisation and continuation estimates, as well as uniqueness.

Carleman [37] proved a lower bound like $Ke^{-|v|^{2+\varepsilon}}$ for hard spheres, with arbitrarily small $\varepsilon > 0$. Much later, unaware of that older work, A. Pulvirenti and Wennberg [151] refined this into a Maxwellian lower bound for all hard potentials. In the case of Maxwellian potentials, there are similar results by Bobylev [23]. The lower bound through bump and maximum principle was used by Desvillettes and I [67] to prove a Gaussian lower bound for the Landau equation with hard potentials. The proof presented here is just a generalisation, obviously suboptimal. Imbert, Mouhot and Silvestre [108] obtain the fine Gaussian lower bound for the Boltzmann equation without cutoff, in great generality, as soon as regularity bounds are available; the key to their analysis is a weak Harnack inequality due to Silvestre [159]. The maximum principle mentioned above, taken from my lecture notes [180], is obviously related to the weak Harnack inequality, and was explored at length in recent research. This maximum principle was also used, for instance, in another work, with Gamba and Panferov [89], to get upper bounds for models of granular media. The general arguments of Pulvirenti–Wennberg and Imbert–Mouhot–Silvestre both use an iterative scheme of spreading type, as did Carleman's proof.

Nash's upper and lower Gaussian bounds are in his famous 1958 paper [144]; such bounds are usually called Aronson estimates. See Fabes and Stroock [78] or Bass [20, Chapter 7] for modern presentations.

Mouhot, Imbert and Silvestre, followed by Cameron, Henderson, Ouyang, Snelson, Turfulea, have worked out an ambitious program showing that the inhomogeneous Landau and Boltzmann equations, with or without cutoff, are regular and well-controlled as long as certain bounds on the macroscopic quantities hold true [34, 35, 103, 108–110, 137, 147]. This is a whole world with the interplay of homogeneous and inhomogeneous features, in which hypoelliptic regularity and diffusion processes regularity are intertwined; Mouhot's lecture at the 2018 International Congress of Mathematicians [140] was surveying the field at a time when it was starting to achieve some success. The most complete results, putting together Imbert–Silvestre [109] and Henderson–Snelson–Tarfulea [103], roughly say that for moderately soft potentials there is global decay, regularity and positivity conditional to uniform upper bounds on two hydrodynamic fields: density $\rho(t, x) = \int f\,dv$ and energy $E(t, x) = (1/2)\int f|v|^2\,dv$. It is natural to ask how far these results can be pushed with very soft potentials: Existing papers prove the regularity conditional to some uniform local L^p bounds, but it would be more satisfactory to assume L^p or L^∞ estimates on the initial datum and conditional bounds for all times on hydrodynamic quantities. In any case, in the past decade a major step has been achieved in our understanding of the Boltzmann equation as it was progressively proven that regularity is basically conditioned to the non-blowup of hydrodynamic fields.

24 Equilibration

Convergence to equilibrium is one of the fundamental predictions of collisional kinetic theory—and the word "prediction" here is not lightly used, since Maxwellian (Gaussian) equilibrium was put forward by Maxwell and Boltzmann even before the existence of atoms and molecules was mainstream. First I shall explain why the quantitative mathematical study of equilibration is closely connected to the problem of Fisher information estimate; this was already mentioned in Sect. 3.

24.1 From Fisher Information Monotonicity to Entropy Inequalities

The main focus of this set of notes was the monotonicity of the Fisher information along solutions of the Boltzmann equation. This is not unrelated to the more classical problem of estimating Boltzmann's classical dissipation functional (entropy production functional). A first way to see the connexion is the following. Consider a collision kernel $\beta(\cos\theta)$ on the sphere. Since the heat and Boltzmann equation with kernel β commute, the decay of Fisher information along the Boltzmann flow is the same as the decay of Boltzmann's entropy production along the heat flow. In other words, if L_β stands for linear Boltzmann equation with kernel $\beta(\cos\theta)$,

$$-\frac{d}{dt}\bigg|_{t=0} D_\beta(e^{t\Delta}F) = -\frac{d}{dt}\bigg|_{t=0} I(e^{tL_\beta}F). \tag{24.1}$$

Here is a relation with the inequalities explored in these notes. Given β, let $L_\#$ be the largest constant such that for all even probability densities on $\mathbb{S}^{d-1}$,

$$\int_{\mathbb{S}^{d-1}} F\,\Gamma_{1,\beta}(\log F) \geq L_\# I(F). \tag{24.2}$$

Note that the left hand side is $(-1/2)(d/dt)I(e^{tL_\beta}F)$. Throughout these notes, we have seen various sufficient conditions for $L_\#$ to be positive, and recipes to estimate this constant. For instance,

- From (17.3), $L_\# \geq (d-2)\Sigma(\beta) > 0$ if $d > 2$;
- From (19.1), $L_\# > 0$ if $\min \beta > 0$;
- From (22.4), if $\beta = \int \mathcal{K}_t \lambda(dt)$, $L_\# \geq (1/2)\int (1 - e^{-2L_* t})\lambda(dt) > 0$.

From (24.2) one may apply Proposition 17.4 to deduce Criterion 13.1 and the monotonicity of the Fisher information for the nonlinear Boltzmann equation. But we may also keep (24.2) and interpret both sides as time-derivatives along S_t: From (24.1) and (1.8), (24.2) is also

$$-\frac{d}{dt}\bigg|_{t=0} D_\beta(e^{t\Delta}F) \geq -2L_\# \frac{d}{dt}\bigg|_{t=0} H(e^{t\Delta}F).$$

Integrating from 0 to $+\infty$ as Stam, it follows the functional inequality

$$D_\beta(F) \geq 2L_\# H(F), \tag{24.3}$$

which itself implies that $H(F(t))$ converges to 0 along the linear Boltzmann flow on the sphere, at least like $O(e^{-L_\# t})$. Actually (24.3) admits two notable limit cases:

- When β is the constant kernel (which corresponds to $t \to \infty$ in (22.4), thus $K^\# \geq 1/2$), this is the spherical linear version of the entropy—entropy production inequality conjectured by Cercignani and proven by Toscani and myself for super hard spheres; and the latter inequality, in turn, is the mother of all quantitative equilibration estimates for the nonlinear Boltzmann equation (homogeneous or not) in the large, when the kernel is not Maxwellian;
- When β concentrates on grazing collisions (which corresponds to the derivative as $t \to 0$ in (22.4), thus $L_\# \geq L_*$), this reduces to the classical logarithmic Sobolev inequality for even functions on the sphere.

In other words, inequality (24.2) acts like a bridge between the theory of logarithmic Sobolev inequalities for diffusion processes, and the theory of entropy—entropy production inequalities in the context of the Boltzmann equation. True, both theories differ significantly when going into details, but at least there is that common core. To summarise and reformulate it: The central inequality is

$$4\Lambda_\# \int_{\mathbb{S}^{d-1}} \Gamma_\beta(F) \leq I(F) \leq \frac{1}{L_\#}\int_{\mathbb{S}^{d-1}} F\Gamma_{1,\beta}(\log F). \tag{24.4}$$

Keeping only the left inequality in (24.4) is a Poincaré problem for L_β, or just the spectral domination of $-L_\beta$ by $-\Delta$. Keeping only the right inequality, and integrating along the heat semigroup, à la Stam, yields the entropic gap inequality $H(F) \le D_\beta(F)/(2L_\#)$, à la Toscani-Villani (except it is on the sphere rather than the whole space). Keeping the whole of (24.4) provides the criterion for decay for the Fisher information along the Boltzmann flow, à la Imbert–Silvestre–Villani. Finally, taking the AGC, the left inequality becomes an equality (with $\Lambda_\# = 1$) and the right one becomes the differential log Sobolev inequality, à la Bakry–Émery, which thus appears as the diffusive limit case of (24.4).

24.2 Rates of Convergence

Now I will try and sketch the overall picture for quantitative equilibration. Again, here I shall focus on the spatially homogeneous case. Without loss of generality, the initial probability density will be assumed to have zero mean ($\int f_0(v)\, v\, dv = 0$) and unit temperature (initial energy $E_0 = d/2$). While the equilibrium is always the same for all collision kernels (the standard Maxwellian, or centered Gaussian with identity covariance matrix), the estimates and expected rates may depend significantly on the features of the collision kernel. In particular,

- In the spectral gap regime ($\gamma + \nu \ge 0$) one expects exponential convergence to equilibrium;
- In the regime with no spectral gap but conditional regularity ($-2 < \gamma + \nu < 0$), convergence should be at best like the exponential of a negative fractional power of time. Such is the case in particular for very soft potentials ($\gamma < -2$).

Whatever the situation, large velocities are one key to the estimates, and obstacles to fast convergence are most often related to velocity tails.

Here is a heuristics for the plausible rate of convergence. In the linearised case, spectral gap is related to the convergence of the nondiagonal part of the covariance matrix, or moments of order 2; so let us consider $Q(v)e^{-|v|^2/2}$, and imagine that there is a distinguished approximate solution of the form

$$\varphi(t, v) = e^{-\lambda t^\alpha} Q(v) e^{-|v|^2/2}.$$

The action of the operator Q or its linearised version will tend to multiply φ by a multiple of $|v|^{\gamma+\nu}$ as $|v| \to \infty$ (each derivative acting on a Gaussian induces a multiplication by v, or kind of). But $\partial_t \varphi$ will be like $-t^{\alpha-1}\varphi$, and this should be homogeneous to $|v|^{(2\alpha)/(1-\alpha)}$ since in the definition of φ, t is homogeneous to $|v|^{2/\alpha}$. So the condition for the good effect of Q to beat the bad effect of the time-derivation, should be $\gamma + \nu \ge 2(\alpha - 1)/\alpha$, or equivalently $\alpha \le 2/(2 - (\gamma + \nu))$. Writing $(\gamma + \nu)_- \ge 0$ for the negative part of $\gamma + \nu$, we just arrived at the

Guess 24.1 *Solutions of the spatially homogeneous Boltzmann equation with fast enough decay at large velocities converge to equilibrium like* $O(\exp(-\lambda t^\alpha))$, *where*

$$\alpha = \frac{2}{2 + (\gamma + \nu)_-}.$$

(So for the Landau–Coulomb equation in dimension 3, the expected rate is $O(\exp(-\lambda t^{2/3}))$.) In a close to equilibrium setting with Gaussian decay, such rate was indeed proven by Guo and Strain for certain cases when $\gamma + \nu < 0$. Starting away from equilibrium, the situation is not as good yet, but with the new estimates for the Cauchy problem we can at least prove that there is convergence to equilibrium, and at least like $O(t^{-\infty})$, that is, faster than any inverse power of time. The main steps will be summarised in the next Subsection.

24.3 Quantitative Entropic Convergence

Start again with the spatially homogeneous Boltzmann equation. From Sect. 23 we already have the control of moments and Sobolev norms of arbitrarily high order like $O(t^\varepsilon)$, with ε as small as desired. And we also have a lower bound like $K(t)e^{-A(t)|v|^q}$ with $\log K(t)$ and $A(t)$ controlled like $O(t^\varepsilon)$. In such a situation there is an entropy—entropy production estimate of the form

$$D_B(f) \geq K t^\varepsilon H(f|M)^{1+\varepsilon}, \tag{24.5}$$

where $H(f|M) = H(f) - H(M) = H_M(f)$ is the relative information of f with respect to its associated equilibrium, $\varepsilon > 0$ is as small as desired, and K depends on d, B, ε, and high order moments. This is a Gronwall inequality on $H(t) = H(f(t)|M)$, of the form $-dH/dt \geq K t^\varepsilon H^{1+\varepsilon}$, which integrates into

$$H(f(t)|M) \leq \left(\frac{1}{H(f_0|M)^\varepsilon} + \left(\frac{\varepsilon K}{1+\varepsilon} \right) t^{1+\varepsilon} \right)^{-\frac{1}{\varepsilon}},$$

and this yields the announced $O(t^{-\infty})$ estimate.

Then, by the Csiszár–Kullback–Pinsker estimate

$$H(f|M) \geq \frac{1}{2} \|f - M\|_{L^1}^2,$$

actually

$$\|f(t) - M\|_{L^1} = O(t^{-\infty}).$$

Interpolating this with the bound

$$\|f(t) - M\|_{L^1_s} \le \|f(t)\|_{L^1_s} + \|M\|_{L^1_s} = O((1+t)^\delta)$$

with s as large and δ as small as desired, we further deduce

$$\|f(t) - M\|_{L^1_\sigma} = O(t^{-\infty})$$

for any $\sigma > 0$. In particular, $\|f\|_{L^1_\sigma}$ remains uniformly bounded as $t \to \infty$, for any σ, and it follows that all decay and regularity bounds are uniform also. Ultimately, it holds

Theorem 24.2 *Let $B(v - v_*, \sigma) = |v - v_*|^\gamma b(k \cdot \sigma)$, where $\gamma \ge -d$, $b(\cos\theta)\sin^{d-2}\theta \ge K\theta^{1+\nu}$, $0 < \nu < 2$. Further assume $-2 < \gamma + \nu < 0$ and the monotonicity criterion for the Fisher information, as in Theorems 10.2 or 13.1. Then if f_0 is a probability distribution with zero mean and unit temperature, with finite moments of all orders and $f_0 \in L^m(\mathbb{R}^d)$ for some $m > d/(d + (\gamma + \nu))$, then the solution $f = f(t, v)$ of the spatially homogeneous Boltzmann equation with collision kernel B and initial datum f_0 becomes instantly smooth and satifies*

$$\forall k \in \mathbb{N}_0, \qquad \forall \kappa > 0, \qquad \forall r > 0, \qquad \forall t \ge t_0, \qquad \left(\frac{|\nabla^k (f - M)|}{\langle v \rangle^\kappa} \right) \le \frac{C}{t^r},$$

where $M = M(v)$ is the centered Maxwellian (Gaussian) with unit temperature, C only depends on t_0, k, κ, r, m, $\|f_0\|_{L^m}$ and $\|f_0\|_{L^1_s}$ for s large enough. In short, f is uniformly bounded in the Schwartz class of rapidly decaying functions, and converges to M in this class with rate $O(t^{-\infty})$.

24.4 Equilibration Through Fisher Information

The decay of I is another approach for the convergence to equilibrium. Actually, the Stam–Gross logarithmic inequality

$$I_M(f) \ge 2H_M(f)$$

shows that convergence in the sense of Fisher information is stronger than in the sense of entropy.

Obviously, the decreasing property of Fisher information along the Boltzmann flow does a good part of the way towards a convergence rate. It turns out that we have all the tools for a good discussion of this. Just by working out again the calculations of Sect. 10, we have

Proposition 24.3 *Consider a kernel $B = B(|v - v_*|, \cos\theta)$, let $\overline{\gamma} = \overline{\gamma}(B, r)$ be defined as in (10.4) and let $K^\#(\overline{\gamma}, r)$ be the largest admissible constant in the functional inequality*

$$\int_{\mathbb{S}^{d-1}} F\,\Gamma_{1,\beta_r}(\log F) \geq \overline{\gamma}(B,r)^2 \int_{\mathbb{S}^{d-1}} \Gamma_{\beta_r}(\sqrt{F}) + K^\#(\overline{\gamma},r)\,I(F), \tag{24.6}$$

required to hold for all even functions $F : \mathbb{S}^{d-1} \to \mathbb{R}_+$. Then, along the spatially homogeneous Boltzmann equation with kernel B, one has

$$-\frac{dI}{dt} \geq 2 \iint_{\mathbb{R}^d\times\mathbb{R}^d} f f_*\, K^\#(\overline{\gamma}, |v-v_*|)\left|\Pi_{k^\perp}\left(\frac{\nabla f}{f} - \left(\frac{\nabla f}{f}\right)_*\right)\right|^2 dv\,dv_*. \tag{24.7}$$

In other words, the dissipation of Fisher information is at least proportional to $D_L(f)$, Landau's dissipation functional, for a Landau equation whose kernel Ψ is proportional to $K^\#(\overline{\gamma}, |v-v_*|)$.

How can we exploit this? First, as in Sect. 24.1, in all the situations discussed in these notes we can find a convenient function $K^\# > 0$; in particular this will be the case for all inverse power law interactions; and then $K^\#$ will be proportional to $|v-v_*|^\gamma$, where $\gamma = (s-(2d-1))/(s-1)$. In other words, there will be $K > 0$ such that

$$-\frac{dI}{dt} \geq K\,D_L^\gamma(f), \tag{24.8}$$

where

$$D_L^\gamma(f) = \frac{1}{2}\iint f f_*\,|v-v_*|^\gamma \left|\Pi_{k^\perp}\left(\frac{\nabla f}{f} - \left(\frac{\nabla f}{f}\right)_*\right)\right|^2 dv\,dv_*. \tag{24.9}$$

Then we may estimate D_L^γ from below, using non-concentration, localisation, smoothness, interpolation. The projector $\Pi_{k^\perp}$ in (24.9) implies a degradation of the estimates, so in general there is a loss of 2 in the exponent of the relative velocity. Assuming, without loss of generality, that f_0 has unit temperature ($E_0 = d/2$), and writing $I(f|M) = I_M(f) = I(f) - I(M)$,

- There is a constant $K = K(d,\gamma,H(f_0))$ such that

$$D_L^2(f) \geq K I(f|M) \tag{24.10}$$

- For any $\varepsilon > 0$ there is a constant $K_\varepsilon = K(d,\gamma,H(f_0),\varepsilon)$ and there are $s = s(d,\gamma,\varepsilon)$ and $k = k(d,\gamma,\varepsilon)$ such that

$$D_L^\gamma(f) \geq K_\varepsilon I(f|M)^{1+\varepsilon}\left(\|f\|_{L^1_s} + \|f\|_{H^k}\right)^{-\varepsilon}. \tag{24.11}$$

This and (24.8), and the moments and smoothness bounds like $O(t^0)$ imply

- for $\gamma = 2$, exponential convergence of f to M in the sense of Fisher information, therefore in entropy and in L^1, and eventually by interpolation, exponential convergence in Schwartz class;
- for any γ (positive or negative, keeping in mind that certain assumptions have to be made to achieve good regularity estimates), convergence of $I(f|M)$ to 0 like $O(t^{-\infty})$, and similarly convergence of f to M like $O(t^{-\infty})$ in Schwartz class.

Remark 24.4 One may conjecture that with a bit more work, it holds $-dI/dt \geq KI$ in the regime of entropic gap, that is, $\gamma + \nu \geq 2$. This may require more specific assumptions on the particular shape of $b(\cos\theta)$; for instance, that b derives from the inverse power laws, or from the fractional Laplace operator on the sphere. It may also require comparison with an integral quantity involving fractional derivatives of higher order than 1.

Remark 24.5 In the end, under the regime considered here, the Fisher information approach yields basically the same rate $O(t^{-\infty})$ as the entropic approach. There are however two notable differences. The first is that the Fisher information approach may require more structure assumptions: to prove the decay we were led to some specific choices and relations between the angular and velocity dependence of B—e.g. power law, or perturbation of fractional diffusion, or high dimension... In the same vein, we cannot just be content with a comparison argument: if $B \geq B_0$ then the associated Boltzmann dissipation functionals can be compared, but for Fisher that is not necessarily true. The second difference is that the Fisher dissipation inequality (24.11) uses moments and smoothness, but does not need any lower bound on f—contrary to (24.5). Both the drawback and the advantage are relatively minor, since on the one hand we saw that the monotonicity for I holds in all the main cases we could think of, and since on the other hand it is often in practice possible to get around the need of lower bound for the convergence to equilibrium, as shown by Carlen, Carvalho and Lu. Of course it might be that for other regimes, the Fisher information approach has some more pronounced advantages.

I will close this section on equilibration with two final, more general remarks. The first one is that even in the entropic approach, Fisher information plays a key role in the backstage, as for general collision kernels nobody knows how to prove a good inequality on the entropy production without going through a Fisher information estimate. The other one is that the possibility to prove and estimate equilibration through the rate of decay of Fisher information, even if it does not yet bring anything fundamentally new with respect to the entropic approach, at least reinforces the consistency of the whole area, and especially the relation between regularity theory, equilibration theory, and information theory.

Bibliographical Notes

The discussion in Sect. 24.1 is taken from [111]. The connexion between the Fisher information decay and the entropy production inequality is implicit in McKean [134] and discussed explicitly by Toscani and I [171].

Entropy–entropy production estimate (24.5) is extracted from my paper [181], which is an improvement of my older work with Toscani [171]. The strategy of convergence in $O(t^{-\infty})$ from slowly increasing bounds is also from another work with Toscani [172]. Carlo Cercignani played an important role in motivating research in this area.

Carlen, Carvalho and Lu [41] were the first to obtain convergence results, although in a weaker sense, for the spatially homogeneous Boltzmann equation with very soft potentials away from equilibrium, refining my strategy from [175]; their great paper is a model of clarity and efficiency to work out relevant conclusions even in a very rough situation. They also showed how to do without lower bound estimates, applying the entropy—entropy production inequality not to the true distribution function but to a modified version like $(1 - e^{-t})f(t, v) + e^{-t}M(v)$. Further see Desvillettes [65] for the spatially homogeneous Landau equation with soft or very soft potentials.

In the 2000's Desvillettes and I worked out a program to extend the $O(t^{-\infty})$ convergence conditionally to regularity bounds, using entropy production bounds, differential inequalities, geometric/analytic functional inequalities, and a lot of interpolation [69–72, 187]. Two main issues remained open after that: To refine the rate from $O(t^{-\infty})$ to exponential or fractional exponential; and to weaken the smoothness assumptions. In both directions, impressive results have been obtained in the following decade.

Begin with the exponential decay, starting with the regime of hard potentials. Of course it had first been proven for linearised versions of the equation, already back in the fifties, see for instance by Grad [94] or Wang Chang and Uhlenbeck [190]. Then it was established for the nonlinear equation in a perturbative homogeneous setting, by Arkeryd with nonquantitative arguments [10, 11], and in a completely quantitative way by Mouhot [138] and Baranger and Mouhot [18]. At that point it had all been for the cutoff case; Mouhot and Strain [142] treated noncutoff kernels. Mouhot [139] used quantitative nonsymmetric spectral theory to prove quantitative exponential convergence for the spatially homogeneous nonlinear Boltzmann equation near equilibrium, under regularity assumptions which are compatible with the nonlinear theory in the large. Gualdani, Mischler and Mouhot [96] extended those results to the spatially inhomogeneous equation, both in a perturbative setting and conditional to global regularity bounds; this involved working out a complete new abstract linear theory—it is actually one nice example of new theory of general interest motivated by problems arising in kinetic theory. Stitching the estimates of [96] (close to equilibrium with the optimal rate) with those that I proved with Desvillettes [72] (far from equilibrium, with $O(t^{-\infty})$ rate, conditional to global regularity) and the conditional regularity program of Imbert–Mouhot–Silvestre, eventually shows that a uniform control on the density and energy (say in the d-

dimensional torus) are enough to achieve a complete theory of existence, decay, regularity and exponential convergence to equilibrium. Of course the spatially homogeneous theory is a very special particular case of that general conditional inhomogeneous theorem.

For soft potentials, Caflisch [33] was the first to obtain decay like $O(\exp(t^{-\alpha}))$ in the cutoff case. This was refined and extended by Guo and Strain [98, 166], who indeed obtained the decay rate of Guess 24.1. They also worked out some less stringent assumptions than Gaussian, e.g. they prove such a decay with a fractional exponential rate, for the Landau–Coulomb equation, if the decay of the distribution function is at least like $e^{-|v|^{1+\varepsilon}}$. Refined results were obtained for the inhomogeneous Landau and noncutoff Boltzmann equations with soft potentials by Cao, Carrapataso, Desvillettes, He, Ji, Mischler, Tristani, Wu [36, 43–45]. All this is in a close-to-equilibrium setting. For data away from equilibrium, in the spatially homogeneous setting the $O(t^{-\infty})$ decay rate for very soft potentials was proven in [111] as a consequence of the new regularity estimates coming from the Fisher information monotonicity—all other ingredients were in place. Stitching together these two theories may be reachable. By the way, the real issue is not so much hard vs soft potentials, than spectral gap vs no spectral gap.

In the special case of Maxwellian kernels ($\gamma = 0$) there are several other ways towards the approach to equilibrium: contracting metrics as studied by Tanaka [168] [183, Section 7.5], Toscani and myself [170], or Carlen–Gabetta–Toscani [40]; central limit theorem, from McKean [134] to Carlen–Gabetta–Toscani again to Dolera–Regazzini [75], or convergence of moments; once again see Bobylev [23]. All those methods eventually lead to exponential convergence if the data is localised enough, see for instance [75], and conversely the convergence can be as slow as desired if the initial datum is not well localised [40].

The new estimates in Sect. 24.4 constitute a natural outgrow of my joint work with Imbert and Silvestre [111]. Inequality (24.10) is from my work with Desvillettes [68] and (24.11) from my work with Toscani [172].

25 Conclusions

Since Torsten Carleman started a sound mathematical study of the spatially homogeneous Boltzmann equation, one century ago, a number of additional ingredients have been progressively incorporated in that program. In addition to the tools of functional analysis and measure theory which Carleman knew so well, over the years the theory has been enriched by moment estimates, harmonic analysis, maximum principles, spectral analysis, information theory, Moser iteration, Harnack inequalities, and the theory of nonlocal diffusive equations. Fisher information was introduced as a useful tool, already sixty years ago by McKean; then in the nineties its impact in the field grew when it was identified as a central tool in the study of quantitative equilibration; and now its impact has grown even more as it has allowed to unlock the stubborn problem of well-posedness for very soft potentials. These recent developments constitute the last brick to date in the theory, and for the very first time we see a consistent picture emerging which encompasses all cases of physical and mathematical interest.

This recent outgrow of the theory echoes and eventually answers most of the key questions on the spatially homogeneous Boltzmann equation which agitated me during my PhD: how to handle singularities in the collision kernel, how to quantify the convergence to equilibrium, and more generally how to sort out the global picture of moments, smoothness, lower bound and convergence to equilibrium. In this jungle I was walking in the footsteps of my senior collaborator Laurent Desvillettes, who had first initiated me to the mysteries of soft potentials.

The theory of the Boltzmann equation comes out from this episode much richer than before and with strengthened consistency. It was also during my PhD that the connexion between entropy—entropy production estimates and information theory was made. This memory is very dear to me, as it was the main outcome of my first scientific stay outside France, in Pavia, on the invitation of Giuseppe Toscani. Our estimate and joint work on Cercignani inequalities stemmed from a shiny coincidence, certainly one of the great lucks of my career. What seemed like a beautiful anomaly or at least a singularity at the time, now seems quite natural and consistent with the way the theory has evolved; so it is a whole architecture which now surrounds this once isolated gem. Finally one may daresay that the collision kernel Q, with its complicated dissipative structure, seems well understood, and this was not the case even just couple of years ago. Frustrating as it may too often be, mathematical physics is at times rewarding.

Several types of further developments may emerge from the current theory.

The most straightforward is a more systematic development of the Cauchy theory for the spatially homogeneous equation in the large, and also for the conditional theory of the spatially inhomogeneous equation. All the ingredients are available, and I sketched most of the estimates, but some of the theorems have only been written under particular assumptions and it would be quite a task to systematically rework the theory with the best available assumptions.

As far as regularity is concerned, now that the spatially homogeneous theory seems to achieve maturity, and that the conditional regularity theory has achieved some impressive successes, the most famous challenges for the Boltzmann equation—hydrodynamic limit, derivation from particle systems—look even more motivating and daunting. Fisher's information may have its role to play there, also. Regularity for the full (inhomogeneous) equation still seems to be the most urgent topic to foray.

But one may also ask questions about the Fisher information itself. Can we be sure that our results are close to optimal, and can one construct a counterexample? that is, a specific collision kernel B for which I is not always decreasing. One may start from the specific kernels yielding a counterexample in Sect. 18, yet things do not seem to come easily from there.

Even if there turn out to be counterexamples for certain kernels, the main striking lesson is the robustness of the Fisher information decay over a very large class of kernels, a striking property which had not been predicted by the community of mathematical physics.

Obviously, this success gives further motivation for the investigation of Fisher's information in the treatment of other equations, either through an integrability a priori estimate, or for the long-term behaviour, or from the point of view of large deviations in Donsker–Varadhan style, a theory which by the way deserves some more quantitative estimates. It also provides a renewed interest for the investigation of the other questions loosely formulated by McKean about higher order behaviour, mentioned at the end of Sect. 3.

One may further enquire whether some of the other functionals known to be monotone for the spatially homogeneous Boltzmann equation with Maxwellian kernel, are actually monotone in greater generality. A case in point is the contractivity property of the Wasserstein distance W_2, proven by Tanaka for Maxwell kernels: Various authors have shown that this distance is exponentially stable along the Boltzmann flow for various kernels; but could it actually be nonincreasing?

Finally, let us pause and reflect on that striking discovery: *The Fisher information is decaying under the action of Boltzmann collisions.* Lyapunov functionals come with a number of features and consequences, and their knowledge is often incorporated in model equations or numerical schemes. Already 25 years ago it was observed that the decay of Fisher's information seems to be associated with the stability of some deterministic numerical schemes, and now that the theory has matured it is time to re-examine that observation which most researchers (including me) had not taken seriously enough. Of course, the proof of decay for the Fisher information is so much more intricate than the proof of increase for the Boltzmann entropy, that it seems absurd to try and force the decrease within the formulation of the code. But still there is food for thought, thinking of the monotonicity property as a test or a guide.

Taking a more distant look, we may also investigate the physical meaning of the decay of the Fisher information. Think of the influence of the H-Theorem, which was mathematically corroborating the already identified Second Law of Thermodynamics, even if just for the specific case of a rarefied gas. Is there another thermodynamical principle which can be formulated with some accuracy and for which a mathematical translation would be precisely the monotonicity of I along the homogeneous Boltzmann equation? In the physics literature there has been some speculation and treatises about the physical meaning of Fisher information, sometimes related to the so-called "thermodynamical length", and its potential role in some of the principles of classical or quantum statistical mechanics. This is consistent with the original meaning of Fisher information related to observation (measurement) and statistics. I do not master the physics literature on this topic well enough to recommend precise references, but at least I can mention the names of two physicists who contributed to those lines of thought: Gavin Crooks (University of Berkeley) and Roy Frieden (University of Arizona). Of course, for systems in which the Fisher information is related to the entropy production (recall that this is the case for the basic linear Fokker–Planck equation), there may be a connection to principles of minimisation of entropy production, explored by Ilya Prigogine (1977 Nobel

Prize in chemistry). But the control of the Fisher information should hold more generally, as its good behaviour under the Boltzmann equation suggests. Nothing comes easily: a major difference between the variations of H and the variations of I is that the former is preserved under transport ($v \cdot \nabla_x$ in the Boltzmann equation), while the latter is not, as transport implies mixing and growing oscillations in phase space, thus a nontrivial behaviour for any quantity involving regularity. And this enormous difficulty is in the order of things: these growing oscillations are associated to a loss of accuracy in the observation and, for instance, the Landau damping phenomenon which has been the subject of a number of mathematical works since my paper with Clément Mouhot. All those remarks justify a study in depth in the hope of finding even richer fruits in the scientific continent discovered by Maxwell and Boltzmann.

References

1. R. Alexandre, Une définition des solutions renormalisées pour l'équation de Boltzmann sans troncature angulaire. C. R. Acad. Sci. Paris Sér. I Math. **328**(11), 987–991 (1999)
2. R. Alexandre, C. Villani, On the Boltzmann equation for long-range interactions. Commun. Pure Appl. Math. **55**(1), 30–70 (2002)
3. R. Alexandre, C. Villani, On the Landau approximation in plasma physics. Ann. Inst. H. Poincaré Anal. Non Linéaire **21**(1), 61–95 (2004)
4. R. Alexandre, L. Desvillettes, C. Villani, B. Wennberg, Entropy dissipation and long-range interactions. Arch. Ration. Mech. Anal. **152**(4), 327–355 (2000)
5. R. Alexandre, Y. Morimoto, S. Ukai, C.-J. Xu, T. Yang, Regularizing effect and local existence for the non-cutoff Boltzmann equation. Arch. Ration. Mech. Anal. **198**, 39–123 (2010)
6. C. Ané, S. Blachère, D. Chafaï, P. Fougères, I. Gentil, F. Malrieu, C. Roberto, G. Scheffer, *Sur les inégalités de Sobolev logarithmiques*. Panoramas et Synthèses , vol. 10 (Société Mathématique de France, 2000)
7. L. Arkeryd, On the Boltzmann equation. Arch. Ration. Mech. Anal. **45**, 1–34 (1972)
8. L. Arkeryd, Intermolecular forces of infinite range and the Boltzmann equation. Arch. Ration. Mech. Anal. **77**, 11–21 (1981)
9. L. Arkeryd, Asymptotic behaviour of the Boltzmann equation with infinite range forces. Commun. Math. Phys. **86**, 475–484 (1982)
10. L. Arkeryd, Stability in L^1 for the spatially homogeneous Boltzmann equation. Arch. Ration. Mech. Anal. **103**(2), 151–167 (1988)
11. L. Arkeryd, On the strong L^1 trend to equilibrium for the Boltzmann equation. Stud. Appl. Math. **87**(3), 283–288 (1992)
12. A.A. Arsen'ev, O. Buryak, On the connection between a solution of the Boltzmann equation and a solution of the Landau–Fokker–Planck equation. Math. USSR Sbornik **69**(2), 465–478 (1991)
13. A.A. Arsen'ev, N. Peskov, On the existence of a generalized solution of Landau's equation. U.S.S.R. Comput. Math. Math. Phys. **17**, 241–246 (1978)
14. D. Bakry, L'hypercontractivité et son utilisation en théorie des semigroupes, in *École d'été de Probabilités de Saint-Flour*. Lecture Notes in Mathematics, vol. 1581 (Springer, Berlin, 1994)
15. D. Bakry, M. Émery, Diffusions hypercontractives, in *Sém. Proba. XIX*. Lecture Notes in Mathematics, vol. 1123 (Springer, Berlin, 1985), pp. 177–206

16. D. Bakry, I. Gentil, M. Ledoux, *Analysis and Geometry of Markov Diffusion Operators*. Grundlehren der mathematischen Wissenschaften, vol. 348 (Springer-Verlag, New York, Berlin, 2014)
17. R. Balescu, *Statistical Mechanics of Charged Particles* (Wiley-Interscience, New York, 1963)
18. C. Baranger, C. Mouhot, Explicit spectral gap estimates for the linearized Boltzmann and Landau operators with hard potentials. Rev. Matem. Iberoam. **21**, 819–841 (2005),
19. A.R. Barron, Entropy and the central limit theorem. Ann. Probab. **14**, 336–342 (1986)
20. R.F. Bass, *Dffusions and Elliptic Operators*. Probability and Its Applications (Springer-Verlag, Berlin, 1998)
21. J.-D. Benamou, Y. Brenier, Weak existence for the semigeostrophic equations formulated as a coupled Monge–Ampère/transport problem. SIAM J. Appl. Math. **58**(5), 1450–1461 (1998)
22. S. Bernstein, Sur les fonctions absolument monotones. Acta Math. **52**, 1–66 (1928)
23. A.V. Bobylev, The theory of the nonlinear, spatially uniform Boltzmann equation for Maxwellian molecules. Sov. Sci. Rev. C. Math. Phys. **7**, 111–233 (1988)
24. A.V. Bobylev, Moment inequalities for the Boltzmann equation and applications to spatially homogeneous problems. J. Statist. Phys. **88**(5–6), 1183–1214 (1997)
25. A.V. Bobylev, C. Cercignani, On the rate of entropy production for the Boltzmann equation. J. Stat. Phys. **94**(3–4), 603–618 (1999)
26. N. Bogoljubov, Problems of Dynamical Theory in Statistical Physics, in *Studies in Statistical Mechanics*, ed. by J. de Boer, G.E. Uhlenbeck (Interscience, New York, 1962)
27. F. Bolley, A. Guillin, C. Villani, Quantitative concentration inequalities for empirical measures on non-compact spaces. Probab. Theory Related Fields **13**(3–4), 287–314 (2007)
28. L. Boltzmann, Weitere Studien über das Wärmegleichgewicht unter Gasmolekülen. Sitzungsberichte der Akademie der Wissenschaften **66**, 275–370 (1872). Translation : Further studies on the thermal equilibrium of gas molecules, in Kinetic Theory 2, 88–174, Ed. S.G. Brush, Pergamon, Oxford (1966)
29. L. Boltzmann, *Lectures on Gas Theory* (University of California Press, Berkeley, 1964). Translated by Stephen G. Brush. Reprint of the 1896–1898 Edition by Dover Publications, 1995
30. F. Bouchut, L. Desvillettes, A proof of the smoothing properties of the positive part of Boltzmann's kernel. Rev. Mat. Iberoamericana **14**(1), 47–61 (1998)
31. L. Boudin, L. Desvillettes, On the singularities of the global small solutions of the full Boltzmann equation. Monatsh. Math. 131, 91–108 (2000)
32. C. Buet, S. Cordier, Conservative and entropy decaying numerical scheme for the isotropic Fokker–Planck–Landau equation. J. Comput. Phys. **145**(1), 228–245 (1998)
33. R. Caflisch, The Boltzmann equation with a soft potential. Commun. Math. Phys. **74**, 71–109 (1980)
34. S. Cameron, S. Snelson, Velocity decay estimates for Boltzmann equation with hard potentials. Nonlinearity **33**(5), 2941–2958 (2020)
35. S. Cameron, L. Silvestre, S. Snelson, Global a priori estimates for the inhomogeneous Landau equation with moderately soft potentials. Ann. Inst. H. Poincaré (C) Anal. Non Lin. **35**(3), 625–642 (2018)
36. C. Cao, L.-B. He, J. Ji, Propagation of moments and sharp convergence rate for inhomogeneous non-cutoff Boltzmann equation with soft potentials. Preprint (2022). arXiv 2204.01394
37. T. Carleman, Sur la solution de l'équation intégrodifférentielle de Boltzmann. Acta. Math. **60**(3), 91–146 (1933)
38. E. Carlen, M. Carvalho, Strict entropy production bounds and stability of the rate of convergence to equilibrium for the Boltzmann equation. J. Stat. Phys. **67**(3–4), 575–608 (1992)
39. E. Carlen, M. Carvalho, Entropy production estimates for Boltzmann equations with physically realistic collision kernels. J. Stat. Phys. **74**(3–4), 743–782 (1994)

40. E.A. Carlen, E. Gabetta, G. Toscani, Propagation of smoothness and the rate of exponential convergence to equilibrium for a spatially homogeneous Maxwellian gas. Commun. Math. Phys. **199**(3), 521–546 (1999)
41. E.A. Carlen, M.C. Carvalho, X. Lu, On strong convergence to equilibrium for the Boltzmann equation with soft potentials. J. Stat. Phys. **135**(4), 681–736 (2009)
42. K. Carrapataso, Quantitative and qualitative Kac's chaos on the Boltzmann sphere. Ann. Inst. H. Poincaré Probab. Stat. **51**(3), 993–1039 (2015)
43. K. Carrapatoso, S. Mischler, Landau equation for very soft and Coulomb potentials near Maxwellians. Ann. PDE **3**(1), 1–65 (2017)
44. K. Carrapatoso, I. Tristani, K.-C. Wu, Cauchy problem and exponential stability for the inhomogeneous Landau equation. Arch. Ration. Mech. Anal. **221**(1), 363–418 (2016)
45. K. Carrapatoso, L. Desvillettes, L. He, Estimates for the large time behavior of the Landau equation in the Coulomb case. Arch. Ration. Mech. Anal. **224**(2), 381–420 (2017)
46. C. Cercignani, H-theorem and trend to equilibrium in the kinetic theory of gases. Arch. Mech. **34**, 231–241 (1982)
47. C. Cercignani, *The Boltzmann Equation and Its Applications* (Springer-Verlag, New York, 1988)
48. C. Cercignani, *Rarefied Gas Dynamics. From Basic Concepts to Actual Calculations* (Cambridge University Press, Cambridge, 2000)
49. C. Cercignani, R. Illner, M. Pulvirenti, *The Mathematical Theory of Dilute Gases* (Springer, New York, 1994)
50. J. Chaker, L. Silvestre, Entropy dissipation estimates for the Boltzmann equation without cut-off. Kinet. Relat. Models **16**(5), 748–763 (2023)
51. Y. Chen, L. He, Regularity of non-cutoff Boltzmann equation with hard potential. Arch. Ration. Mech. Anal. **201**(2), 501–548 (2011)
52. Y. Chen, L. Desvillettes, L. He, Smoothing effects for classical solutions of the full Landau equation. Arch. Ration. Mech. Anal. **193**, 21–55 (2009)
53. T.M. Cover, J.A. Thomas, *Elements of Information Theory*. A Wiley-Interscience Publication (John Wiley & Sons Inc., New York, 1991).
54. P. Degond, B. Lucquin-Desreux, The Fokker–Planck asymptotics of the Boltzmann collision operator in the Coulomb case. Math. Models Methods Appl. Sci. **2**(2), 167–182 (1992)
55. J.-L. Delcroix, A. Bers, *Physique des Plasmas* (InterEditions/CNRS Editions, Paris, 1994)
56. A. Dembo, O. Zeitouni, *Large Deviations Techniques and Applications*, 2nd edn. (Springer-Verlag, New York, 1998)
57. A. Dembo, T.M. Cover, J.A. Thomas, Information theoretic inequalities. IEEE Trans. Inform. Theory **37**(6), 1501–1518 (1991)
58. L. Desvillettes, Entropy dissipation rate and convergence in kinetic equations. Commun. Math. Phys. **123**(4), 687–702 (1989)
59. L. Desvillettes, Convergence to equilibrium in large time for Boltzmann and BGK equations. Arch. Ration. Mech. Anal. **110**(1), 73–91 (1990)
60. L. Desvillettes, On asymptotics of the Boltzmann equation when the collisions become grazing. Transport Theory Statist. Phys. **21**(3), 259–276 (1992)
61. L. Desvillettes, Some applications of the method of moments for the homogeneous Boltzmann equation. Arch. Ration. Mech. Anal. **123**(4), 387–395 (1993)
62. L. Desvillettes, About the regularizing properties of the non-cut-off Kac equation. Commun. Math. Phys. **168**(2), 417–440 (1995)
63. L. Desvillettes, Regularization for the non-cutoff 2D radially symmetric Boltzmann equation with a velocity dependent cross section. Transport Theory Stat. Phys. **25**(3–5), 383–394 (1996)
64. L. Desvillettes, Regularization properties of the 2-dimensional non radially symmetric noncutoff spatially homogeneous Boltzmann equation for Maxwellian molecules. Transport Theory Stat. Phys. **26**(3), 341–357 (1997)
65. L. Desvillettes, Entropy dissipation estimates for the Landau equation in the Coulomb case and applications. J. Funct. Anal. **269**, 1359–1403 (2015)

66. L. Desvillettes, C. Mouhot, Stability and uniqueness for the spatially homogeneous Boltzmann equation with long-range interaction. Arch. Ration. Mech. Anal. **193**(2), 227–253 (2009)
67. L. Desvillettes, C. Villani, On the spatially homogeneous Landau equation for hard potentials. I. Existence, uniqueness and smoothness. Commun. Partial Differ. Equ. **25**(1–2), 179–259 (2000)
68. L. Desvillettes, C. Villani, On the spatially homogeneous Landau equation for hard potentials. II. H-theorem and applications. Commun. Partial Differ. Equ. **25**(1–2), 261–298 (2000)
69. L. Desvillettes, C. Villani, Entropic methods for the study of the longtime behavior of kinetic equations. Transport Theory Stat. Phys. **30**(2–3), 155–168 (2001). The Sixteenth International Conference on Transport Theory, Part I (Atlanta, GA, 1999)
70. L. Desvillettes, C. Villani, On the trend to global equilibrium in spatially inhomogeneous entropy-dissipating systems: the linear Fokker–Planck equation. Commun. Pure Appl. Math. **54**(1), 1–42 (2001)
71. L. Desvillettes, C. Villani, On a variant of Korn's inequality arising in statistical mechanics. ESAIM Control Optim. Calc. Var. **8**, 603–619 (2002) (electronic)
72. L. Desvillettes, C. Villani, On the trend to global equilibrium for spatially inhomogeneous kinetic systems: the Boltzmann equation. Invent. Math. **159**(2), 245–316 (2005)
73. R.J. DiPerna, P.-L. Lions, On the Cauchy problem for the Boltzmann equation: global existence and weak stability. Ann. Math. (2) **130**, 312–366 (1989)
74. E. Dolera, On the computation of the spectrum of the linearized Boltzmann collision operator for Maxwellian molecules. Bolletino U.M.I. (9) **IV**, 47–68 (2011)
75. E. Dolera, E. Regazzini, Proof of a McKean conjecture on the rate of convergence of Boltzmann-equation solutions. Probab. Theory Related Fields **160**, 315–389 (2014)
76. M. Donsker, S.R.S. Varadhan, Asymptotic evaluation of certain Markov process expectations for large time, I. Commun. Pure Appl. Math. **28**, 1–47 (1975)
77. T. Elmroth, Global boundedness of moments of solutions of the Boltzmann equation for forces of infinite range. Arch. Ration. Mech. Anal. **82**, 1–12 (1983)
78. E. Fabes, D. Stroock, A new proof of Moser's parabolic harnack inequality using the old ideas of Nash. Arch. Ration. Mech. Anal. **96**, 327–338 (1986)
79. R.A. Fisher, Theory of statistical estimation. Math. Proc. Camb. Philos. Soc. **22**, 700–725 (1925)
80. E. Fontenas, Sur les minorations des constantes de Sobolev et de Sobolev logarithmiques pour les opérateurs de Jacobi et de Laguerre. Séminaire de probabilités (Strasbourg) **32**, 14–29 (1998)
81. N. Fournier, Uniqueness of bounded solutions for the homogeneous Landau equation with a Coulomb potential. Commun. Math. Phys. **299**, 765–782 (2010)
82. N. Fournier, H. Guérin, On the uniqueness for the spatially homogeneous Boltzmann equation with a strong angular singularity. J. Stat. Phys. **131**, 749–781 (2008)
83. N. Fournier, H. Guérin, Well-posedness of the spatially homogeneous Landau equation for soft potentials. J. Funct. Anal. **256**(8), 2542–2560 (2009)
84. N. Fournier, C. Mouhot, On the well-posedness of the spatially homogeneous boltzmann equation with a moderate angular singularity. Commun. Math. Phys. **289**, 803–824 (2009)
85. N. Fournier, M. Hauray, S. Mischler, Propagation of chaos for the 2D viscous vortex model. J. Eur. Math. Soc. **16**(7), 1425–1466 (2014)
86. M. Frazier, B. Jawerth, G. Weiss, *Littlewood–Paley Theory and the Study of Function Spaces*. CBMS, vol. 79 (American Mathematical Society, Providence, 1991)
87. C. Frye, C. Efthimiou, *Spherical Harmonics in p Dimensions* (World Scientific, Singapore, 2014)
88. I. Gallagher, L. Saint-Raymond, B. Texier, *From Newton to Boltzmann: Hard Spheres and Short-Range Potentials* (European Mathematical Society, Helsinki, 2014). Erratum available at https://ems.press/books/zlam/206/contents
89. I.M. Gamba, V. Panferov, C. Villani, Upper Maxwellian bounds for the spatially homogeneous Boltzmann equation. Arch. Ration. Mech. Anal. **194**, 253–282 (2009)

90. F. Golse, M. Gualdani, C. Imbert, A. Vasseur, Partial regularity in time for the space-homogeneous Landau equation with Coulomb potential. Ann. Sci. Ec. Norm. Super. (4) **55**(6), 1575–1611 (2022)
91. F. Golse, C. Imbert, A. Vasseur, Local regularity for the space-homogenous Landau equation with very soft potentials. Preprint (2022). arXiv:2206.05155
92. F. Golse, C. Imbert, L. Silvestre, Partial regularity in time for the space-homogeneous Boltzmann equation with very soft potentials. Preprint (2023). arXiv:2312.11079
93. T. Goudon, On Boltzmann equations and Fokker–Planck asymptotics : influence of grazing collisions. J. Stat. Phys. **89**(3–4), 751–776 (1997)
94. H. Grad, Principles of the kinetic theory of gases, in *Flügge's Handbuch des Physik*, vol. XII (Springer-Verlag, Berlin, 1958), pp. 205–294
95. L. Gross, Logarithmic Sobolev inequalities. Am. J. Math. **97**, 1061–1083 (1975)
96. M. Gualdani, S. Mischler, C. Mouhot, *Factorization of Non-symmetric Operators and Exponential H-Theorem*. Mémoires de la SMF, vol. 153 (Société Mathématique de France, Marseille, 2017)
97. N. Guillen, L. Silvestre, The Landau equation does not blow up. Preprint (2023). arXiv:2311.09420
98. Y. Guo, R. Strain, Exponential decay for soft potentials near Maxwellian. Arch. Ration. Mech. Anal. **187**, 287–339 (2008)
99. T. Gustafsson, L^p-estimates for the nonlinear spatially homogeneous Boltzmann equation. Arch. Ration. Mech. Anal. **92**(1), 23–57 (1986)
100. M. Hauray, S. Mischler, On Kac's chaos and related problems. J. Funct, Anal. **266**(10), 6055–6157 (2014)
101. L. He, Well-posedness of spatially homogeneous Boltzmann equation with full-range interaction. Commun. Math. Phys. 447–476 (2012)
102. C. Henderson, S. Snelson, A. Tarfulea, Local solutions of the Landau equation with rough, slowly decaying initial data. Ann. Inst. H. Poincaré Anal. Non Linéaire **37**(6), 1345–1377 (2020)
103. S. Henderson, S. Snelson, A. Tarfulea, Self-generating lower bounds and continuation for the Boltzmann equation. Calc. Var. Partial Differ. Equ. **59**(6), No.191 (2020)
104. C. Henderson, S. Snelson, A. Tarfulea, Classical solution of the Boltzmann equation with irregular initial data. Preprint (2022). arXiv:2207.0349
105. C. Henderson, S. Snelson, A. Tarfulea, Decay estimates and continuation for the non-cutoff Boltzmann equation. Preprint (2023). arXiv:2311.02235
106. R. Holley, D.W. Stroock, Logarithmic Sobolev inequalities and stochastic Ising models. J. Stat. Phys. **46**(5–6), 1159–1194 (1987)
107. E. Ikenberry, C. Truesdell, On the pressures and the flux of energy in a gas according to Maxwell's kinetic theory. I. J. Ration. Mech. Anal. **5**, 1–54 (1956)
108. C. Imbert, C. Mouhot, L. Silvestre, Decay estimates for large velocities in the Boltzmann equation without cut-of. J. Éc. Polytech. Math. **7**, 143–184 (2020)
109. C. Imbert, L. Silvestre, Global regularity estimates for the Boltzmann equation without cut-off. J. Am. Math. Soc. **35**(3), 625–703 (2022)
110. C. Imbert, C. Mouhot, L. Silvestre, Gaussian lower bounds for the Boltzmann equation without cutoff. SIAM J. Math. Anal. **52**(3), 2930–2944 (2020)
111. C. Imbert, L. Silvestre, C. Villani, On the monotonicity of the Fisher information for Boltzmann equation. Preprint (2024). arXiv 2409.01183
112. H. Jeffreys, An invariant form for the prior probability in estimation problems. Proc. R. Soc. Lond. Ser. A **186**(1007), 453–461 (1946)
113. S. Ji, Bounds for the optimal constant of the Bakry–Émery Γ_2 criterion inequality on RP^{d-1}. Announced at the Mathemata Conference, Crete, July 2024. Preprint (2024)
114. R. Jordan, D. Kinderlehrer, F. Otto, The variational formulation of the Fokker–Planck equation. SIAM J. Math. Anal. **29**(1), 1–17 (1998)
115. M. Kac, Foundations of kinetic theory, in *Proceedings of the Third Berkeley Symposium on Mathematical Statistics and Probability, 1954–1955*, vol. III (University of California Press, Berkeley/Los Angeles, 1956), pp. 171–197

116. L. Landau, Die kinetische Gleichung für den Fall Coulombscher Wechselwirkung. Phys. Z. Sowjet. **10**, 154 (1936) Translation : The transport equation in the case of Coulomb interactions, in D. ter Haar, ed., Collected papers of L.D. Landau, pp. 163–170. Pergamon Press, Oxford, 1981
117. C. Le Bihan, Effets de bord et comportement en temps long en théorie cinétique collisionnelle. Ph.D Thesis, ENS Lyon, 2023
118. M. Ledoux, On an integral criterion for hypercontractivity of diffusion semigroups and extremal functions. J. Funct. Anal. **105**(2), 444–465 (1992)
119. M. Ledoux, Differentials of entropy and Fisher information along heat flow: a brief review of some conjectures. Preprint (2023). Available online on the personal page https://perso.math.univ-toulouse.fr/ledoux/files/2024/09/Entropy-conjectures.pdf
120. E.H. Lieb, M. Loss, *Analysis*, 2nd edn. (American Mathematical Society, Providence, 2001)
121. E.M. Lifshitz, Pitaevskiĭ, L. P. *Course of Theoretical Physics ["Landau–Lifshits"]*, vol. 10 (Pergamon Press, Oxford, 1981). Translated from the Russian by J. B. Sykes and R. N. Franklin
122. D. Lindley, *Boltzmann's Atom: The Great Debate That Launched A Revolution In Physics* (Simon and Schuster, New York, 2001)
123. Y.V. Linnik, An information-theoretic proof of the central limit theorem with the Lindenberg condition. Theory Probab. Appl. **4**, 288–299 (1959)
124. P.-L. Lions, Compactness in Boltzmann's equation via Fourier integral operators and applications. J. Math. Kyoto Univ. **34**(2), 391–427, 429–461 (1994)
125. P.-L. Lions, On Boltzmann and Landau equations. Philos. Trans. R. Soc. Lond. Ser. A A **346**, 191–204 (1994)
126. P.-L. Lions, Regularity and compactness for Boltzmann collision operators without angular cut-off. C.R. Acad. Sci. Paris **326**(1), 37–41 (1998)
127. P.-L. Lions, C. Villani, Régularité optimale de racines carrées. C. R. Acad. Sci. Paris Sér. I Math. **321**(12), 1537–1541 (1995)
128. X. Lu, A result on uniqueness of mild solutions of Boltzmann equations. Transport Theory Stat. Phys. **26**(1–2), 209–220 (1997),
129. X. Lu, Conservation of energy, entropy identity, and local stability for the spatially homogeneous Boltzmann equation. J. Stat. Phys. **96**(3–4), 765–796 (1999)
130. A. Lunardi, *Interpolation Theory*, 3rd edn. (Scuola Normale Superiore, Pisa, 2018)
131. G. Matthes, G. Toscani, Variation on a theme by Bobylev and Villani. C.R. Acad. Sci. Paris, Ser. I **350**(1–2), 107–110 (2012)
132. J.C. Maxwell, On the dynamical theory of gases. Philos. Trans. R. Soc. Lond. Ser. A **157**, 49–88 (1867)
133. J.C. Maxwell, On stresses in rarefied gases arising from inequalities of temperatures. Philos. Trans. R. Soc. Lond. Ser. A **170**, 231–256 (1879)
134. H.J. McKean, Speed of approach to equilibrium for Kac's caricature of a Maxwellian gas. Arch. Ration. Mech. Anal. **21**, 343–367 (1966)
135. S. Mischler, C. Mouhot, Kac's program in kinetic theory. Invent. Math. **193**(1), 1–147 (2013)
136. S. Mischler, B. Wennberg, On the spatially homogeneous Boltzmann equation. Ann. Inst. H. Poincaré Anal. Non Linéaire **16**(4), 467–501 (1999)
137. C. Mouhot, Quantitative lower bounds for the full Boltzmann equation. i. Periodic boundary conditions. Commun. Partial Differ. Equ. **30**(4–6), 881–917 (2005)
138. C. Mouhot, Explicit coercivity estimates for the linearized Boltzmann and Landau operators. Commun. Partial Differ. Equ. **31**, 1321–1348 (2006)
139. C. Mouhot, Rate of convergence to equilibrium for the spatially homogeneous Boltzmann equation with hard potentials. Commun. Math. Phys. **261**, 629–672 (2006)
140. C. Mouhot, De Giorgi—Nash—Moser and Hörmander theories: new interplays, in *Proceedings of the International Congress of Mathematicians, Rio de Janeiro 2018. Vol. III. Invited lectures* (World Scientific Publication, Hackensack, 2018), pp. 2467–2493
141. C. Mouhot, DeGiorgi–Nash–Moser theory for kinetic equations, in *Proceedings of the Mathemata Conference, Chania, Crete, Greece* (Springer-Verlag, 2024). To appear

142. C. Mouhot, R. Strain, Spectral gap and coercivity estimates for linearized Boltzmann collision operators without angular cutoff. J. Math. Pures Appl. (9) **87**(5), 515–535 (2007)
143. C. Mouhot, C. Villani, Regularity theory for the spatially homogeneous Boltzmann equation with cut-off. Arch. Ration. Mech. Anal. **173**(2), 169–212 (2004)
144. J. Nash, Continuity of solutions of parabolic and elliptic equations. Am. J. Math. **80**, 931–954 (1958)
145. F. Otto, The geometry of dissipative evolution equations: the porous medium equation. Commun. Partial Differ. Equ. **26**(1–2), 101–174 (2001)
146. F. Otto, C. Villani, Generalization of an inequality by Talagrand and links with the logarithmic Sobolev inequality. J. Funct. Anal. **173**(2), 361–400 (2000)
147. Z. Ouyang, L. Silvestre, Conditional L^∞ estimates for the non-cutoff Boltzmann equation in a bounded domain. Preprint (2023). arXiv: 2305.02392
148. J. Perrin, *Les Atomes*, ed. by A. Fuchs (CNRS Editions, Paris, 2014). Reedition of the 1913 original, with foreword and comments
149. A.J. Povzner, The Boltzmann equation in the kinetic theory of gases. Am. Math. Soc. Transl. **47**, Ser. 2, 193–214 (1965)
150. A. Proutière, *New Results of Regularization for Weak Solutions of Boltzmann Equation*. MAPMO, vol. 23 (University of Orléans, Orléans, 1998). Unpublished work, Preprint
151. A. Pulvirenti, B. Wennberg, A Maxwellian lower bound for solutions to the Boltzmann equation. Commun. Math. Phys. **183**, 145–160 (1997)
152. P. Quittner, P. Souplet, *Superlinear Parabolic Problems: Blow-up, Global Existence and Steady States*, 2nd edn. (Birkhäuser, Basel, 2019)
153. F. Rezakhanlou, *Lectures on the Large Deviation Principle* (Department of Mathematics,, University of California, Berkeley, 2017). Available online
154. O. Rothaus, Hypercontractivity and the Bakry–Emery criterion for compact Lie groups,. J. Funct. Anal. **65**, 358–367 (1986)
155. E. Rutherford, The scattering of α and β particles by matter and the structure of the atom. Philos. Mag. Series **6**, 21 (1911)
156. L. Saloff-Coste, Precise estimates on the rate at which certain diffusions tend to equilibrium. Math. Z. **217**, 641–677 (1994)
157. R. Schilling, R. Song, Z. Vondraček, *Bernstein Functions. Theory and Applications*. Studies in Mathematics, vol. 37 (de Gruyter, Berlin, 2010)
158. C.E. Shannon, W. Weaver, *The Mathematical Theory of Communication* (The University of Illinois Press, Urbana, 1949)
159. L. Silvestre, A new regularization mechanism for the Boltzmann equation without cut-off. Commun. Math. Phys. **348**, 69–100 (2016)
160. L. Silvestre, Collision kernel (2024). GitHub repository, https://github.com/luissilvestre/collisionkernel
161. S. Snelson, C. Solomon, A continuation criterion for the Landau equation with very soft and Coulomb potentials. Preprint (2023).
162. Y. Sone, *Kinetic Theory and Fluid Dynamics*. Modeling and Simulation in Science, Engineering and Technology (Birkhäuser, Basel, 2002)
163. H. Spohn, *Large Scale Dynamics of Interacting Particles*. Texts and Monographs in Physics (Springer-Verlag, Berlin, 1991)
164. A. Stam, Some inequalities satisfied by the quantities of information of Fisher and Shannon. Inform. Control **2**, 101–112 (1959)
165. E. Stein, *Harmonic Analysis: Real-Variable Methods, Orthogonality, and Oscillatory Integrals* (Princeton University Press, Princeton, 1995)
166. R.M. Strain, Y. Guo, Almost exponential decay near Maxwellian. Commun. Partial Differ. Equ. **31**(1–3), 417–429 (2006)
167. S.J. Szarek, D. Voiculescu, Shannon's entropy power inequality via restricted Minkowski sums, in *Geometric Aspects of Functional Analysis*. Lecture Notes in Mathematics, vol. 1745 (Springer, Berlin, 2000), pp. 257–262

168. H. Tanaka, Probabilistic treatment of the Boltzmann equation of Maxwellian molecules. Z. Wahrsch. Verw. Gebiete **46**(1) (1978/1979), 67–105
169. G. Toscani, New a priori estimates for the spatially homogeneous Boltzmann equation. Cont. Mech. Thermodyn. **4**, 81–93 (1992)
170. G. Toscani, C. Villani, Probability metrics and uniqueness of the solution to the Boltzmann equation for a Maxwell gas. J. Stat. Phys. **94**(3–4), 619–637 (1999)
171. G. Toscani, C. Villani, Sharp entropy dissipation bounds and explicit rate of trend to equilibrium for the spatially homogeneous Boltzmann equation. Commun. Math. Phys. **203**(3), 667–706 (1999)
172. G. Toscani, C. Villani, On the trend to equilibrium for some dissipative systems with slowly increasing a priori bounds. J. Stat. Phys. **98**(5–6), 1279–1309 (2000)
173. C. Villani, On the Cauchy problem for Landau equation: sequential stability, global existence. Adv. Differ. Equ. **1**(5), 793–816 (1996)
174. C. Villani, Fisher information estimates for Boltzmann's collision operator. J. Math. Pures Appl. **77**(8), 821–837 (1998)
175. C. Villani, On a new class of weak solutions to the spatially homogeneous Boltzmann and Landau equations. Arch. Ration. Mech. Anal. **143**(3), 273–307 (1998)
176. C. Villani, On the spatially homogeneous Landau equation for Maxwellian molecules. Math. Models Methods Appl. Sci. **8**(6), 957–983 (1998)
177. C. Villani, Regularity estimates via the entropy dissipation for the spatially homogeneous Boltzmann equation without cut-off. Rev. Mat. Iberoamericana **15**(2), 335–352 (1999)
178. C. Villani, Contribution à l'étude mathématique des collisions en théorie cinétique. Master's Thesis, University of Paris-Dauphine, France, 2000
179. C. Villani, Decrease of the Fisher information for solutions of the spatially homogeneous Landau equation with Maxwellian molecules. Math. Models Methods Appl. Sci. **10**(2), 153–161 (2000)
180. C. Villani, A review of mathematical topics in collisional kinetic theory, in *Handbook of Mathematical Fluid Dynamics*, vol. I (North-Holland, Amsterdam, 2002), pp. 71–305
181. C. Villani, Cercignani's conjecture is sometimes true and always almost true. Commun. Math. Phys. **234**(3), 455–490 (2003)
182. C. Villani, Propriétés qualitatives des solutions de léquation de Boltzmann. Technical Report, Cours Peccot, Collège de France, 2003
183. C. Villani, *Topics in Optimal Transportation*. Graduate Studies in Mathematics, vol. 58 (American Mathematical Society, Providence, 2003)
184. C. Villani, Convergence to equilibrium: Entropy production and hypocoercivity, in *Rarefied gas dynamics : 24th International Symposium on Rarefied Gas Dynamics, Monopoli (Bari), Italy, 10-16 July 2004: RGD 24*, vol. 58, ed. by M. Capitelli. Harold Grad Lecture (2005)
185. C. Villani, Hypocoercive diffusion operators, in *Proceedings of the International Congress of Mathematicians (Madrid, 2006)*, vol. III (European Mathematical Society, 2006), pp. 473–498
186. C. Villani, Entropy dissipation and convergence to equilibrium, in *Entropy Methods for the Boltzmann Equation*, ed. by S.O.F. Golse. Lecture Notes in Mathematics, vol. 1916 (North-Holland, New York, 2008), pp. 1–70. Lectures from a special semester at the Centre Émile Borel, Institut H. Poincaré, Paris, 2001
187. C. Villani, *Hypocoercivity*. Memoirs of the American Mathematical Society (AMS, Providence, 2009)
188. C. Villani, *Optimal Transport, Old and New*. Grundlehren des Mathematischen Wissenschaften, vol. 338 (Springer-Verlag, Berlin, New York, 2009)
189. D. Voiculescu, Lectures on free probability theory, in *Lectures on Probability Theory and Statistics (Saint-Flour, 1998)*. Lecture Notes in Mathematics, vol. 1738 (Springer, Berlin, 2000), pp. 279–349
190. C.S. Wang Chang, G.E. Uhlenbeck, J. de Boer, The heat conductivity and viscosity of polyatomic gases, in *Studies in Statistical Mechanics*, vol. II (North-Holland, Amsterdam, 1964), pp. 241–268

191. F. Weissler, Local existence and nonexistence for semilinear parabolic equations in L^p. Indiana Univ. Math. J. **29**, 79–102 (1980)
192. B. Wennberg, On moments and uniqueness for solutions to the space homogeneous Boltzmann equation. Transport Theory Stat. Phys. **24**(4), 533–539 (1994)
193. H. Zaag, Sur la description des formations de singularités pour l'équation de la chaleur non linéaire. Ph.D Thesis, University of Cergy-Pontoise, France, 1998

Part IV
Music and Mathematics, a Gentle Bridge

Microtones, 31-TET and Just Intonation

Franck Jedrzejewski

At the beginning of the twentieth century, composers sought to enrich their palette of sounds and timbres by working on sound frequencies through the use of micro-intervals. Some used untempered sounds in a natural way, others used the divisions of the current tempered system as did Christiaan Huygens and Adriaan Fokker for the 31-TET (Tone Equal Temperament, or 31-EDO, Equal Division of the Octave), dividing the octave into 31 equal parts, still others sought to develop compositional theories and worked on the creation of dedicated instruments. Today, micro-intervals are increasingly present in many musical spheres, whether contemporary classical, ambient music, dub, rock, metal or jazz. Microtones pose many problems, both historical and theoretical, that we will try to present succinctly.

1 First Microtonal Experiments

At the beginning of the twentieth century, the idea of using microintervals (*microtones* or *microintervals* are musical intervals that are not multiples of the tempered semitone, in other words any untempered frequency) reappeared in different places on the planet (Russia, Germany-Czechoslovakia, France, Mexico, USA). Here are some historical elements that have upset the imperial order of the 12-tone scale.

In Lichterfelde, near Berlin, at the end of the nineteenth century, Georg August Behrens-Senegalden laid the foundations of an *achromatic piano*, which has twice the number of notes of a traditional piano. In India, the British composer John Foulds wrote a string quartet based on quarter tones (1896) that is now lost. In Berlin, Willi Möllendorff published a brochure *Musik mit Vierteltönen* [28] and filed

F. Jedrzejewski (✉)
CEA/Paris Saclay University, Paris, France

© The Author(s), under exclusive license to Springer Nature Switzerland AG 2026
N. D. Alikakos, C. Villani (eds.), *Analysis Techniques for Mathematical Physics and Geometry – and Music*, Lecture Notes in Mathematics 2390,
https://doi.org/10.1007/978-3-032-11396-2_8

a patent for a bichromatic harmonium. His colleague, Jörg Mager, also published a brochure [26] accompanied by a work in quarter tones *Die Flucht der Armen* [*The flight of the poor*]. Interested in instrument making, he developed one of the first micro-interval instruments, the *Elektrophon*, renamed *Sphärophon* around 1926. Supported by Prince Emile Schlösschen, who founded the Society for the *Studiengesellschaft für elektro-akustische Musik* [*Study of Electroacoustic Music*] in 1929, Mager developed and perfected a new instrument, the *Partiturophon*. For their part, the Futurists, under the pen of Nikolai Kulbin, published in the Almanach *Die Blaue Reiter* [*The Blue Rider*] an article *The Free Music* in which he considered the use of microtones up to the eighth tone [23].

In 1922, the Italian composer Ferruccio Busoni published in *Melos*, a visionary article on third-tone music, *Bericht über Dritteltöne* [*Communication on third-tones*] [5].

In Mexico, violinist and composer Julián Carrillo experimented with the "sonido 13", a symbolic complement to the 12 traditional sounds that opened the way to sixteenth tones. In Prague, Alois Hába composed in microtempered universes and in 1923 opened a microtonal music class at the Prague Conservatory, training many students. In 1927 in Leipzig, he published a *Neue Harmonielehre des diatonischen, chromatischen, viertel, drittel, sechstel und zwölfteltonsystems* [*New Course in Harmony in the Half-, Third-, Quarter-, Sixth-, and Twelfth-Tone Systems*] [14]. In the United States, Charles Ives composed *Three Quarter-Tone Pieces* for two pianos tuned a quarter-tone apart. The first two pieces (*Largo* and *Allegro*) are thought to date from 1923–24 and the third (*Chorale*) was written between 1903 and 1914. Charles Ives analyzed them in his study *Some Quarter-Tone Impressions* [17]. His *Fourth Symphony*, also called *Universe Symphony* (1909–16) is an incomplete score which also uses microtones. It was completed and created by several conductors including Johnny Reinhard.

In the 1930s, Hans Barth (1897–1956) experimented with microtonal resources. With George Weitz, he perfected a portable quarter-tone piano (1928), an instrument for which he composed numerous pieces. His *Concerto for quarter-tone piano and strings* was premiered in Philadelphia in 1930 by the composer at the piano under the direction of Leopold Stokowski.

In Saint Petersburg, Ivan Wyschnegradsky developed his own microtonal theory. In Paris, where he settled permanently in 1920, he wrote *La loi de la pansonorité* [*The law of pansonority*], a fundamental work that developed the theory of non-octavian spaces and which was not published until 43 years later [31]. He contacted piano makers. At his request, Pleyel adapted a double piano with pneumatic transmission capable of playing quarter tones. But this piano did not satisfy him. Before his Parisian exile, in Saint Petersburg, he frequented the circle of ultrachromatists, led by Gueorgui Rimsky-Korsakov, Nikolai's grandson. Nikolai Rimsky-Korsakov had 7 children (4 sons and 3 daughters). His second son Mikhail had a son, Georgy, and his third son Vladimir also had a son, Andrey Vladimirovich, who together with Aleksander Ivanov was the inventor of the *emeriton* (the name *emeriton* is an acronym for "electric musical instrument", followed by the initials (*RI*) of its inventors, A.V. Rimsky-Korsakov and A.A. Ivanova, and the word *tone*). The first version of the emeriton was built in 1935, and the latest version dates from the 1950s. In 1932 Georgy composed an *Octet for two emeritons and small ensemble*, while in Moscow Arseny Avraamov developed a theory of ultrachromatism [2]. In 1917, together with Yevgeny Sholpo, he founded the *Leonardo da Vinci Association* and tried to synthesize sound spectra into graphic scores. It must be said that at that time there was a great deal of interest in this question. In 1923, Vladimir Baranov-Rossiné had filed a patent for an *optophonic piano* and the Czech Zdeněk Pešánek had built a colour keyboard, the *spectrophone*, in collaboration with the pianist Erwin Schulhoff. Raoul Hausmann wrote an article *Optophonetics* [16] for Lázlo Moholy-Nagy's *MA* magazine and filed a patent for his instrument, the *optophone*. For his part, Sholpo built an electro-optical synthesizer, the *variophone*, and Arseny Avraamov presented his experiments on a mechanical orchestra made of modified pianos, harmoniums and noisemakers: the *Symphony of the Sirens*, which was created in Baku in 1922 and revived in Moscow the following year. In 1938, Boris Yankovsky met Yevgeny Murzin, a young designer with a passion for sound synthesis. Together, they decided to build a new synthesizer based on the principles of the variophone, the first of which, which has now disappeared, was not built until 1957. A second version of this electro-optical synthesizer, named *ANS* (from the initials of the composer Alexander Nikolayevich Scriabin), was built in 1964. It has since been exhibited at the Glinka Museum in Moscow.

Microtones were also used in large-scale works, and not only in chamber music. In 1931, Alois Hába's quarter-tone opera *Matka* [*The Mother*] was premiered by Hermann Scherchen in Munich. It was revived after the war in 1947 under the direction of Karel Ancerl.

The influence of folk music, which naturally employs microintervals, has been significant for some composers. However, they did not pursue their microtonal research, and the use of microtones remained limited. Examples include emblematic works such as Georges Enescu's *Violin Sonata No. 3*, written in 1926, and his opera *Oedipe*, completed in 1931, which employ quarter tones, as well as the original version of Bartok's *Sonata for Solo Violin*, written in 1944, from which the quarter tones were removed at the suggestion of Yehudi Menuhin to make the Sonata easier to play. Peter Bartok and Nelson Dellamaggiore restored the original piece. In another area, and although he was not influenced by folk music, Pierre Boulez wrote a microtonal canon in the first version of *Le Visage Nuptial* in 1947, and a tone row of 24 quarter tones in *Polyphony X* for 18 instruments (1950–1951), but later abandoned microintervals [4].

Early composers and theorists considered microintervals to be refinements of the 12-tone tempered scale that were divided into equal parts. Things changed radically with the viewpoint of the American Harry Partch, who published *Genesis of a Music* [30] in 1934. It was no longer a question of equally dividing the octave but of rethinking all acoustic relationships according to natural resonance. The fractional numbers representing musical intervals had to be as simple as possible. Just intonation was born. It would lead to a vast compositional movement that was not limited to the borders of the United States. This movement was accompanied by rich discoveries and theoretical speculations, including those of Erv Wilson,

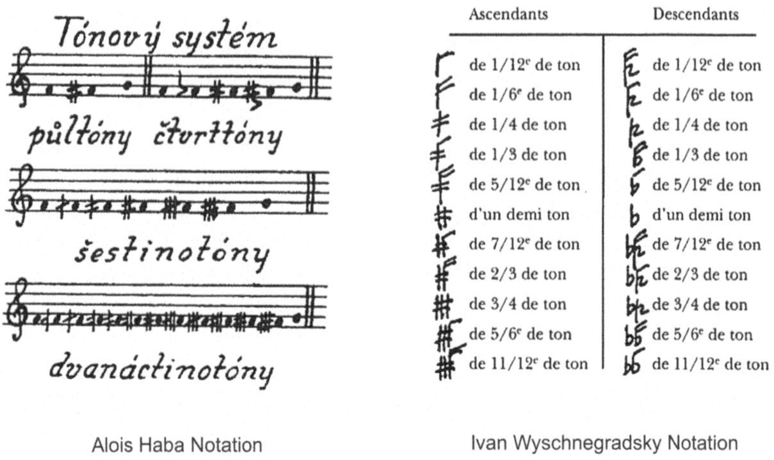

Fig. 1 Some microtones notations

who extended Partch's teaching and whose theories were presented in Terumi Narushima's book, *Microtonality and the Tuning Systems of Erv Wilson* [29].

In the Netherlands, the physicist Adriaan Fokker was interested in the works of Christiaan Huygens [6] and in harmonic accuracy. Taking up Huygens' theories, he spent his life promoting the 31-tone temperament, founding the *Stichting Nauwluisterendheid* [*Society for Listening to Harmonic Intuition*], which was renamed the *Stichting Huygens-Fokker* [*Huygens-Fokker Society*] on 17 March 1966. Fokker commissioned many composers to write for this new temperament. In 1966, he published *Neue Musik mit 31 Tönen* [*New music with 31 notes*] [9], building a bridge both historically between the meantone temperament and the 31-tone temperament, and theoretically between the equal division of an octave and its conception in just intonation.

In Paris in 1955, then at the Universal Exhibition in Brussels in 1958, the *Carrillo metamorfoseadores pianos* were presented, tuned from 1/2 to 1/16th of a tone, one piano for each temperament.

Two major microtonal currents emerged: one based on the equal division of the tempered system, the other based on just intonation. But new directions have arised: systems based on the division of intervals other than the octave (2) such as the tritave (3) with Bohlen-Pierce and Wendy Carlos (Fig. 1).

2 Microtonal Instruments

Designed for the 12-tone temperament, traditional keyboards were not adapted to the new tuning systems. This problem had already been encountered in the sixteenth century, when the search for perfect fifths and thirds in all keys encouraged makers and theorists to produce keyboards with broken steps and feints. One only has to leaf through Mersenne's *L'harmonie universelle* to realize the diversity of solutions proposed. The specificity of tuning systems also imposed multiple solutions: the enharmonic harmoniums of the nineteenth century, such as those of Colin Brown or Robert Bosanquet, sought the greatest variety of playing possibilities. Special temperaments such as that of Adriaan Fokker with its 31 tones per octave, required a complete recomposition of the layout of the keyboard keys. It had already been discussed in the sixteenth century by Nicola Vicentino, Francisco Salinas and designed by Fabio Colonna for his archicembalo. With the introduction of the discursive functions of microtonality, the problematic of keyboards evolved. The representation of tuning systems according to a network, which since Hugo Riemann, has been designated by the German name of *Tonnetz*, influenced the design of keyboards that began to imitate this tonal network or tonnetz. More compact keyboards appeared, with diamond-shaped or hexagonal keys. When the tuning of the instrument is no longer based on the octave, as is the case in some temperaments such as those of Wendy Carlos or Bohlen-Pierce, the periodicity is displaced and with it the entire design of the keyboard or the keying of wind instruments. Elaine Walker is an American composer. A student at Berklee College of Music, she founded the electronic music band ZIA and designed various keyboard instruments for the Bohlen-Pierce temperament.

> I performed *Stick Men* and a newer piece called *Love Song*, at the same symposium [Bohlen-Pierce Symposium], with Dr Boulanger on the sonome (the hexagon keyboard invented by Peter Davies) rearranged for the Bohlen-Pierce Scale. I built the other two keyboards especially for this performance. The one I'm playing is a "Vertical Keyboard", arranged for Bohlen-Pierce, in Dur II mode [C, C#, D, E, F, F#, G, H, H#, J, A, B, B#,C]. I call it the "BP-tar". The other keyboard, which Dan Sedgewick is playing, is a Korg Poly-800 rearranged for Bohlen-Pierce, in Lambda mode [C, C#, D, E, F, F#, G, H, H#, J, A, A#, B, C]. Dan's wife, Marji Gere, is playing a BP-tuned violin. (Elaine Walker, Program Note).

As can be seen in this example, tuning an existing instrument is often a solution, but it is done to the detriment of the completeness of the system: some notes are omitted or shifted over the total range of the instrument. The introduction of electronic devices made it possible at the beginning of the twentieth century to approach the question of the realization of keyboards for micro-interval instruments differently.

Several instruments have left their mark on electronic instrument making in an original way and have achieved a certain notoriety: this is the case of the *Theremin* or the *Ondes Martenot*. This theremin invented in the 1920s by Léon Thermine (real name Lev Termen) is a melodic instrument that is manipulated by gesture. The pitch of the sound rose when the user's right hand approached an antenna connected to a variable capacitor. The further the hand moved away, the lower the sound. With the left hand, the instrumentalist adjusted the volume. It was originally an entirely haptic instrument, just like Nikolai Oboukhov's *Croix sonore*, whose functionalities would be taken up much later by the adaptation of sensors. The success of the theremin was due in part to its marketing by RCA in 1929. That was the year Joseph Schillinger's *First Airphonic Suite for theremin and orchestra* was created in Cleveland. The instrument produced a continuum of sounds, but did not allow fine control of the frequencies produced in isolation. This was also the case for the *ondioline* (or *ondiola*), an instrument invented by Georges Jenny in 1941, whose micro-intervals Giacinto Scelsi, who owned two examples, liked to hear. However, this instrument marked a real advance in the control of timbres by electronics. The ondioline is presented in the form of a three-octave keyboard that actually covers six octaves with a set of keys. As with the Ondes Martenot, lateral pressure on the keys allows vibrato effects to be obtained. On the front panel, potentiometers modify the sound parameters. A knee lever and a pedal regulate the intensity. Geneviève Robert premiered the *Concerto Taj-Mahal for ondioline and orchestra* by Darius Cittanova.

With the *trautonium*, or its renovated version the *mixturtrautonium*, Friedrich Trautwein's instrument, presented for the first time in Berlin in 1930, a new step was taken in the way of working with timbres. However, and although marketed by the Telefunken firm in 1932, the instrument did not have the expected success. Even if famous composers such as Paul Hindemith or Harald Genzmer wrote concertos for trautonium, or composers such as Werner Egk, Paul Höffer, Helmut Riethmüller or Richard Strauss wrote more or less ambitious pieces for this instrument, the works for trautonium remained confidential. Dedicatee and performer of many works, Oskar Sala spent his life defending the instrument. Pressure on the slider allows the sound to be modulated and micro-intervals to be produced, but their accuracy remains as on a stringed instrument under the control of the player's ear.

A student of Milton Babbitt and Roger Sessions, John Eaton is an American composer born in 1935. He was passionate about micro-intervals and built with an engineer named Paul Ketoff, an instrument adapted to his acoustic research, the *SynKet* (*Synthesiser Ketoff*). He composed a set of works for different tuning systems including quarter and sixth tones, but reserved pieces for large groups for the Synket such as the *Concert Piece for SynKet and orchestra* (1966) or *Mass for soprano and synthesizer ensemble* (1969). It is one of the first synthesizers capable of producing micro-intervals, which, due to its prototype character, remained the work of a single composer, even if it was used in some film music by Ennio Morricone.

An emblematic example of micro-intervallar construction, the Fokker organ is an example of an instrument that was designed in the early years as an acoustic instrument and became, mutatis mutandis, an electronic instrument, which has retained its traditional construction through new achievements.

The development of the MIDI standard created a small revolution in the world of instrument construction. The Siemen Terpstra compagny developed different systems for the 31-tone per octave temperament from the 1980s. On the advice of the composer Joel Mandelbaum, assisted by Dylan Howarth and Garnet Willis, and on the principles of previous keyboards, the *Terpstra Generalized Keyboard* was created by Johnny Reinhardt in 2007.

Recently, and reconnecting with past research from the sixteenth century, Swiss musicians, makers and musicologists have undertaken, as part of the *Studio 31 project*, to reconstruct Nicola Vicentino's archiorgano, with today's means and knowledge. Led by musicologist Martin Kirnbauer, composer Caspar Johannes

Walter and harpsichordist Johannes Keller, the project reconstructed the instrument from plans and the few surviving representations. The organ is tuned to the trentunisonic system. It has two manuals, 132 keys and 126 pipes.

In Europe, Franz Richter Herf, rector of the Mozarteum in Salzburg and composer, decided to compose only in the twelfth tone system. Throughout his life, he developed the concept of ekmelic music. In order to produce twelfth tones, he had an electronic organ, the *ekmelic organ*, built in 1974. With the help of Helmut Baron and Georg Eberl, both from Salzburg, Franz Richter Herf was able to experiment with his own music in November 1974. A second version of this organ was built in 1981 by engineer Reinhard Hanel for the Mozarteum University in Salzburg; the three keyboards had 6 octaves, producing 438 different sounds.

Another experiment was conducted by the German organist Hans-André Stamm as part of a research project with Professor Martin Vogel of the University of Bonn. Together, they built an *enharmonische Pfeifenorgel* [*enharmonic pipe organ*] capable of producing just chords as well as eighth or quarter tones.

Aaron Andrew Hunt's company H-Pi Instruments, specializing in the development of micro-interval solutions, both in software (Microsynth) and in microtonal boxes and keyboards (tonal plexus) marketed the *Tuning box*. One of the advantages of electronic instruments is that they are able to memorize pre-programmed tunings and reconfigure the keyboard instantly. Often having several colors, the keyboard helps the instrumentalist to situate himself in the frequency ranges of the keyboard: the base is often yellow and the green, blue and purple deviations give increasingly sharp or ascending sounds, while the orange and red colors give increasingly descending sounds. Dolores Catherino's *Towards the Continuum*, composed for the tonal plexus, uses a temperament of 106 notes per octave.

3 From Euler and Huygens to Fokker

The first mentions of a division of the harmonic line into 31 equal parts date from the sixteenth century. Francisco de Salinas (1513–1590), a Spanish organist and music theorist, attributes it to Nicola Vicentino (1511–1575 or 1576). In his 1555 treatise, *L'antica musica ridotta alla moderna prattica,* Vicentino divides the tone into five equal parts and the semitone into three parts, so that the octave is divided into $5 \times 5 + 2 \times 3 = 31$ parts. But unlike Christiaan Huygens, the octave is not thought of by Vicentino as a whole divided into parts.

3.1 Tunings and Temperament

This division of the tone into 5 parts is confirmed by Francisco Salinas (1513–1590), a Spanish organist monk, who was only 10 years old when Pietro Aaron (1489–1545) wrote his *Toscanello della Musica* (Venice, 1523) [1]. In this text, he lays the first theoretical foundations of the meantone temperament. Since one can never complete the cycle of natural fifths (3/2), the idea of this *meantone temperament* is to seek to complete the cycle of thirds.

In fact, it is well known that it is impossible to close the cycle of pure fifths (of ratio 3/2). Otherwise, it would be necessary that by evolving from fifth to fifth (C–G–D–A, etc.), one could fall back on an octave. But this is impossible because the equation

$$2^p = \left(\frac{3}{2}\right)^q$$

has no integer solution (other than the trivial solution). When we go through 12 fifths, we end up in the vicinity of 7 octaves. The difference defines the Pythagorean comma:

$$\text{Pythagorean comma} = \frac{12 \text{ fifths}}{7 \text{ octaves}} = \frac{3^{12}}{2^{19}}$$

which represents 23.46 cents when the octave is divided into 1200 cents. In equal temperament tuning, it is this Pythagorean comma that is distributed to decrease each pure fifth by one twelfth of a Pythagorean comma.

As the cycle of fifths could not close on the octaves, it was considered to take not fifths, but pure major thirds (of ratio 5/4) and it was sought to construct a scale having the most pure major thirds. But whether major (5/4) or minor (6/5), natural thirds, no more than perfect fifths, cannot close the cycle. It is therefore necessary to resolve to compromises, taking care if possible to construct a system in which as many perfect fifths and thirds as possible appear. In the Pythagorean system, the major third (C–E obtained from the fifths cycle C–G–D–A–E, $\frac{3^4}{2^6} = \frac{81}{64}$) differs from the pure major third (5/4). The difference between the Pythagorean major third and the pure major third defines the syntonic comma:

$$\text{syntonic comma} = \frac{\text{Pythagorean major third}}{\text{pure major third}} = \frac{81}{80}$$

which is about 21.51 cents. Instead of reducing the fifths by a fraction of a Pythagorean comma, the idea was to reduce the fifths by a fraction n of a synthonic comma. In other words, the idea was to impose the equation:

$$2^p = \left(\frac{3}{2}\left(\frac{81}{80}\right)^{-n}\right)^q$$

which defines the meantone temperament. Various values of n have been proposed. The first was that of Pietro Aaron (c. 1480—after 1545), in 1523. Aaron's solution, which became widespread in the sixteenth century, is to diminish the Pythagorean just fifths (of ratio 3/2) by a quarter of a syntonic comma ($n = 1/4$). This leads to a temperament, the *quarter-comma meantone*, in which no fifth is just but all except one are diminished by 5 cents, 8 major thirds are just but no minor third is just, and 9 minor thirds are diminished, like the fifths, by a quarter of a comma.

3.2 Meantone Temperament and 31-edo

Christiaan Huygens (1629–1695) studied the division of the tone into 5 parts and the semitone into 3 parts and therefore of the octave into 31 equal parts. He noted that this temperament included an approximation of the meantone temperament, which justified in his opinion the use of the tricesimoprimal system. It was hardly used, although some organists or composers such as the Dutch organist Quirinus van Blakenburg (1654–1739) praised its merits, at least from a theoretical point of view (see Fig. 2).

It was not until the physicist Adriaan Fokker (1887–1972) that the trentunisonic temperament took off.

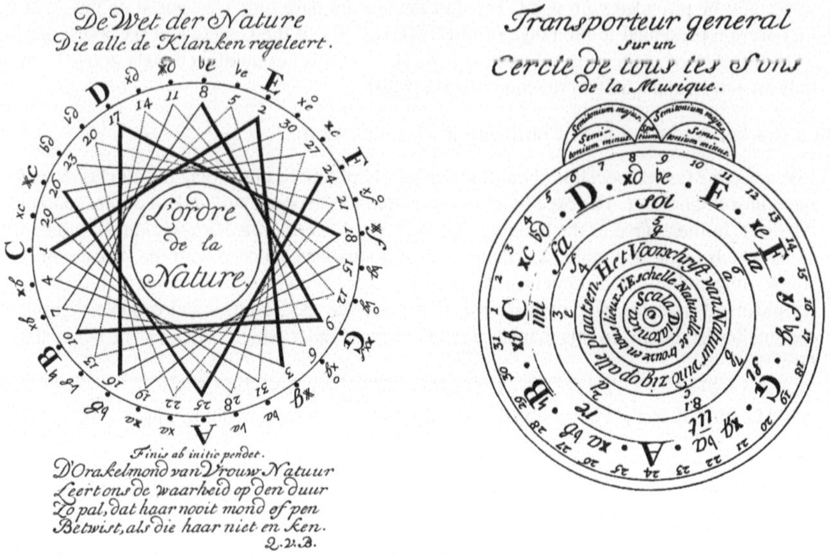

Fig. 2 Quirinus Blankenburg, *Elementa musica*, Gravenhage,1739

3.3 The 31-tone Organ and the Stichting Huygens-Fokker

Huygens' theories were taken up in the 1950s by Adriaan Fokker who developed a harmonic theory based on Euler genera. Fokker had an organ built with 31 tones per octave and founded what is today called the *Stichting Huygens-Fokker* in 1960 and placed commissions to composers. Fokker was interested in the theories that Leonhard Euler had expounded in his *Tentamen novae theoriae musicae*, [8].

In order to hear and musically realize Euler's trentunisonic system and genera, an instrument was needed. This is how the Euler organ was built. Leonhard Euler had invented new tuning systems that were generated by a simple geometric process. First, Fokker had an organ built that could play Euler's first ten genres (3^2, $3^2.5$, 3.5^2, 5^3, $3^2.7$, $3.5.7$, $5^2.7$, 3.7^2, 5.7^2, 7^3). He relates:

> Shortly before the war, the *Geluidstichting* was founded in the Netherlands, foundation which dedicated itself, in addition to the fight against noise, to other acoustic problems. The Geluidstichting granted the necessary funds for the building of a small organ, on which the ten Genera of three numbers each cited above could be played. The organ was constructed by Van Leeuwen Brothers, organ builders in Leiderdorp, and installed in the Teyler Museum at Haarlem. It had ten stops, corresponding to the Genera. Each stop had twelve labial pipes, corresponding to twelve equal-sized white keys of the keyboard. By means of knobs each stop could be provided with wind. Three of the Genera have only four notes; on the organ they were represented in the range of three octaves. Six of the Genera had six notes; they were represented in the range of two octaves. The tenth Genus numbered eight notes; for it only an octave and a half were required. [10, p. 20]

Then Fokker had the idea of building a 31-tone organ.

> I sketched a 31-step special keyboard, which would permit both the play of full chords and rapid movement [. . .]. The execution was entrusted to the organ builder B. Pels en Zoon in Alkmaar. In the course of the building, the *Stichting Nederlandse Muziekbelangen* granted sums which made it possible for us to furnish a smaller keyboard with the conventional seven while and five black keys in addition to the large keyboard with 31-notes to the octave. By means of switches we could choose 12 pipes from the 31 pipes and connect them to the smaller keyboard. In this manner the 9 Genera of Euler which contained 12 notes, the Genera
> $$(3^3.5^2),(3^2.5^3),(3^3.7^2),(3^2.7^3),(5^3.7^2),(5^2.7^3),(3^2.5.7),(3.5^2.7),(3.5.7^2)$$
> could be played on the conventional keyboard with the conventional technique. [10, p. 23–24]

Through the association he had just created, Fokker placed orders with Dutch composers to use his instrument or to compose for the trentunisonic system, which he was trying to popularize. He became friends with the composer and organist Anton De Beer, who mastered the repertoire of Fokker's organ. But De Beer was not content with being a brilliant performer; he developed a new electronic instrument, the *archiphone*, whose name was a tribute to the archicembalo or arciorgano given by Vicentino. The archiphone had the same keyboard as Fokker's organ, a system of hexagonal keys derived from the tonnetz, but had forty different timbres. In 1970, at a concert in the Teyler Museum in Haarlem, Anton de Beer inaugurated a six-manual archiphone, for which he wrote the opening piece simply entitled *Intrada 1st November 1970.*

In total, four instruments were built: the first two are at the Huygens-Fokker Foundation in Haarlem (Netherlands), the third is owned by the composer William Coates (Australia) and the fourth was at Webster College in St. Louis (Missouri, USA) for a long time before returning to its owner.

Several composers have written for the archiphone, including Ton de Kruyf who composed a radio play *Quahquahtinchan in den vreemde* (1971) and Joel Mandelbaum who wrote *The Dybbuk* and *Three songs for soprano, two violins and archiphone* on a text by Judy Berman.

4 New Music Theory

From Alois Hába to Georg Friedrich Haas, from Ivan Wyschnegradsky to Alain Bancquart, microtonal evolution has been accompanied by the publication of numerous music theory manuals, introducing new concepts adapted to microintervals. Finding, theorizing, classifying new scales, transposing existing notions in the domain of 12 tones such as the 12-tone row or the modes of limited transposition to microinterval domains, reinventing diatonic scales in a scale of N tones, weaving networks of acoustic relationships such as Hugo Riemann's tonnetz, Partch's tonality diamond or Ben Johnson's networks, reinventing modulations as Alain Bancquart will do by substituting what he calls *mutations* [3], it is a whole world of new concepts that has emerged with microtonality. Some theoretical aspects of these new concepts are presented in this section, leaving aside areas such as microtone notation (see Fig. 1).

4.1 Chords Enumeration

The first lessons in harmony textbooks consist of classifying chords and considering the most common chords such as major and minor triads, then seventh chords, and so on. In a microtonal universe, the question arises of classifying chords and scales. How many chords are there up to transpositions (and inversions) if you have not 12, but N notes in an octave? This is an important question. There have been (at least) two main ways in history to classify chords:

1. Up to transpositions. This was proposed by Anatole Loquin (1834–1903), then later by Edmond Costère (pseudonym of the famous magistrate of the French Supreme Court Edouard Coester (1905–2001), for the most part unkown, in *Lois et styles des harmonies musicales* (1954) [7]. Two chords A and B of same cardinality are equivalent if it exists n, such that $B = T_n(A)$ where T_n is the translation $T_n(x) = x + n \mod N$.
2. Up to transpositions and inversions. This was proposed by Howard Hanson (1896–1981) in *Harmonic Materials of Modern Music* (1960) [15], and taken

up by Allen Forte (1926–2014) in *The Structure of Atonal Music* (1973) [11]. Two chords A and B of same cardinality are equivalent if it exists n, such that $B = T_n(A)$ or $B = I_n(A)$, where I_n is the inversion $I_n(x) = -x + n \mod N$.

One might think that it is easy to count the number of k-chords in a temperament equal to N degrees, and that this number is simply the number of k-combinations in a set of N elements. But this is wrong. There is an obstruction to this, due to the existence of the sets of limited transposition (SLT). Let's see how this works on some simple examples.

- If $N = 12$, $k = 2$, there is only one SLT: the tritone wich has 6 transpositions, the other 5 intervals have 12 transpositions

$$\binom{12}{2} = 66 = 5 \times 12 + 1 \times 6$$

  So there are only $5 + 1 = 6$ intervals, up to transpositions.
- If $N = 12$, $k = 3$, there is only one SLT: the triad $\{0, 4, 8\}$ which has 4 transpositions.

$$\binom{12}{3} = 220 = 18 \times 12 + 1 \times 4$$

  So there are $18 + 1 = 19$ triads, up to transpositions.
- If $N = 12$, $k = 4$, there are 3 SLT: $\{0, 3, 6, 9\}$ with 3 transpositions, $\{0, 2, 6, 8\}$ with 6 transpositions, $\{0, 1, 6, 7\}$ with 6 transpositions.

$$\binom{12}{4} = 495 = 40 \times 12 + 1 \times 3 + 1 \times 6 + 1 \times 6$$

  So there are $40 + 1 + 1 + 1 = 43$ 4-chords up to transpositions.

In fact, we will see that the enumeration is facilitated by the use of action groups and combinatorial results due to Brunside and Polya, and does not require knowledge of SLT.

In the first half of the twentieth century, Olivier Messiaen introduced the concept of limited transposition sets [27], but he only used some of them in his own compositions. He called these sets *modes of limited transposition*. Some sets were already known before Messiaen presented his theory. The *whole tone* scale $\{0, 2, 4, 6, 8, 10\}_{12}$ was used by many composers. Claude Debussy wrote a prelude (*Voiles*) using only this mode. Messiaen's third mode $\{0, 1, 2, 4, 5, 6, 8, 9, 10\}_{12}$ is the enneaphonic scale that Scriabin uses in his *Piano Prelude, opus 74*. An analysis of the combination reveals that there is in the 12-Tone equal temperament, 16 classes of limited transpositions sets. A complete enumeration of limited transposition sets has been made for the quarter tone universe, [19].

4.2 Loquin-Costère Theory

If we limit ourselves to considering chords up to transpositions, major triads and minor triads belong to two different classes. This classification is well suited to the analysis of classical music, in which major and minor characters must be distinguished.

An application of one of Polya's theorems to the cyclic group allows us to construct the generating polynomial and prove the following result. Proofs of the enumeration results can be found in *A Compendium of Musical Mathematics* [21, p. 19–28].

Theorem 1 *Under the action of the cyclic group (up to transpositions), the number of scales of k degrees are the coefficients of x^k in :*

$$P_{C_N}(x) = \frac{1}{N}\sum_{d|N} \varphi\left(\frac{N}{d}\right)(1 + x^{N/d})^d$$

For example, for $N = 12$, the polynomial

$$P_{C_{12}}(x) = 1 + x + 6x^2 + 19x^3 + 43x^4 + 66x^5 + 80x^6$$
$$+66x^7 + 43x^8 + 19x^9 + 6x^{10} + x^{11} + x^{12}$$

says that there are 6 intervals, 19 triads, 43 tetrachords, 66 pentachords and 80 hexachords classes up to transpositions.

The number of modes of k degrees in the N-tet is

$$M_{N,k} = \binom{N-1}{k-1} = \frac{(N-1)!}{(N-k)!(k-1)!}$$

and the total number of modes is given by the formula:

$$M_N = \sum_{k=1}^{N} M_{N,k} = \sum_{k=0}^{N-1} \binom{N-1}{k} = 2^{N-1}$$

4.3 Hanson-Forte Theory

If we consider the classification of chords up to transposition and inversions, in which major triads and minor triads belong to the same class, the application of a theorem of Polya to the dihedral group leads to the following result:

Theorem 2 *Under the action of the dihedral group $\mathcal{D}_N$, (up to tranpositions and inversions), the number of k-chords are the coefficients of x^k in :*

$$P_{\mathcal{D}_N}(x) = \begin{cases} \frac{1}{2}P_{C_N}(x) + \frac{1}{2}(1+x)(1+x^2)^{(N-1)/2} & \textit{if } N \textit{ is odd} \\ \frac{1}{2}P_{C_N}(x) + \frac{1}{4}((1+x)^2(1+x^2)^{(N-2)/2} + (1+x^2)^{N/2}) & \textit{if } N \textit{ is even} \end{cases}$$

For exemple, if $N = 12$, the polynomial is

$$P_{\mathcal{D}_{12}}(z) = 1 + z + 6z^2 + 12z^3 + 29z^4 + 38z^5 + 50z^6$$
$$+38z^7 + 29z^8 + 12z^9 + 6z^{10} + z^{11} + z^{12}$$

This polynomial says that they are 6 intervals, 12 triads, 29 tetrachords, 38 pentachords, 50 hexachords classes.

4.4 Sets of Limited Transposition

With these results, it is possible to establish the complete enumeration of limited transposition sets in an N-tone universe. We prove the following result:

Theorem 3 *The number of sets of limited transposition is given by*

$$L_N = P_N + \frac{2}{N}(K_N - M_N)$$

where P_N is the total number of collections

$$P_N = \frac{1}{N}\sum_{d|N}\varphi(\frac{N}{d})$$

M_N is the number of modes $M_N = 2^{N-1}$, and K_N is given by

$$K_N = \frac{1}{2}\sum_{k=1}^{r}(-1)^{k+1}\sum_{i_1=1}^{r-k+1}\sum_{i_2=i_1+1}^{r-k+2}\cdots\sum_{i_k=i_{k-1}+1}^{r} 2^{\frac{N}{p_{i_1}p_{i_2}\cdots p_{i_k}}}$$

where N is decomposed in prime factors $N = p_1^{k_1}p_2^{k_2}\ldots p_r^{k_r}$, with $p_i > 1$, $k_i > 0$, $(i = 1,\ldots,r)$ et $r > 0$.

In particular, for the usual microtonal universes:

Ton	*N*	#SLT	#Chords Classes (up to transpositions)
1/3	18	69	14601
1/4	24	381	699251
1/5	30	2300	35792567
1/6	36	14939	1908881899
1/8	48	703331	5864062367251
1/12	72	1909580799	65588423374144427519
1/16	96	5864196582931	825293359523589782053586451

The number of classes of limited transposition sets is growing exponentially. For the 96-edo (equal division of an octave), aka space of sixteenth tones, which is now common practice (the Sauter firm having manufactured a sixteenth tone piano whose 96 keys constitute a single octave), harmonic theory seems beyond human reach, just like composition which requires the use of a computer.

4.5 New Diatonicity

When we go from a tempered universe of 12 tones to a universe of N tones, the question arises as to what happens to scales and chords, sharps and flats. In our current system, the seven white keys of the piano are representative of the diatonic scale. The black keys, sharps or flats, facilitate transpositions of the diatonic scale to different pitches while maintaining the same interval structure. In other words, the question is what will be the white and black keys in an N-tone universe? How are they distributed between the white and black keys?

A simple way to consider this problem is to go from the classical space of 12 tones to the space of quarter tones, and to extend the tetrachordal structures from one universe to the other. This is what was done by Ivan Wyschnegradsky in the preface to the *24 preludes for piano, opus 22*. He defines a scale that he calls the "13-tone diatonicized chromatism". By numbering the chromatic tones of the quarter tones $0 = C$, $1 = C + 1/4$ tone, $2 = C + 1/2$ tone, etc. this scale is written: (The index 24 indicates that we are in a universe of 24 tones per octave.)

$$\{0, 1, 3, 5, 7, 9, 11, 12, 14, 16, 18, 20, 22\}_{24}$$

Another way to look at the problem is to start with a singular property of diatonic scales. When we move from the C major scale to the G major scale, we see that a single note changes, *F* becomes *F sharp*. This property is repeated when we move from one scale to another by transposition to the fifth (or fourth). By looking for scales that have this property in microtonal universes, we define scales that were invented by Alain Louvier [25]. He calls them échelles à transposition progressive [*Progressive transposition scales*]. For example, the scale with the structure 4,334,343 (i.e. in which the notes are spaced 4 quarter tones, then 3 quarter tones, then 3 quarter tones,

Fig. 3 Sets of progressive transposition by Alain Louvier

etc.) corresponds to the scale of $F_{24} = \{10, 14, 17, 20, 0, 3, 7\}_{24}$. By transposing this scale 7 quarter tones higher (by adding 7 to the elements of F_{24}) we obtain a new scale $Ab^{-}_{24} = \{17, 21, 0, 3, 7, 10, 14\}_{24}$ which differs from the previous one by the note 21 which replaces the note 20. This can be seen in Fig. 3.

Other models of diatonicity use the notion of a well-formed scale and maximally evenness. Let us observe a simple property. The elements of the scales are obtained by adding the same number v (the same transposition). Starting from 20 and adding $v = 7$ modulo 24, we construct the scale $\{0, 3, 4, 7, 10, 11, 14, 17, 20, 21\}_{24}$:

$$20 - 3 - 10 - 17 - 0 - 7 - 14 - 21 - 4 - 11$$

Similarly, the classical diatonic scale of C $\{0, 2, 4, 5, 7, 9, 11\}_{12}$ is obtained by starting from 5 and adding v=7 modulo 12:

$$5 - 0 - 7 - 2 - 9 - 4 - 11$$

Comparing the distribution of sharps and flats, or, thinking of the keyboard of white and black keys, we see that in the 12-note system as in the 24-note system, there are never three consecutive white keys and never two consecutive black keys. Which puts us on the trail of a definition of the diatonic scale. Several solutions are possible and the problem is reduced to the choice of the transposition index m. We must therefore give ourselves sufficient criteria so that the diatonic scale exists for all values of N. From this diatonic scale, it is possible to construct (if it exists) one (or more) major scales and its relative minor. To do this, we must generalize the notion of tritone, which is the heart of tonal theory. The simple case, which we will assume here, is to take N even. In this case, the generalized tritone (which is not always a space of 3 tones) is the pair $\{0, N/2\}$ and its transpositions. It is in fact the only diad with limited transposition. In this model, when N is odd, the major scale does not always exist (for instance for $N = 15, 21$ etc).

Let us assume here that m is given and that N is even. Then, it is possible to construct a diatonic scale that coincides with the major scale:

$$\text{Dia}_N = \text{Maj}_N = \{0, m, 2m, 3m, 4m, \ldots, (N/2-1)m, N-m\}_N$$

It is presented according to the following diagram in which the notes are linked horizontally by transposition index m. For $N = 12$, $m = 5$, the tritone is the pair $\{1, 7\}$. We then have a representation of the $A\flat$ major scale. For the quarter tones, we must take $N = 24$, $m = 11$, and the tritone is the pair $\{1, 13\}$. We then have a major scale of 13 tones, which corresponds to Wyschnegradsky's "Position G" of the *24 Preludes*.

$$\begin{array}{llllllll} 0 \,— & N-m & & & & & & \\ & \quad | & & & & & & \\ & (N/2-1)m & — \cdots — & 4m & — & 3m & — & 2m \; — \; m \end{array}$$

From this major scale, we construct the *relative minor scale* by replacing the note $3m$ with the note $2m + N/2$ forming a generalized tritone with the note $2m$.

$$\text{Min}_N = \{0, m, 2m + N/2, 4m, 5m, \ldots, (N/2-1)m, N-m\}$$

The relative minor scale is thus formed of two tritones according to the following diagram. For $N = 12$, $m = 7$, we obtain the F minor scale.

$$\begin{array}{llllll} 0 \,— & N-m & & & 2m + N/2 & \\ & \quad | & & & \quad | & \\ & (N/2-1)m & — \cdots — & 4m \; — & 2m & — \; m \end{array}$$

The case where N is odd is more difficult, because the major character then results from the existence of a set of limited transposition playing the role of a generalized tritone which has more than two notes (see [21]).

4.6 Homometric Sets

Homometry is a property that appeared in the middle of the twentieth century in crystallography. It was transposed to the musical domain in the following way. Given a chord, or rather a pitch class set (pcset for short), that is to say a chord up to transposition and inversion, its Interval class content is a list of every possible interval occurring in this pitch-class set. Two pcsets A and B of the same cardinal are

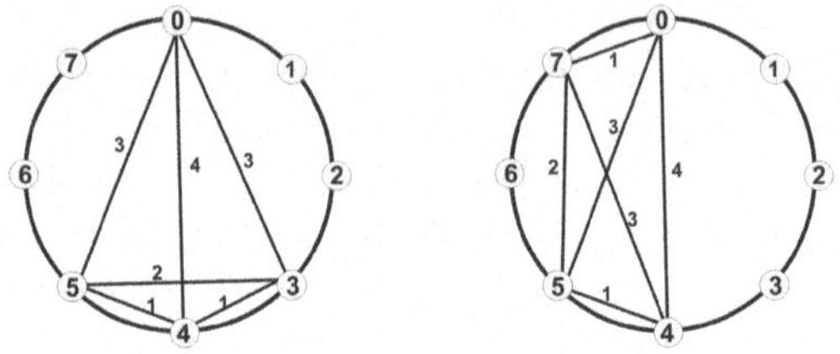

Fig. 4 Homometric sets in 8-edo

Z-related or homometric in the N-edo if they have the same interval class content.

$$(A\ \mathcal{Z}_N\ B) \Leftrightarrow ic(A) = ic(B)$$

One of the first known homometric sets was found by physicist Lindo Patterson in 1944 (see Fig. 4). He showed that the following sets are homometric.

$$\{0, 3, 4, 5\}\ \mathcal{Z}_8\ \{0, 4, 5, 7\}$$

Since then, we have known all the homometric sets of the 12-note equal temperament. For example, the sets noted 6Z24 in the Forte classification and its complement set class 6Z46 are homometric

$$(0, 1,\ 3, 4, 6, 8)\ \mathcal{Z}_{12}\ (0, 1, 2, 4, 6, 9)$$

Moreover, all sets and their complement are homometric. In the classical system of 12 tones, Allen Forte has shown that there is only one homometric pair of tetrachords $\{0, 1, 4, 6\}$ and $\{0, 1, 3, 7\}$, three pairs of pentachrods {0, 1, 2, 4, 7} and {0, 1, 3, 5, 6}; {0, 1, 2, 5, 8} and {0, 1, 3, 8, 9}; {0, 1, 2, 5, 9} and {0, 1, 3, 4, 8}; and 15 pairs of hexachords, such as 6-Z29 {0, 1, 3, 6, 7, 10} and 6Z-50 {0, 1, 3, 6, 8, 9}.

Composer Tom Johnson has intensively studied Z-related sets applied to rhythms. He tackled homometries in 2013 and since then has continued to deal with it. *Intervals, for flexible instrumentation* (2013) connects in different ways the 48 transforms of homometric sets {0,1,3,7} and {0,1,4,6}. The *Slight Variations* (2015) for piano, composed two years later, are built on the Z-related hexachords {0,1,3,4,5,7} and {0,2,3,4,5,8}. More recent is the piece composed for two wood-blocks *Homometry 108* (2019), based on 108 homometric chords and three permutations.

For the first values of N, homometric sets go in pairs. But for $N \geq 16$, as David Lewin noticed, multiplets appear, except for N prime. In his article, *On extended Z-Triples* [24], he gave the three triples of Z-associated hexachords mod 16. In the quarter-tone system, homometric sets are very rich. The 24-edo has a homometric triplet of pentachords:

$$\{0, 1, 4, 8, 13\}_{24}, \{0, 1, 4, 13, 20\}_{24}, \{0, 1, 5, 9, 12\}_{24}$$

one sextuplet of 7 notes, one octuplet of 9 notes, and three very strange dodecuplets of 12 notes. This data was used by the French composer Stéphane de Gérando, with whom I worked and whose collaboration led to the writing of *Homometric Attractors 2 for main computer in real time, instrumental ensemble and two dancers*, based on all homometric quarter-tone sets. It was premiered in Paris on December 16, 2014.

5 New Compositions

Serialism and microintervality have always had a tumultuous relationship. The twelve-tone and serial theories developed at the beginning of the twentieth century by Arnold Schönberg were poorly adapted to microtonal universes. While in classical space, a series requires memorizing 12 notes, in a third-tone space it requires 18 notes and in a quarter-tone space, 24 notes. Yet composers such as Henri Pousseur, Karlheinz Stockhausen, or more fundamentally, Alain Bancquart, Jean-Étienne Marie, Ben Johnston, Bogusław Schaeffer, have written new compositional theories that intersect microtones and series. While Pousseur's *Bulle de Temps* (1995) uses a series of 17 tones, Stockhausen's *Himmelfahrt* a series of 24 tones, and Jean-Étienne Marie's *Ecce Ancilla Domina* a series of 24 tones, Alain Bancquart has turned to theorizing incomplete series, using, for example, series of 13 tones in quarter-tonal universes. Details can be found in [20, p. 230–241].

For several decades, the number of microtonal compositions has continued to grow. Today, all aesthetic trends are represented. From the historical microtonal avant-garde, with composers such as the Norwegian guitarist Bjørn Fongaard or the French composer Alain Louvier, works that explore the space in 31-edo, vigorously defended by the keyboardist Ere Livonen and the Scale ensemble, creating pieces such as *Solar Neon* (2010) by Ned McGowan, or *Zercheinen* (2010) by Steffen Krebber, to composers of spectral music, such as Tristan Murail, Gérard Grisey, who work on microtonal approximations of spectra to serve as material for the composition. Microtones are not only ornaments, but become real constituent elements of timbral research. Here we must acknowledge the theoretical and compositional work of George Friedrich Haas [12, 13], whose piece *In Vain* (2000) for orchestra will remain emblematic of microtonal music.

It is undoubtedly this question of timbre that is at the root of the attraction of certain composers towards the Balinese gamelan and Javanese scales such as the pelog and the slendro, in which these composers found the serenity of just intonation. Lou Harrisson is undoubtedly the most famous composer who, with his friend William Colvig, studied with Jody Diamond and Wasitodipuro, director of the court gamelan of Jogjakarta. They both worked on the creation of the first American gamelan. They built many instruments inspired by the Indonesian tradition and influenced composers such as Bill Alves, Douglas Leedy, Greg Schiemer. David Doty composed for Other Music's American gamelan, an association he had founded. Daniel Schmidt founded the Berkeley Gamelan, and Wayne Vitale founded the Bay Area-based Gamelan Sekar Jaya ensemble. Gamelan had become for some the place to experiment with just intonation.

At the same time, electroacoustic music trends mixed with minimalist trends and those of microtonal drone music. Some microtonalists were influenced by Morton Feldman, Alvin Lucier and Pauline Oliveros. An influence that was not limited to classical music composers, but also reached electronic music, house, dub and ambient music. Emblematic of this trend is the English composer Sevish who self-produced more than thirty albums, declining micro-intervals in all varieties of equal tuning, but also Bohlen-Pierce tuning, based on the tritave.

Composers like David Hykes or Michael Harrison were influenced by the minimalism of La Monte Young and Marian Zazeela. Others like Dean Drummond followed in the footsteps of Harry Partch, others were influenced by the works of James Tenney or Marc Sabat, or the unjustly lesser-known works of Gloria Coates, who composed symphonies and string quartets in abundance.

In rock music and related trends, fretless guitars and microtonal guitars play a leading role. They were introduced by musicians like Jon Catler, who founded the band *The Microtones* with his brother Brad in the 1980s. At that time, the Japanese band *Syzygys* used an organ inspired by Partch's chromolodeon based on a 43-tone scale. Since then, experimental rock bands like the Australian band *King Gizzard & The Lizard Wizard* from Melbourne produced an album *Flying Microtonal Banana* (2017) using microtonal guitars, as did the band *The Mercury Tree* from Portland, Oregon in their album *Spidermilk* (2020). Psychedelic or experimental metal bands have taken over like the band *Asteroidi Esadecafonici* founded by Fabrizio Fulvio Fausto Fiale and Fiorenzo Vicari, or like *Cryptic Ruse* founded by Jason Yerger.

Among jazzmen, the first attempts at using micro-intervals date back to the 1960s. Don Ellis had a quarter-tone trumpet built by the Frank Holton Company in 1965, which he first used at the Monterey Jazz Festival on September 20, 1966. The trumpet section plays quarter-tones on the recording of the *New Don Ellis Band Goes Underground*. In 1975 he published a textbook of quarter-tone composition (*Quarter Tones. A text with musical examples, exercises and etudes*).

The greatest microtonalist jazzman is undoubtedly Joe Maneri. Professor of music at the New England Conservatory of Music, he spent his life working in the world of twelfth of a tone. He founded a quartet with his son and violinist Mat Maneri, Randy Peterson drums and John Lockwood bass that recorded many microtonal albums. He is also the founder of the Boston Microtonal Society.

Steve Coleman made several microtonal attempts that can be found on his albums *The Ascension to Light* (2001) or *Lucidarium* (2003) but Coleman did not develop a microtonal theory, unlike Hafez Modirzadeh who developed the concept of *chromodality*. In 1992, he defended a thesis *Chromodality and the cross-cultural exchange of musical structure* at Wesleyan University. He has recorded with the best jazz musicians: with trumpeter Amir El-Saffar and pianist Vijay Iyer in *Post-Chromodal Out!* (2012), with three eminent pianists Craig Taborn, Kris Davis and Tyshawn Sorey on *Facets* (2021).

Claiming microtonal expressionism, the German saxophonist Philipp Gerschlauer met Johnny Reinhard who explained to him his theory of 8th octave overtone tuning. This meeting was decisive for his microtonal orientation. In 2015, at Berklee College he met David Fiuczynski with whom he recorded *Mikrojazz. Neue Expressionistische Musik* (2017).

In free jazz and improvised music, Jon Rose and Veryan Weston have been interested in the question of temperaments and have produced several albums where they pose the question of the search for the Pythagorean comma.

On the other side of the Atlantic, Austrian keyboardist Georg Vogel has built several instruments including the *M-Claviton*. He founded *Dsilton*, a band dedicated to music in 31-edo.

Among recent creations of contemporary classical music, a marked interest in microtonal pianos, including the new use of Carillo pianos, appears. Kyle Gann's piece *Hyperchromatica* composed between 2012 and 2021 is written for three retuned, computer-driven pianos, Taylor Brook's *Virtutes Occultae* (2021) uses six retuned pianos, and Georg Friedrich Haas's older limited approximations is written for 6 micro-tonally tuned pianos and orchestra (2010). Interest in instrument making is still very much alive. Let us mention Bastien David's *metallophone* and his piece written especially for this instrument *Les Métamorphoses* (2022).

To conclude this necessarily incomplete panorama of microtonal music, which can be completed by consulting [18, 22], we must acknowledge the work of bassoonist Johnny Reinhard, who has organized hundreds of concerts, played and created a large number of works and who has just set up an extraordinary research and discussion center, the Virtual Microtonal University.

References

1. P. Aaron, *Toscanello in musica*, 1st editvspace@ion 1523 (Marchio Sessa, Venice, 1539)
2. A. Avraamov, « Jenseits von Temperierung und Tonalität ». Melos **I**, 131–134, 160–166, 184–188 (1920)
3. A. Bancquart, Musique: imaginaire, combinatoire. Essai sur la composition en micro-intervalles. *Les Cahiers du CIREM*, vols. 28–29 (Presses Universitaires de Tours, Tours, 1994), pp. 9–64
4. P. Boulez, Propositions, in *Polyphonie*, vol. 2, (Richard Masse éditeur, Paris, 1948), pp. 65–72
5. F. Busoni, Bericht über Dritteltöne [Communication on third-tones]. Melos **3**, 198–199 (1922)
6. R. Caddeo, X. Hascher, F. Jedrzejewski, A. Papadopoulos, *Christiaan Huygens, Écrits sur la musique et le son* (Éditions Hermann, Paris, 2021)
7. E. Costère, *Lois et styles des harmonies musicales* (Presses Universitaires de France, Paris, 1954)
8. L. Euler, *Tentamen Novae Theoriae Musicae Ex Certissimis Harmoniae Principiis Dilucide Expositae* (ex typographia Academiae Scientiarum, Saint Petersburg, 1739)
9. A. Fokker, *Neue Musik mit 31 Tönen*, Düsseldorf, Verlag der Gesellschaft zur Förderung der systematischen Musikwissenschaft (1966)
10. A. Fokker, *New Music with 31 Notes*. Series of Monographs on Basic Questions in Music, vol. 5 (Verlag für Systematische Musikwissenschaft, 1975). English Translation by Leigh Gerdine, Bonn-Bad Godesberg, Orpheus
11. A. Forte, *The Structure of Atonal Music* (Yale University Press, London, 1973)
12. G.-F. Haas, Fünf Thesen zur Mikrotonalität. Positionen **48**, 42–44 (2001)

13. G.F. Haas, Mikrotonalitäten. *Musik der anderen Tradition: Mikrotonale Tonwelten*, ed. by H.-K. Metzger, R. Riehn, Musik-Konzepte, edn. (Text+Kritik, München, 2003), pp. 59–65
14. A. Hába, *Neue Harmonielehre des diatonischen, chromatischen, viertel, drittel, sechstel und zwölfteltonsystems* (Fr. Kistner & C.F.W. Siegel, Leipzig, 1927)
15. H. Hanson, *Harmonic Materials of Modern Music* (Irvington Pub., Irvington, 1960)
16. R. Hausmann, « Optophonétique », *Revue Ma*, 1922, Repris dans *Sensorialité excentrique – Eccentric Sensoriality* (Les Presses du réel, Dijon, 2002)
17. C.E. Ives, Some 'Quarter-Tone' Impressions. *Pro Musica Quarterly*, vol. 6 (Franco-American Musical Society, 1925), pp. 24–33. Reprint in *Essays Before a Sonata and Other Writings*, Howard Boatwright (ed.), W.W. Norton, New York, 1962
18. F. Jedrzejewski, *Dictionnaire des musiques microtonales* (L'Harmattan, Paris, 2014)
19. F. Jedrzejewski, *Hétérotopies musicales. Modèles mathématiques de la musique* (Hermann, Paris, 2020)
20. F. Jedrzejewski, *La musique dodécaphonique et sérielle: une nouvelle histoire* (Brepols, Turnhout, 2021)
21. F. Jedrzejewski, *A Compendium of Musical Mathematics* (World Scientific, Singapore, 2024)
22. F. Jedrzejewski, *The Palgrave Dictionary of Microtonal Music* (Springer-Macmillan, 2026, forthcoming)
23. N. Kulbin, Die freie Musik. *Der Blaue Reiter* (Berlin, 1911). Self-published, Moscow, 1915. Reprint with additions, (Klaus Mankheit Verlag, München, 1984), pp. 125–131
24. D. Lewin, On extended Z-triples. Theory Pract. **7**, 38–39 (1981)
25. A. Louvier, Recherche et classification des modes dans les tempéraments égaux. Musurgia **IV**(3), 119–131 (1997)
26. J. Mager, *Vierteltonmusik* (Franz Kuthal, Aschaffenburg, 1915)
27. O. Messiaen, *The Technique of my Musical Language* (Leduc, Paris, 1944)
28. W. Möllendorf, Musik mit Vierteltönen (F.E.C. Leuckart, Leipzig, 1917)
29. T. Narushima, *Microtonality and the Tuning Systems of Erv Wilson* (Routledge, Milton Park, 2018)
30. H. Partch, *Genesis of a Music. Monophony: the Relation of its Music to Historic and Contemporary Trends; its Philosophy, Concepts, and Principles; its Relation to Historic and Proposed Intonations; and its Application to Musical Instruments* (University of Wisconsin Press, Madison, 1949). Second enlarged edition *Genesis of a Music: An Account of a Creative Work, its Roots and its Fulfillments*. Da Capo Press, New York, 1974
31. I. Wyschnegradsky, *La loi de la pansonorité (Manuscript, 1953)* (Contrechamps, Geneva, 1996). Preface de Pascale Criton, Édité par Franck Jedrzejewski

From Bird Feather to Digits: Some Thoughts About Olivier Messiaen

Claude Ledoux

Mathematics/Music: a complex binomial. Indeed, the second term cannot exist without the first, as the world of sound is subject to certain numerical laws that enable numerous logical operations. But what about reciprocity? To what extent could mathematics be subject to the laws of music? Ultimately, could the problem not be posed differently? Our culture postulates the separation of emotions (which would generate the Arts) and cognition (which would generate the Sciences). Albert Moukheiber, a brain specialist, criticizes this position: [...] *these separations between emotions and reason, between scientific and [artistic] subjects, between creativity and analysis, are so deeply rooted in our language and our culture that getting rid of them is probably going to be difficult.*[1] The author deduces from his own research that emotions and reasoning are highly correlated mental states. *One cannot exist without the other. All functions are involved at the same time, for everything, but modulated differently.*

Mathematics and music are part of the same desire to represent the world, or a world beyond through some abstract projection. Poetry is always present. To express this, it is revealing to recall those great scientists who were also masters of their musical instruments. Think of violinist Albert Einstein, pianists Max Planck and Werner Heisenberg, or, closer in time, Bertrand Maury, a remarkable mathematician and keyboard virtuoso. It's also worth remembering, from the other shore, how composers combine their sense of intuition with the use of quantitative

[1] Albert Moukheiber, *Neuromania*, Alary Éditions, 2024.

C. Ledoux (✉)
ARTS2 Mons (Royal Conservatory of Music), Brussels, Belgium

CNSM, Paris, France

Royal Academy of Belgium, Fine Arts class, Brussels, Belgium

© The Author(s), under exclusive license to Springer Nature Switzerland AG 2026
N. D. Alikakos, C. Villani (eds.), *Analysis Techniques for Mathematical Physics and Geometry – and Music*, Lecture Notes in Mathematics 2390,
https://doi.org/10.1007/978-3-032-11396-2_9

and qualitative logic to elaborate their musical discourse. The musical analysis that follows will unveil a few aspects of research on the subject, taking as an example the case of Olivier Messiaen, a twentieth-century French composer who constantly wandered between more or less complex numerical speculations and the tenderness of caressing sonorities for the most common of listeners, when he wasn't focusing on birdsong, those little intermediaries between man and God.

Let's turn to Messiaen. His musical work is a paradox. The music often speaks for itself, and doesn't necessarily require in-depth knowledge of contemporary art to be appreciated by many music lovers. Could there be some symbolic clue linking the compositional act of this great composer to our personal, intimate apprehension of music, or even of the world's sound experience? It's an intriguing question, and Messiaen's theoretical writings are far from providing an answer. More than a genuine analytical exploration, his treatises tell us about the master's extraordinary erudition and reveal a few technical procedures, but tell us little about the tangible nature underlying the composing reality at work in his music. As such, these lines will modestly attempt to lift a small corner of the veil, as others have done previously, to offer the elements of a very rough sketch of Messiaen's thinking based on a few technical rudiments, a few examinations of materials that will allow us to imagine the motivations that enabled Messiaen to initiate one or other "musical project".

And to open this session, we give the floor to one of the most famous students of the generation of composers who attended Olivier Messiaen's post-war classes. In the early 1950s, Stockhausen was introduced to the great Parisian teacher through the young Belgian composer Karel Goeyvaerts,[2] who had already introduced his German counterpart to the French artist's most contemporary scores. In a lecture given in Brussels in 1965, Stockhausen, who had stepped in at the last minute to replace an indisposed Messiaen, evoked some memories of this prolific period:

> *As I stand in for Olivier Messiaen this evening, I'm very happy to start by offering you a few ideas about his work. I remember the year 1951, when I arrived in Paris. Messiaen was lecturing on Hindu rhythms, on the rhythmic models of Plain-chant. He was analyzing Mozart's concertos, examining his exploration on rhythmic cells. At the same time, I was invited by Pierre Boulez, who was working in the musique concrète studios in Paris, to work there myself.*
>
> *Messiaen achieved a certain synthesis between Webern's research, historical sources—even from other traditions—and his own French tradition. We have seen that the different paths that led to Messiaen were not so different, whether they came from the Debussy school or the Viennese school. So there was a basic thought that proposed—after a secular language that had concentrated on musical characters, themes and motifs—to create works from what Messiaen called "cells", from neumes that were not at first characterized by behavior in a form, but only—as we used to say at the time—by parameters (durations, pitches, intensities, timbres). In a small piano piece,* Mode de valeurs et d'intensités, *which has since become famous, Messiaen reached a point where all musical figures and*

[2] Delaere, Marc (2016). The Stockhausen-Goeyvaerts Correspondence and the Aesthetic Foundations of Serialism in the Early 1950s. In The Musical Legacy of Karlheinz Stockhausen: Looking Back and Forward, edited by M.J. Grant and Imke Misch, 20–34. Hofheim: Wolke Verlag.

> *characters disappeared. All that remained was a structure. He had worked with one mode for pitches, another for durations and a third for attacks, to create a fabric—if I can call it that—of music; a fabric of sound. It was a fundamental change in musical language. The music suddenly sounds very decorative. It has no definite beginning or end. You get the impression that the music has already begun and could go on forever. All of a sudden, we had arrived at sound alone, i.e. "punctual music", because we were concentrating only on points. There were no more figures, no more "Gestalten", no more musical characters, no more musical motives as a motivation for something.*[3]

Interesting as it is, this intervention tells us as much about the author as it does about Messiaen himself. Although the concept of *Punkte* now belongs to Stockhausen's exclusive verbal prerogative, the fact remains that this *Mode de valeurs et d'intensités*—this "little thing" composed in 1949 as an operative exercise in complex musical principles, and whose success eluded its author's comprehension as this piece remained a meteor in his musical output—had to strike many avant-garde minds of the time. At the risk of focusing exclusively on the "letter", to the detriment of the "spirit". As a result, the work was viewed solely in terms of its material and structure, without regard for the concept of "musical project", an important notion in the eyes of composers.

How, moreover, are we to reconcile *a-priori* antinomic propositions, i.e. the objectivity of structuralist thought and the subjectivity of analytic interpretation? The exercise is frightening, but could prove fascinating in more ways than one. And to return to Stockhausen to tell us about his first appreciation of the piano piece in question: "music from the stars".[4] A totally subjective cry from the heart from the one who, like his French mentor, was to associate music and theological thought in the history of twentieth-century music.[5] This is perhaps an important point of reference for understanding Messiaen's motivation in writing such a work. But before jumping to a hasty interpretation, perhaps it would be worthwhile to engage in a critical reading of the short excerpt from Stockhausen's lecture.

Starting with the young German aficionado's plural use of the notion of mode, an incongruity given the title of the composition in question: "Mode...", in the singular! The young composer's intention is entirely understandable in view of the emphasis placed on the treatment of the multiple parameters of musical sound—in the present context: pitch, duration, intensity, timbre and space—by the proponents of generalized serialism of which he was a part. The confusion between "modes" and "series" seems to be active, so much so that Messiaen's work has often been described as the first attempt at generalized serial composition. However,

---

[3] Karlheinz Stockhausen's lecture at the Institut de Sociologie, Brussels, March 4, 1965. RTBF sound archives. Transcribed by the author of this article.

[4] Richard TOOP, "Messiaen/ Goeyvaerts, Fano/ Stockhausen, Boulez", in *Pespectives of New Music*, Vol. 13, No 1. (1974), pp. 141–199.

[5] Let us not forget how Stockhausen associated technological creation with spirituality. Thus, from his first electroacoustic attempts, the composer imagined composing an *Electronic Mass* for Köln Cathedral, a project immediately refused by the rather conservative religious institutions. The project, however, would later take the form of an "electronic motet", the *Gesang der Jünglinge* premiered in Köln on December 15, 1956.

this terminology misapplies here, because in Stockhausen's sense, parameters can be ordered and manipulated independently of one another, creating results that stimulate the composer's imagination both structurally and perceptually. The Frenchman's point of view is quite different: no independence of parameters! On closer examination, the score takes us on a journey that could be defined as a subtle exploration of a reservoir of thirty-six frozen "musical objects", each with its own identity (defined by a specific pitch, duration, intensity and articulation), a profile[6] determined once and for all, which will never change in the course of the work. Messiaen, in his introduction to his publication, does not hesitate to mention this "mode" divided into three sequences spread over the score's three staves:

Is there any point in differentiating the concept of "series" from that of "mode"? To answer this question, let's remember that Messiaen comes from a lineage of composers passionate about modal issues. What's more, as Stockhausen mentioned, the composer had broadened his horizons by studying non-European music. India, so appreciated by the Parisian composer and professor, could reveal more of the subtlety of the modal fact and make us aware of the impact of the concept.[7] As Robin Cooper pointed out in an article from the 1960s,[8] modality is not immediately apparent in certain ragas. It is only as the form meanders and insists that the qualities of a specific modal orientation gradually reveal themselves to the listener. In fact, this modality is not intended to be scalar (build on abstract scales), but formular (defined by melodic formulas), in such a way that it can only be fully understood by examining the distribution of its components in relation to one another. The work thus acts as an organism in (trans)formation, refusing any framework for any kind of development, but responding to the needs of a genetic code of "mode", thus becoming a receptacle for a revelation of the intrinsic qualities of the initial material, renewing the temporal conception of music, since this genetic code thus incriminated comes from itself and generates itself ad infinitum. No beginning, no real end, as Stockhausen said of Messiaen's work. In order to grasp its musical reality, it remains to analyze the organization of this distribution of punctual elements, a sound constellation conceived in the image of the myriad stars that any human being can observe on a dark night, a multiplicity of points of varying brilliance whose logical organization escapes our perception. However, over time, prolonged observation of this cluster reveals one or another pattern, linking certain points according to different luminosity weightings (aren't these lines the origin of the common cartography of the sky?).

[6] As Boulez defined it in his analysis of the score in « *Le temps musical*: Pierre Boulez présente et dirige Boulez, G. Ligeti, O. Messiaen, E. Carter—Introduction à une histoire du temps musical de Guillaume de Machaut à nos jours », cassettes Radio France/Ircam, K5002, Paris, 1988.

[7] Messiaen was largely inspired by the *Encyclopedia of Music, Dictionnaire du Conservatoire* by Albert Lavignac, published in 1921, more particularly from Volume I, p. 257–376, including the article "India" by J. Grosset.

[8] Robin COOPER, Propositions pour un modèle transformationnel de description musicale, in *Musique en jeu*, no. 28, 1967, pp. 70–87.

The score proves difficult to read for many musicians, especially pianists. This complexity lies in a singular topography that betrays Messiaen the organist: three-voice writing reminiscent of the historical layout of a *trio sonata*, characterized by a translation of two-manual and pedal writing (and its corollary, the partial loss of the vertical spatial reference of the bass/treble relationship) into the world of the piano. It also reveals Messiaen's knowledge of medieval music, of polyphonic writing with three voices, each acting in a different, proportional temporality. An examination of the durations (Example 1) reveals a 1:2:4 relationship between divisions I, II (double time expansion) and III (quadruple time expansion) of the mode, updating the research of the *Ars Mensurabilis* of the thirteenth century. Of course, this is hardly perceptible in the first few bars. To the average listener, they express chaos rather than any comprehensible organization. No coherent logic seems to explain the distribution of the musical objects defined above in the musical space. The punctual phenomenon seems to act at the level of the sound particle, locally insignificant to the listener, but nevertheless inscribed in a universe of musical events calling for a global listening—to paraphrase Xenakis—of clouds in a galaxy of sounds. This last allusion is poetic, and already calls for a desire to transcend our materiality—an understandable aspiration for that post-war time, with its desire to radically distance ourselves from those immanent expressions (sentimentality, nationalism, narcissism, apology for death, post-romantic excess, etc.) that led mankind towards the tragedies of the twentieth century, replacing them with an impulse towards spirituality and its corollary, abstraction.[9] The music thus created eludes us, fleeing our traditional sensory apprehension to reveal the order of an elsewhere that we sense without understanding. Didn't the Middle Ages, to which Messiaen refers, make the same postulate? By gradually integrating this order into a dialectic between "perfect time" (*tempus perfectum*), infinite, elusive, transcendent and divine, and the time of our imperfect humanity (*tempus imperfectum*), directional, binary by our physical constitution, by our walking movement and unconscious of this end of time of a beyond that we could only represent symbolically (Example 2).

Listening to this *Mode de valeurs et d'intensités*, it's clear that this dialectic of "transcendent (beyond our understanding)/immanent (with reference to our senses)" takes shape as the discourse (trans)forms, This is evidenced by the presence of the work's lowest note, a vigorously accented C# in the fortissimo nuance, which punctuates the score three times, thus marking out a sound space which, while infinite, externalizes one of the limits of its ambitus, and thus expresses a very human index of finitude. But if this sonic space is delimited by an ostentatious low pitch/frequency, the same cannot be said of the high-frequencies domain, all the more so as the dynamics assigned to the mode's highest pitches are confined to an indescribable sound expression.

[9] Paul Klee, Personal diary 1915: "The more terrible this world turns out to be, the more art seeks to be abstract, while a happy world produces art turned towards here below"

Example 1 Mode defined by Messiaen in the introduction to the score. (© Éditions Durand)

Since reason is lost in conjecture at this stage, it would be a good idea to return to a more rigorous, structural analysis, as Stockhausen invited us to do in his lecture. To begin with, we need to take a close look at the mode itself, and try to understand its intrinsic potential. For this purpose, it would be useful to propose a linear calibration of the 36 musical identities, focusing more specifically on pitches and dynamics (with below a numerical order defined for the three divisions of the mode of Example 1).

This layout immediately reveals a number of properties. There are four implicit layers characterized in their register, four profiles capable of guiding perception in a specific case of proximity distribution (Example 3):

A. The first eight musical objects (eight pitches with their dynamic and duration) offer an extremely capricious profile in terms of dynamic and interval variability. However, it is interesting to note its strong chromatic emphasis, as well as the existence of structural tritones, themselves in semitone relationships (Eb-La/D-Lab/G-C#).
B. The next segment, made up of interpenetrations of the three divisions, contrasts its ten or so values with the previous profile through its quasi-linear *forte* dynamics (significant presence of *ff* and *mf)*, associated with a a diatonic curve revealing a traditional church mode. The pitch B natural could be added as the culmination of a structural diminuendo, adding a little ambiguity to the modal aspect.
C. Subsequent identities are again contrasted by the strong chromatization of their succession and the relevance of a low-amplitude dynamic. This segment results from the interpenetration of two divisions.
D. The last section of this mode is initiated by a continuation of the chromatic movement, this time in a strong nuance. This is followed by an extreme fall

Example 2 First page of *Mode de valeurs et d'intensités*. (© Éditions Durand)

> of objects in the low register of the instrument, in an extremely strong and accentuated nuance, in parallel with an arithmetical and important extension of the durations (could this be a rhetorical act, a variety of contemporary *catabasis*[10] to lead us to this capital C# of the score?)

This last segment is particularly interesting. It is made up of low pitches, with long durations, marked accentuation and often exaggerated dynamics. In a certain way, Messiaen imagined his "mode" as the sum of materials that can fit into a model close

[10] a rhetorical figure widely used in early music, a motif consisting of a series of notes taken in a descending movement, expressing a geographical shift or, figuratively speaking, manifesting a shift towards our human condition.

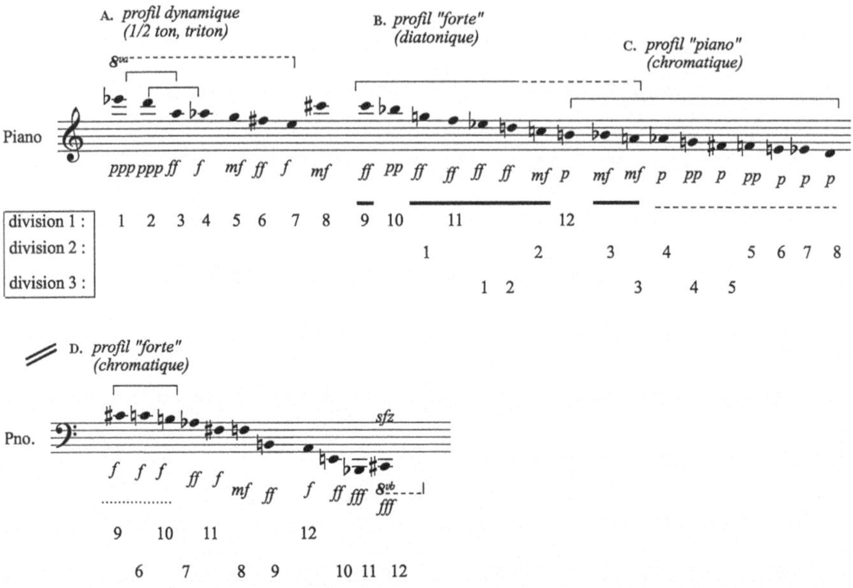

Example 3 Linear distribution of the complete mode

to natural resonance, such as that of a piano sound. Example 4 offers an acoustic analysis via a graphical representation (a sonogram) of the timbre of the lowest Bb of this instrument (the fundamental and the space encompassing the next 80 harmonics), along with a few frequency reference points delimiting the acoustic formants, which have been transcribed into musical notes.

This three-dimensional representation of the sound spectrum (time in seconds on the x-axis, frequencies on the y-axis, with intensity varying according to the degree of grayness of the frequency lines) shows that low frequencies (the fundamental and first harmonics) have the longest durations and the strongest intensities. The higher the harmonic ranks, the more temporally the partials diminish and the less powerful they become (although this loss of energy is not totally linear with the exploration of the high registers). This graphic representation makes clear the extent to which the phenomenon of resonance was at the source of the initial organization of Messiaen's material, i.e. the elaboration of his "mode". A number of writings have sketched out the problem of timbre in the construction of Messiaen's discourse, and in particular addressed the question of the "harmonie-timbre" (a sonic ambiguity between harmony and timbre) so dear to Debussy.[11] This idea of a vertical construction akin to natural resonance also seems to contribute to making

[11] Klaus SCHWEIZER, "Olivier Messiaens Klavieretude "Mode de valeurs et d'intensites", in *Archiv für Musikwissenschaft*, 30. Jahrg., H. 2. (1973), pp. 128–146. Julian ANDERSON, "Dans le contexte", in *Entretemps*, 8, 1989, pp. 13–23.

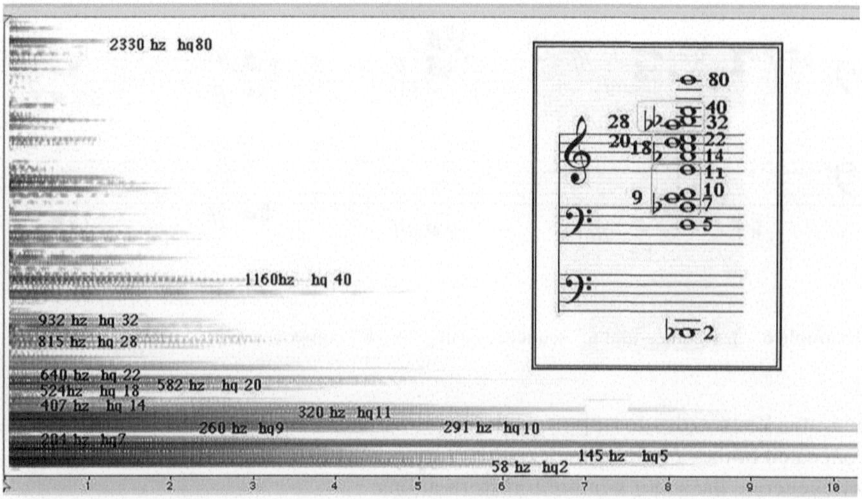

Example 4 Sonogram of a low Bb0 on a Yamaha C3 piano

Example 5 some occurrences of the low Bb and of the fragmented construction of a referential harmony-timbre

sense as you listen to this *Mode de valeurs et d'intensités.* In fact, it's enough to compare a few appearances of low notes other than the C# we've already discussed to grasp its importance. For example, listening to the occurrences of low Bb allows us to realize how polyphony seems to organize and progressively reveal a context of harmonics evoking a certain momentary stability, a transformation that stimulates our *a-priori* zones of understanding and knowledge, all the more so as it is intimately related to what resembles a crackling exploration of the components of a natural resonance model, as the following example demonstrates (Example 5).

If this notion of "harmonie-timbre" proves to be congruent, then it enables us to make sense of the last identities constituting this "mode". In view of the polyphonic constructions Messiaen gradually elaborates on the basis of his low notes, we can imagine that these were selected on the basis of a sequence of "harmonies-timbre" (Example 6, to which it would be interesting to add the corresponding durations

Example 6 "harmonies-timbre" sequences using "mode" components

and attacks in order to approach the acoustic model of reference), a proposition concealed in the "mode" which becomes more than ever a kind of genetic code that will generate the score's future transformations:

Chaos, ancient modality, "harmonies-timbre": these are all concepts that seem to have nothing to do with each other, except that they are potentially present in the "mode" examined in this paragraph. However, a simple technique, that of permutation, is used to bring them together. It's worth recalling that Messiaen uses three divisions of this "mode", each made up of 12 elements represented in this analytical framework by the numbers 1 to 12 (cf. Example 3). It is then up to the composer to *com-pose* using an often abstract numerical organization.

An example of conceptual permutation par excellence lies in starting with both extremities of the numerical series in order to approach the center of this order of twelve values:

1 - 2 - 3 - 4 - 5 - 6 - 7 - 8 - 9 - 10 - 11 - 12

1 – 12 – 2 – 11 – 3 – 10 – 4 - 9 – 5 – 8 – 6 – 7

This arrangement is rigorously applied in division I by the composer from bar 24 onwards, in order to bring in the low C#.

A related procedure (here in groups of three events) occurs from the last eighth note of bar 29 in division II, a sequence built on the resonance of low C#:

1 – 2 – 3 / 12 – 11 – 10 / 5 – 4 – 6 / 9 – 8 – 7.

A final, similar example affects division III as early as bar 61:

1 –	7 –
2 –	8 –
3 –	9 –
4 –	10 –
5 –	11 –
6 –	Bringing 12 (low C#)

These few arrangements demonstrate the meticulous application of a conceptual procedure that doesn't overly affect hearing, but which, when strategically located in the score, generates a chaotic halo by way of premonition or environment of a crucial aural landmark. This type of numerical manipulation is reminiscent of magic squares and other mathematical games (later over-exploited by the adepts of the generalized series, led by Boulez), which enabled their authors to create an analogy with a hidden order in the world, a disposition of spiritual essence, a complexity which, for a theological mind, would reflect the arcana of a universe created by a divine entity, once again beyond human comprehension; yet another opportunity to return to our earlier reflection on temporality and its underlying dialectic.

Numerical operations may seem simple, but their application in the perceptive field multiplies their effects. Admittedly, the inversion of a numerical sequence transpires from its mathematical banality. However, adapting it to the realm of sounds and sound objects, to its psychoacoustic implications, reveals the unheard (*inouï* in French, meaning never heard). In this way, Messiaen attempts to focus attention, bringing music back to its perceptive dimension. Thus, the rather free permutations—the choice of the order of permutation types is intuitive, but quite perceptible in its written result—of Divisions I and II will focus on the first sequences, as demonstrated by Example 7a, b.[12] From the insistence on the first events (1 - 2 - 3 - 4) emerges an audible behavior, a thematization without motive, or rather an insistence that will function as a background *Gestalt* (to contradict Stockhausen!), a symbolic expression of a chaos that is in reality only divine order. On the other hand, the journey between 1 and 12 and its retrogradation, from 12 to 1, allows us to glimpse the idea of temporal dilations-contractions in terms of duration, a metaphor for the subliminal breathing of the musical universe.

To remain in the perceptual field, from bar 46 onwards, Messiaen abandons the instrument's low notes for ten bars or so, depriving himself of the qualities of either aural cue (such as C#) or "harmonie-timbre", and confining himself to the middle register. In this way, the score reveals, by means of new specific permutations, the diatonic segment B previously mentioned, enveloped by a background of fine ethereal chromatic lace, at the very limit of audibility, or highlighted by figures belonging to segment A, distant echoes of a chaos increasingly distanced from the *ff* or *mf* nucleus described in Example 8. At this particular moment in the discourse, it's up to the composer to divulge a new musical dimension close to an ancient conception of modality—sounding like a Gregorian chant -, all the more stabilized in that its agogic is inscribed in a relative striated time (to use Messiaen's terms), audible for the first time in the score.

Chaos, modal revelation, exacerbation of a fragmented "harmonie-timbre"—these are the milestones of a discourse that is the opposite of the "decorative impression" Stockhausen spoke of in his lecture. In fact, these three stages are merely the result of a program contained in the initial genetic code, the fruit of a lib-

[12] With the two divisions facing each other, the numerical work is read from left to right by division, with a specific typographic adaptation to reveal its organization in ascending or descending order.

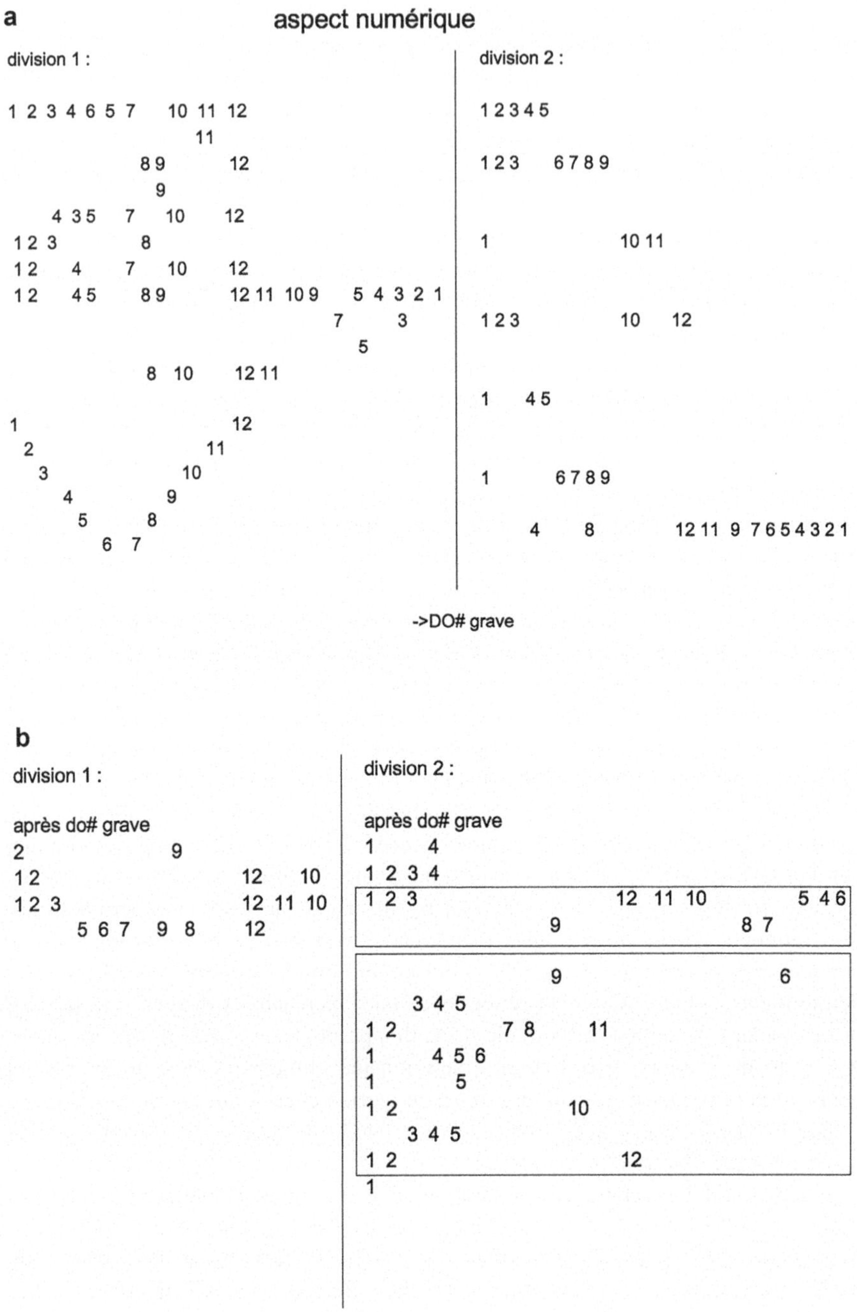

Example 7 (**a**) Numerical permutations of divisions I and II, bars 1–28. (**b**) Numerical permutations of divisions I and II, after bar 28. Read all the examples of numerical series by division, from left to right, from the first line to the last

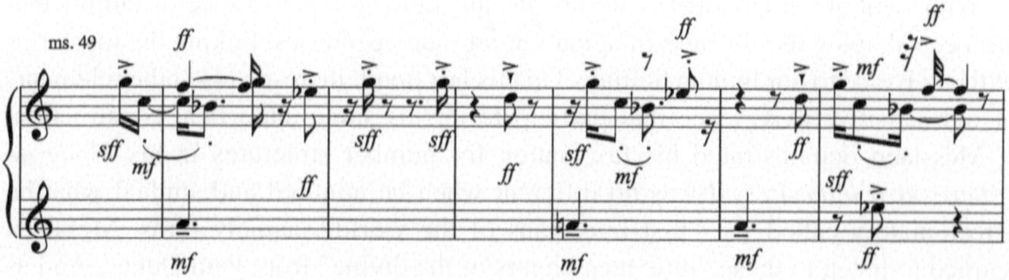

Example 8 Modal core of bars 49 to 55

eration of the potentialities integrated into the fragments of the "mode" that gives the work its title. This variety of genes from the "musical chromosome" will initiate (in its initiatory sense) the discourse through its exploration of the first musical objects from division I—identities lost in the initial chaos of complex polyphony—towards the last objects from division III—strong personalities from which will emerge the understanding of a surrounding world made of Gestalten organized in crackling «harmonies-timbres»—passing through the associations of profiles coming from the interweaving of the three divisions, the time to create an audible modal link, like the aftertaste of a plain-chant close to our human sensibility. The stakes are high, as analytical interpretation, after careful examination of materials and techniques, brings us back to the idea of a compositional project whose motivation would be to renew the dialectic between spirituality and materiality. Chaos confronts the listener with the existence of an unintelligible higher order, and offers a simple experience of fascination—that famous "music of the stars" proposed by Stockhausen. However, the emergence of reference points (including the low C#) marks out the space and reactivates the dimension of a humanity confirmed by a fragrance of modality that is pervasive in the senses; a veritable bubble—as in Hieronymus Bosch's work, echoed by Messiaen in his *Cinq Rechants*—of palpable sensitivity in the midst of chaos, which in turn becomes the source of a reconstruction of the universe made up of "harmonies-timbre", an infinite space, not enclosed by a relevant upper register limit, as it is open to the resonances of innumerable high-ranking harmonics, derived from the sonorities of the black-and-white instrument set in vibration for this masterly work. It should be noted that other poetic interpretations could prove possible in the light of certain epistemological investigations, all the more so as the work was composed at a time of manifest renewal, of hope for a better world after the chaos of a destructive war.

The fact remains that the implementation of such a discourse was only possible thanks to the application of meticulous, extremely refined work on numerical permutations, thus making it possible to update the problematic of modal distributionalism (a world away from serial thinking), and beyond that, to divulge the potentialities contained in the original "mode", confirming its status as a "forme en devenir" (form in progress).

With symbols and axioms as arguments, the digital refers to a specific culture that has been shaping the Western imagination for many centuries, linking the mysteries of the universe to our human finitude. On this last point, there are also other elements of our sensitive experience that participate in this same dialectical construction. If Messiaen demonstrated his fascination for number structures in his *Mode de valeurs et d'intensités*, it was no different when he admired and studied what he affectionately called the "first musicians of the world", namely birds. Messiaen learned to listen to these "little messengers of the divine" from Paul Dukas. And as François-Bernard Mâche, a specialist in the composer's relationship with nature[13], reminds us:

> *The bird embodies total freedom in metrical pulsation, absolute certainty in improvisation. It also has a spiritual connotation, dear to Olivier Messiaen: birds [...] enjoy a providential infallibility of instinct. They were the companions of Saint Francis of Assisi. Paradoxically, Messiaen was more interested in the formalism of bird song than in its mystical aspect [...].*
>
> *Olivier Messiaen combined formalism and extreme realism: this was the focus of his research in the 50s. He tried to strike a balance between accepted and invented forms, between the natural and the linguistic.*[14]

Birdsong has been a constant theme in the composer-pedagogue's work, from the allusions in *Quatuor pour la fin du temps* to the innumerable pieces from the 1950s onwards, whose musical project faithfully follows the modeling[15] of these animal sounds. But what exactly is this representation? The artist worked "by ear". No computer, no acoustic analysis tools. Simply bird songs taken as dictation, or recorded with a tape recorder to keep a memory trace. All that remained was to transcribe musically, with the help of a pencil and a sheet of paper with staves, not only the inflections and other melismas of these passerines, but also the specific color of these much-loved sounds. Much was made clear at this level, as Messiaen attempted to make audible the sonic qualities of these birdsongs (today we would speak of their spectral qualities) in order to include them in a larger musical project:

> *In my* Catalogue d'Oiseaux, *one could point to a large number of innovations, because reproducing the timbre of birds forced me to constantly invent chords, sonorities, sound combinations and sound complexes, resulting in a piano that doesn't sound "harmonically" like other pianos.*[16]

---

[13] François-Bernard MÂCHE, *De la musique, des langues et des oiseaux*. Interview with Bruno Serrou. éditions de Maule, Institut national de l'audiovisuel, 2007.

[14] excerpt from article on the Internet: Olivier Messiaen ornithologist, http://www.canalacademie.com/Olivier-Messiaen-ornithologue.html

[15] As a reminder, acoustic musical modeling is an attempt to use traditional instruments to represent sounds from our life experiences (transposing the spectral content of a sound event via a symbolic notation—a score—enabling instrumental mimicry and interpretative reproduction). Through its symbolic inscription, modeling enables the manipulation of its multiple parameters. Since the last quarter of the twentieth century, modeling has been facilitated by the detailed analysis of a sound model with the help of computers.

[16] Anik LESUR, Claude SAMUEL, *Messiaen, le livre du centenaire*, Éditions Symétrie, Lyon, 2008

To help us understand this musical reproduction of birdsong, it would be interesting to transpose this quotation to the world of the orchestra, and place under the analytical microscope a simple instance of this type of modeling performed by ear. In this perspective, let's take a look at one of the composer's last works, whose musical project is easy to grasp: *La Ville d'en-haut*, for solo piano and small orchestra, composed in 1987.

From the outset, Messiaen asserts his theological perspective by highlighting two texts from Sacred Scripture:

- "Seek the things that are above, where Christ is" (Saint Paul, Epistle to the Colossians, chp.3, v.1)
- "And I saw the Holy City coming down out of heaven from God" (Revelation, chp.21, v.2)

In this case, transcendence is the subject of a discourse punctuated by numerous birdsongs, including that of the Melodious Warbler (Hypolaïs polyglotte) (Example 9).

From the outset, the question arises of the relationship of this modeling within the score to its original model. A sonogram of this bird song will demonstrate the infallibility of Messiaen's ear in this field, and the subtlety with which the composer transposed the qualities of the song of this little musician of nature into the instrumental realm.

Surprisingly, the rhythmic transcription corresponds exactly to the rhythm proposed by Messiaen in his score! When it comes to understanding the model in terms of pitches, the sonogram is invaluable, as it reveals a number of frequency qualities (indicated in hertz for the first sound segment in Example 10), essential to understanding the timbre of this Melodious Warbler (Hypolaïs Polyglotte). So what can we deduce? First of all, the first segment (a) involves two events (a1, a2), the components of which fall within extremely narrow frequency bandwidth (with a frequency contraction from a1 to a2), resulting in a very rough timbre and interference due to the proximity of the frequencies. The transcription in musical terms could be noted as follows (with a quarter-tone approximation), in order to better understand the components of timbre of the bird in (a):

As for the succession of subsequent events, the sonogram reveals its essential characteristic, a powerful argument for understanding the spectral evolution of the bird's timbre. Indeed, the spectral representation shows a progressive extension of the frequency bandwidth ambitus, so as not only to enrich the timbre, but also to attenuate the effect of the liminal roughness (Example 11).

It remains to examine how Messiaen brought out these spectral qualities. Example 12—a one-staff reduction of the chords from Example 9—will help us to better grasp the space of pitches that have become components of "harmonies-timbre" (since the composer's aim is not to have us hear complete and strict spectral harmonies, as in Example 11, but rather aggregates likely to recall the sonic qualities of birdsong):

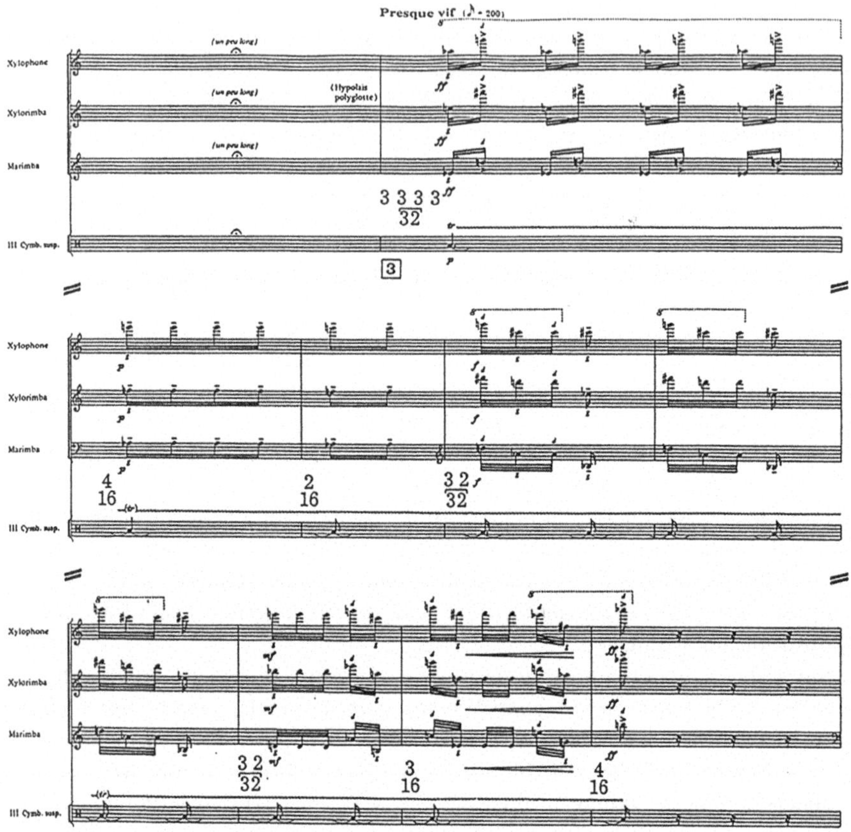

Example 9 Modelled song (pitches, rhythms and timbre) of the Melodious Warbler (Hypolaïs polyglotte), using a percussion ensemble in *La Ville d'en-haut.* (© Éditions Alphonse Leduc)

Reducing these chords to a close position (according to the postulate of the *set theory*) will help us better understand the sonic implications of the succession of such arrangements:

This example makes it clear that Messiaen does not attempt to reproduce the exact frequency components of the bird's song; indeed, this would be difficult for him, given the absence of microtonality in his writing, and the impossibility of discriminating by ear between extremely small, tightly-packed interval components (which can now be calculated by computer). As a result, it attempts to express, through the selection of its aggregates and its overall curve, the essential qualities of its spectral evolution: a rough first "harmonie-timbre" (like a1—cf. Example 11), an aggregate also parasitized by the noisy attack transients of keyboard percussion instruments, immediately undergoing contraction (like a2), and subsequently proposing interval expansion, thus reducing harmonic tension—as demonstrated by the vectors in Example 13.

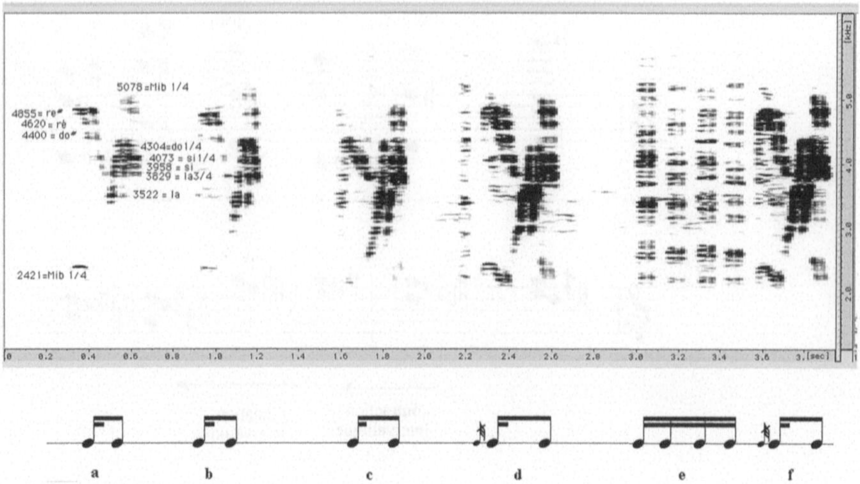

Example 10 Sonogram of a song by the Melodious Warbler (Hypolaïs Polyglotte) and rhythmic transcription of its musical events. (Sonogram based on a sound sample taken from: Bertel BRUUN, *The Hamlyn guide to birds of Britain and Europe*, Hamlyn Publishing Group Limited, London, 1978 (Book + Cassettes))

Example 11 Modeling of event (a) confined to the height field

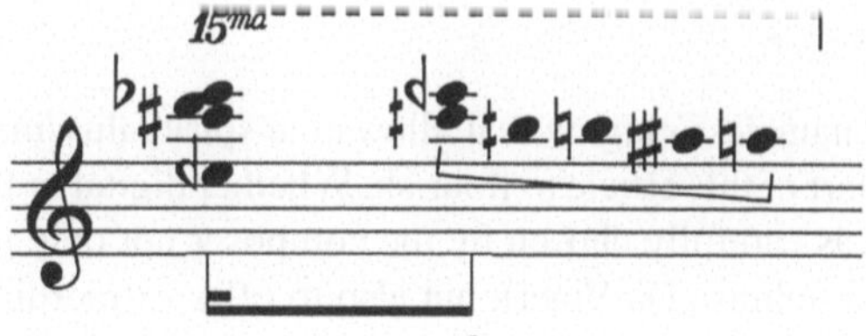

Example 12 reduction of chords at number 3 from the score of *La Ville d'en-haut*

This small example demonstrates, once again, how Messiaen travels between his formalist preoccupation and his desire to express materiality, the better to give us a glimpse of a higher order that eludes our immediate comprehension. The songs of this bird, mediator between heaven and earth, titillate our implicit knowledge of the world. Its modeling acts symbolically as an essential reference point for our personal integration into this world, which can be extended to the universe. But its quasi-magical treatment calls up a mystery that analysis can never fully reveal,

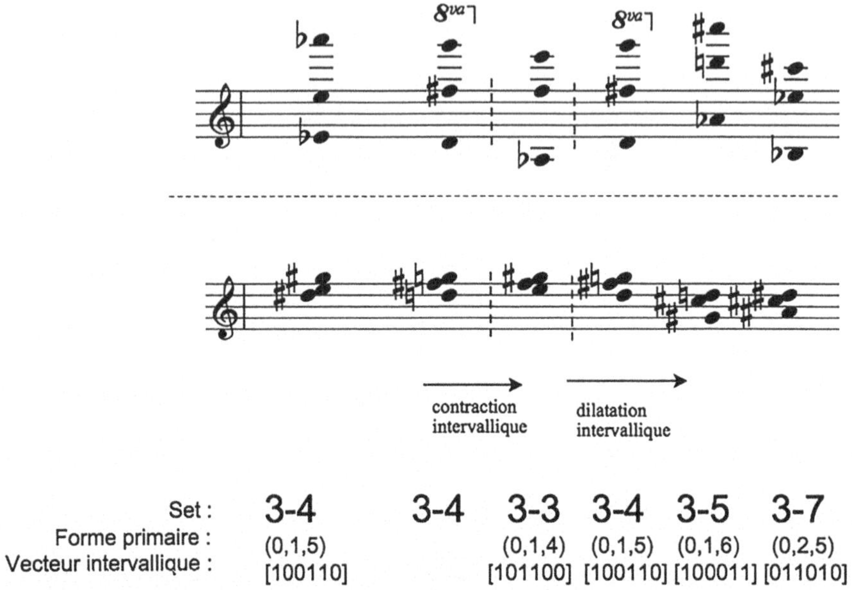

Example 13 Set theory analysis of "harmonies-timbre" from Example 12

a nameless enigma that allows the spirituality/materiality dialectic envisaged at the start of this article to flourish. Whether digital or birdsong, each of these information was carefully chosen by the composer not only to open a window on the world (to paraphrase Da Vinci), but also to offer extraordinary access to a universe declaring its uniqueness, in which not only God finds his place, but also every creative spark whose force projects the listener towards an overwhelming unknown.

References

1. J. Anderson, Dans le contexte. Entretemps **8**, 13–23 (1989)
2. P. Arnault, N. Darbon, *Messiaen, les sons impalpables du rêve*, Collection Compositeurs de notre temps, Millénaire III Éditions (2006)
3. B. Bruun, *The Hamlyn Guide to Birds of Britain and Europe* (Hamlyn Publishing Group Limited, London, 1978) (Livre + Cassettes)
4. R. Cooper, Propositions pour un modèle transformationnel de description musicale. Musique en jeu **28**, 70–87 (1967)
5. M. Delaere, The Stockhausen-Goeyvaerts correspondence and the aesthetic foundations of serialism in the early 1950s, in *The Musical Legacy of Karlheinz Stockhausen: Looking Back and Forward*, ed. by M.J. Grant, I. Misch, (Wolke Verlag, Hofheim, 2016), pp. 20–34
6. A. Lavignac, Encyclopédie de la musique, Dictionnaire du Conservatoire, Librairie Delagrave, Paris (1921)
7. A. Lesur, C. Samuel, *Messiaen, le livre du centenaire* (Éditions Symétrie, Lyon, 2008)

8. F-B Mâche, *De la musique, des langues et des oiseaux*. Entretien avec Bruno Serrou. éditions de Maule, Institut national de l'audiovisuel (2007)
9. O. Messiaen, *Technique de mon langage musical* (Alphonse Leduc, Paris, 1944)
10. O. Messiaen, *Traité de rythme, de couleur et d'ornithologie*, Tomes 1 à 7, Alphonse Leduc (2000)
11. A. Moukheiber, *Neuromania*, Alary Éditions (2024)
12. K. Schweizer, Olivier Messiaens Klavieretüde "Mode de valeurs et d'intensites. *Archiv für Musikwissenschaft*, 30. Jahrg., H. 2. (1973), pp. 128–146
13. R. Toop, Messiaen/Goeyvaerts, Fano/Stockhausen, Boulez. Pespectives of New Music **13**(1), 141–199 (1974)